From Nuclei to Stars

Festschrift in Honor of Gerald E Brown

From Nuclei to Stars

Festschrift in Honor of Gerald E Brown

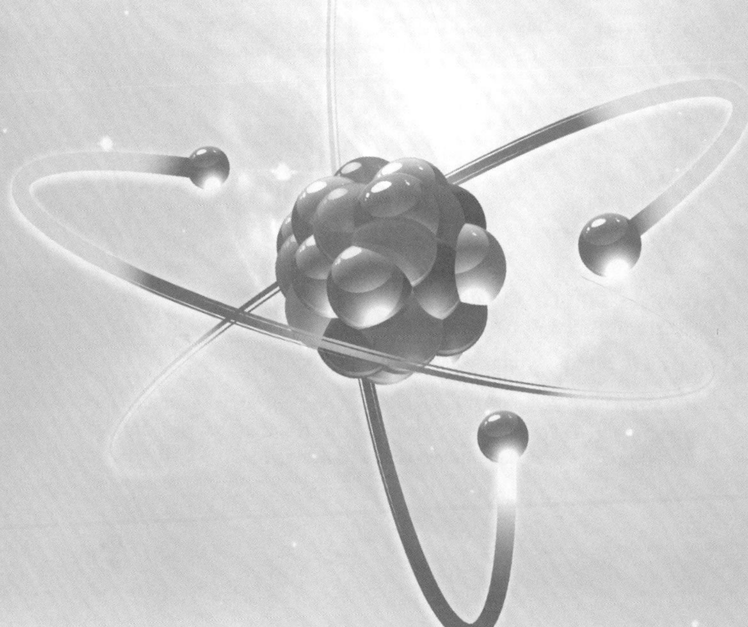

Editor

Sabine Lee

University of Birmingham, UK

World Scientific

NEW JERSEY • LONDON • SINGAPORE • BEIJING • SHANGHAI • HONG KONG • TAIPEI • CHENNAI

Published by

World Scientific Publishing Co. Pte. Ltd.
5 Toh Tuck Link, Singapore 596224
USA office: 27 Warren Street, Suite 401-402, Hackensack, NJ 07601
UK office: 57 Shelton Street, Covent Garden, London WC2H 9HE

British Library Cataloguing-in-Publication Data
A catalogue record for this book is available from the British Library.

FROM NUCLEI TO STARS
Festschrift in Honor of Gerald E Brown

ISBN-13 978-981-4329-08-8
ISBN-10 981-4329-08-8

Typeset by Stallion Press
Email: enquiries@stallionpress.com

Printed in Singapore by Mainland Press Pte Ltd.

Professor Gerald E Brown

PREFACE

What is the Universe?

In one way or another, Gerry Brown has been concerned with questions about the universe, about its vast expanse as well as about its most miniscule fundamental constituents of matter, ever since he was a young boy growing up as the youngest of six children in the intellectually stimulating environment of the household of his mathematician father, Dean George Brown. This volume, a collection of articles dedicated to Gerry on his 85th birthday, contains discussions of many of the issues which have attracted his interest over the years. The contributions are written by his former students, co-authors, colleagues and admirers strongly influenced by Gerry's own scientific tastes. With this compilation we want to express our respect, admiration and gratitude; we want to celebrate Gerry's scientific and scholarly achievements, the inspirational quality of his teaching and the enthusiasm which he himself displayed in his research and which stimulated so many of his students and colleagues over the decades.

Wherever Gerry studied and worked, at Wisconsin, Yale, Princeton, Birmingham, Copenhagen or Stony Brook, he left a mark on the scientific community within which he was operating. As evident in the contributions to this Festschrift, his influence went way beyond the scientific impact; it often resulted in long-lasting friendships and life-long collaborations.

The idea for this volume arose out of my many extensive conversations with Gerry about his own personal and scientific development, He saw this as an outcome of his intense and constructive working relationships with his three main teachers, his "eagles": Gregory Breit, Rudolf Peierls and Hans Bethe. By the time I first met Gerry about a decade ago, he himself had become an immensely influential eagle himself, having guided and continuing to guide many successful scientists on their way to discovery and academic success. Gerry has always been keen to communicate his gratitude to his teachers and he has always felt a strong sense of responsibility to give back to the scientific community something similar to the support he felt fortunate to have received from his eagles All contributors to this Festschrift share the experience of having been at the receiving end of Gerry's enthusiastic support of their academic endeavours. We unite in thanking, you, Gerry, for your encouragement and assistance and we wish you and Betty well for the future.

Sabine Lee
University of Birmingham
July 2011

A NOTE FROM PROFESSOR CHEN NING YANG

One of my early contributions to Stony Brook was to convince Gerry Brown to move from Princeton to Stony Brook.

I had first gotten to know Gerry in 1964 when he joined the Physics Department of Princeton University. It happened that Nina Byers, one of my collaborators then, was visiting the Institute for Advanced Study in 1964–1965. She was a good friend of Gerry's in Birmingham and the three of us often got together in Princeton to talk shop. It was in these conversations that I learned to appreciate Gerry's good instincts about physics.

After I moved to Stony Brook in the summer of 1966, I knew one of my responsibilities was to help recruit the best physicists for the Department. To describe what happened, let me quote Gerry from an article[a] he wrote in 1992:

> "In 1967 Frank made a short visit back to Princeton. I was sitting at my desk in Palmer Physical Laboratory when he came in and asked me to go for a walk with him. Although Linwood Lee had set up a tandem Van der Graaf in Stony Brook, there was no nuclear theory, and here I was, sitting as professor at the world center of research, where I'd built up one of the leading nuclear theory groups. I was caught by Frank's vision, and because of my profound admiration of him, I resigned at Princeton and took up my present position in Stony Brook in 1968. This was the best of several good decisions I made in my life. I have helped Frank build up one of the 12 best physics departments in the country (as listed in *Changing Times*, November, 1983 edition, based on a five-volume study published by the National Academy Press in 1982. An amusing article 'Where the Academic Elite Meet to Inbreed' in Physics

---

[a]G.E. Brown, "C.N. Yang's influence on my life and research", in *Chen Ning Yang — A Great Physicist of the Twentieth Century*, ed. by C.S. Lin and S.T. Yau, International Press, 1975.

> Today, January 1989, p. 15 finds that, relative to other universities, fewer elite inbreed in Stony Brook than in the other top departments.) Because of the outstanding eminence and dignity given the department by Frank, our faculty has not indulged in squabbles, and we have had a helpful attitude towards each other in building up not only an excellent, but also harmonious research and teaching environment."

Gerry built up in Stony Brook in a short time a small world famous centre of nuclear theory. Later, with his interests broadening to astrophysics, Stony Brook also attracted many visiting astrophysicists. One side effect of those visitors I shall never forget: When Bethe visited in 1980, it happened that BBC was airing a documentary on Oppenheimer. At a party I asked Bethe his reaction to the documentary's description of the Los Alamos days and he said emphatically, "Excellent".

I take this opportunity to say once more how much I personally am grateful that Gerry had agreed in 1967 to take that walk with me.

C.N. Yang
Tsinghua University, Beijing
May 31, 2011

CONTENTS

Chapter 1

What is the Universe? G.E. Brown — His Life and Work

Sabine Lee*

I first met Gerry when, as a young historian, I was researching the life and work of one of his mentors, Rudolf Peierls. In a conversation with Sir Rudolf, in the spring of 1995, he intimated that the colleague who probably knew him best was Gerry Brown and, if I could, I should talk to him about the finer details of any biographical work on Peierls. As will be clear from the paper below, this assessment was entirely correct. In 2004, on the occasion of the naming of the Rudolf Peierls Centre for Theoretical Physics in Oxford, I finally had the opportunity to make Gerry's acquaintance. It was the beginning of a fruitful collaboration during which I learnt much about physics, its history and many of its key players in the 20th century.

In this paper I am tracing Gerry Brown's own background, personal and scientific. It is as much an attempt to tell a story of Gerry's life and work as it is an attempt to tell the stories which I believe Gerry would like to hear told about his family, his friends, his colleagues and, above all, his teachers. My reading of events and developments is based on Gerry's own autobiographical writing, most notably "Fly with Eagles" [1] but also on recollections gathered in other written works such as the *Formation and Evolution of Black Holes in the Galaxy* [2] and *The Nucleon-Nucleon-Interaction* [3], as well as on numerous conversations we have had over the years. It was in these discourses that I developed a strong sense of his gratitude to his teachers, his three "Eagles," Gregory Breit, Rudolf Peierls and Hans Bethe. Therefore much of the account below will highlight facets of Gerry's life, which demonstrate what he himself often emphasised, namely the immense influence of his three "Eagles" on his personal and professional life. As much of Gerry's work on the nucleon-nucleon interaction, his work done at Princeton and also later at Stony Brook is described in detail or alluded to in various other contributions in this volume, these will feature less prominently in this biographical essay.

*Department of Modern History, University of Birmingham, Edgbaston Birmingham B15 2TT, UK. E-mail: s.lee@bham.ac.uk

1. Family Background and Early Life

Gerald E. Brown was born in Brookings, South Dakota, on July 22, 1926. His father had come to South Dakota State College in 1897, when it was little more than an "Aggie" School, gathering would-be farmers over the winter to teach them a variety of subjects. But the so-called Land Grant Colleges, which were given land by the federal government, played a leading role in instructing farmers with experts in agriculture, hybrid seeds, etc. travelling around the state to advise them.

His paternal grandfather, John Brown, was born in Pennsylvania and moved to Ohio in 1837. In 1854 he moved on to Illinois, enlisted in the Union Army, August 13, 1862 and served for 3 years. After the war, June 3, 1866, he sold his farm and went to Missouri by covered wagon, settling in Appleton City. While in the Army, he had marched about 700 miles and decided that — if he lived — he would return to Missouri and make this place his home. He was one of the first settlers there and named the new community "Round Prairie".

Gerry's father, George Lincoln Brown, was born in 1869. Grandfather John had a farm on which he could bring up his four sons, and they remained relatively poor farmers. Gerry's father left the country school at age 17 to become an accountant, enrolling at the University of Missouri. He turned out to be a good mathematician and was encouraged to stay on at the University. His attendance was irregular, because he had to go back home to help get the harvests in and share other chores on the home farm. Nevertheless, he received the $50 scholarship prize awarded each year for the best junior student in general science, and during his senior year was a teaching fellow in the Department of German, teaching a language he had learned in the University earlier. He graduated in 1892 with the degree of Bachelor of Science. Returning to the University as a teaching fellow in mathematics, George Brown pursued advanced work leading to a Master of Science which he received in 1893. He then moved on to the University of Chicago, a new institution and a new mathematics institute that had only opened in 1892.

It was an exciting time in the development of American Universities in general and American mathematics in particular. The second half of the 19th century had seen American mathematics transformed from the academic backwaters of earlier periods, and the last quarter of the century witnessed rapid progress in the creation of a mathematical research community. J.J. Sylvester, who had joined the newly-founded John Hopkins University in Baltimore in 1876, had established the first American school of mathematical research, and thereby laid the foundations of the rapid advance of the subsequent decades. When Sylvester "retired" from Baltimore to become Savilian Professor of Geometry in Oxford, John Hopkins University tried to interest Felix Klein, then Professor at Leipzig, to take up the Chair in mathematics. But the appointment negotiations remained unsuccessful, and Klein, instead, moved to Göttingen in 1886. Nevertheless he became an increasingly significant influence on the development of American mathematics, as many young American mathematicians (Haskell, Thompson, Snow, Tyler, White) embarked on travelling apprenticeships across the Atlantic in order to study with Klein in Göttingen. Furthermore, some of Klein's German students later settled in America and played a leading role in the progress of mathematics as a research discipline [4], not least in the university where George Brown studied for his doctorate and which became one of the leading forces in American physics at the turn of the century: the University of Chicago.

The first Chair of the Mathematics Department there was Eliakim Hastings Moore,[a] who had previously been an associate professor at Northwestern. Moore immediately appointed Oskar Bolza[b] and Heinrich Maschke,[c] and the three of them became the core of the department in its initial decade and a half. Gerry's father was among the first of a cohort of 39 students who, between 1892 and Maschke's death in 1908, graduated from Chicago with doctoral degrees in mathematics.

Between 1894 and 1896 George Brown was a fellow in Mathematics at Chicago, where he received the degree of Doctor of Philosophy. E.H. Moore was the leading mathematician in the Chicago group, but Heinrich Maschke was George's thesis supervisor. The results on collinear groups which Maschke published from 1897 onwards were based on Moore's work, but they also followed closely on the heels of George's dissertation research. At Moore's suggestion, he had undertaken an analysis of the structure of the ternary collineation group of order 360 some time prior to the

[a]Eliakhim Hastings Moore (1862–1932), had studied at Yale where he obtained his Ph.D. in 1885; after research in Berlin under Weierstrass and Kronecker he taught at Northwestern before moving to Chicago.

[b]Oskar Bolza (1857–1942), had studied physics at the University of Berlin under Helmholtz and Kirchoff before turning to mathematics, which he studied under Christoffel and Reye at Strasbourg, under Schwarz at Göttingen and under Weierstrass at Berlin. In 1886 he completed his doctorate, supervised by Felix Klein, at Göttingen.

[c]Heinrich Maschke (1853–1908), studied mathematics at Heidelberg and Berlin under Königsberger, Weierstrass, Kummer and Kronecker. He received his doctorate from Göttingen, where he was supervised by Felix Klein.

summer of 1896. His 15-page Ph.D. thesis was entitled *The Ternary Linear Transformation Group* $G_{3.360}$ *and its Complete Invariant System.*

There was no denying the continuing influence of Felix Klein on Maschke. Klein himself, together with some of his students, had been studying the equation of the sixth degree [5]. Klein had made various attempts to employ quaternary linear groups in the study of the sextic equation. Then, in 1889, Siegfried Valentiner had found a simple linear ternary group of order 360, but as he had published his result in Danish it was not widely noticed. Klein's student Andres Wiman, however, drew attention to it and developed the idea further, drawing, among others, on work of Heinrich Maschke [6].

Writing to Felix Klein on 28 July 1896, "Heinrich Maschke proudly reported that his own student, George Brown had not only established the isomorphism of this group with the alternating group on six letters but also, under Maschke's guidance determined its complete invariant system."[d] In reporting his student's success he enquired whether these results should not be published, and — knowing of Klein's influence on academic journals — Maschke suggested to write up an abstract of the work for inclusion in either the *Mathematische Annalen* or the *Göttinger Nachrichten.* But George's research and Maschke's attempts to see his student's work published and thereby recognised widely, came too late. The next issue of the *Mathematische Annalen* which arrived in Chicago a little later contained an article by Wiman which duplicated Brown's results [7].

It was one of the cruel unfortunate cases of important discoveries being made independently and concurrently, without either of the researcher being aware of the other's efforts. As Maschke clarified in a letter to Klein: "Brown spoke on G_{360} last Friday, that is, before the number of the *Annalen* arrived. Naturally, my question concerning publication has now become superfluous. Our Brown, however, will not have done his work in vain; he will publish it in one of the American journals with reference to Wiman's work."[e] Yet, this apparently never happened. On earning his doctorate, he returned to South Dakota. In 1896, he was appointed teacher of mathematics in the Rock Island High School. He left after three months to accept a position as Professor of Mathematics at South Dakota State College.

[d]Heinrich Maschke to Felix Klein, 28 July 1896, see [4], p. 401.
[e]Heinrich Maschke to Felix Klein, 24 August 1896, see [4], p. 401.

Gerry's father sought out culture, although he had little background from his home environment. He loved opera, and acted as a stage hand in the Chicago Opera. In particular he had a strong liking for Wagner, partly because it was sung in German. In Brookings, South Dakota, George Brown met Geraldine Winifred Louchs, born in 1879, and married her in 1898. They had three children together, Cecil, Elisabeth and Florence. In 1908, Geraldine died of a heart attack. Her younger sister, Anna York Louchs, born in 1885, who had taken a degree in literature at South Dakota State College, came to Brookings to look after the children. George Brown married her in 1910 and they had four more children, Winifred, Charlotte, George Jr. and Gerry.

The Brown family lived in an unusual isolated intellectual environment, with their main friends being the Hutton and Hume families. Professor and Mrs Hutton had moved to the US from Germany, and Prof. Hume, who taught animal husbandry, had taken his degree and got his training in Germany. The older children in Gerry's family had counterparts in the Hutton and Hume families, and they always celebrated holidays, Thanksgiving and Christmas together. Since George Brown was relatively old, 57, when his youngest son was born, Gerry had no counterparts and, as he would later recall, was thoroughly spoilt by everyone!

Members of Dean Brown's family (George had advanced to being Dean of General Sciences) were expected to excel in school work, and Gerry's father was noticeably disappointed when Winifred only tied for valedictorian in high school. After his junior year in high school, Gerry's brother George went to the University of Minnesota for a summer school and took five courses in chemistry, physics and mathematics and got A's in all of them. When he came home, his father looked at his report card and sniffed: "That's all right, if they don't give A+'s." George did not return to high school to get a degree, instead he attended South Dakota State College, and later took his Ph.D. in chemistry at Brown University. Teachers who had been at the local high school for a long time often called Gerry Cecil after his half brother; those who had come later called him George. Gerry later mused whether some of his somewhat revolutionary activities, which will be described below, were expressions of a belated rebellion against his upbringing as Dean Brown's youngest son.

Gerry's six brothers and sisters all received their bachelors' degrees from South Dakota State College (and his brothers went on to study for their Ph.D.'s at Brown University), but Gerry decided very early that he wanted to go to a better university (although in retrospect his brothers all did very well in science). His father's salary was not high, and he asked Gerry how he was going to finance college away from home. Gerry's pragmatic response was to take a number of jobs, starting with delivering newspapers in the early morning, even in winter when the temperature dropped to nearly $-40°$. But his main job was writing for the *Brookings County Press*. This was a weekly newspaper playing second fiddle to the *Brookings County Register*, which came out twice a week, and which had a much wider circulation and a greater number of subscriptions.

The *Press* was a historical paper, which had existed in Brookings for a long time, and probably should have been shut down long before Gerry began writing for it. A man named Turner, whose ancestors had run the paper practically forever (that is in a town which was consolidated in the late 1800's) was editor. The *Press* was run in archaeological fashion. Turner had hired an American Indian to run the linotype machine which dropped molten lead into molds, and then had to be remelted, a task that Gerry had to help with on the Fridays after the newspaper came out. As for local news, Gerry covered all events, deaths, weddings (for which he had made out a form for the families to fill out, describing the dress of the bride in detail.). He spent nearly every afternoon there, and wrote a good fraction of the locally oriented events for the paper.

Since Mr Turner did not close down unpaid subscriptions, Gerry had to collect many from the subscribers with which he paid his salary, not really knowing how Mr Turner himself lived. As the paper had been around for a long time, the *Press* carried advertisements of the businesses on the one Main Street of Brookings, and Gerry could also collect money from them. He was still owed a few hundred dollars when he left the *Press*, but he had saved enough money to go off to college.

In the summer of 1943 Gerry went to summer school at South Dakota State College, including accelerated courses of the Army Special Training Program (ASTP). He took differential and integral calculus, which at the time was usually taught in the first and second year of the state college. He did not go back to his senior year in high school, and instead just continued in college. After he had received his Ph.D. and five honorary D.Sc.'s, Brookings High School gave him an honorary high school degree which he treasures to this day.

By Christmas 1943 Gerry had taken all the courses in physics and mathematics that were offered during those war years, plus some literature and history in order to broaden his education. He wrote to the University of Minnesota and to the University of Iowa requesting admission as a sophomore. The University of Minnesota declined on the grounds that Gerry did not have a high school degree; but the head of the physics department at Iowa University, George Washington Carver, who knew Gerry's father agreed to his joining the department and so Gerry went after Christmas 1943.

Gerry had started college at 16 and when he was 17 he knew that he would be taken into the services. Army life did not appeal to him, and he reasoned that life on ships was likely to be much cleaner. So he tried to enlist in the navy for officers' training. Gerry twice went to Sioux Falls, where the navy selection process took place. The first time he failed the physical exams already at the weighing machine for being underweight (125 lbs) and having flat feet even though he could never understand why this would have been important for the navy. The weight issue, however, he could improve on and after the consumption of gallons of malted milk and plenty of hamburgers he made it further the second time, when he was turned down for colour blindness and an overbite.

So instead Gerry went to Iowa City to begin his sophomore year at the University of Iowa in January, 1944. He knew that he wanted to study physics and had known that since he was 6 years old when he asked, to the amusement, Prof. Hume, "**What is the Universe**?" As is clear from numerous contributions to this volume, this questions has stayed with him all his life.

To his great luck, as it would turn out, Professor Carver at Iowa studied colour, and in the physics department library, Gerry found the same Japanese tests for colour blindness that he had failed when he was trying to join the navy. They consisted of 50 pages, each with a large circle, filled with large dots of various colours forming the numbers. Gerry was determined never again to fail a test for colour blindness, so he took the book of colour plates home and memorized the number he should say when he saw another one. Armed mentally and having revised the colour plates thoroughly, he took the Eddy test to qualify as a recruit for the naval radio technician program. As the day when he received his test results was the final day of recruiting for some time, Gerry enlisted then and there and with the clothes he had on and the books he was studying he immediately went by train to Arnes Iowa, for a physical exam, which — to his great relief — he passed without problems. That same evening he was off to boot camp at the Great Lakes Naval Training Center, north of Chicago, on Lake Michigan. Having passed the Eddy test he was a first class seaman and the 120-some other recruits were apprentice seamen, so he was in charge of the group.

As one would expect, the training was gruelling. It was not so much the marching, calisthenics, rifle practice that posed the problems, nor — generally — the long hours of physical exercise. The real challenge was the round the clock watch set by one of the older seamen, but even this eventually ended, and near the end of the boot camp, the recruits were given the possibility of taking written exams for officers' training. Gerry found them easy, one on physics, one on mathematics and one on English, mostly testing vocabulary. Following the exams he was interviewed by an officer. Gerry was quite sure that he had written a perfect exam, and the officer seemed very impressed by him. So Gerry was sent for a physical exam, supposedly the toughest the navy gave, the same as for navy pilots. Of course he knew why he passed the colour test, but he had no idea how he passed the physical examination given all his other defects. One can only assume that since he had done so well in the written exams officers turned a blind eye to his physical shortcomings and pushed him through. He reasoned that after his induction into the navy they looked at Gerry's body and decided he would be no good at fighting, and thought that they would do better to make an electrical engineering officer out of him. This war had been going badly, and, as Gerry later recalled, the general thinking was that it would go on for a long time and they were ready to invest several years in soldiers' training. Therefore Gerry was sent to the University of Wisconsin and began taking classes in the autumn term of 1944.

2. The First Eagle: Gregory Breit

Gerry was clumsy in the electrical engineers' laboratory practices — his supervisor had a fit when he put wires into winding posts the wrong way. Luckily students were graded on the basis of written exams, which he had no trouble with. But for the young sailor, university was not only about academic study. He and his fellow sailors were in competition with the other V12 programs at universities across the country for physical fitness, and push ups, pull ups, sit ups etc. were still very much part of the daily routine. Having to defend a record of being second in the nation in physical fitness there was a strong sense of commitment to do the required exercises in order not to let one's mates down. The mind was willing and the body had to follow. Even Gerry's! The day started with calisthenics early in the morning on the top floor of the armory. Then he had to hold on to the railing and falter down to the ground floor, walk up and down Bascom Hill to get over to his classes.

After the first autumn in Wisconsin, Gerry was given the option of becoming a physics major, and he signed up, among other subjects, for the graduate course in theoretical physics that Gregory Breit was teaching. Gerry would arrive about ten minutes late to his class, having gone up and down Bascom Hill after calisthenics, and collapse into his chair. He often wondered what his tutor thought of the strange conduct of the by then 18 year-old sailor. Normally Gregory Breit expected first-class behaviour of students and Gerry clearly was unable to fulfil this expectation because of his aching joints.

At the end of term, Gerry took his final exam and received a mark of 40 points out of 100. The course had been a difficult one, and Breit was a convoluted lecturer. It was hard to follow his line of thought. Yet, 40 was a far cry from the 100s he had been getting throughout his earlier education, and as expectations of high academic achievement had been drummed into him from an early age, disappointment was great. To his great relief, Gerry soon found out that his score was second highest in the class, only one graduate student, Johnny Powell,[f] being (way) ahead of him.

Little did Gerry know at this stage how his fate would be tied up with Breit's for years to come. Little, also, did he understand that the small (but powerfully built) man, so seemingly unassuming, was one of the best nuclear physicists in the world at the time, and certainly one of the best in the U.S.

It was not until much later, too, that Gerry learnt about his professor's background, and learnt to appreciate how this might have affected what was generally perceived to be an "odd" character. Gregory Breit was born on 14 July 1899 [8] in Nikolaev, in Southern Ukraine, a city built on the estuary of the Yuzhny Bug River, about 40 miles from the Black Sea. Founded as a shipyard in 1789, the site was chosen specifically for the purpose of repairing naval ships serving in the Russo-Turkish War. In 1862 a commercial harbour was opened and in 1873 a railway was built to the port. Gregory's parents, Alfred Schneider, and his wife Alexandra Smirnova Breit Schneider, had owned a textbook business. After Alexandra's death, in 1911, Alfred sold the business and emigrated to the United States in 1912. Gregory and his sister Lubov were left in the charge of a governess for three years, attending

[f] John L. Powell, (1920–2004) was a graduate of Reed College who would earn his doctoral degree from the University of Wisconsin in 1948; he joined the faculty of the University of Oregon in 1955 after working on the Manhattan Project, at MIT and the University of Chicago and teaching at Wisconsin 1948–1953.

schools in Nikolaev, until Alfred instructed them to join him in the United States in 1915. Upon his arrival, at the age of sixteen, Gregory enrolled at John Hopkins University where, over the subsequent six years he continued his formal education, collecting three degrees in the process: his B.A. in 1918, his M.A. in 1921, and (at the age of 22!) his Ph.D. His doctoral dissertation on the distributed capacity of inductive coils [9] seemed — at first sight — a little trivial for an advanced research project, but as Breit's biographer McAllister Hull pointed out, it was "a masterly piece of applied mathematics, a skill Breit demonstrated for the rest of his professorial life."[g]

Having been supported by scholarships since his arrival in the US, following his Ph.D. Breit could finally move into a paid position. He moved back to Europe to take up a National Research Council fellowship at the University of Leiden (1921–1922). His subsequent appointments established him firmly in the American physics community, with another short-term appointment at the University of Minnesota (1923–1924), and longer spells at the Carnegie Institute of Washington (1924–1929 as mathematical physicist and 1929–1944 as research associate), at New York University (1929–1934) and at the University of Wisconsin (1934–1947).

It was as assistant professor at the University of Minnesota that Breit met Merle A. Tuve with whom he would publish significant research papers of impact well beyond the physics community. Tuve had graduated at the University of Minnesota in physics in 1922 and obtained a master's degree in 1923. Following a year at Princeton, where he was an instructor, Tuve went to Johns Hopkins University to work for his doctorate. While at Minnesota, Merle and Breit developed a close friendship, which continued after Breit moved to the Department of Terrestrial Magnetism of the Carnegie Institute of Washington in 1924. After Tuve's arrival at Johns Hopkins, Breit sought his collaboration in a possible effort to study the ionosphere. To demonstrate the existence of the ionosphere it would be necessary to find evidence that radio signals arrived over at least two paths, a ground wave and a sky wave. Tuve devised the necessary detecting equipment and Breit and Tuve were able to use a Naval Research Laboratory oscillator as their source of radiation. The experiment was a success. Verification of the existence of the ionosphere opened an important field of research and suggested the practicability of radar [10].

Even before publishing these ground breaking papers on the ionosphere, Gregory Breit had turned his attention to the nascent field of quantum

[g]See [8], p. 31.

mechanics, which would become one of the focal points of his research for years to come. Having been the first American physicist to realize the advantages of inducing nuclear reactions with artificial sources, he published his early ideas on nuclear reactions in 1929 [11]. In the mid-1920s already he had begun his studies of dispersion relations [12]. This was followed by several other papers culminating in his substantial review articles in 1932–1933 [13]. In 1928, he had discussed the interpretation of Dirac's Theory of the Electron [14], and he used this understanding of the Dirac equation in the following years when developing the theory of two-electron interactions.[h] It was one of the finer points of the theory developed by Breit before the war, the fine structure calculations of helium that he and Gerry would return to after the war when studying the effect of the finite size of the proton on the fine and hyperfine structure of hydrogen spectra.

Students could be forgiven for overlooking the exceptional qualities of a world-class researcher, not least because Breit found it difficult to translate them into effective teaching. To Gerry, Breit looked like any professor in South Dakota State College, except that the latter would have had better offices. It was hard to imagine that here he was in the presence of one of the leading although immensely unappreciated physicists of the time. That Gregory Breit's contributions did not find the appreciation that they deserved at the time was at least in part due to his many personal peculiarities which, as Gerry would comment later, were his own fault, or rather, resulted from the circumstances of his growing up and maturing mostly in personally very enclosed environments.

The first significant seminar paper that Gerry gave in Breit's weekly seminar was on the so-called Pinocchio experiment (thus named by Niels Bohr), named with reference to the authors who had presented the results of the experiment: M. Conversi, E. Pancini & O. Piccioni [15]. The Italians stopped positive and negative mesons in 3 cm of iron as an absorber by using magnetized iron plates to keep mesons of the same signs while keeping away mesons of opposite signs. Delayed coincidences of the disintegration electrons from the mesons were detected. The importance of the experiment was that several centimetres of iron were required to stop the mesons, so that they were clearly not the strongly interacting mesons that Yukawa had

[h]Gregory Breit published widely on various aspects of the theory of two-electron interactions, including the separation of angles in the two-electron problem, the calculation of fine structure and hyperfine structure of helium, about isotope shift in hyperfine structure to name but a few.

suggested as the mesons the exchange of which gave interactions between nucleons. They turned out to be muons. Gerry had found out in electrical engineering that the magnetic field inside of iron **B** was much greater than the magnetic field in free space **H** because of what is called the magnetic susceptibility. In the seminar paper that Gerry gave, Breit drew him out to establish the main points by asking questions to which he knew Gerry would know the answers, because they had discussed them before. Breit had instructed him carefully about the physics contained in the papers. Gregory Breit had a reputation for what Gerry would refer to as being "a tough cookie" and some young women who attended the seminar said afterwards: "You sure handled Gregory well!" It was at that point that Gerry was beginning to realize that Breit took great interest in his training, and he began to get an idea of what research was like.

Breit's name is best known from the Breit-Wigner theory of nuclear resonances. For a time later in life, namely 1964–1968, Gerry was a colleague of Eugene Wigner's in Princeton, and it was at that time that he heard a story in connection with the Breit-Wigner theory which — like few others — demonstrated the unpredictable character of Breit in his personal dealings with colleagues and students alike. Wigner was an expert in the early study of nuclear forces and received a Nobel Prize for his work on symmetry. Yet, when he had been in Princeton in the 1930's, it was not clear that Princeton would promote him to tenure. So he accepted an offer from the University of Wisconsin. Upon arriving there, Gregory Breit greeted him and said: "Now we have two people, so we need a leader for our group." Wigner was a man of strong opinions and apparently he replied that he did not think they needed a leader. Breit, flushed with anger, floored him with a left to the jaw. His temper sometimes got the better of him, but he always regretted it later. He was a man of strong, often hidden, emotions. Gerry himself experienced some of this during his time with him in Wisconsin.

After the end of the war, Gerry signed out of the V12 program because he would have had to spend several years in the navy, had he become an Ensign. He was sent out to Bremerton ship yards, just west of Seattle. Essentially all of the electrician technician mates had been college students and the navy would not have had any left, had they discharged all of them immediately. Gerry had no training in the electrician technician program, and therefore he had no experience with the radar. But he was assigned to the radio shack, and ended up scraping paint off the floor and repainting it with chromium paint thus "mothballing" it. To keep himself intellectually occupied at Bremerton, he took his pocket cards with German, French and Russian words and phrases. Books in the base library taught him how they should be pronounced, but years later, when he tried out the languages, no natives understood him. Nonetheless, at least Gerry would be able to read books in these languages for decades to come.

He wrote back to the University of Wisconsin explaining his previous academic career with them and following his detailed exposition the authorities agreed on his having accrued sufficient credits for a bachelor's degree. Yet, Gerry had taken only one foreign language, German, and they required two for the award of a bachelor of science degree. However, helpfully, they dug up from their archives a degree, bachelor of philosophy, which only required one language and sent along a scroll attesting to Gerry's graduation.

In the autumn of 1946 Gerry returned to the University of Wisconsin to continue his studies with Gregory Breit in order to obtain a Ph.D. Until then he had seen none of the peculiarities of Gregory described above. While in the Navy, Breit had always been very kind to Gerry. On his return to Wisconsin, in 1946, for the first time he experienced Breit's less pleasant side in a canoeing incidence which has been described elsewhere and which many acquaintances of Breit would have regarded as rather typical.[i] Breit had a canoe, and in September 1946 he invited a German student and Gerry to go canoeing with him across Lake Mendota. They fetched his canoe from storage and first he put the German boy in front position to steer the canoe, Gerry in the middle as passenger and Breit as aft paddler. Breit was small but powerful, and he paddled with great force. The stern paddler essentially steers the canoe so that the front man has to work hard from keeping the boat from turning. The German boy could not keep the boat straight and once they got across the lake, he "remembered he had an appointment", leapt out of the boat and took a bus back. Gerry had seen what Breit had been doing and silently took his position in the front part of the boat. Try as Breit would to turn the boat, Gerry kept the canoe straight, because the physical fitness he had acquired in the Navy had made him stronger than his physique would have suggested. Without a word the two physicists carried the canoe back into storage, and Gerry sensed that Breit had new respect for him. Many years later, when Breit had moved to Buffalo after retiring from Yale, he phoned Gerry on a Saturday afternoon, as he often did. That time he asked: "Do you remember when I took you canoeing when you came back to Wisconsin? I knew that you had some stuff in you then." This was the first and last time they ever discussed the canoe trip.

---

[i]See [1], p. 5.

Despite some problems in the interpersonal relationship, Gerry always treasured his time with Breit, who taught him a lot. He always held that he could not have been given a more appropriate thesis problem — and this, of course, was the most important stepping stone on the way to the successful academic career that was to follow. For Gerry, Breit was the right man at the right time for advising him on his doctoral studies.

In January 1947, Breit was offered a professorship at Yale, and Gerry, alongside other most other members of Breit's research group, followed him there. Before the move, at Christmas 1946, he went to visit his sister Florence. She and her husband Caroll Bottum, who was professor of agricultural economy in Purdue, lived not so far from Madison in Lafayette, Indiana. During the visit, the siblings heard that their father had had a stroke, and they took the train to Brookings, South Dakota. Dean Brown was affected in his speech, but he kept repeating how proud he was that Professor Breit was taking his son to Yale. Gerry's father lived for several years as an invalid and died while Gerry was in England in the 1950's (as did his mother). True to his promise, Gerry had cost his parents nothing for his education. He had paid his own way to Iowa and the U.S. Government took over from there. Yet, his mother said that if he was going to Yale, he should have a suit and insisted that he took $100 from her to pay for it. Gerry did buy a suit in Yale, but he never told his mother that he only paid $25 for it. Friends of his in the clothing business in New Haven stayed at the Yale University post office during lunchtime and bought up the expensive suits that the many wealthy undergraduates were happy to sell for ready money and Gerry bought a nice grey flannel suit. But, as would become evident throughout his time at Yale, nothing could hide the fact that he was not a typical Yalee.

Although Yale University was very grand, there was no one of Breit's stature in the physics department and Gerry was disappointed in the graduate courses. The Yale physics department was really rather old-fashioned. First year graduate students had to take a course taught by Lyman Page, from 9:00 to 10:40 every day of the week. He taught from his books and students called the lecture "Page by Page". They were given the opportunity of taking the qualifying exam just before the beginning of the fall term and had Gerry not passed it he would have had to take the first part of the course the following autumn. Therefore he worked late into the night all summer and passed the qualifying exam. With this, he could finally begin his research.

One of Breit's important papers in the 1930's was to show the isospin dependence of the nucleon-nucleon force; i.e. when Coulomb effects were removed from the proton-proton scattering, the scattering when considered in the same states as neutron-proton scattering, was the same for the two systems [16]. Only later, Nicholas Kemmer of Cambridge put the statement into invariance under rotation in isospin [17]. The Coulomb wave functions in the proton-proton scattering were hypergeometric functions and were miserable to work with. Breit had formulated them before the war; in fact he had employed John Wheeler one year as a postdoc to work with them in the early 1930's at Wisconsin [18]. The WPA[j] had tabulated in books tables of mathematical functions. Breit wanted complete tables of hypergeometrical functions, which his students worked out on electric calculators, Marchand or Monroematics. The professor had organized most of his students into computational projects, especially computing Coulomb wave functions for nuclear scattering processes. Gerry was not enamoured with punching the hand computers of that time, and he made no secret of it. His eye sight had given him some trouble already, and in this particular instance, he was quite relieved to be able to point to his personal defect. At the time it was not possible for Gerry to get glasses that would have enabled him to read in long stretches — he would get migraines, headaches where his whole vision would black out and he would get sick. Only later in Birmingham, from an oculist in the British National Health Service, he found out that his eyes were too far apart. Once he was given glasses that helped him focus at the distance he stopped getting the migraines that he had had throughout college and graduate school. But at the time, at Yale, he simply could not punch the machines for too long, and hence he told Breit that he needed a research problem that did not require lengthy computing. Not only that; he also told him that he did not want to do the calculations, which he considered mindless (although he did not tell him that!) and that he was going to go back to Wisconsin if this was what Breit considered theoretical physics. To give a more accurate impression of the exchange, Gerry did not stand up proudly to his teacher; according to his later recollections, he was actually (silently) shaking in his boots when delivering his little speech. Breit asked, "How can I pay you if you don't work?" to which Gerry replied, "You're not paying me. I'm on the GI bill."

[j]Works Progress Association, founded by F. D. Roosevelt to provide employment for out-of-work intellectuals.

Somewhat crestfallen, Breit reconsidered Gerry's comment and then told him that he had a very good problem for him. And indeed, Gerry agreed that his assignment was an attractive one: he was tasked to treat the electron and proton more or less on the same footing relativistically, the combined hydrogen atom wave function being described by the product of Dirac-like four-component relativistic ones, making up the Breit 16-component wave function. The idea was then to give the proton some structure, a finite radius, and to see what influence the proton structure would have on the fine and hyperfine structure of hydrogen. Having only just started studying quantum mechanics, Gerry had only really begun grasping the one-component Schrödinger equation — never mind any higher orders. So he dug out Dirac's 1928 paper on his relativistic theory of the electron, the wave function of which had four components with the proton an external potential and studied it [19]. The next step was to reduce it to the two-component Pauli equation and to derive the Schrödinger equation as its nonrelativistic limit. It was a rather complicated and cumbersome way to learn Schrödinger nonrelativistic quantum mechanics! But in some respects it was characteristic of Gerry's approach to study at that stage — manipulating quantities before completely grasping their meaning. Not work-shy by any stretch of the imagination and independent in his studying, his calculations and manipulations of the 16 component Breit equations put him far ahead of anyone, except Breit, in the then relatively archaic theoretical physics department at Yale. At the time, Gerry did not expect anyone ever to measure the small effects from finite proton size on hyperfine and fine structure of hydrogen, but now these have been measured to many decimal places. In fact, at the time the experiments had not been carried out to show that the proton had a finite size.

Although research took up much of Gerry's time during this period, he had a number of other interests. He was fascinated by Russia, having taught himself how to read Russian the year he was scraping paint for the purpose of mothballing ships. As he found out quickly, there was a Ukrainian Hall in New Haven, where the Ukrainians gathered regularly. The girls were very attractive and he joined the Ukrainian folk dance troop. A professional instructor came up to teach them. Later, during the year, they visited other cities with large Ukrainian populations, for instance Trenton and Erie, the latter having a lot of iron workers, which included many Ukrainians. The troop from each city would perform, one after another. Then they would all have supper together on long tables with lots of vodka and Ukrainian culinary specialties. After supper, the old folks would socialize while the dancing groups mixed and danced with each other. Being relatively tall, Gerry's usual part of the routine was to escort a girl out from one side of the stage, show her off as we went across the stage and then run around the back and come out with another girl. He was too tall for the individual dances the men did alone, going down on the knees and kicking the "saratchka"; but being acutely aware of the large number of beautiful Ukrainian girls, Gerry did not mind at all that his role involved dancing in partnership with them.

The Yale undergraduates mostly came from affluent homes, and they were often conservative. The graduate students, many like Gerry himself, who had come back from the military, as a rule, were more liberal. Gerry was taken by the third party movement, in which Henry Wallace ran for President. Indeed, he went to the Third Party Convention in Chicago as the Yale student delegate. In Wisconsin he had had a lot of political discussions and he had been delegated to meet the cosmopolitan (compared with South Dakota) students, many from New York City. In the physics building he had shared an office with other graduate students. He had been particularly impressed by Paul Zilsel, an Austrian Jew, who had fled to England at the onset of the Nazi occupation in 1938 and later emigrated to the U.S. Paul was the son of Edgar Zilsel, who, as a philosopher, connected Marxist views to the positivistic perspectives of the Vienna Circle, working specifically on the social and historic conditions of the development of modern science [20]. Paul had done his undergraduate degree at the College of Charleston (South Carolina) and a master's at Wisconsin. Like Gerry, he had accompanied Gregory Breit to Yale. Paul loved music, art, literature, good food and drink. He spoke many languages, had an exceptional breadth and depth of knowledge and a terrific sense of humour. All this made a great impression on Gerry, and the two had long discussions. As his father Edgar, whose left-wing political views had been held with conviction and had ultimately prevented him from pursuing a university career in his native Austria, Paul was a courageous political activist. He was never afraid to express his views and to act upon his convictions, which caused him problems with the authorities repeatedly, especially during the McCarthy Era. Gerry's own Third Party interests moved steadily leftwards in discussions with Paul, so that he ended up as a member of the Communist Party in 1947. Although by his own admission he felt pretty embarrassed going from door to door in the poorer areas of New Haven to sell *The Daily Worker*, the communist newspaper, which he had found to be badly written, he regarded such duties as necessary. Mrs Hulme in Brookings had gathered from his letters that he had been moving left, and wrote him a letter of advice, warning him that his activities would make life more difficult for me. She was right, as the next few years would prove.

In the summer of 1948 the Communist Party held their state convention in the Ukrainian Hall. The foreign affairs member of the national communist party came up from New York City as the representative of the Central Party. All of the other graduate students — there were no undergraduates in the Yale branch of the Communist Party — were away, so Gerry was asked to attend. This was just after the time that the Yugoslav Party had been expelled from the Comintern. There had been many animated discussions about this in the Yale branch, and the members there could not understand how the different parties could call each other friends and then, from one day to the next, be at each other's throats. A standard requirement for membership in the Communist Party was "self-criticism" which actually meant criticizing other members. As became obvious from the report of the Yale group that Gerry delivered at the party convention, toeing the line was not the prime characteristic of the students who — elsewhere — had learnt to think independently. Not surprisingly, the representative from the central party took issue with the report. Reminiscing about the occasion, Gerry recalled: "The party representative got up and said: 'In a vote of 22 to one, which side do you think would be correct?' I answered: 'In 1943 when the communist party voted to dissolve itself because they wanted to give Frankin Delano Roosevelt their full support, in the 20 (or so) vs 3 vote to dissolve the party, whom would you have thought correct?' "[k] The central party representative was furious, but held his counsel until the end of the meeting when he summarized it. In those days party speakers thought they should emulate Stalin, who gave several hour speeches. The representative spent the better part of an hour excoriating Gerry and the intellectuals in general for their bourgeois views.

This period was of course the time that witnessed a right wing swing figure-headed by McCarthy. There was a lot of real aggression. Gerry later referred particularly to violence that took place upstate in Poughkeepsie. The Veterans of Foreign Wars had taken what they considered to be left-wing books, or liberal books which they deemed to be left wing out of some library, probably that of the communist party headquarters, and they made a big show of burning them on the street. This was too reminiscent of the book burning by the Nazis during the Third Reich. So the New York

[k]The dissolution had later been realized to be a mistake and the party was reconstituted.

State communist party organized a massive meeting of several thousand people at a meadow outside of Peekskill, close to Poughkeepsie. Paul Robeson,[l] the famous black singer, was to sing at a concert. The leather workers in New York City, who were mostly left-wing, were to form a cordon around the meadow to protect the public. All of the Yale left-wing people, not just communists, were called upon to attend. It was when the protestors wanted to leave that the violence started. The Poughkeepsie police force had deputized a lot of the Veterans of Foreign Wars and other right-wing groups to be officers during the day. They led all of the cars out along a small country road which went out of the meadow and then up a hill into Poughkeepsie where it met the highway. En route, "patriotic people" were greeting the cars with stones and windscreens got smashed. The violence convinced Gerry that the US were heading towards fascism. It seems unreasonable in retrospect, but one must remember that during the time it seemed that there was no one willing to confront McCarthy. The left-wing around New York, where it was strongest, was demolished by Peekskill. Ironically, at the time, Gerry, together with Paul Zilsel, was expelled by the Communist Party for left-wing deviationism. Careerwise, this may have been a blessing in disguise for Gerry, as he hid himself in his research which was going well!

Gregory Breit had been getting suspicious of Gerry's political activities, since FBI personnel had visited the head of the physics department, Willie Watson, and inquired about Gerry's political activities. He asked Gerry what organizations he belonged to, and the latter truthfully admitted being a member of the Methodist Young People organization! In fact, by then, he really had had enough of politics. Around that time Hans Freistadt, who was a young American theoretical physicist was denied his passport for being a member of the Communist Party.[m] Although this withholding of travel documents was later judged to be illegal by the Supreme Court, Gerry

[l]Paul Leroy Robeson (1898–1976), American singer and actor, gained notoriety for his political radicalism and activism in the civil rights movement.
[m]Hans Freistadt, a native of Austria and son of the Jewish radical intellectual Bruno Frei, openly acknowledged his Communist Party membership, debated political questions with other students, and edited the local Communist Student Bulletin. Led by radio commentator Fulton Lewis, conservatives attacked the Atomic Energy Commission (AEC) in 1949 for awarding Freistadt a graduate fellowship to study physics at Chapel Hill. AEC chair David Lilienthal was forced to resign, and the AEC banned Communists from its programs. Freistadt lost his fellowship but continued to support the Party and its programs for social change. In 1949, he attacked the university for refusing to admit an African American to the law school. He finished his degree in 1950 and pursued a career

decided that he had better get out of the country. This meant finishing his thesis fast. As, at the time, English was the only language in which he could converse with ease, he chose England as a starting point for his job search. He still had the G.I. bill support through the summer of 1950, so he was asking for no financial support. He wrote to the physics departments at the Universities of Manchester, Glasgow and Birmingham. From Manchester, the head of the physics department, Patrick Blackett replied that they did not have any room. From Glasgow, Max Born replied that he was too old to supervise research and from Birmingham, Rudolf Peierls immediately sent him a three-penny folded airmail saying: "Come ahead." So Gerry accepted, and as he would often repeat later on, this was a life-changing choice. But, as Gerry realized, the first and foremost task to be achieved before boarding any ship in order to pursue academic physics was to complete his doctoral research.

His thesis, eventually, would consist of five papers. The first one was on the "Effect of Nuclear Motion on the Fine Structure of Hydrogen" [21]. The problem was re-examined from the point of view of the Hamiltonian applicable to two charged particles and accounting for effects of order v^2/c^2 where v is the velocity. The work differed from the earlier work of Bechert and Meixner [22] in that the energy of Dirac's electron in a central field was taken as the reference point. The solution was carried out in terms of an eight-component rather than a four-component approximation to the 16-component wave function. The result is the same for practical purposes as that of Darwin for the prequantum-mechanical problem [23] and of Bechert and Meixner for the four-component approximation. The energy formula is affected only by the original Bohr reduced mass correction to the term value and by a term which is independent of the particular fine structure component and depends only on the principal quantum number. Bethe's electrodynamic shift [24] is, therefore, not obscured to within terms of relative order $\alpha^2/(m/M)$ effect of nuclear motion. Here α is the fine structure constant $\alpha = 1/137$, and m/M is the ratio of electron mass to proton mass, $n/M \sim$ .00054, very small. This paper was received by the *Physical Review* July, 26, 1948, just after Gerry's 22nd birthday. So he felt that he was making progress! This was soon followed by "The Effect of Nuclear Motion on the Hyperfine Structure of Hydrogen" [25]. When George Arfken and Gerry took the manuscript for the latter to Breit's

as a physicist, although his radicalism hampered his advancement. After the controversy, university trustees instituted a loyalty oath for employees.

office for him to look at, they did not know that one of his postdocs had just resigned. This came as a real blow to Breit. Public recognition for his work had always been slight, a fact for which Breit himself was somewhat to blame, because his papers were written in a convoluted style. In Niels Bohr's complementarity between truth and clarity, Breit was the most truthful, and therefore most obscure theoretical physicist! But he usually had a rapport with this graduate students and postdocs, so the resignation of a postdoc, in Gerry's recollection, really affected his mentor's pride. He decided to take it out on Gerry and George. It did not help that preparing a paper for publication for Breit was a great chore. Circles in blue or red were drawn around each Greek symbol such as α and then a line with a ruler was drawn over to the side and labelled "Greek alpha, l.c." Recalling the episode when their paper was finalized for publication, Gerry later recounted that he had forgotten to label in such a way the alpha on the first page of the manuscript. "Gregory asked: 'What is that symbol?' I said, 'Oh Professor Breit, I only labelled it on the second page; I forgot to do it on the first.' George saw that Gregory was losing his temper by seeing the blood vessels expanding on the back of his neck and he made it to the door. I was not so agile, and a chair got in my way. Gregory grabbed the back of my belt and threw me into a chair, swearing at me and finishing with 'when I talk to you, you listen to me!' I was really pretty upset by this and left for a week. I would like to think that I was philosophical and said to myself that Gregory could not help himself and, after all, I had been accepted by a world's expert. In any case how in the world could I be working with a 16-component relativistic equation at age 22; I wasn't any genius. Whatever I thought I had no inclination to leave Gregory and when I got back to Yale, he behaved very kindly towards me."[n]

Gerry wrote another paper with George Arfken, "Effects of the Proton Radius on Nuclear Motion Correction for the Hyperfine Structure of Hydrogen" [26]. Then he went home to South Dakota for the summer of 1949 to write his thesis, in which he also included a nice relation he had discovered, regarding Dirac's theory of the electron [27]. When he had his thesis typed back at Yale, Breit read this note and came rushing up to his office at night, admitting that he had not realized that Gerry had worked on this. Breit himself had published a similar short note on a different topic in the *Proceedings of the National Academy*, and to Gerry's delight offered to propose this note for publication there, too.

[n]Private communication.

Gerry returned from Christmas vacation in 1949/50 to Yale and announced that he would defend his thesis in late January. The entire department was supposed to attend a thesis defence. Gerry had chosen a room that had blackboards around the side and he had them all filled with equations involving the 16-component Breit wave functions. One by one the faculty members stuck their heads inside the door, looked around and disappeared to their offices, presumably taken aback by what covered the blackboards. Only Gregory Breit and his students remained!

Towards the end of January 1950, Breit asked Gerry to assemble all of the numerical calculations he had done, so that they could be used by the other students "to build on". This was a frightening prospect as Gerry only found two sheets of paper where, in his initial work on the hypergeometric functions, he had carried the JWKB approximations to second order. He put these in a clip folder and gave them to his mentor. Breit looked unbelievingly at his meagre array of calculations and exclaimed: "Do you mean that this is *all* you did during my 3 1/2 years with me? I have a mind to keep you another year so that you learn how to work." As he recalled later, Gerry summoned up all of his courage and for the second time during his Yale years he confronted his tutor: "Well, Professor Breit, I defended my thesis and deposited it in the library. I published 5 papers. I have arranged a postdoc position in England, and I have a boat ticket for tomorrow afternoon. I don't see what you can do about it." On hearing this, Breit called all of his students in, 12 of them, and sat them around a table. He disdainfully held up the two sheets Gerry had given him and said: "I want to show you what I do NOT expect of you when you leave."

Of course Gerry was upset. But despite the unsatisfactory valedictory experience, Gerry felt strongly that he had learnt more physics than he could have learnt anywhere else in the US, not least because Gregory Breit had given him a physics problem that suited him down to the ground. However, his experiences with his second formative influence, the second Eagle on the other side of the Atlantic, the German-born British physicist Rudolf Peierls, were diametrically opposed — yet equally stimulating, if not more so.

3. The Second Eagle: Rudolf Peierls and Birmingham

On February 4, 1950, Gerry arrived in Southampton on the USS America. On the train up to Birmingham, he shared the compartment with two businessmen from London. They kindly advised him that he should write the address of his digs on a piece of paper, as the taxi drivers in Birmingham would not understand him. This seemed a rather ominous start to his first visit abroad, especially given that Gerry had chosen his destination on the grounds that this would be one of the few countries where he would be able to make himself understood without problems.

In fact, he found his new home in Harborne without problems. Jens Lindhard, who was on leave from the Institute of Theoretical Physics in Copenhagen to spend some time at Peierls' department, had found Gerry accommodation in a large boarding house near the university. Mrs Bedford, the landlady, was a good cook, and Jens and Gerry enjoyed their breakfasts and suppers there. What they did not enjoy was the fact that the only heating in the house was an open fire, lit on the ground floor at supper time. Mrs Bedford had heard about Americans' problems with adjusting to the British heating habits and offered Gerry two extra blankets for the first nights. So Gerry's first impressions were that life in England, indeed, was very different; in fact, as he would point out repeatedly later, he felt that everything was different.

Yet, what he had embarked upon were a wonderful ten years during which he was brought into the extended Peierls family and he learned from his second Eagle, or — as he was known affectionately among his students and postdocs — "Prof" his way of doing theoretical physics. In his autobiographical essay, Gerry — displaying his appreciation of German literature — quotes Eugen Roth who wrote "Der Mensch schaut in die Zeit zurück und sieht, sein Unglück war sein Glück."[o] Indeed, what may have been perceived as a difficult circumstance, effectively being forced to leave one's home country and resettle elsewhere to avoid political prosecution, turned out to be a stroke of good luck. While for many of his friends their political activities in the 1950's would become real stumbling blocks on their way to academic success, Gerry, in stark contrast and by sheer coincidence, ended up in what by many would have been regarded the most stimulating theoretical physics group in the world: Prof.'s Group.[p]

They came from all over the world to England's sprawling industrial town of Birmingham, where one had to go into the centre of town three miles away from the university at Edgbaston, if one wanted a warm meal after 5 p.m. With Freeman Dyson's arrival at Birmingham in the autumn of 1949,

[o]Man looks back in time and realises that what he perceived as being misfortune, in fact, was good luck (S.L.).

[p]This subchapter is based on a talk which Gerry Brown gave at Oxford in October 2004 on the occasion of the naming of the Theoretical Physics building as the Rudolf Peierls Centre for Theoretical Physics.

quantum electrodynamics arrived at Prof.'s institute, as Dyson taught students, staff and Prof. alike initially about the Feynman/Schwinger/Dyson results and later about the new formulations of the old theory. Freeman Dyson was sharing an "office" with Jens Lindhard, a small room in army barracks, technically "Quonset Huts". Accommodation within the physics department was notoriously limited, and huts were employed to deal with the problem. They were hardly luxurious. The heat went off at 5 p.m. and dark air rose through the holes in the wooden floors. Comfortable it was not, but it was conducive to a clear sense of "being in it together" that developed among the young scientists from all over the globe.

The Australian Dick Dalitz was finishing his Ph.D. thesis, having moved, for want of people to talk with, to Birmingham from Bristol where he had spent the preceding year rederiving the formulae in Wolfgang Heitler's book on the quantum theory of radiation [28]. Wladik Swiatecki was finishing his thesis on the surface energy of nuclei. Geoff Ravenhall, the exception to British education, in that he had come to the university from a secondary modern school rather than through the traditional grammar school, had just finished his Ph.D. thesis on the electron-electron interaction in atoms. The 6'6" Australian Fred Barker, who studied nuclear interactions towered over all of them. Stuart Butler also came from Australia and had already done some original work on deuteron stripping. Hugh McManus was a lecturer having written his thesis in Birmingham on the classical extended relativistic model of the electron, and Max Krook, George Kynch and Gerry Field were lecturers in this Department of Mathematical Physics that Prof. had put together after returning from Los Alamos where he had played an important role in the theoretical division of the Manhattan Project. As Gerry later commented "they came like bees searching for a favourite location to build an edifice, and they did just that, forming Prof.'s group with Rudi Peierls' wife Genia the queen bee." No matter where they came from, they were taken into the Peierls home, fed and entertained and they were freed of the usual problems of life outside of physics at least from making difficult decisions: Genia made them for them.

The sheer lack of comfort of life in Birmingham, the rationing of coal that was still going on in 1950, the twelve pence of meat they could buy each week (although liver and kidney could be got outside this ration if the butcher was so kind as to sell them to you) united Prof.'s group in camaraderie, and made them all highly appreciative of Genia's cooking. Moreover, it made everybody look forward to the Peierls' parties (in their tremendous leasehold house) which lasted into the small hours of the morning, the last hours punctuated by Genia's screams of delight as the opposing teams in charades neared their guessing goals, helped by a complex detailed sign language such as patting one's rear to add letters onto the part of the answers already guessed. In short, Prof.'s group were one large family, thrown inwards on each other by the then lack of outside distractions in this large industrial city.

The group had its daily routines. Starting with work in the huts in the morning followed by a stroll over to the main building for coffee in mid-morning to talk with the experimentalists. They passed the "Poynting Vector", a high campanile which they jokingly called that name, Poynting having been professor of physics at Birmingham. They had long lunches in the university refectory of overcooked cabbage and mashed potatoes, shepherd's pie (mashed potatoes with bits of hamburger meat hidden in yet another plate of mashed potatoes, all for twelve pence.) then coffee again in the common room, and then, finally, tea with the experimentalists in the afternoon.

Entertainment was limited, and so many in the group took part in the tennis played at the courts just below the huts, running back into the huts when the rain began, dumping their sweaty clothes over the heating radiator — until Genia came and enforced more sanitary arrangements. And, as many would later testify, they were happy! They loved research; and they knew that this group provided as good a setting for theoretical physics as any other anywhere else.

Ever since entering the extended Peierls family, Gerry would claim (with no disrespect for his first wife Traudl or, indeed, his second wife Betty) that Genia Peierls was the most remarkable woman he had ever known. He would classify her with Berthold Brecht's *Verfremdungseffekt*. This was a technique, essentially a scene in his plays, to withdraw the audience from the emotional drama of the play to give them time to "regroup", i.e. to look at the situation dispassionately and consider it in the cold light of reason. Most people are prisoners of habit; as they march ahead they put one foot ahead of the other and follow the path they began earlier without much contemplation. Genia could step aside, survey the situation, and come up with a completely novel solution to a problem that no one else had thought about. Of course, the person presenting a problem had to accept Genia's solution — Genia would check by long telephone calls that the subject had understood her solution and then came to check that he or she was carrying it out. Like many of members of the extended Peierls family, Gerry found that life was not always easy with Genia, but she was immensely kind. She had a voice which rocketed through the Peierls' leasehold mansion, such as: "Gerry, Breakfast!" from the corner of consummation to the opposite one of Gerry's bedroom, and one could feel the whole house shudder.

Gerry later vividly recalled his first few weeks and months in the Peierls household. After supper he would usually go out for a stroll in the evening to get fresh air. Before going to bed he would take a long bath, filling the bathtub with hot water. Often this was the only time of the day that he was sufficiently warm as to not shiver. The bathroom was next door to the bedroom, and Prof. and Genia would generally go to bed soon after the dinner. They would converse in Russian about everything that had gone on during the day and Gerry could hear Mrs Peierls exploding in delight as Prof. told her something that amused her. Gerry later commented that he had never seen a stronger love than existed between them. Prof. was certainly "man for her".

It was during those early years in Birmingham that Gerry first met Prof.'s old friend from his student days in Munich, Hans Bethe. The two had been graduate students under Arnold Sommerfeld and they remained life-long friends, occasional scientific collaborators and guardians of the Birmingham-Cornell pipeline, an incredibly successful informal scientific exchange set-up in which the two physicists sent each other their most promising doctoral and post-doctoral researchers to the benefit of their respective institutions and the scientific community more generally. Little did Gerry know at that time that more than quarter of a century he would begin a fruitful and long-lasting collaboration with Prof.'s closest friend, when, as will be described below, in the late 1970s they began working on astrophysical problems.

When Peierls was preparing to go on sabbatical to Princeton, he asked Gerry to take his senior year undergraduate course in hydrodynamics in the spring term of 1952. Gerry told him that he had never even studied a course in this subject, and Prof. replied that teaching it would be an excellent way to learn the material. In fact, the course evolved quickly into compressible flow, shock waves, etc. and Prof. had Hans Bethe's lecture notes from Los Alamos. So Gerry already in those early Birmingham days learnt Hans Bethe's way of handling shock waves, long before he got to collaborate with him decades later.

During the winter he had been going to the overseas' club in the middle of Birmingham, a social centre run by the city of Birmingham so that young people, students, au pair girls, etc. from abroad, could have tea and enjoy social life. Gerry had met a girl from Alsace and tried to improve his French. When he wanted to learn German, his fate was to meet Waltraud (Traudl) Kunick in his quest to improve the language which he had studied one year at the University of Wisconsin. Genia soon formed an idea of Gerry's evening social activities, but did not say much about them. Gerry invited Traudl to the Peierls Christmas Dinner in 1951. She arrived by bicycle several hours late. Genia was enthralled by this careless attitude, because she always told Gerry that he took life, including appointments, too seriously. One weekend in March, Genia confronted him with: "Gerry, you are aesthetic mute! You should get married. If you are married before first weekend in April you can save tax. You can save 10 or 15 pounds!... I will give you wedding reception." Just the same weekend Traudl told Gerry that she and her friend, who were both au pair girls taking care of the house keeping for a dormitory of nurses who worked at the Selly Oak Hospital, had been fired, and that she was going back to Germany. Gerry said: "Nonsense. We will get married." So Genia got out the invitations to the wedding reception: "Professor, in absentia, and Mrs Peierls invite you to the wedding reception of their boarder, Dr. Gerry Brown, on April 1 at their home at 40 Carpenter Road, Birmingham." Gerry had been pretty discreet, especially with Traudl's friends, and several of them thought that the whole business was an April Fool's joke, and did not come. But the nurses from the hospital came and Genia laid on a colourful affair. (One did not have the money at the time to make it lavish.)

When Gerry arrived in Birmingham, Freeman Dyson was giving crystal clear lectures on quantum electrodynamics, an area in which Gerry wanted to work. Other years Prof. lectured on solid state physics, Dalitz on particle physics and Max Krook on mathematical problems. Everybody attended those lectures: graduate students, postdocs and lecturers. In every subject they were brought up, if they paid attention to the course, to a level that allowed them to do research. And most of them later became professors at good universities, mostly working on completely different problems to the ones they had written their theses on. Prof. seemingly knew everything; criticized everything. He worked on a plaster wall, painted black to mimic a blackboard, a somewhat partially hollowed long piece of wood below to hold chalk. Any time he was not sitting or writing or on the phone — there was only one phone in the entire hut, in the secretary's office with extension to Prof. — colleagues could go in and try out a new idea on him, getting a reaction; severely criticized for the most part. At first it was unnerving that he knew everything. Then one went away, revised the concepts, and improved calculations. One did not have to wait long for response. They all argued with each other, invented new ways of doing each other's problems, went to the seminars, talked with many visitors, all of this organised by Prof. on what today would be considered a shoestring. As a reminder, the yearly budget for the office work of mail, preprints (which they produced by typing on a paper with wax on it, then tying it around a cylinder which went through the machine, ink coming out of the places that had been typed onto the paper.) came to £ 175 per year.

In Yale Gerry had tried to learn from Schwinger's papers [29]. He began telling Prof. about Schwinger's correlation functions. He immediately wrote them down on the blackboard-painted plasterboard and started calculating with them. Gregory Breit would never have gone about matters that informally. With him one would have had to have made an appointment to discuss the matter at some time in the future and given him copies of the topic one wanted to discuss. In fact, Gerry had kept the Schwinger papers hidden in his desk drawer at Yale, reading them secretly. He had been on the second floor of the tower in Sloane Physics Lab and Breit on the first floor. When he heard Breit's footsteps on the staircase he would put the papers back in his desk drawer and seem to be working on calculations with his 16 component wave function. Breit did, however, take an interest in Schwinger's work. Whereas Schwinger calculated from the QED the vacuum fluctuation correction to the charge of the electron in hydrogen, Breit published a paper showing that Dirac theory left this addition as a possibility [30]. Schwinger then showed that there was an error in Breit's work, the only error that Gerry ever had been aware of, even though he had had to read almost all of Breit's papers on theoretical physics. Breit was very unhappy about the error, pointing out to Gerry that he had had to make his calculation while standing over his desk because he had had a painful boil on his neck. Such human matters are usually washed away by history.

Gerry was totally amazed both by Prof.'s lack of formality, and by his ability to delve immediately to the central core of any given problem. The first day Peierls showed him why the Feynman correlation functions were so much easier to use than Schwinger's, because of their boundary conditions. As mentioned above, Freeman Dyson had already in the autumn of 1949, lectured about the Feynman, Schwinger and Dyson results in quantum electrodynamics, for which the first two, plus Tomonaga, later received the Nobel Prize. Dyson was a Royal Society Research Fellow at the time. He had beautiful lecture notes [31] of which Edwin Thompson Jaynes once remarked: "For a generation of physicists they (the lecture notes) were the happy medium: clearer and more motivated than Feynman and getting to the point faster than Schwinger." But better than having just his lectures Gerry had Freeman, the man himself! If Gerry's frequent interruptions ever annoyed Dyson, he was too polite to mention it, but Gerry later remarked that Freeman was something like an honorary short-term Eagle between Gregory Breit and Rudolf Peierls, because he considered himself to be his postdoc during his first two years in Birmingham — at least from the vast amount he learned from Dyson. The one thing he did not learn, however, was how Dyson himself actually did research. Gerry later remarked that he seemed a little like C.N. Yang, whom he met at Princeton later and who would eventually bring him to Stony Brook.[q] Freeman and Frank were similar in that they arrived at "gems", brilliant and glowing, but they seemed to arrive at them by sheer genius more than work, although Freeman hinted to Gerry at one point that he had worked very hard at some research some time.

In May of 1950, three was an international meeting of Theoretical Physics in Paris. This had, for many reasons, a profound influence on Gerry's professional life. Only just over 100 theoretical physicists attended, and those included most of the famous theoreticians from all over the world: Niels Bohr, Wolfgang Pauli, and many others. As common in those days the conference covered virtually all subjects in theoretical physics. In those "olden" days the good theoretical physicists followed almost everything in every subject; at least they were broadly interested.

One evening Gerry was invited by the Peierls to have supper with them and Richard Feynman. He went to the Peierls hotel room, which was on the 3rd floor of a French hotel. Prof. was ready to go, but Mrs. Peierls was in the bathroom, in fact, according to her, locked in the bathroom. Prof. shouted through the door: "Turn the key the other way." She said that she had turned the key the other way. After a while it was clear that they would be late for dinner. The bathroom was in a room built out perpendicular from the bedroom, so that the window of the bathroom was accessible from the one in the bedroom. The only problem was that there was the possibility of the three storey drop to the pavement below. Gerry put one of his feet on the hotel room, where Prof. held his hand, stepped across with his other foot to the sill of the open bathroom window, and then Genia pulled him across. He turned the key "the other way" and the bathroom door opened. Gerry always believed that it was his chivalry (perhaps complemented by his love of gardening) that led the Peierls to invite him to stay with them in their house as a boarder from the autumn of 1950.

[q]See paper by C.N. Yang in this volume.

The Brown-Ravenhall "Disease"

Freeman Dyson taught his Birmingham colleagues that *the* problem to be solved was the bound state problem for two electrons, the Lamb Shift which arose from the vacuum fluctuations having been solved. Gerry knew about the Breit electron-electron interaction, which had been used in calculating the atomic spectra of helium. When he arrived in Birmingham, he found that Geoff Ravenhall had just finished his thesis on the electron-electron interaction problem, using Feynman's covariant result and that he had derived a new term, additional to that of Breit, arising from the scalar and longitudinal interactions.

Gerry was unhappy with his result and argued with him a lot. Yet, somehow, Prof. catalyzed a collaboration which led to a real gem in atomic physics, although it is known, rather unfortunately, as the "Brown-Ravenhall Disease". The collaboration leading to this discovery was somewhat typical for the way research was carried out at Prof.'s group. Prof., Geoff Ravenhall, and Gerry spent a morning looking at the two electrons interacting in helium, on the blackboard. They had not filled the electronic negative energy states; they were looking at only two electrons. One of them (most likely Prof.) realised that the wave function was unstable. The two electrons could interact, one of them dropping into a negative energy state and the other one going off in a positive energy state, conserving energy. So the wave function would vanish in a short time. Having recognised this, they filled the negative energy electron states with electrons, and began by projecting the two interacting electrons with positive energy states. They then introduced pair production, but it was cut down considerably because the intermediate states had high energies, since they included the electron and positron masses, the positrons being the holes in the otherwise filled negative energy electron sea. Beginning from a wave function with only (in plane wave presentation) positive energy states, they then showed that only small corrections were needed to take care of the infinite sea of negative energy states. The result of these calculations *à trois* were published as "On the interaction of two electrons" [32]. The paper had an interesting history which is still continuing more than half a century later, although no-one paid much attention to it until 24 years after its initial publication. Atomic physicists began putting two-body correlations into Hartree–Fock calculations, the latter being mean field theories without any two-body correlations. They found that, upon iterations, their wave functions disappeared on them. Joe Sucher, professor in particle physics at the University of Maryland, realized what was happening. In 1985, he constructed a simple model to show that continuum dissolution occurs if the negative energy states are left empty [33]. Sucher ended the abstract to the publication of his results with the sentence: "A simple model is constructed to illustrate the validity of this "folk theorem", first noted by Brown and Ravenhall in 1951." Sucher cast the problem by labelling the uncorrected (without projection onto positive energy states) as continuum dissolution. This dissolution, of course was what had first struck Gerry, Ravenhall and Prof. However, the need for projection onto positive energy states remedied a problem with the Breit interaction [34].

$$I = \frac{e^2}{r_{12}} - \frac{e^2}{r_{12}}(\vec{\alpha}^{(1)} \cdot \vec{\alpha}^{(2)} + \vec{\alpha}^{(1)} \cdot \vec{r}_{12}\alpha^{(2)} \cdot r_{12}).$$

The α's here are Dirac operators for the velocity (actually 4×4 matrices), but one can think of their expectation values as v/c. In calculating spectroscopic correction in energy to the electronic states in helium (which has two electrons) Hans Bethe had iterated I to second order which meant taking sums of matrix elements of I between the ground states and intermediate eigenstates, squaring these matrix elements, dividing by $E_0 - E_i$ where E_0 is the ground state and E_i is the (common) energy of the intermediate state and the summing over i using closure [35].

With his results, Bethe found large contributions from second-order terms that were in disagreement with experiment. Gregory Breit was piqued by this, saying that he had arrived at his two-electron interaction by analogy with the classical interaction, and it should not be used to higher order. Indeed, this was what Ravenhall and Gerry had found. The matrix elements of the Dirac operators $\vec{\alpha}$ for the velocity, when taken between positive and negative energy states, were of order unity. Ravenhall and Brown showed that these should be excluded, by using projection operators which limited the interaction to operate only between positive energy states. When negative energy states were included, in higher order, the denominators in the expressions had to contain the energy of the electron-positron pair. Sucher's paper organized this so that it could be used in the electron many-body problem.

The fact that if one simply went ahead in, for instance, a relativistic Hartree-Fock calculation, without special treatment of negative energy states, then the wave function would disappear, came to be known as the Brown-Ravenhall disease, which was not particularly complimentary to Geoff and Gerry. It had taken 24 years until anyone had noticed their paper, because in the Hartree-Fock (mean field) calculation only one particle is allowed to change state at a time. Before the "disease" sets in, one of the electrons has to go off in a positive energy state, the other in a negative energy state, so the two-particle interactions must be put in before the disease "breaks out". The correct way to handle the negative energy states is to treat them as an electron-positron loop correction. Without the special treatment of the negative energy states, the wave function would disappear. The Brown-Ravenhall "disease" is well known in the atomic problems and taken care of by introducing pair theory; in contrast, the question of stability of quantum mechanical solutions, i.e. the problem that they might "fall through" any preset lower limit, has occupied mathematicians more or less continuously. In fact, there is a thin but steady stream of papers being published even today dealing with this issue [36].

Electron-electron interaction in heavy atoms

The next piece of work that Gerry engaged in at Birmingham was on the derivation from quantum electrodynamics of a Hamiltonian describing the interaction of two fast-moving electrons in a heavy atom to order $a = e^2/\hbar c(\alpha \cong 1/137)$. The case considered was exemplified by the interaction of the two K-electrons in a heavy element, where v/c, the ratio of the electron velocity to that of light, is of order Z_α, where Z is the nuclear charge of the atom. His paper "Electron-Electron Interaction in Heavy Atoms" was only six pages long [37], but the result surprised many at the time and has been used frequently since then. The main point of the paper is that if the wave functions of the electrons are determined by the nuclear potential, which couples with Z_α, α being the fine structure constant $\alpha = 1/137$, then for large Z the electrons are in *stationary* states, so that the usual retardation correction between interaction at different times in electrodynamics is zero.

The end result of the electron-electron interaction, to within order $1/Z$ corrections for heavy atoms is

$$H_{\text{int}} = \frac{e^2}{r_{12}}(1 - \underline{\alpha}_1 \cdot \underline{\alpha}_2)$$

where the α's are the Dirac operators for the velocities.

Dick Dalitz immediately pointed out that the H_{int} would have a complex factor in excited states of the atom. Evolution of the matrix elements of this factor gives, upon squaring it, the time dependent life-time of the state. However, applied to a state above the ground state, in which all of the lower single-particle states are filled, the imaginary parts from the various levels cancel out between each state, and the system is stable. In conclusion, the idea that a strongly bound state of the two innermost electrons should have no retardation in the interaction between the two electrons was really easy to work out, but nobody had done it, largely because nobody had realized that a stationary state in quantum mechanics was really stationary. (Classically, the faster the electrons move the greater the retardation.)

As with the "Brown–Ravenhall disease" and with the Raleigh Scattering, which will be discussed below, these were all problems where the solution contradicted "accepted wisdom". Gerry later frequently expressed himself fortunate to have had advisers with the knowledge and confidence of Rudolf Peierls, and later Hans Bethe, who would immediately see what other investigators had missed, and who would similarly give him confidence in his research. This gave him the courage to uphold even the most counterintuitive results. The general reaction of many of Gerry's colleagues to results that contradicted accepted wisdom was that his work had to be wrong, as otherwise someone else would have come to a similar conclusion before him. However, Prof.'s encouragement, occasionally to believe in the acceptance of "surprises" in theoretical physics served Gerry well in standing firm in the incidences when he stumbled across such unexpected or counterintuitive result.

The interaction in the above equation differs from the Breit interaction through having no retardation in the interaction between electrons; i.e. it has the two electrons interacting instantaneously. At first sight this may seem strange, because in a heavy nucleus like lead or uranium, the two electrons run around with nearly the speed of light c, whereas in the lighter nuclei where Z is smaller, they do not travel as fast. The point is, however, that in a heavy nucleus like lead with $Z = 82$, the innermost electrons (so-called K) electrons are locked into the innermost state of the Pb nucleus. To move them to another state by interaction with the other K-electron, (one of them has spin up and the other has spin down) will involve a change in wave function of order $\alpha/Z\alpha = 1/Z$ where α is the fine structure constant. If $Z \gg 1$ then $1/Z$ is small and so is the change in wave functions. Thus, the K-electrons are locked into what are called "stationary states" in quantum mechanics, and nothing is moving; therefore there is no retardation.

Gerry found a good application for his 1952 paper by using it in the context of lattice gauge calculations of the energies of the pseudoscalar π and the scalar σ mesons as the temperature T moves downwards through Tc, the critical temperature for transition from quarks and gluons to hadrons. This is the transition the early universe went through some seconds after the big bang, here the transition from quarks and gluons at Tc is to hot matter involving nucleons, also, which build up the elements found on earth. Lattice gauge calculations enable one to solve the Yang-Mills equations which describe the quark-gluon system above Tc. In Yang-Mills, in the region not far above Tc, the main interactions between quarks are just Coulomb interactions. They change with temperature in a different way than the Coulomb interactions discussed above. Since the Yang-Mills theory (quantum chromo dynamics), which has the property of what is called asymptotic freedom, the Coulomb interactions (and other ones) go to zero as $T \to \infty$. However, in the region of temperature just above Tc, they behave like normal Coulomb interactions.

The π and σ particles are each made up out of bound states of a quark and antiquark, the antiquark having the opposite charge to that of the quark. In this sense, they are like positronium, which is made up out of a bound state of electron and positron. In a number of papers, Hatsuda and Kunihiro [38] showed that the π and σ meson masses went to zero at Tc. A proviso must be added. This statement is true in what is called the "chiral limit". However, the π-meson gets a small mass from coupling to a heavy, so-called "Higgs" particle which comes at a much higher energy scale than QCD. This small mass of the pion goes through Tc, but is neglected in the present argument. Somewhat surprisingly, in the neighbourhood of Tc the Coulomb interaction not only predominates over the other interactions, but behaves exactly like the Coulomb interactions between electrons and electrons and positrons.

The latter calculations determine the Coulomb potential between quarks and antiquarks for the case of heavy quarks. These find for T near Tc that $\alpha_s \cong 1$. The quark and antiquark masses are ~1 GeV so in order to make the bound states massless, one needs a binding energy of ~2 GeV. The lattice gauge calculations find for the Coulomb potential $V = -\alpha_s/r =$ with $\alpha_s = 7$, that the binding is 1 GeV. This is the same as in the non-relativistic hydrogen atom that the net binding energy is 1/2 of the total potential energy, half of the latter being cancelled by the kinetic energy. This 50% cancellation results from what is called the virial theorem. Thinking more deeply about the question one might wonder why the non-relativistic Schrödinger energy applies to what is clearly a relativistic situation. In the lattice calculation one has to regulate the singular behaviour of the short range potential. In doing so, the short-range part of the potential is rounded off, not going to ∞ as $r \to 0$. This regularization procedure removes the relativistic aspects so one can use arguments for the hydrogen atom of the non-relativistic Schrödinger equation. The fact that results are checked by chiral invariance justify this direction. Thus

$$H_{\text{int}} = \frac{2e^2}{r_{12}}$$

and the binding energy is doubled, bringing the bound state energy of π and σ, composed of quark and antiquark, to zero. In the way described here, one brings the masses of quark and antiquark, which make up the π and σ mesons, to zero and transforms the quarks in lattice calculations from those of mass ~1 GeV to massless ones [39].

The above is an example of what will be mentioned below as "Brown-Rho Scaling", in this particular case the meson masses which come from breaking chiral invariance going to zero as chiral symmetry is restored at Tc. Either density or temperature can act as background fields to scale towards chiral symmetry restoration. In a practical sense, the lattice gauge calculators seemed to be pretty happy that Gerry corrected their heavy quarks to nearly massless quarks in the region near Tc where they were calculating.

Surprises in theoretical physics

As mentioned above, Prof. encouraged his students and younger colleagues to have the courage of their convictions and not to shy away from standing up for counterintuitive or "surprising" results of their scientific research. Peierls assembled such surprises in two very popular books: *Surprises in Theoretical Physics* and *More Surprises in Theoretical Physics* [40].

One piece of Gerry's research contained such a surprise and was duly included in the first volume. If an atom is exposed to radiation of a frequency much higher than the resonance frequency of even K electrons, the dominant processes are Compton effect and photo-ionization. However, for some purposes one is interested in the coherent scattering, in which the atom remains in its ground state, although the cross section of this is rather small.

If one uses Dirac's equation for describing the atomic electron, the scattering of radiation is a two-stage process in which first the incident proton is absorbed, and then the secondary photon emitted, or vice versa. The contribution to the scattering amplitude of the first process is proportional to

$$\sum_n \frac{\langle \varphi_0 | e'_\mu \gamma_\mu e^{-ik\cdot r} | \varphi_n \rangle \langle \varphi_n | e_\nu \gamma_\nu e^{ik\cdot r} | \varphi_0 \rangle}{E_0 + \hbar c\omega - E_n}$$

where e_μ, e'_ν are the polarization vectors and k, k' the wave vectors of the incident and scattered photon, respectively, γ_μ the Dirac matrices, φ_0 the atomic wave functions for a K electron, and φ_n the wave function of any excited state. The summation over n includes integration over the continuous spectrum. So far this result is exact, except for radiative corrections of the relative order of magnitude $\alpha = e^2/\hbar c$, if one includes also the similar term in which the emission precedes the absorption.[r]

For high k, it is reasonable to neglect the potential energy and to replace the scattering states by plane waves. Rather "surprisingly", it turns out that the result for the scattering amplitude this way is wrong and gets worse the higher the photon energy and the lighter the atom.

This had been shown by Gerry Brown and Woodward [41] when they calculated also the next term obtained, by adding to the plane waves for the intermediate states the first order Born correction for the scattering of a free electron by the Coulomb field, finding the ratio for the first-order term to the approximation which uses only plane waves.

$$\frac{A_1}{A_0} = \frac{\Delta p}{\pi Z \alpha mc} - \frac{1}{1 + \frac{8mc}{Z\alpha\Delta p}}$$

with $\Delta p = \hbar|k' - k|$ being the momentum transfer.

The reason for this surprising result is that the "leading term" becomes extremely small. In more physical terms, as argued by Brown-Woodward, the momentum transfer Δp must ultimately come from the atomic nucleus. In the approximate expression A_0 the effect of the nucleus is felt only in the ground state wave function φ_0. Therefore one relies on the momentum fluctuations, contained in that function because of the interaction with the Coulomb field. In the correction term one allows the electron to exchange momentum with the nucleus while it is in the excited state, and the comparison shows that it is easier for the electron to pick up momentum when it is already in a state of high energy.

When Gerry was engaged in the calculations for this paper, he had to deal with three-dimensional equations, which could be expanded in

[r]See [40], p. 40.

a series of angular momentum eigenstates, which converged well if the wavelength $2\pi/k$ was not too small compared with the range of the atomic wave function φ_0. The equations could be conveniently solved on electronic computers. Computing, in the 1950s, however, was a rather different experience to today. There were no computers in Birmingham. The EDSAC I, the computer in Free School Lane in Cambridge, the first university electronic computer, though a real step forward from the Marchands and Monroematics that Gregory Breit's students had been using, was still primitive compared with even the most basic of modern computers. EDSAC I consisted of a set of long mercury tubes in which numbers were stored to the base 10, a blip travelling down a tube comparing the number, and instructions were put in by a punched tape which had to be programmed in binary. Initial instructions took up the first approximately 10% of the total storage of 1024. If high accuracy was required, the storage bins could be joined together two by two. Gerry was given the privilege to use the machine, but only as a partially qualified user, which meant that he could turn the machine off, but he was not allowed to turn it on!

At the time, Gerry had no grant to support the work. His salary was £ 500 per year, when he started at Birmingham. This was adequate but not generous, and money tended to be tight. Therefore, in the beginning, once lectures were finished on a Friday, or sometimes sooner, he hitchhiked cross country from Birmingham to Cambridge. The Cambridge eateries were novel, compared to Birmingham, and Gerry usually made use of the luxury of a good meal for half a crown (two shillings, six pence). Then he would go up to Free School Lane to the EDSAC and punch his program on a spool of paper tape. Towards midnight the fully qualified users would have gone home, and he finally got to work on the computer himself. But by that time, connections of some of the vacuum tubes, which ran the machine, were loose. Gerry's high-tech approach to this problem was to take his shoe off and gently hit the vacuum racks until they operated well again. Then he would run his spool, but usually, there were still errors in it. With no support programs to help find the errors, Gerry would go back to Birmingham and turn over the entire programme in his head, trying to deduce where the errors could be. If the machine stopped or if he could get no further, or if some runs really seemed to work, he would work through the night until early morning, shutting the machine off. Then he would sleep on a cart in the Ladies' restroom where a blanket was conveniently left, and resume work Sunday morning, assuming a fully qualified user had come in and turned the computer on.

In 1953 his work was substantially lightened when Gerry took on Sheila Brenner as a student. She had finished Part III of the Mathematics Tripos at Cambridge, being a whiz in the classical mathematics that Cambridge excelled in. As a Cambridge graduate she was allowed to be a fully qualified user. Furthermore, both her parents were doctors, her father a heart specialist in Queen Elisabeth Hospital in Birmingham, and both had cars, one of which she could borrow on the weekend. Also, Sheila knew many of the older students and younger researchers in Cambridge, many of whom lived with their wives at home, so that their rooms at College were empty. This meant that Gerry could finally do away with the overnight stays in the Ladies' restrooms exchanging them for the marginally more comfortable college rooms.

Of course, Sheila also helped Gerry with the calculations, and as will be described below, showed him in some detail how to do many of them. The result of all this computing was an impressive array of four papers in the *Proceedings of the Royal Society* [42].

First steps to becoming an eagle himself? Gerry's students

One way of grasping some of the features of Gerry's work, especially in quantum electrodynamics (QED) is by looking at the work of the students he advised and supervised at Birmingham. All those students were very remarkable, each in his or her own way.

Prof. had so many students coming to do their Ph.D.'s with him that he was happy for Gerry and other more advanced staff to suggest research problems. He would immediately join in critically and make sure that students worked them out properly. What Gerry did not realize initially, when advising students, was that he himself wanted to know the answers to so many problems that he needed all the help he could get. Remarkably, not only did he not run out of ideas for Ph.D. research — more significantly, he never ran out of doable projects. None of his seventy odd doctoral students got stuck (at least not to the point of despair), and they all finished their theses.

Gerry's first student, Michael Radcliffe, worked on the Bethe-Salpeter equation. It was the first covariant two-body equation, a generalization of Feynman's one-body one, which is still very difficult to solve. Radcliffe tackled solutions for special cases. He learned the new Feynman version of quantum chromodynamics and worked out the decays of the singlet state of positronium which decays into two photons, and the triplet state which decays into three photons.

The second student, Sheila Brenner, has already been mentioned above. She was extremely well-versed in mathematics. Her adviser in Cambridge had been Lady Jeffreys (née Swirles), the wife of mathematician and statistician Sir Harold Jeffreys. Sheila's connections in Cambridge proved to be very useful. Among others, she introduced Gerry to Douglas Hartree. He had made his name at the University of Manchester by the Hartree-self consistent fields for atoms, and then moved to Cambridge. He helped Gerry and Sheila programming EDSAC I and even checked some of the results by calculations with a hand computer. He had used the latter in Manchester to do the self-consistent field calculations.

An interloper into Gerry's quantum electrodynamics crew was the Greek Thermis Kanellopoulos who had completed a Ph.D. in mathematics in Athens, but wanted to learn physics. His English initially was not very good, so Gerry had to design a thesis problem which was chiefly mathematics with little need for language. His thesis consisted essentially of the two papers Gerry and he published together on proton-proton scattering [43]. Birmingham possessed at that time (briefly) the highest energy proton 1GeV accelerator in the world, built by the then Professor of Physics, Mark Oliphant. Gerry assumed that some of those experiments could be carried out locally, although the energy of the accelerator was really too high for the theory he and Thermis used. Furthermore the experiments proceeded only slowly at what Gerry would later refer to as the "British pace". One example of this was that once the machine was built, scientists had to wait a month for experiments to begin running because the inside of the building had to be painted.

With the Canadian Glen Schaefer Gerry began preparing to work out the Lamb Shift in heavy atoms. Schaefer did the mathematical preparation for this in a paper on expansions in angular momentum in bound-state perturbation theory [45], and his thesis was based on this, on another joint paper with Gerry and a third joint paper with another of Gerry's students, Jim Langer. In his first year with Gerry, Langer had worked on a problem in atomic physics, and in the second year he contributed to the calculations leading to this joint paper which developed a method for calculating the Lamb shift of an inner electron in a heavy atom [46]. In 1959, Jim Langer wrote a paper on nuclear physics with Gerry and Cyrano Dominicis, on leave from Saclay in France [47]. Even at that time, Gerry remarked that the width and breadth of Jim Langer's knowledge at that early stage in his career was staggering. Jim went back to Carnegie Mellon University, where he had done his first degree, and later became a permanent member of the Institute of Theoretical Physics at the University of California, Santa Barbara, where remains to this day.

Another American who joined the Birmingham department around that time was Elliot Lieb. He began his research with Gerry on a topic which was a natural spin-off from earlier work on the Lamb Shift of a tightly-bound electron [48]. Subsequently, he went on to spend two years working with Sam Edwards, his doctoral supervisor. While Lieb's physics, even then, was memorable in its own right, as memorable, if not more so, was his stay for its linguistic features. Prof. had allocated Elliot an office to be shared with a visiting Japanese postdoc who did not speak fluent English, with the idea that he would learn from Elliot. He did learn some things. Gerry later recalled hearing him say about his work one day: "I am up shit creek without a paddle." The flip side of the linguistic collaborative effort was Elliot's acquisition of Japanese. After his spell at Birmingham he went on a Fulbright grant to Kyoto!

A particularly close friendship developed between Gerry and the particle physicist of the department, Dick Dalitz. Dick had received his B.A. and B.Sc. degrees in mathematics and physics in 1944 and 1945 from the University of Melbourne. In 1946 he moved to England for postgraduate studies at Cambridge and then worked at the University of Bristol before coming to Birmingham in 1949. There he had written up his doctoral thesis under Nicholas Kemmer on "Zero-zero transitions in nuclei", which he had calculated in the relativistic Feynman theory. Gerry knew from his discussions with him that in Bristol Dick had worked carefully through Heitler's field theory book [49]. It must have pleased him immensely to see how much simpler and more elegant the Feynman theory was. Dick was extremely forceful, and one of the leaders in developing particle physics. In 1951, pondering a new photographic image of an unusual cosmic ray event, it occurred to him that it could be due to a newly discovered unstable particle called the neutral pi-meson decaying into one photon plus an electron and an anti-electron [50]. He calculated that this must happen, on average, once for every 80 decays, and this prediction was subsequently verified. The technique of measuring these so-called Dalitz pairs was used to measure the parity of the pion and the τ meson. Parity deals with the relation of something to its mirror image and at the time it was believed to remain unchanged; unbeknown to him, Dalitz's work was about to overthrow this received wisdom.

Among the strange particles were two -τ and θ- that appeared to be identical twins but for one feature: τ decayed into three pions, θ into two. As the law of parity implied that particles could decay into either an even or an odd number of pions, it was believed τ and θ were different. In 1954, Dalitz looked at decays of τ into three pions and so introduced into physics what he called a "phase space plot", a map-like representation known since as the Dalitz plot. In the half century since their discovery, Dalitz plots have led to the discovery of numerous ephemeral particles. Dalitz then investigated the seemingly different τ and θ particles musing that perhaps the law of odds and evens, "parity", was not true. Prof. discouraged him from pursuing those ideas and it was left to T.D. Lee and C.N. Yang to solve the riddle. They pointed out that the law of parity had been tested only for the strong and electromagnetic forces. For the weak force, the scientific verdict had not yet been given, and it was the weak force that was at work in the $\theta - \tau$ case. They were proved to be right in establishing this connection; yet it was the application of Dalitz's plot that set the sequence in train. Much of the research on this topic had been done when Dick spent 1953–1955 in the U.S., with research positions at Cornell and Stanford, visiting also Princeton and Brookhaven National Laboratory. He returned to England as a Reader in Mathematical Physics to the University of Birmingham for a year before becoming Professor of Physics at the Enrico Fermi Institute for Nuclear Studies and the Department of Physics at the University of Chicago in 1956. He moved to Oxford in 1963 as a Royal Society Research Professor, just in time to welcome Prof., who was about to move to the Wykeham Chair there.

However, Gerry's friendship with Dalitz was established in the early Birmingham days. He was confronted with not only the genius of Dyson, but also the power and knowledge of Dick. After being an outstanding Ph.D. student at Yale, Gerry found himself hopelessly unprepared, by his own admission, to compete in this new environment, but Rudi Peierls took care of that getting all his junior colleagues into close interaction but not into competition, which was one of the great strengths of the department. Dick, as Gerry would later remember, was immensely hard working and serious, and a man of relatively few words in social situations (especially compared with Gerry's own more extrovert way). He and his family, like the Browns, lived on the Bourneville Estate, and the couples saw a lot of each other. In fact, Dick and Valda owned a square box of a car on wheels that they paid £20 for and which Dick kept running, thanks to his ability with machinery. They were all poor, and rationing was still severe. But the joint trips to the central market in Birmingham to get bags of Brussels sprouts, etc. in order to make what they called "Eintopf", a vegetable stew, on weekend refined by some meat, forged a friendship that would last a lifetime.

It might sound unreal to say that Prof. had time for research, but he did, and he did it in the Niels Bohr way. He had ideas on how to work a problem and then went around and talked with all of his students and postdocs. As a result his junior colleagues benefitted, and so did Prof. himself. Yet, occasionally, Prof. would work on "his own" project. Freeman Dyson wrote about the early 1950s at Birmingham [51]: "During my two years in Rudi's group the most important discovery was made by Rudi himself, who found a new and elegant way of deriving the quantization rules for any classical field theory [52]." Prof. began with constructing relativistic Poisson brackets. These were then quantized by analogy with the ordinary Poisson brackets, the commutator between two field quantities being given by

$$\lfloor \Phi_\alpha(x_1), \Phi_\beta(x_2) \rfloor = i\hbar\{\Phi_\alpha(x_1), \Phi_\beta(x_2)\}$$

the quantity in curly brackets being the commutator. All operators are understood to be in the Heisenberg representation. Reasoning to justify this was given by Freeman Dyson. Of course, knowing the commutation laws, one can construct the operators.

The Peierls quantization procedure has proved to be very useful in quantumchromodynamics. In 1953 Prof. published, together with Max Chrétien a paper on properties of form factors in non-local theories [53]. Soon after they published a paper on gauge-invariant non-local interactions [54]. Gerry later admitted to a certain feeling of jealousy when Sam Edwards arrived in Birmingham as postdoc from Manchester. He immediately enveloped Prof. in functional calculations. Prof. was completely immersed, as far as his own research was concerned, by field equations in functional forms [55]. As mentioned above, Gerry had tried to battle through the Schwinger formalism for calculations in quantum electrodynamics while working with Gregory Breit. Once after reading some work of Gamow, he had questioned Gregory about Gamow's complex eigenvalues in beta decay, the imaginary part describing the lifetime, not knowing about Gregory's aversion to complex eigenvalues. Gregory turned to face him and solemnly asked: "Have you been reading a book?" Clearly, Prof. had no such hang-ups — quite the contrary, he encouraged his students to "read books", and on occasion Gerry felt that he had missed out on this experience in his earlier training.

It seemed to him that using Schwinger's formalism for functional analysis, with functional differentiation to obtain the equations of motion was combining complications. In addition, he felt that he simply could not master the formalism. One thing that he did understand was that the actual calculations were carried out by completing the square, with the result that one did not really need to understand the formalism in detail. About half a century later, at the opening of the Rudolf Peierls Centre for Theoretical Physics in Oxford in October 2004, Gerry met Sam Edwards again. In the meantime, Sam had worked on many such problems and trained students, all over physics and all over the world, but especially in thermodynamics and related subjects. Gerry told him that all he had understood of his work with Peierls was that one had to complete the square. To his amazement, Sam Edward agreed: "Yes", he said. "That is all one has to do!"

Peierls was always deeply interested in conservation laws and made sure that in his own work they were satisfied. In 1941, he had co-authored a paper on Lorentz invariance in quantum theory [56]. Eddington had claimed that the normal quantum mechanical treatment of the hydrogen atom could not be reconciled with relativity. Prof.'s paper with P.A.M. Dirac and M.H. Pryce justified the standard treatment. In the 1950's Aage Bohr and Ben Mottelson developed the collective model for nuclear rotations. J.P. Elliott gave a group theory description of quadrupole rotations using a schematic quadrupole-quadrupole interaction [57]. The fundamental question was how to handle the fact that there should be three collective variables describing the collective rotations. Then the total number of variables should be $3N$ where N is the number of rotation particles, but how does one handle the redundancy, because the number of degrees of freedom of the non-rotating shell model is already $3N$. Thus Prof. and J. Yoccoz had a great problem, which they solved by beginning from the approach of Hill and Wheeler [58] the solution of which Yoccoz presented in their joint paper [59].

David Thouless took over the work on collective degrees of freedom. He had been a graduate student of Hans Bethe's. Hans had assigned him the job of determining what properties of the Brueckner G-Matrix (effective interaction) would produce the empirically known shell $VSM(r)$. In 1959 he joined the Peierls department in Birmingham [60]. Peierls' papers with Yoccoz and Thouless treated rotational states of nuclei by a variational method. As trial function the first paper used a projection of the independent particle wave function or angular momentum eigenstates. However, in the analogous but soluble problem of translation this procedure loses the Galileo invariance and gives a wrong answer for the centre-of-motion. This is corrected by a "double representation which mixes states of different velocity." A wrap up of the entire problem with its solution is found in the paper by Peierls and Thouless [61]. Here, wave packets of momentum (angular momentum) and of linear (angular) velocities were taken for the original wave functions, and then the desired momentum and angular momentum were projected out of this. The treatment of rotations was continued by Peierls and J.N. Urbano in Oxford, in their paper on Villas Formalism for Nuclear Reaction and in Peierls single-authored paper on perturbation theory for projected states [62].

Before the background of this wide variety of cutting edge research around Prof., it is not surprising that Gerry made Birmingham his home in this significant phase of his career, especially given the above-mentioned difficulties which he assumed he would have had in the US for political reasons. Nevertheless, like most young scientists he felt that input from elsewhere would provide additional stimulus, and in 1958 he planned a one-term sabbatical at the Institute of Theoretical Physics in Copenhagen. A small problem in contemplating Copenhagen was that he did not possess valid travel documents. Initially, in 1949, he had received a passport just a few months before Joe McCarthy had come on the scene. This was valid only for England and only for one year. When he turned it in for renewal after one year, the American authorities kept it. This caused problems because the British wanted to stamp it yearly when they gave Gerry permission to live in the country. The net result was that he had neither a passport nor permission to stay in England for several years. He spent many a Saturday planning strategies, appealing his case all along the line to American authorities and then writing to the British Home Office, telling them what he had done. They always replied, asking him to let them know the outcome of his appeal. Evidently no one in the British Home Office really wanted to make a decision. Finally, they asked Gerry to reapply for his US passport and tell them the results. In due course, the British Consulate in Birmingham forwarded him the request from Washington that he submit an affidavit about his activities in the Communist Party. Peierls thought that the British Home Office would consider this a reasonable request, so Gerry prepared the affidavit. Some time later a request came back for another affidavit supplying names of everyone Gerry had known in the Communist Party. Even though he had long been expelled, many of the members he had known were highly idealistic people. Gerry did not wish to involve others. He told the US Consul in Birmingham that such an affidavit, like the first, was a very serious matter and he needed to consider it carefully; could he please be given a copy of the request? Since the matter was highly sensitive, the consul banished his British secretary from his office and copied the request by hand for Gerry who took it back and gave it to Peierls to lock up in his safe. Gerry knew that the British would never expel him from the country for refusing to comply with such a request. Indeed, Secretary Burton of the University wrote to the British Home Office, saying that Gerry's request for permission to live in England had been in limbo for several years and demanding that a decision be taken. After some time, he was granted leave permission. In 1956 Dr. Nathan, friend of Albert Einstein's and executor of his will, had carried a case similar to Gerry's through the US Supreme Court. The Court ruled that the State Department could not deny passports to US citizens. So when Gerry reapplied he got his passport back just in time to go on sabbatical to Copenhagen in 1958.

This was a decisive phase in Gerry's career. More mobile after the post-McCarthy sorting out of his passport issues, he was naturally keen to see his scientific achievements awarded by promoted posts. He had been put forward by Prof. for promotion in 1957, but on the occasion lost out to more senior colleagues. By 1958, Gerry's international standing had increased to such a level that Prof. took the unusual step of putting him forward for promotion to a professorship although even promotion to a senior lectureship had not been finalised at that point. The subsequent episode surrounding Gerry's first academic promotions tell us as much about his excellent scientific reputation as they tell us about his mentor's determination to see bright young people rewarded for their excellence (and they also tell us how cumbersome university politics can serve to hinder such a reward). In response to Peierls' initial request, the Vice Chancellor indicated that Gerry's age (he was 31 at the time) and the fact that he held only an "ordinary" lectureship would make the conferment of the professorial title all but impossible as would the fact that if "such a dramatic promotion were made the door would be opened to an embarrassing number of claims on behalf of other lecturers or readers in the university." [63] Rather incensed by the university's attitude towards academic achievements, Peierls took up the fight, pointing out the more enlightened attitude of other outstanding research institutions such as Harwell or the ETH in Zurich which did not consider young age as an obstacle to academic brilliance and which were prepared to acknowledge these without prejudice.[s] Gerry finally received notice of his first promotion: not to professor, nor to reader but to senior lecturer at the bottom of the pay scale.[t] Disappointing as this was, it was — rather unusually — followed by a further promotion later that year to reader. This was a vast improvement over his previous position, but given Gerry's

[s]See [63], p. 576–578.
[t]See [63], p. 588.

excellence, it was still likely that he would further improve his position by moving away from Birmingham, which was exactly what happened in 1960.

He had qualms about leaving Birmingham, where — following the offer of a professorship from the Nordic institute for Theoretical Physics (NORDITA) — he had been offered promotion to a professorship, too. It was a difficult choice between Birmingham and Copenhagen, because the theoretical work in Birmingham was of such high quality. However, Aage Bohr and Ben Mottelson at NORDITA were at the centre of the world's experimental data, and this was a huge attraction for Gerry and in the end tilted the balance in favour of Bohr's institute.

Lamb shift of a tightly bound electron

Gerry wanted to end his work on quantumelectrodynamics with the greatest of challenges, a problem, though, that he was well equipped to deal with because of his experience in calculating the two-point Coulomb Green's function for Rayleigh scattering and because of what he had learnt about renormalization in order to obtain a finite answer. As mentioned above, he had some excellent collaborators in his students, Glen Schafer, Jim Langer and also in David Mayers, whom Douglas Hartree "lent" to him in order to carry out the numerical calculations. Despite this, the problem initially looked insurmountable. Not only did the two-point Green function have to be evaluated for each of a high number of angular momenta l, which then had to be summed over, but the task of renormalization looked formidable. In the case of the Lamb Shift of hydrogen, the whole subject of covariant renormalization was only a few years old. This work had been carried out in perturbation theory, there $Z\alpha$ was the same as α (for hydrogen). Therefore, one could easily Lorentz transform the equations or change variables, except that in infinite terms the change of variables could cause erroneous changes in finite terms. But these could be kept track of.

In the case of the Lamb Shift of, for instance, Pb or Uranium, there was a natural frame of reference, the one in which the nucleus was at rest, so one was essentially restricted to working in that frame. One great help was that just as Gerry was beginning, important work on the one point Coulomb Green function (where $x_1 = x_2$) had been carried out by Norman Kroll and Eyvind Wichmann at Columbia [64]. They calculated the corrections to the Uehling term in the vacuum polarization, and looking at their work, one could imagine that the generalization from the one-point to the two-point calculation required for the Lamb Shift would be very

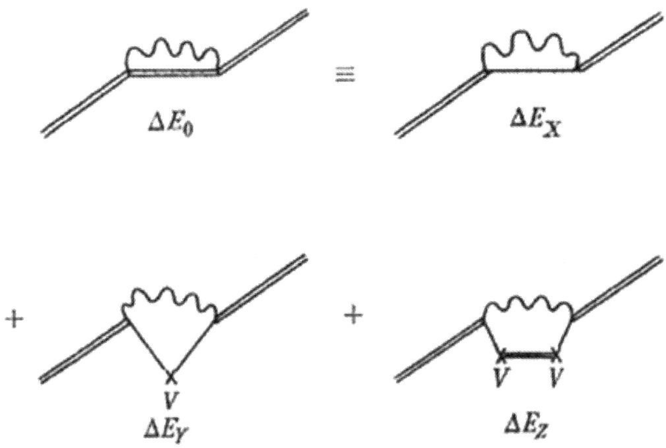

Fig. 1. The double line indicates propagation with the propagator in the external field, $S_F^{(V)}(x, y)$, where x and y are four-vectors but can be taken to have equal times. The single line indicates propagation with the free propagator $S_F^{(0)}$ and the wave line represents the photon propagator. The single line indicates propagation with the free propagator $S_F^{(0)}$ and the wave line represents the photon propagator [66].

difficult. The expression for the self-energy ΔE_0 was the same as given by Kroll and Pollock in 1952 [65]. Gerry's procedure was to expand the bound-state propagator between the two points x and y, which took into account the effects of the (external) Coulomb field to all orders, in an expansion such as to explicitly exhibit the infinite terms which must be included as renormalizations of the finite terms.

This is exhibited graphically in Fig. 1. The procedure of taking into account the Born approximation to lowest order was suggested by the earlier work on Rayleigh scattering, and then the (finite) term with the two V's picked up the exact remainder. The double lined propagator was the one in the so-called Furry representations. The strategy was to put the first two terms on the right hand side of Fig. 1 into formally covariant form, so that they could be renormalized; they contained all of the divergences. It turned out that by analytic continuation, the integration over the three-vector photon momentum **k** could be carried out first, leaving the integration over the energy component ω. This was just the opposite order to that in which the integration with Feynman graphs was usually carried out. It was known since Karplus and Kroll [67] that if M_x is evaluated covariantly, it contains mass and charge normalization parts, i.e.

$$M_x = X_A + X_B + X_C$$

where X_A corresponds to mass renormalization, X_B to charge renormalization and X_C is a finite part which should be included in the Lamb Shift. Similarly,

$$M_F = -X_B + Y_C$$

where Y_C is a finite part to be included in the Lamb Shift. The term Z is finite and gives no trouble. If M is evaluated covariantly, then the correct finite part is $-(M - X_{Ax})/2\pi i$, the factor $2\pi i$ coming in relating time to energy. Alternatively, the correct finite part is

$$\Delta E_c = -\frac{1}{2\pi i}\{(M - M_X) + X_B + X_C\}.$$

Although this alternative form appears to be clumsy, $(M - M_X)$ has only a logarithmic divergence, whereas M has a linear one, and it is clear that in the last form the calculation converges with increasing ω much more rapidly. It is, in fact, just the combination $M - M_X$ that was calculated non-relativistically in the first calculation of the Lamb Shift by Bethe in 1947 [68].

The formalism was outlined in a paper by Gerry and David Mayers [69]. Most of the calculations were actually carried out by David Mayers in Aldermaston, the British Defence Establishment, where he had a job between his posts in Cambridge and Oxford. Gerry did not have access to Aldermaston, so he had to figure out various tests which the calculations had to satisfy and then check that the work done by Mayers were consistent with these.

In the sum over angular momenta l they had to evaluate each l contribution. Their asymptotic expansion showed that the difference with l of the terms went as l^{-2} for large values in l. So even though the convergence of the series in l was not rapid, they could end up by adding the asymptotic behaviour. The final result for the Lamb Shift was 41 Rydbergs, about the size of 38 Ry found by Kroll and Wichmann. Other contributions to the k absorption edge of order mc^2 and αm^2 where $\alpha = 1/137 - e\hbar c$ had been noted by Brenner and Brown. The net energy of the K-electron was found to be -6138.31 Ry. The best experimental value for the K-energy was, at the time, that of Saxon [70]: 6107.7 ± 0.6 Ry. Other terms than the Lamb Shift were calculated and included by David Mayers [71] who found a theoretical value of 6099 Rydberg as compared with the experimental value of 6107.7 ± 0.6 Ry, showing that the inclusion of the Lamb Shift markedly improved the agreement.

In 1960 Gerry received a letter from Walter Johnson, an atomic theorist in Notre Dame. He pointed out an error in the calculations. Most investigators would be very upset to find out they had made what turned out to be a simple mistake. Walter, however, generously credited Gerry with having worked out the formalism and Gerry himself was happy that someone else in the world was looking at effects from quantumelectrodynamics in heavy atoms, because he had the subject almost completely to himself during his ten years at Birmingham. For decades, his theory was generally used to calculate the Lamb Shift in heavy atoms [72], and of course with modern computers the calculations became easier, especially compared with Gerry's own experiences with the Cambridge EDSAC I!

Having received the offer of a permanent position at NORDITA in the summer of 1959, Gerry spent that summer at the University of Minnesota in Minneapolis, supported by their accelerator contract. Although he had shifted into theoretical nuclear physics at Birmingham in the late 1950's, the research at the University of Minnesota catapulted him head on into nuclear structure physics. The trigger was a Gordon Research Conference. Initiated by Neil Elbridge Gordon, who had expanded an intermittent set of summer meetings at the chemistry department of John Hopkins University in the late 1920s into annual summer schools from 1931, the Gordon Conferences, over the years developed into a large non-profit organization managed by and for the benefit of the entire scientific community, aiming to provide an international forum for the presentation and discussion of frontier research in sciences and related technologies. From the humble beginnings in 1931, the Gordon Conferences evolved into a forum that attracted in excess of 20,000 researchers annually to its more than 200 conferences.

In 1959, under the auspices of Peter Axel and A. Hanson, the first conference on photonuclear reactions took place. Many experiments were carried out with betatrons and other accelerators. The betatron was particularly simple because it whirled electrons around, in that way accelerating them. The accelerated electrons radiated gobs of gamma rays and the latter could be scattered off nuclei. Generally, it was hard to get clean data, but one characteristic stuck out from the many experiments: the scattering of the gamma rays took place by nuclear energy levels which were at almost twice the energy at which the single particle excitations, i.e. the excitations of a particle from a filled level to an empty level, were in the shell model. These excitations were the lowest-lying ones, the so-called dipole excitations. In stark contrast to modern conferences with hundreds or thousands of participants, these Gordon Conferences were attended by only 30 or 40 experimentalists and a few theorists, who had some interest in photonuclear physics. The photonuclear conference in 1959 had **one** theorist, Joe Levinger, who was devoted to photonuclear physics. Gerry, by his own admission, was spending the summer of 1959 in Minnesota, largely because he liked to swim. The added advantage was that the university had a cyclotron and active experimental physicists, especially Norton Hintz. Gerry had an invitation not only to attend the electromagnetic probes Gordon Conference, but to give the theoretical talk, as Joe Levinger was unable to attend. When discussing the talk with Mark Bolsterli, whom he had known from his visit to Birmingham, he remarked that he really would like something entirely new for his lecture. Elliott and Flowers [73] had calculated the dipole states in ^{16}O. There were five shell model transitions between the last filled shell and the first empty shell. Each of these had its own (unperturbed) transition, the one it would have were it the only transition in the shell model. But these were all connected in the matrix which one had to diagonalize in order to obtain the possible eigenfunctions by what Gerry termed the dipole-dipole interaction. A lot of angular momentum algebra was involved. The Legendre Polynomials had, by convention, phase factors which made them either positive or negative in their derivation. The dipole transition involved the calculation of matrix elements of the dipole operator

$$D = \frac{1}{2} \sum \tau_{3i} z_i$$

where τ_{3i} was the operator for isospin +1 for a neutron, −1 for a proton, between the initial and final shell model states. Looking at the matrix to be diagonalized by Elliott and Flowers, one only saw a randomness in sign, + and −. The first things Gerry did was to change the convention in the Legendre Polynomials so that all of the dipole matrix elements between the filled and unfilled states were positive. With the new conventions, every matrix element in the Elliott and Flowers paper was positive!

This meant that all of the matrix elements in the nucleus would cooperate in pushing the column vector, which described the components of the dipole state, up in energy. Mark Bolsterli and Gerry constructed a schematic matrix by using zero-range forces, not calculating any matrix elements but taking them proportional to the product of the square roots of the degeneracies (the number of projections of the angular momentum of the relevant particles) of the initial and final particles. This built in the general property that the matrix elements were larger for the higher angular momentum initial final states. The latter differed by only one unit, the dipole transition involving the lowest possible change in angular momentum. Bolsterli and Gerry actually constructed the model and then showed the solution pictorially in a single morning in June 1959. They did not know it at the time, but the model had the same ingredients as that of plasma oscillations in an electron gas, the latter being simpler because the matrix elements were all the constant matrix element which was the Fourier transform of the Coulomb potential $V(G) = 4\pi G$.

At the time, there was not much interaction between the solid state physicists and nuclear physicists. As Gerry later recalled, the nuclear work was often so tied up in angular momentum algebra, that the physics was hidden. In any case, the highest shell model level was pushed up in their schematic model and almost all of the dipole strength flowed into it. Furthermore, most of the sum of diagonal elements went into further raising this state. Thus, it looked to observers as if the giant dipole state lay well above the shell model transitions and indeed it did.

Gerry went to the Gordon Conference, with his two or three sheets on which he had derived the giant dipole state in a cardboard roll. The chairman Joe Wesener, told him that he had 50 minutes for his presentation plus an additional ten minutes for discussion. Gerry went to the blackboard and completed his talk in about 10 minutes. There was complete silence. Everyone understood that he and Bolsterli had solved the problem of the Giant Dipole State. Joe Wesener called for a 50 minute intermission during which they could walk around the small hills surrounding the school. Gerry's old teacher Gregory Breit was well impressed and was finally proud of his Ph.D. student.

Nucleon-nucleon interaction

The nucleon-nucleon interaction was beginning to occupy physicists' interests, with Ernest Lawrence's Berkeley cyclotron operational and measurements of all kinds of scattering and correlations of nucleons hitting each other at 200 MeV centre of mass energy emerging from there. Of course people knew of the pion of mass 140 MeV, but its exchange did not do much to explain the scattering. In particular, the proton-proton scattering was very nearly isotropic. This would be the case were it purely S-wave, but the cross section was too large for this, indicating that many partial waves were involved.

Abraham Pais, at the Institute for Advanced Study in Princeton, suggested that a large spin-orbit force was present in the higher particle waves, which would fill in the scattering at 90° and would tend to make the scattering isotropic. Gerry quite liked this idea and was off to calculate it on the Cambridge EDSAC computer, but Prof. warned him about getting into too heavy calculations. At the time, the short-range nucleon-nucleon interaction was described by a vertical hard core of ~0.5 fm in radius. Gerry knew that in atoms, where the electric interaction between electron and nucleus was attractive, there was a spin-orbit interaction which was repulsive. The spin-orbit force in the nucleon-nucleon interaction was known to have the opposite sign to that in atoms. But the hard core interaction had the opposite sign to the central interaction in atoms. The atomic spin orbit interaction was known to result from the relativistic effect of the spin orbit coupling. The nuclear spin-orbit term had opposite sign, but it arose from putting the repulsive hard core into the Dirac equation, rather than having the attractive Coulomb potential as in hydrogen atoms.

Prof. thought Gerry's idea of spin-orbit term arising from the repulsive core in the relativistic Dirac equation to be a good one. Dan Sprung from Mc Master University had joined the Birmingham group and, Prof. suggested to him that he work out the nuclear spin-orbit problem. It took him some time to do it, but he wrote his doctoral thesis on it, and Sprung and J.B. Willis wrote a paper about it in 1960 [74]. Around that time, the vector mesons, now known as ω and ρ were discovered. The former has no isospin dependence. It was known that the exchange of a photon gave an attractive electron-atom potential. So it was natural, as Gregory Breit and S.N. Gupta suggested independently [75], to introduce a massive photon, the ω into the nucleon-nucleon interaction. It took over the effect of the vertical hard core. The relativistic correction gave rise to Gerry's spin-orbit term, and this term gave the necessary large angle proton-proton scattering, isotropic at the Berkeley energies.

As Gerry later reminisced, the entire development of the meson exchange giving rise to the nucleon-nucleon interaction was torturously slow, largely because investigators thought of the nucleon-nucleon interaction as a strong interaction problem. They were hesitant to make any kind of expansion which might be viewed as perturbative. The covariance bothered them; they were also hesitant as to how to handle the time component of the momentum exchanged between the two nucleons.

The vertical hard core had been introduced by Bob Jastrow in order to obtain nuclear saturation. In the work of Euler, Heisenberg's student in 1937, there was no repulsion in the S-wave states and it was suggested that the angular momentum states would saturate the nuclear matter. They turned out not to give enough repulsion. So the infinite repulsive core was introduced and, at the time, it was suggested to contain all sorts of mesons, etc. that were too complicated to handle explicitly [76].

4. Copenhagen, Princeton and Stony Brook

When Gerry moved to Copenhagen in the summer of 1960, it was quite natural that he should take up the problem of constructing the nucleon-nucleon interaction to be used between nucleons in shell-model states. Copenhagen was at its zenith in nuclear physics, with the Bohr–Mottelson collective model explaining more and more nuclear phenomena, especially excitations, and more and more data flowing in daily.

Bohr and Mottelson had built up the extremely successful "Copenhagen school" in nuclear physics (the more famous pre-war Copenhagen school still permeated the buildings of the Bohr Institute and NORDITA), but Gerry wanted to understand how the nucleon-nucleon interactions in the nucleus built up the structure and excitations. It was clear that this was to be a thankless job, because the schematic interactions, delta function plus quadrupole-quadrupole interaction, worked so well in describing nuclear phenomena that Gerry knew he would get no credit for a "microscopic" derivation. Nonetheless, he wanted to give the work a more "fundamental" foundation. As a result, quite often he found himself in a minority of one playing "devil's advocate", or as he put it in his autobiographical essay, "loyal opposition",[u] questioning each ad hoc assumption and adding liveliness to the discussions. His recollections of the Monday morning meetings at the institute, when theorists and experimentalists met for discussions, demonstrate his position vividly. Forever questioning the way in which experimentalists connected their data, after one such session a colleague remarked, "We had a very heated debate today, as heated as it could be with odds of 25 to 1!"

Nuclear theory distinguished itself from most other many-body theory by the extremely strong repulsion felt by two nucleons at short distances, often taken to be an infinite hard-core interaction, which we now know to originate from the exchange of massive vector mesons. Moszkowski and

[u]See [1], p. 12.

Scott had proposed a delightfully simple way to handle the short-range interaction [77]. This was to integrate the Schrödinger equation, with the wave function starting from zero at the hard-core radius, out to the radius at which the phase shift would be zero, if the potential outside of it were thrown away. But instead of throwing it away, the long-range remaining piece of the interaction was to be used as the effective interaction. This "Moszkowski–Scott separation" worked well because nuclei and nuclear matter are rather dilute, so that if two particles are close together, a third particle is unlikely to be in the vicinity. At Copenhagen, Gerry encouraged a Norwegian student, Kristofer Kolltveit, and a Finnish student, Alpo Kallio, to tackle this problem. Their solution, the Kallio–Kolltveit potential provided a real step forward and became extremely successful, considering its simplicity [78].

Being able to travel freely, Gerry was now in a position to spend his sabbaticals back in the US. During a year of leave from NORDITA, in 1962–1963, he went to MIT, where he continued to work on the many-body theory of nuclear matter, realizing that although the two-body tensor force contributed a lot to the binding energy in second order, higher-order effects were small. Having just returned to Europe from his MIT sabbatical, in 1963 Princeton offered Gerry a professorship and he decided to take it up in 1964. As in Copenhagen, the view muted at Princeton was that the nucleon-nucleon interaction in nuclei resulted from such strong interactions that it was impossible to work it out, starting from the elementary nucleon-nucleon interaction measured in scattering experiments. When Gerry arrived at Princeton in early September, 1964, he met Eugene Wigner, who enquired about his research plans. When Gerry replied that he was planning on working out the nucleon-nucleon interaction in nuclei, Wigner replied that it would probably take someone cleverer than him. Gerry's response was that the two probably had a different understanding what it meant to "work out". Gerry wanted to achieve a working knowledge, sufficiently good to be able to work out problems in nuclear physics. He almost certainly was envisaging something less accurate what Wigner himself would have in mind.

Gerry, in his efforts of working out the nucleon-nucleon interaction was building on an already formidable body of scholarly work: Brueckner's theory dealing with the strongly repulsive hard core potential which was later understood as originating from the strongly repulsive potential from the ω meson exchange, the ω being a massive photon. Furthermore, there was Bethe's effort of trying to organize the Brueckner theory to form a framework amenable to calculations in finite systems such as nuclei. Also, significantly, Scott and Moszkowski developed a method that allowed one to handle off-shell effects in a trivial way, using simple physical arguments. Given the amount of effort put in earlier to handle these effects, one might be suspicious about the simplicity, but as was shown later, an elegant renormalization group approach and effective field theory justify the Moszkowski–Scott approach.[v]

Many discussions concerned the question how to treat the energy of the nucleon, when it was off its energy shell; i.e. how to quantify its energy of interaction with other nucleons when the nucleon itself was in an excited state. Since hard core potentials of infinite height were used frequently, the nucleon much of the time was in an intermediate state of very high momentum. The off-shell effects were generally treated in Rayleigh–Taylor perturbation theory and the treatment seemed ugly in appearance compared with the covariant treatment of the fields, essentially in the interaction representation by Feynman and Schwinger, where they were given their time dependence from the free-particle Hamiltonian. The major work on the nucleus as a many body system, on the structure of the nucleus and on the spectra, nature of excited levels etc. was led the Copenhagen School, and so Gerry was intimately familiar with the results [79].

Many years were spent in using the G-matrix to tame the infinitely repulsive interactions as used in the nuclear matter problems. It was natural to try it in finite systems. A very important effort applied to finite nuclei was carried out by Dawson, Talmi and Walecka [80] who solved the Bethe-Goldstone equation not only for the ground state but also excited states and then looked at the spectra of ^{18}O. Igal Talmi then systematized the properties of subsystems of particles in partly filled shells of the shell model [81]; e.g. the calcium isotopes where n particles are in the $f_{\frac{7}{2}}$ shell. He would call these f^n where $j = \frac{7}{2}$. Talmi realized that any G-matrix model like that of Dawson, Talmi and Walecka would be such that the force between the nucleons in the j_n configuration would be attractive such that ω would be negative; i.e. that in the quadratic term there would be attraction between the nucleons. However, in the shell model nuclei he had looked at α was positive. This argument disabled essentially all of the models using the G-matrix as interaction to date, and spurred Gerry and his colleague Tom Kuo at Princeton to go on to the next higher order.

[v]See [3], Chapter 2.

The Kuo–Brown calculation

Brown and Tom Kuo had the Moszkowski–Scott method with its implication of a real division between short and long range interaction at their disposal, which put all of the physics into the well-measured long-rage part of the interaction; furthermore, Talmi's work had led to the formulation of the requirements that a two-body interaction had to satisfy. Gerry knew from Copenhagen that effects on the interaction from nucleons polarizing the nuclear medium were important. His first morning at Palmer Physical Laboratory,[w] he listened to a suggestion which Ben Bayman, then assistant professor there, had proposed to George Bertsch, a young graduate student. Gerry showed him that the idea probably would not work out. George had sat listening, then in the middle of the discussion suddenly walked out without saying anything. Noticing Gerry's puzzled reaction, Ben explained: "George saw that it wouldn't work out, so why waste any more time on it?" When George, the following morning, asked Gerry for a problem to work on, he proposed that George work out, in ^{16}O, the correction to the two-body interaction that came through polarization of the nucleus, to lowest order, essentially including the nucleon particle-hole bubble. The calculations involved a lot of angular momentum recoupling, and Gerry estimated that George would be kept busy for months. To his astonishment he brought in the results the next week. The polarization correction had all the right properties to satisfy Igal Talmi — the quadratic term was repulsive — and later they showed that most of the Copenhagen quadrupole-quadrupole interaction came from this polarization bubble diagram [82]. He had worked out the lowest-order core polarization and showed by including it, that the physically simplest requirement by Talmi was now satisfied with the inclusion of the core polarization.

Bethe, Brandow and Petschick [83] took care of additional complications namely the large effects from the energy being off-shell in the intermediate states as well as the effect of blocking intermediate states by the Pauli Principle. They put a substantial gap into the single particle spectrum for intermediate states right at the Fermi surface which served to cut down the interaction. Bethe [84] and Moszkowski [85] suggested that the intermediate states might well be represented by free particles. This would leave a gap in the energy, because the initial states of the particles would be bound in the shell model. Gerry and Tom Kuo adapted plane wave intermediate states

[w]See [1], p. 14.

which have quite commonly been used since then; i.e. by being off-shell the intermediate states lost most of their originally attractive energy.

They used the Moszkowski–Scott method for the states that would be bound or nearly bound, the $^1S_0$ and $^3S_1$ states. The states of higher l were calculated in the Reference Spectrum approximation. Gerry and Kuo showed that the different P-states contribute with different signs to the energy, so they are not very important, and the D-states are not large, which is the reason for their results being quite accurate. As Gerry recalled later, the response of the physics community was mixed: particle physicists considered the method too messy; nuclear physicists remained unconvinced about accuracy; and the Copenhagen gurus Bohr and Mottelson were sceptical that it was possible to arrive at a quantitative solution without including their vibrational induced interaction, the bubble sum, to all orders. Divided as opinion was, the paper became a citation classic, and about a decade and half after the Brown–Kuo results had been formulated, Tom Kuo summarised the significance of their contribution as follows:

> What are the uses of such a microscopic nuclear structure theory? ... if we have a fundamental nuclear structure theory with which we can not only understand known nuclear properties but also predict new nuclear properties, it can indeed be very useful. We are still quite far from this ambitious goal. The tangible contribution of the Kuo–Brown paper is probably in providing some evidences and hopes for such a theory. Using quantum mechanics and an internucleon interaction based on two-nucleon scattering experiments, we found that the calculated nuclear properties of nuclei ^{18}O and ^{18}F were in unexpectedly good agreement with experiments. It is probably the first major nuclear structure calculation where one does *not* treat the effective internucleon interactions as free parameters and adjust them so as to give good results. The application of this theory to many other nuclei yielded equally encouraging results. Nevertheless, our approach is by no means free from defects. It employed a number of approximations. Unlike the Coulomb and gravitational interactions which are weak, the nucleon-nucleon interactions are strong and our knowledge about them and how to handle them in many-body theories is still rather preliminary. Much has been done since this paper, and much more remains to be done. Let us hope that physicists will keep on working in this direction and complete before long what we were not able to complete [86].

Indeed, while many questions about the nucleon-nucleon interaction remain unsolved, progress in this field has been remarkable, which is evident from some of the contributions to this volume. But undoubtedly, there is more to come.

While Princeton was certainly intellectually stimulating, Gerry's family much preferred living in Copenhagen, and with the Princeton rules at the time preventing frequent periods of leave except for government service, it was difficult to see how the family requirements would be squared with Gerry's work commitments, although NORDITA was a cooperative institute involving five governments. As described elsewhere in this volume, in 1967 C.N. Yang, who had moved from the Institute for Advanced Study to Stony Brook, asked Gerry to join him there. Gerry had learnt to ignore little local difficulties, such as rationing, unpleasant living conditions and the like during his spell at Birmingham in the 1950s and he had come to realize that the right environment for academic progress does not necessarily involve having access to the most pristine laboratory. At the time, Stony Brook looked like a construction site. Yet, Gerry took Yang's invitation seriously. He worked out a scheme in which he divided his time between Stony Brook and NORDITA, taking leave without pay from one institution while being at the other. In 1968 he moved to Stony Brook officially, taking his graduate students back and forth with him for the next 18 years until he retired from NORDITA.

Stony Brook gave Gerry the resources to build up a nuclear theory institute, and C.N. Yang was helpful at every stage in his endeavours. The lessons learnt at Birmingham and Copenhagen about group dynamics, critical mass of students and the essential input of distinguished visitors came in useful in his project of building up a dynamic group. How productive the group has been and continues to be is evident in the many contributions of former and current members of the Stony Brook theory group to this volume. As the stories of scientific endeavour and progress are best told by the participants in these successes, they will be taken up by the protagonists of the events in the core of this book, and this essay will focus on Gerry's last Eagle and long-term collaborator: Hans Bethe.

5. The Third Eagle: Hans Bethe

Gerry first met the Bethes in 1951, while lodging in the Peierls household. But it was not until many years later that he started to collaborate with Hans. The story of the beginning of their most fruitful collaboration has been described in detail elsewhere [87], but it is just as well to recall it here briefly, because rarely has there been a collaboration of "mature" researchers that has commenced so late in their lives, dealing with cutting-edge science, leading to such a prolific output with such remarkable impact.

On April 1, 1978 (yet again that ominous date) Gerry picked the Bethes up at the airport in Copenhagen. Hans had sprained his ankle on Mount Pion in Turkey, an occurrence which Gerry read as a bad omen for his planned work on the pion-nuclear many-body problem, where each higher-order term calculated was as large as all of the lower-order terms put together. Instead, Gerry suggested that they worked on the theory of supernovae. Although Bethe claimed not to know anything about them, the challenge must have attracted him, as Gerry pointed out to him that he believed that the work in this field was full of errors waiting to be corrected.

Gerry had already published a paper with James Lattimer and Ted Mazurek [88] which demonstrated that many of the excited states of nuclei had essentially been left out in works up until then. But Gerry felt that he needed an expert on explosions to make a breakthrough, and Bethe clearly was as much of an expert in this field as one could hope for, given that he had been head of theory in the atomic bomb program in Los Alamos under Oppenheimer.

As a welcome gift, Gerry left a computer printout of a large star that Stan Woosley had evolved up to the point of collapse with Hans Bethe. He did not disappoint — when Gerry went to see him, he explained that the entropy of the iron core was very low, less than 1 (in units of k_B^{-1}) per nucleon, which meant that the iron core would collapse without being held up until the iron nuclei merged into nuclear matter. There was not enough entropy for them to break up. However, all supernova calculations on the supercomputers since the war had shown that the collapse was held up at about 1/1000 of nuclear matter density. When Gerry pointed this out to Bethe, the latter simply remarked: "They're wrong!"

As it happened, in May of 1978 a topical meeting on the collapse of stars and resulting supernovae took place in Copenhagen, and everyone began tabulating the entropy. It was obvious by then that before Gerry's paper with Lattimer and Mazurek no one had included sufficient excited states in nuclei, which soak up the entropy without producing pressure.

The first published result of the remarkable collaboration of Gerry and Hans Bethe was BBAL [89], named after the initials of the authors, or "Babble" as Willy Fowler would refer to it. The paper explained the collapse of stars. Gerry later said that he felt like a graduate student (and Hans treated him like one during working hours) taking a crash course in thermodynamics, while avidly soaking up all of Hans's tricks for simplifying calculations. Among the other by-products of the Brown-Bethe project was that all of a sudden astrophysics became an important activity at

NORDITA, and also at Stony Brook, where Jim Lattimer and Madappa Prakash joined Gerry to form a strong nuclear astrophysics group. Having settled the collapse of large stars, Gerry and Hans Bethe thought they would understand the subsequent explosion quickly. But even now, more than three decades later, no satisfactory successful supernovae calculations exist.

After BBAL, for well over twenty years, Bethe and Gerry for the month of January shared an apartment or condominium on the west coast, at Santa Barbara, Santa Cruz, or later most years at Caltech. They lived and worked together, having long, interesting discussions and conversations. In the end, they knew each other well enough so that one of them only had to begin a sentence for the other one to know what would follow.

Gerry has described these annual research trips in some detail [90]. Housed in university accommodation, Gerry was responsible for the housekeeping and for satisfying the culinary needs by doing the shopping and cooking. In the evening, he would supply a problem that he wanted Bethe to think about the next morning. This Bethe would usually do during a long morning bath, which was followed by an equally substantial breakfast, consumed while explaining the "line of attack" to be used on the problem at hand together with an estimated research target to be reached by lunchtime, a meal that would usually be consumed in the refectory. Following another work session after lunch the late afternoon would be used to discuss the results, with the evenings reserved for the roast dinner which Gerry would have prepared for them. The published output of this collaboration is astounding, even if much of the later work on black holes is not generally accepted by astrophysicists and astronomers. What began with "Equation of State in the Gravitational Collapse of Stars", the above mentioned "babble" in 1979 was concluded eventually in 2003 with 'Evolution and Merging of Binaries with Compact Objects', published posthumously in Bethe's Centenary volume of *Physics Report* [91]. Gerry chose to print a copy of the manuscript of Bethe's original notes of their collaborative effort, and this neatly sums up what Gerry would often refer to as the "Bethe Way" of doing physics. In Gerry's words: the Bethe way "when confronted with any problem was to sit down with his stack of paper and fountain pen and his slide rule and calculate the problem in the most obvious way. Most importantly, not to be deterred by anyone who tells you — as most colleagues will — that the situation is much more complicated than you think... he would begin working out whatever problem he had planned to do, remembering all constants and would fill the white sheets at a nearly constant rate. He headed straight towards the light at the end of the tunnel. If he ran into a barrier he would go around, over or under it, filling more of the paper."[x] The manuscript page, page 5424 of the pages thus filled, bears witness to this method (as do quite literally tens of thousands of other papers held in the Bethe Archives at Cornell [92]). Although Gerry never drew the comparison, his own approach of "solving the nucleon-nucleon interaction", attempted when he first went to Princeton, bore resemblance with the undeterred determination of tackling a problem no matter how difficult and complicated it is deemed to be by other colleagues.

Gerry's and Bethe's collaboration was prolific. It defied the notion that physics is a young person's pastime. At the time of their last publication the two scientists had a combined ago of well over a century and a half. In terms of scientific interests they stayed young at heart, and they were not shy to venture into new territories. In 1985 they wrote a Scientific American article, "Theory of Supernovae [93]". In January 1987 they went on to start working on the theory of relativistic heavy-ion collisions. That year, when the two scientists left Caltech after their January research spell, Gerry speculated that it was now time for a supernova explosion. And on February 23 one occurred! Just in time to supply the two physicists with a wealth of new related problems. Neutrinos were detected from the explosion for about 12 seconds; the optical display occurred, and Gerry and Bethe waited expectantly to "see" the neutron star. Each January Gerry would ask observers at Caltech when they could see it. They always replied that it would take another year for the debris to clear out. Gerry decided that he should be able to work out the time when they should observe it from the ambient matter accreting onto it, with its binding energy converted into X rays. For this he joined forces with Joe Weingartner [94], and they found that they should have seen the neutron star in 1988.

With the numerical work completed, Gerry phoned Hans Bethe one evening and told him, that he believed that 1987A had gone into a black hole. The latter disagreed, listing all the reasons why a neutron star was there; however, after a lengthy discussion, Bethe was finally persuaded. The result of these calculations and the ensuing discussion formed the basis of yet another paper published 1994 [95], suggesting that 1987A had gone into a black hole because it exceeded the maximum neutron star mass of $1.5\,M_{\odot}$.

In the autumn of 1993 Gerry attended a workshop on hot QCD with a companion program on dense stellar systems. It led him to think about

[x]See [90], pp. 16, 18.

the common envelope evolution of a neutron star in the envelope of an expanding red giant, which he found interesting, because the neutron star could accrete mass from its companion. In response to his paper with Bethe giving the maximum neutron star mass as $\sim 1.5\,M_{\odot}$, people had agreed that neutron stars are formed with masses $\lesssim 1.4\,M_{\odot}$ in their evolution. However, in binary evolution leading to double neutron stars, after the first neutron star is formed, it must go through common envelope evolution with the expanding (evolving) companion. Gerry reasoned that there, in different situations, it could accrete varying amounts of mass. Thus, if one did not see neutron stars of masses greater than $\sim 1.5\,M_{\odot}$ they must have gone into black holes, as Gerry and Hans Bethe has assumed SN1987A to have done.

In researching earlier work on binary evolution, Gerry noted that common assumptions of accretion rates could not account for the production of any significant increase in neutron star mass during common envelope evolution. However, according to Chevalier's work [96] larger accretion rates ($\dot{M} > 10^4\,\dot{M}_{Edd}$) would lead to matter accreting onto the neutron star trapping the outgoing radiation and carrying it back onto the neutron star in the adiabatic inflow. Gerry had encountered a similar situation in the collapse of stars, where neutrinos are trapped (at much higher densities) and confined to the star. Therefore, he immediately investigated envelope densities in the red giant companion and saw that accretion would proceed at a sufficiently great accretion rate, a rate sufficiently large to send the neutron star into a black hole. Thus, the standard scenario for evolving binary neutron stars ended up with a low-mass black hole of mass $\sim 2.4\,M_{\odot}$ and a neutron star mass of $1.4 M_{\odot}$ formed in the final explosion. Gerry wrote up his results in 1994 and submitted the paper to the *Astrophysical Journal* [97], where the referee commented that the results were likely to have a major impact on the evolutionary models of X-ray binaries and recycled pulsars. With such a positive evaluation of his first steps into binary evolution research, it is hardly surprising that the topic would keep Gerry occupied for some time to come.

In the year of Hans Bethe's 90th birthday, initiated by a query from Kip Thorne about the input of merging black-hole and neutron-star binaries on gravitational waves, he and Gerry embarked on their first joint project on binary evolution, a topic Bethe knew little about at the time. As "introductory reading", Gerry sent him an article by Meurs and van den Heuvel [98] that developed standard binary evolution. A few days later Gerry received a fax from Bethe, in which he had rederived their mass results, which had taken up 20 pages in the original paper, in less than two pages. After this initial "comprehension and summary" exercise, Gerry and Hans Bethe who both had a reputation of being good at calculating phenomena that could not be seen, went on to work out "Evolution of Binary Compact Objects that Merge [99]". They did this analytically, with simple mathematics leading the referee to comment that this was "an interesting paper, making a useful point in an unusually simple and elegant manner".[y] By their own admission, it was the most complimentary referee report that either Gerry or Hans Bethe had ever received. Coming from two eminent scientists who between them had been awarded with the Nobel Prize, the Max Planck Medal, the Henry Draper Medal, the Benjamin Franklin Medal, the Eddington Medal, the Enrico Fermi Award, the Lomonosow Medal, the Bruce Medal, the National Medal of Science, the Rumford Prize, the Hans Bethe Medal, the John Prize Wetherill Medal and many others, this was saying something about the quality of their research at a period in their lives when most people would rest on their laurels.

Not unlike his own teachers, Gerry thrived through the work done in collaboration, both with his Eagles and with his research students and junior collaborators. At the last count, he had supervised around seventy Ph.D. students, many of whom are now professors in their own right in many parts of the globe and many of whom have contributed to this volume. They were and are gifted researchers, and Gerry saw his main contribution to their success in starting them off on promising and exciting projects and problems. He had learnt from his Eagles that a strong, active and dynamic research group was an essential ingredient in building a group and fostering young and promising talent. Not only did he help gather a critical mass of exceptional talent around him; not only did he devote time and energy to this students and colleagues; not only did he create an inspiring atmosphere in which scientific work flourished and in which his students were supported well beyond the call of duty; as important as all this is the fact that be became a true friend to his students, post-docs and colleagues alike. It is this friendship, above all, that is to be celebrated with this *Festschrift*.

References and Endnotes

[1] G.E. Brown, "Fly with Eagles", *Annual Review of Nuclear and Particle Science* **51**, 1–22 (2000).

[y] See [87], p. 152.

[2] H.A. Bethe, G.E. Brown and C.-H. Lee (eds.), *Formation and Evolution of Black Holes in the Galaxy. Selected Papers with Commentary*, Singapore: World Scientific Press, 2003.
[3] G.E. Brown, T.T.S. Kuo, J.W. Holt and S. Lee (eds.), *The Nucleon-Nucleon Interaction and the Nuclear Many-Body Problem*, Singapore: World Scientific Press, 2010.
[4] For details about the development of American mathematics in the late 19th century and the role of Felix Klein, see Karen Hunger Parshall and David E. Rowe, *The Emergence of the American Mathematical Research Community, 1876–1900: J.J. Sylvester, Felix Klein and E.H. Moore*, Providence: Amer. Mathematical Society 1994.
[5] Wilibald Reichardt, "Ein Beitrag zur Theorie der Gleichungen sechsten Grades", *Leipziger Berichte* **37**, (1885); and Wilibald Reichardt, "Über die Normierung der Borchardtschen Moduln der hyperelliptischen Funktionen vom Geschlecht p = 2", *Mathematische Annalen* **28**, 84–98 (1886); Frank Nelson Cole, "A contribution to the theory of the general equation of the sixth degree", *American Journal of Mathematics* **8**, 265–86 (1885–86).
[6] H. Maschke, "Aufstellung des vollen Formensystems einer quartären Gruppe von 51840 linearen Substitutionen", *Mathematische Annalen* **33**, 317–44 (1889).
[7] Anders Wiman, "Über eine einfache Gruppe von 360 ebenen Collineationen", *Mathematische Annalen* **47**, 531–56 (1896).
[8] Many of the details used in this chapter are taken from the only detailed account of Gregory Breit's life, McAllister Hull's Biographical Memoir, written for the National Academy of Sciences. See Hull, McAllister, "Gregory Breit", *Biographical Memoirs of the National Academy of Sciences* **74**, 27–56 (1998).
[9] G. Breit, "The Distributed Capacity of Inductive Coils", *Phys. Rev.* **17**, 649–77 (1921).
[10] G. Breit and M.A. Tuve, "Radio Evidence of the existence of the Kennelly Heaviside layer", *J. Wash. Acad. Sci.* **16**, 98 (1926); G. Breit and M.A. Tuve, "A Test of the existence of the Conducting Layer", *Phys. Rev.* **28**, 554–575 (1926).
[11] G. Breit, "On the possibility of nuclear disintegration by artificial sources", *Phys. Rev.* **36**, 817–818 (1929).
[12] G. Breit, "The quantum theory of dispersion", *Nature* **114**, 310 (1924).
[13] G. Breit, "Quantum theory of dispersion. Parts I–V", *Rev. Mod. Phys.* **4**, 504–576 (1932) and G. Breit, "Quantum theory of dispersion. Parts VI–VII", *Rev. Mod. Phys.* **5**, 90–140 (1932).
[14] G. Breit, "An Interpretation of Dirac's Theory of the Electron", *Proceedings of the National Academy of Sciences of the United States of America* **14,** 553–559 (1928).
[15] M. Conversi, E. Pancini and O. Piccioni, "On the Disintegration of Negative Muons", *Phys. Rev.* **71**, 209–210 (1947).
[16] G. Breit, "The Effect of Retardation on the Interaction of Two Electrons", *Phys. Rev.* **34**, 553–573 (1929).

[17] N. Kemmer, "Zur Theorie der Neutron-Proton-Wechselwirking", *Helv. Phys. Act.* **10**, 47–67 (1937).

[18] John A. Wheeler and G. Breit, "Li^+ Fine Structure and Wave Functions near the Nucleus", *Phys. Rev.* **44**, 948 (1933); G. Breit and John A. Wheeler, "Collision of Two Light Quanta", *Phys. Rev.* 46, 1087–1091 (1934); F.L. Yost, John A. Wheeler and G. Breit, "Coulomb Wave Functions in Repulsive Fields", *Phys. Rev.* **49**, 174–189 (1936).

[19] P.A.M. Dirac, "The Quantum Theory of the Electron", *Proc. Roy. Soc.* **A117**, 610–624 (1928).

[20] Edgar Zilsel, *The Social Origins of Modern Science*, Dordrecht: Kluwer Academic Publishers, 2003.

[21] G. Breit and G.E. Brown, "Effect of Nuclear Motion on the Fine Structure of Hydrogen", *Phys. Rev.* **74**, 1278–1284 (1948).

[22] K. Bechert and J. Meixner, "Structure of the hydrogen lines", *Annalen der Physik* **22**, 525–36 (1935).

[23] C.G. Darwin, "Wave equations of the electron", *Proc. Roy. Soc.* **A118**, 654–80 (1928).

[24] H.A. Bethe, "The electromagnetic shift of energy levels", *Phys. Rev.* **72**, 339–41 (1947).

[25] G. Breit, G.E. Brown and G.B. Arfken, "The Effect of Nuclear Motion on the Hyperfine Structure of Hydrogen", *Phys. Ref.* **76**, 1299–1304 (1949).

[26] G.E. Brown and G.B. Arfken, "Effects of the Proton Radius on Nuclear Motion Correction for the Hyperfine Structure of Hydrogen", *Phys. Rev.* **76**, 1305–1306 (1949).

[27] G.E. Brown, "Note on a relation in Dirac's theory of the electron", *Proc. Nat. Acad. Sci. U.S.A.* **34**, 15–17 (1950).

[28] W. Heitler, *The Quantum Theory of Radiation*, Oxford: Oxford University Press, 1936.

[29] J.S. Schwinger, "On quantum electrodynamics and the magnetic moment of the electron". *Phys. Rev.* **73**, 416–441; J.S. Schwinger, "Quantum electrodynamics. I. A covariant formulation". *Phys. Rev.* **74**, 1439–1461 (1948); J.S. Schwinger, "Quantum electrodynamics. II. Vacuum polarization and self-energy". *Phys. Rev.* **75**, 651–679 (1949); J.S. Schwinger, "Quantum electrodynamics. III. The electrodynamic properties of the electron", *Phys. Rev.* **76**, 790–817 (1949).

[30] For details about Schwinger's relationship with Breit see: Kimball A. Milton, "In Appreciation: Julian Schwinger: From Nuclear Physics and Quantum Electrodynamics to Source Theory and Beyond", *Phys. Perspect.* **9**, 70–114 (2007).

[31] F. Dyson, *Advanced Quantum Mechanics*, transcribed by David Derbes, Singapore: World Scientific Press, 2007.

[32] G.E. Brown and D.G. Ravenhall, "On the Interaction of two electrons", *Proc. Roy. Soc.* **A208**, 552–559 (1951).

[33] J. Sucher, "Continuum Dissolution and the Relativistic Many Body Problem. A Solvable Model", *Phys. Rev. Lett.* **55**, 1033–1035 (1955).

[34] G. Breit, "The Effect of Retardation on the Interaction of Two Electrons", *Phys. Rev.* **34**, 553–573 (1929).
[35] H.A. Bethe and E.E. Salpeter, *Quantum Mechanics of One- and Two-Electron Atoms, New York:* Academic Press, 1957.
[36] For recent examples see e.g. Heinz Siedentrop, "The spectrum of relativistic atoms according to Bethe and Salpeter and Beyond", *J. Comput. Appl. Math.* **208**, 155–163 (2007); Sergey Morozov and Semjon Vugaltor, "Stability of Atoms in the Brown-Ravenhall Model", *Annales Henri Poincaré* **7** (4), 661–687 (2006).
[37] G.E. Brown, "Electron-Electron Interaction in Heavy Atoms", *Phil. Mag.* **43**, 467–471 (1952).
[38] For a review see the discussion of "soft nodes" by T. Hatsuda and T. Kunihiro, "QCD phenomenology based on a chiral effective Lagrangian", *Phys. Rep.* **247**, 221–367 (1994).
[39] This behaviour of the lattice gauge calculations is described in G.E. Brown, B.A. Gellman and Mannque Rho, "Matter formed at the BNL relativistic heavy ion collider", *Phys. Rev. Lett.* **96**, 132301 (2006).
[40] Rudolf Peierls, *Surprises in Theoretical Physics*, Princeton: Princeton University Press, 1979; Rudolf Peierls, *More Surprises in Theoretical Physics*, Princeton: Princeton University Press, 1991.
[41] G.E. Brown and J.B. Woodward, "Coherent Scattering of Gamma Rays by Bound Electrons", *Proc. Phys. Soc.* **A65**, 977–980 (1952).
[42] G.E. Brown, R.E. Peierls and J.B. Woodward, "The Coherent Scattering of γ-rays by K-electrons in Heavy Atoms I. Method", *Proc. Roy. Soc.* **A227**, 51–58 (1954); S. Brenner, G.E. Brown and J.B. Woodward, "The Coherent Scattering of γ-Rays by K-electrons in Heavy Atoms II. The Scattering of 0.32 $Mc^2\gamma$-Rays in Mercury", *Proc. Roy. Soc.* **A227**, 59–71 (1954); G.E. Brown and D.F. Mayers, "The Coherent Scattering of γ-Rays by K-electrons in Heavy Atoms III. The Scattering of 0.64 $Mc^2\gamma$-Rays in Mercury", *Proc. Roy. Soc.* **A234**, 387–390 (1956); G.E. Brown and D.F. Mayers, "The Coherent Scattering of γ-Rays by K-electrons in Heavy Atoms IV. The Scattering of 1.28 *and* 2.56 $Mc^2\gamma$-Rays in Mercury", *Proc. Roy. Soc.* **A242**, 89–95 (1957).
[43] T.V. Kanellopoulos and G.E. Brown, "Polarization Experiments in Proton-Proton Scattering", *Proc. Phys. Soc.* **A70**, 690–702 (1957); G.E. Brown and T.V. Kanellopoulos, "Analysis in Coincidence in Proton-Proton Scattering", *Proc. Phys. Soc.* **A70**, 703–707 (1957).
[44] W. Heiter, *The Quantum Theory of Radiation*, Oxford: Oxford University Press, 1936.
[45] G.E. Brown and G.W. Schaefer, "Expansion in Angular Momentum in Bound-State Perturbation Theory", *Proc. Roy. Soc.* **A233**, 527–536 (1956).
[46] G.E. Brown, J.S. Langer and G.W. Schaefer, "Lamb Shift of a Tightly Bound Electron. I. Method", *Proc. Roy. Soc.*, A **251**, 92–104 (1959).
[47] G.E. Brown, C.T. De Dominicis and J.S. Langer, "Perturbation Theory in Nuclear Reactions", *Ann. Phys.* **16**, 209–220 (1959).

[48] E.H. Lieb, "Second Order Radiative Corrections to the Magnetic Moment of a Bound Electron", *Phil. Mag.* **46**, 311–316 (1955).
[49] W. Heitler, *The Quantum Theory of Radiation*, Oxford: Oxford University Press, 1936.
[50] R.H. Dalitz, "On the alternative decay process for the natural π-meson", *Proc. Phys. Soc.* **A64**, 667 (1951).
[51] S. Lee, *Sir Rudolf Peierls: Selected Private and Scientific Correspondence*, vol. 1, Singapore: World Scientific Publishers, 2007, ix.
[52] R.E. Peierls, "The Commutation Laws of Relativistic Field Theory", *Proc. Roy. Soc.* **A214**, 143–157 (1962).
[53] M. Chrétien and R.E. Peierls, "Properties of Form Factors in Non-Local Theories", *Nuovo Cimento* **10**, 668–76 (1953).
[54] M. Chrétien and R.E. Peierls, "A study of gauge-invariant non-local interactions", *Proc. Roy. Soc.* **A223**, 468–81 (1954).
[55] S.F. Edwards and R.E. Peierls: "Field equations in functional form", *Proc. Roy. Soc.* **A244**, 24–32 (1954).
[56] P.A.M. Dirac, R.E. Peierls and M.H.L. Pryce, "On Lorentz-Invariance in the quantum theory", *Mathematical Proceedings of the Cambridge Philosophical Society* **38,** 193–200 (1942).
[57] J.B. Elliott, "Collective Motion in the Nuclear Shell Model. I. Classification Schemes for States of Mixed Configurations. I", *Proc. Roy. Soc.* **A245**, 128–145 (1958); J.B. Elliott, "Collective Motion in the Nuclear Shell Model. I. Classification Schemes for States of Mixed Configurations II. The Introduction of Intrinsic Wave Functions", *Proc. Roy. Soc.* **245**, 562–581 (1958).
[58] D.H. Hill and J.A. Wheeler, "Nuclear Constitution and the Interpretation of Fission Phenomena", *Phys. Rev.* **89**, 1102–1145 (1953).
[59] R. E. Peierls and J. Yoccoz, "The Collective Model of Nuclear Motion", *Proc. Phys. Soc.* **A70**, 381–387 (1957).
[60] D.J. Thouless, "Stability Conditions and nuclear rotations in the Hartree-Fock theory", *Nucl. Phys.* **21**, 225–32 (1960).
[61] R.E. Peierls and D.J. Thouless, "Variational Approach to Collective Motion", *Nucl. Phys.* **38**, 154–176 (1962).
[62] R.E. Peierls and J.N. Urbano, "The Villars Formalism For Nuclear Reaction", *Journal of Physics* **A1**, 1–10 (1968) and R.E. Peierls, "Perturbation Theory for Projected States", *Proc. Roy. Soc.* **A333**, 157–170 (1978).
[63] S. Lee, *Sir Rudolf Peierls: Selected and Private and Scientific Correspondence*, Vol. 2, Singapore: World Scientific Press, 2009, 575.
[64] Eyvind H. Wichmann and Norman M. Kroll, "Vacuum Polarization in a Strong Coulomb Field", *Phys. Rev.* **96**, 232–234 (1954) and Eyvind H. Wichmann and Norman M. Kroll, "Vacuum Polarization in a Strong Coulomb Field", *Phys. Rev.* **101**, 843–859 (1956).
[65] N.M. Kroll and F. Pollock, "Second Order Radiative Corrections to Hyperfine Structure", *Phys. Rev.* **86**, 876–888 (1952).
[66] G.E. Brown, J.S. Langer and G.W. Schaefer, "Lamb Shift of a Tightly Bound Electron. I. Method". *Proc. Roy. Soc.* **A251**, 92–104 (1959); figures on p. 94.

[67] R. Karplus and N.M. Kroll, "Fourth-Order Corrections in Quantum Electrodynamics and the Magnetic Moment of the Electron", *Phys. Rev.* 77, 536–549 (1950).

[68] H.A. Bethe, "The Electromagnetic Shift of Energy Levels", *Phys. Rev.* **72,** 339–341 (1947). See also article by Freeman Dyson, "Hans Bethe and Quantum Electrodynamics" (in G.E. Brown and Chang-Hwan Lee (eds.), *Hans Bethe and his Physics*, Singapore: World Scientific Press, 2006, 157–164) which gives an elegant discussion of how Hans Bethe calculated this.

[69] G.E. Brown and D.F. Mayers, "Lamb Shift of a Tightly Bound Electron. II. Calculation for the K-electrons in mercury", *Proc. Roy. Soc.* **A251**, 105–109 (1959).

[70] D. Saxon, University of Wisconsin, unpublished Ph.D. thesis (1954).

[71] D.F. Mayers, G.E. Brown and E.A. Sanderson, "Lamb Shift and Energy Level of a Tightly Bound Electron", *Phys. Rev.* **3**, 90 (1959).

[72] K.T. Cheng, W.R. Johnson and J. Sapirstein, "Screened Lamb Shift Calculations for lithiumlike uranium, sodiumlike platinum and copperlike gold", *Phys. Rev. Lett.* **66**, 2960–2963 (1991) especially references 11–15.

[73] J.P. Elliott and B.F. Flowers, "The Odd-Parity states of ^{16}O and ^{16}N", *Proc. Roy. Soc.* **A242**, 57–80 (1957).

[74] D.W.L. Sprung and J.B. Willis, "Spin-orbit potential in proton-proton scattering", *Proc. Phys. Soc.* **76**, 539–544 (1960).

[75] G. Breit, "Nucleon-Nucleon Spin Orbit and the Repulsive Core", *Phys. Rev.* **120**, 287–292 (1960); S.N. Gupta, "Pion Theory of Nuclear Forces with Nucleon Recoil", *Phys. Rev.* **117**, 1146–1151 (1960).

[76] For details see Brown *et al.* (eds.), *The Nucleon-Nucleon Interaction and the Nuclear Many-Body Problem*, Singapore: World Scientific Publishers, 2010, chapter 1.

[77] S.A. Moszkowski and B.L. Scott, "Nuclear Forces and the Properties of Nuclear Matter", *Ann. Phys.* **11**, 65–115 (1960).

[78] A. Kallio and K. Kolltveit, "An application of the separation method in shell model calculation", *Nucl. Phys.* **53**, 87–96 (1964).

[79] Aage Bohr and Ben R. Mottelson, *Nuclear Structure*, (vol. I Single Particle Motion; vol. II Nuclear Deformations), New York: W.A. Benjamin, 1969.

[80] J.F. Dawson, I. Talmi and J.D. Walecka, "Calculation of the level spectrum of O^{18} from the free two-nucleon potential", *Ann. Phys.* **18**, 339–66 (1962).

[81] I. Talmi, "Effective Interactions and Coupling Schemes in Nuclei", *Rev. Mod. Phys.* **34**, 704–722 (1962).

[82] G.F. Bertsch, "Role of core polarization in two-body interaction", *Nucl. Phys.* **74**, 234–240 (1965).

[83] H.A. Bethe, B.H. Brandow and A.G. Petschek, "Reference Spectrum Method for Nuclear Matter", *Phys. Rev.* **129**, 225–64 (1963).

[84] H.A. Bethe, "Three-Body Correlations in Nuclear Matter", *Phys. Rev.* **138**, B804–B822 (1965).

[85] S.A. Moszkowski, "Short-Range Correlations in Nuclear Matter", *Phys. Rev.* **140**, B283–B286 (1965).

[86] T.T.S. Kuo, CC1, January 1980; http://www.garfield.library.upenn.edu/classics1980/A1980HX93600001.pdf.
[87] H.A. Bethe, G.E. Brown and C.H. Lee, *Formation and Evolution of Black Holes in the Galaxy, Selected Papers with Commentary.* Singapore: World Scientific Publishers, 2003, 1.
[88] T.J. Mazurek, J.M. Lattimer and G.E. Brown, "Nuclear forces, partition functions, and dissociation in stellar collapse", *Astrophys. J.* **229**, 713–727 (1978).
[89] H.A. Bethe, G.E. Brown, J. Applegate and J.M. Lattimer, "Equation of state in the gravitational collapse of stars", *Nucl. Phys.* **A324** 487–533 (1979).
[90] Gerald E. Brown and Chang-Hwan Lee (eds.), *Hans Bethe and His Physics*, Singapore: World Scientific Press, 2006, 18ff.
[91] H.A. Bethe, G.E. Brown and C.-H. Lee, "Evolution and Merging of Binaries with Compact Objects", *Physics Report* **442**, 5–22 (2007).
[92] Guide to the Hans Bethe Papers, Collection Number: 14-22-976, Division of Rare and Manuscript Collections Cornell University Library; http://rmc.library.cornell.edu/ead/htmldocs/RMA00976.html.
[93] H.A. Bethe and G.E. Brown, "How a Supernova Explodes", *Scientific Amer.* **252** (5), 60–68 (1985).
[94] G.E. Brown and J.C. Weingartner, "Accretion onto and radiation from the compact object formed in SN 1987A", *Astrophys. J.* **436**, 843–847 (1994).
[95] G.E. Brown and H.A. Bethe, "A Scenario for a Large Number of Low-Mass Black Holes in the Galaxy", *Astrophys. J.* **423**, 659 (1994).
[96] R.A. Chevalier, "Neutron star accretion in a stellar envelope", *Astrophys. J.* **41**, L33—L36 (1993).
[97] G.E. Brown, "Neutron star accretion and binary pulsar formation", *Astrophys. J.* **440**, 270–279 (1995).
[98] E.J.A. Meurs and E.P.J. van den Heuvel, "The number of evolved early-type close binaries in the Galaxy", *Astron. Astrophys.* **226**, 88–107 (1989).
[99] H.A. Bethe and G.E. Brown, "Evolution of binary compact objects that merge", *Astrophys. J.* **506**, 780–789 (1998).

Chapter 2

Toward a Fully Relativistic Theory of Quantum Information

Christoph Adami*

Information theory is a statistical theory dealing with the relative state of detectors and physical systems. Because of this physicality of information, the classical framework of Shannon needs to be extended to deal with quantum detectors, perhaps moving at relativistic speeds, or even within curved space-time. Considerable progress toward such a theory has been achieved in the last fifteen years, while much is still not understood. This review recapitulates some milestones along this road, and speculates about future ones.

1. Preface: From Nuclei to Quantum Information

I am sure I am one of the more junior contributors to this volume celebrating Gerry Brown's 85th birthday, and still I've known him for 25 years. I arrived as a young graduate student at Stony Brook University in 1986, and Gerry immediately introduced me to every member of his Nuclear Theory group, ending with his postdoc Ismail Zahed. He pointed to a chair in Ismail's office, said: "You guys talk", and left. I started to work with Ismail that day, and when he was promoted to Assistant Professor I became his first graduate student. Gerry and I only started to work together closely within the last two years of my Ph.D., and the collaboration intensified when he took me on his yearly Spring visits to the Kellogg Radiation Laboratory at the California Institute of Technology. There, I had the opportunity to meet Hans Bethe, who visited Caltech every Spring to work with Gerry. Over the following years, Hans and I became good

*Department of Physics and Astronomy and Department of Microbiology and Molecular Genetics, Michigan State University, East Lansing, MI 48824, USA.

friends and Hans's influence on my growth as a scientist would end up rivaling the influence that Gerry had on me [1]. In particular Hans was always very interested in my shifting interests from nuclear and high energy theory first towards quantum information theory and the foundations of quantum mechanics, and then to theoretical biology. At the same time, Gerry and Hans's collaboration on the physics of binary stars and in particular black holes continued to intrigue me. I ended up staying at Caltech for 12 years.

While I spend most of my time now working in biology, I still sometimes return to work in physics. People like Gerry and Hans have reinforced to me the fun that comes with attempting to understand the universe's basic principles, and when lucky enough, unravel a few of them. Perhaps it is not a coincidence that one of the striking applications of quantum relativistic information theory that I describe below is to the physics of black holes. Gerry and I discussed black holes and binary stars endlessly on walks in the mountains adjacent to Caltech, and on the phone (often on Sunday mornings) when he was back in New York. Why did I store away article after article on black holes in the 1990s when I wasn't nearly working on the subject? I am sure it was Gerry's influence, who taught me to go after your gut instinct, and not worry if you are called crazy. I've been called crazy in many a referee's review, and I've come to realize that this usually signals that I am on to something. Thus I dedicate this article to you Gerry: there are crazy things buried in here too.

2. Entropy and Information: Classical Theory

Since Shannon's historical pair of papers [2], information theory has changed from an engineering discipline to a full-fledged theory within physics [3]. While a considerable part of Shannon's theory deals with communication channels and codes [4], the concepts of entropy and information he introduced are crucial to our understanding of the physics of measurement, and turn out to be more general than thermodynamical entropy. Thus, information theory represents an important part of statistical physics both at equilibrium and away from it.

In the following, I present an overview of some crucial aspects of entropy and information in classical and quantum physics, with extensions to the special and general theory of relativity. While not exhaustive, the treatment is at an introductory level, with pointers to the technical literature where appropriate.

2.1. *Entropy*

The concepts of entropy and information are the central constructs of Shannon's theory. They quantify the ability of observers to make *predictions*, in particular how well an observer equipped with a specific measurement apparatus can make predictions about another physical system. Shannon entropies (also known as *uncertainties*) are defined for mathematical objects called *random variables.* A random variable X is a mathematical object that can take on a finite number of discrete states x_i, where $i = 1, \ldots, N$ with probabilities p_i. Now, physical systems are not mathematical objects, nor are their states necessarily discrete. However, if we want to quantify our uncertainty about the state of a physical system, then in reality we only need to quantify our uncertainty about the *possible outcomes of a measurement* of that system. In other words, an observer's maximal uncertainty about a system is not a property of the system, but rather a property of the measurement device with which the observer is about to examine the system. For example, suppose I am armed with a measurement device that is simply a "presence-detector". Then the maximal uncertainty I have about the physical system under consideration is 1 bit, which is the amount of *potential information* I can obtain about that system, given this measurement device.

As a consequence, in information theory the entropy of a physical system is undefined if we do not specify the device that we are going to use to reduce that entropy. A standard example for a random variable (that is also a physical system) is the six-sided even die. Usually, the maximal entropy attributed to this system is $\log_2(6)$ bits. Is this all there is to know about this system? What if we are interested not only in the face of the die that is up, but also the angle that the die has made with respect to due North? Further, since the die is physical, it is made of molecules and these can be in different states depending on the temperature of the system. Are those knowable? What about the state of the atoms making up the molecules? All these could conceivably provide labels such that the number of states to describe the die is in reality much larger. What about the state of the nuclei? Or the quarks and gluons inside those?

This type of thinking makes it clear that we cannot speak about the entropy of an isolated system without reference to the coarse-graining of states that is implied by the choice of detector (but I will comment on the continuous variable limit of entropies below). So, even though detectors exist that record continuous variables (such as, say, a mercury thermometer), each detector has a finite resolution such that it is indeed appropriate to consider only the *discrete* version of the Shannon entropy, which is given in terms of the probabilities p_i as[a]

$$H(X) = -\sum_i^N p_i \log p_i \,. \tag{1}$$

For any physical system, how are those probabilities obtained? In principle, this can be done both by experiment and by theory. Once I have defined the N possible states of my system by choosing a detector for it, the *a priori* maximal entropy is defined as

$$H_{\rm max} = \log N. \tag{2}$$

Experiments using my detector can now sharpen my knowledge of the system. By tabulating the frequency with which each of the N states appears, we can estimate the probabilities p_i. Note, however, that this is a biased estimate that approaches the true entropy Eq. (1) only in the limit of an infinite number of trials [5]. On the other hand, some of the possible states of the system (or more precisely, possible states of my detector interacting with the system) can be eliminated by using some knowledge of the physics of the system. For example, we may know some initial data or averages that characterize the system. This becomes clear in particular if the degrees of freedom that we choose to characterize the system with are position, momentum, and energy, i.e. if we consider the *thermodynamical entropy* of the system (see below). In this respect it is instructive to consider for a moment the continuous variable equivalent of the Shannon entropy, also known as the *differential entropy*, defined as [4]

$$h(X) = -\int_S f(x) \log f(x) dx \,, \tag{3}$$

with a probability density function $f(x)$ with support S. It turns out that while $h(X)$ is invariant with respect to translations $[h(X+c) = h(X)]$, it is not invariant under arbitrary coordinate transformations: the entropy is *renormalized* under such changes instead. For example,

$$h(cX) = h(X) + \log |c|. \tag{4}$$

In particular, this implies that if we introduce a discretization of continuous space (e.g. via $p_i = \int_{i\Delta}^{(i+1)\Delta} f(x)dx$) and consider the limit of the discretized

[a]From now on, I shall not indicate the basis of the logarithm, which only serves to set the units of entropy and information (base 2, e.g. sets the unit to a "bit").

version as $\Delta \to 0$, we find that

$$H_\Delta[p_i] = -\sum_i p_i \log p_i \underset{\Delta\to 0}{\longrightarrow} h(X) - \log \Delta. \tag{5}$$

Thus, as the resolution of a measurement device is increased, the entropy is renormalized via an infinite term. Of course, we are used to such infinite renormalizations from quantum field theory, and just as in the field theory case, the "unphysical" renormalization is due to an unphysical assumption about the attainable precision of measurements. Just as in quantum field theory, differences of quantities do make sense: the shared (or mutual) differential entropy is finite in the limit $\Delta \to 0$ as the infinities cancel.

2.2. *Conditional entropy*

Let us look at the basic process that reduces uncertainty: a measurement. When measuring the state of system X, I need to bring it into contact with a system Y. If Y is my measurement device, then usually I can consider it to be completely known (at least, it is completely known with respect to the degrees of freedom I *care* about). In other words, my device is in a particular state y_0 with certainty. After interacting with X, this is not the case anymore. Let us imagine an interaction between the systems X and Y that is such that

$$x_i y_0 \to x_i y_i \quad i = 1, \ldots, N, \tag{6}$$

that is, the states of the measurement device y_i end up reflecting the states of X. This is a perfect measurement, since no state of X remains unresolved. More generally, let X have N states while Y has M states, and let us suppose that $M < N$. Then we can imagine that each state of Y reflects an *average* of a number of X's states, so that the probability to find Y in state y_j is given by q_j, where $q_j = \sum_i p_{ij}$, and p_{ij} is the joint probability to find X in state x_i and Y in state y_j. The measurement process then proceeds as

$$x_i y_0 \to \langle x \rangle_j y_j \tag{7}$$

where

$$\langle x \rangle_j = \sum_i p_{i|j} x_i. \tag{8}$$

In Eq. (8) above, I introduced the *conditional probability*

$$p_{i|j} = \frac{p_{ij}}{q_j} \tag{9}$$

that X is in state i *given* that Y is in state j. In the perfect measurement above, this probability was 1 if $i = j$ and 0 otherwise (i.e. $p_{i|j} = \delta_{ij}$), but in the imperfect measurement, X is distributed across some of its states i with a probability distribution $p_{i|j}$, for each j.

We can then calculate the *conditional entropy* (or remaining entropy) of the system X given we found Y in a particular state y_j after the measurement:

$$H(X|Y = y_j) = -\sum_i^N p_{i|j} \log p_{i|j}. \tag{10}$$

This remaining entropy is guaranteed to be smaller than or equal to the unconditional entropy $H(X)$, because the worst case scenario is that Y doesn't resolve *any* states of X, in which case $p_{i|j} = p_i$. But since we didn't know anything about X to begin with, $p_i = 1/N$, and thus $H(X|Y = y_j) \leq \log N$.

Let us imagine that we did learn something from the measurement of X using Y, and let us imagine furthermore that this knowledge is permanent. Then we can express our new-found knowledge about X by saying that we know the probability distribution of X, p_i, and this distribution is *not* the uniform distribution $p_i = 1/N$. Of course, in principle we should say that this is a conditional probability $p_{i|j}$, but if the knowledge we have obtained is permanent, there is no need to constantly remind ourselves that the probability distribution is conditional on our knowledge of certain other variables connected with X. We simply say that X is distributed according to p_i, and the entropy of X is

$$H_{\text{actual}}(X) = -\sum_i \log p_i \log p_i. \tag{11}$$

According to this strict view, all Shannon entropies of the form (11) are conditional if they are not maximal. And we can quantify our knowledge about X simply by subtracting this uncertainty from the maximal one:

$$I = H_{\max}(X) - H_{\text{actual}}(X). \tag{12}$$

This knowledge, of course, is *information*. We can see from this expression that the entropy $H_{\max}$ can be seen as *potential information*: it quantifies how much is knowable about this system. If my actual entropy vanishes, then all of the potential information is realized.

2.3. *Information*

In Eq. (12), we quantified our knowledge about the states of X by the difference between the maximal and the actual entropy of the system. This was a special case because we assumed that after the measurement, Y was in state y_j with certainty, i.e. there is no remaining uncertainty associated with the measurement device Y (of course, this is appropriate for a measurement device). In a more general scenario where two random variables are correlated with each other, we can imagine that Y (after the interaction with X) instead is in state y_j with probability q_j (in other words, we have reduced our uncertainty about Y somewhat, but we don't know everything about it, just as for X). We can then define the *average conditional entropy* of X simply as

$$H(X|Y) = \sum_j q_j H(X|Y = y_j) \tag{13}$$

and the information that Y has about X is then the difference between the unconditional entropy $H(X)$ and Eq. (13) above,

$$H(X:Y) = H(X) - H(X|Y). \tag{14}$$

The colon between X and Y in the expression for the information $H(X:Y)$ is conventional, and indicates that it stands for an entropy *shared* between X and Y. According to our strict definition of unconditional entropies given above, $H(X) = \log N$, but in the standard literature $H(X)$ refers to the actual uncertainty of X given whatever knowledge allowed me to obtain the probability distribution p_i, that is, Eq. (11). In the case nothing is known *a priori* about X, Eq. (14) equals Eq. (12).

Equation (14) can be rewritten to display the symmetry between the observing system and the observed:

$$H(X:Y) = H(X) + H(Y) - H(XY), \tag{15}$$

where $H(XY)$ is just the joint entropy of both X and Y combined. This joint entropy would equal the sum of each of X's and Y's entropy only in the case that there are *no correlations* between X's and Y's states. If that would be the case, we could not make any predictions about X just from knowing something about Y. The information (15), therefore, would vanish.

2.3.1. *Example: thermodynamics*

We can view Thermodynamics as a particular case of Shannon theory. First, if we agree that the degrees of freedom of interest are position and momentum, then the maximal entropy of any system is defined by its volume in phase space:

$$H_{\rm max} = \log \Delta\Gamma, \tag{16}$$

where $\Delta\Gamma = \frac{\Delta p \Delta q}{k}$ is the number of states within the phase space volume $\Delta p \Delta q$. Now the normalization factor k introduced in (16) clearly serves to coarse grain the number of states, and should be related to the resolution of our measurement device. In quantum mechanics, of course, this factor is given by the amount of phase space volume occupied by each quantum state, $k = (2\pi\hbar)^n$ where n is the number of degrees of freedom of the system. Does this mean that in this case it is not my type of detector that sets the maximum entropy of the system? Actually, this is still true, only that here we assume a quantum mechanical perfect detector, while still averaging over certain internal states of the system inaccessible to this detector.

Suppose I am contemplating a system whose maximum entropy I have determined to be Eq. (16), but I have some additional information. For example, I know that this system has been undisturbed for a long time, and I know its total energy E, and perhaps even the temperature T. Of course, this kind of knowledge can be obtained by a number of different ways. It could be obtained by experiment, or it could be obtained by inference, or theory. How does this knowledge reduce my uncertainty? In this case, we use our *knowledge of physics* to predict that the probabilities $\rho(p,q)$ going into our entropy

$$H(p,q) = -\sum_{\Delta p \Delta q} \rho(p,q) \log \rho(p,q) \tag{17}$$

is given by the canonical distribution[b]

$$\rho(p,q) = \frac{1}{Z} e^{-E(p,q)/T}, \tag{18}$$

[b]We set Boltzmann's constant equal to 1 throughout. This constant, of course, sets the scale of thermodynamical entropy, and would end up multiplying the Shannon entropy just like any particular choice of base for the logarithm would.

where Z is the usual normalization constant, and the sum in (17) goes over all positions and momenta in the phase space volume $\Delta p \Delta q$. The amount of knowledge we have about the system according to Eq. (12) is then just the difference between the maximal and actual uncertainties:

$$I = \log \frac{\Delta \Gamma}{Z} - \frac{E}{T}, \tag{19}$$

where $E = \sum_{\Delta p \Delta q} \rho(p, q) E(p, q)$ is the energy of the system.

3. Quantum Theory

In quantum mechanics, the concept of entropy translates very easily, but the concept of information is thorny. John von Neumann introduced his eponymous quantum mechanical entropy as early as 1927 [6], a full 21 years before Shannon introduced its classical limit! In fact, it was von Neumann who suggested to Shannon to call his formula (1) "entropy", simply because, as he said, "your uncertainty function has been used in statistical mechanics under that name" [7].

3.1. *Measurement*

In quantum mechanics, measurement plays a prominent role, and is still considered somewhat mysterious in many respects. The proper theory to describe measurement dynamics in quantum physics, not surprisingly, is quantum information theory. As in the classical theory, the uncertainty about a quantum system can only be defined in terms of the detector states, which in quantum mechanics is a discrete set of *eigenstates* of a measurement operator. The quantum system itself is described by a wave function, given in terms of the quantum system's eigenbasis, which may or may not be the same as the measurement device's basis.

For example, say we would like to "measure an electron". In this case, we may mean that we would like to measure the position of an electron, whose wave function is given by $\Psi(q)$, where q is the coordinate of the electron. Further, let the measurement device be characterized initially by its eigenfunction $\phi_0(\xi)$, where ξ may summarize the coordinates of the device. Before measurement, i.e. before the electron interacts with the measurement device, the system is described by the wave function

$$\Psi(q)\phi_0(\xi). \tag{20}$$

After the interaction, the wave function is a superposition of the eigenfunctions of electron and measurement device

$$\sum_n \psi_n(q)\phi_n(\xi). \tag{21}$$

Following orthodox measurement theory [8], the classical nature of the measurement apparatus implies that after measurement the "pointer" variable ξ takes on a well-defined value at each point in time; the wave function, as it turns out, is thus *not* given by the entire sum in (21) but rather by the single term

$$\psi_n(q)\phi_n(\xi). \tag{22}$$

The wave function (21) is said to have collapsed to (22).

Let us now study what actually happens in such a measurement in detail. For ease of notation, let us recast this problem into the language of state vectors instead. The first stage of the measurement involves the interaction of the quantum system Q with the measurement device (or "ancilla") A. Both the quantum system and the ancilla are fully determined by their state vector, yet, let us assume that the state of Q (described by state vector $|x\rangle$) is unknown whereas the state of the ancilla is prepared in a special state $|0\rangle$, say. The state vector of the combined system $|QA\rangle$ before measurement then is

$$|\Psi_{t=0}\rangle = |x\rangle|0\rangle \equiv |x,0\rangle. \tag{23}$$

The von Neumann measurement [9] is described by the unitary evolution of QA via the interaction Hamiltonian

$$\hat{H} = -\hat{X}_Q\hat{P}_A, \tag{24}$$

operating on the product space of Q and A. Here, $\hat{X}_Q$ is the observable to be measured, and $\hat{P}_A$ the operator *conjugate* to the degree of freedom of A that will reflect the result of the measurement. We now obtain for the state vector $|QA\rangle$ after measurement (e.g. at $t = 1$, putting $\hbar = 1$)

$$|\Psi_{t=1}\rangle = e^{i\hat{X}_Q\hat{P}_A}|x,0\rangle = e^{ix\hat{P}_A}|x,0\rangle = |x,x\rangle. \tag{25}$$

Thus, the pointer variable in A that previously pointed to zero now also points to the position x that Q is in. This operation appears to be very much like the classical measurement process Eq. (6), but it turns out to be quite different. In general, the unitary operation (25) introduces quantum *entanglement* between the system being measured and the measurement apparatus, a concept that is beyond the classical idea of correlations.

That entanglement is very different from correlations becomes evident if we apply the unitary operation described above to an initial quantum state which is in a quantum *superposition* of two states:

$$|\Psi_{t=0}\rangle = |x + y, 0\rangle. \tag{26}$$

Then, the linearity of quantum mechanics implies that

$$|\Psi_{t=1}\rangle = e^{i\hat{X}_Q\hat{P}_M}(|x, 0\rangle + |y, 0\rangle) = |x, x\rangle + |y, y\rangle. \tag{27}$$

This state is very different from what we would expect in classical physics, because Q and A are not just correlated (like, e.g. the state $|x + y, x + y\rangle$ would be) but rather they are *quantum entangled.* They now form *one* system that cannot be thought of as composite.

This nonseparability of a quantum system and the device measuring it is at the heart of all quantum mysteries. Indeed, it is at the heart of *quantum randomness*, the puzzling emergence of unpredictability in a theory that is unitary, i.e. where all probabilities are conserved. What is being asked here of the measurement device, namely to describe the system Q, is logically impossible because after entanglement the system has grown to QA. Thus, the detector is being asked to describe a system that is *larger* (with respect to the possible number of states) than the detector, because it includes the detector *itself.* This is precisely the same predicament that befalls a computer program that is asked to determine its own halting probability, in the famous *Halting Problem* [10] analogue of Gödel's famous Incompleteness Theorem [11]. Chaitin [12] showed that the self-referential nature of the question that is posed to a computer program written to solve the Halting Problem gives rise to randomness in pure Mathematics: the halting probability $\Omega = \sum_{p \text{ halts}} 2^{-|p|}$, where the sum goes over all the programs p that halt and $|p|$ is the size of those programs, is random in every way that we measure randomness [13]. A quantum measurement is self-referential in the same manner, since the detector is asked to describe its own state, which is logically impossible. Thus we see that quantum randomness has mathematical, or rather logical, randomness at its very heart.

3.2. *von Neumann entropy*

Because of the uncertainty inherent in standard projective measurements, measurements of a quantum system Q are described as *expectation values*, which are averages of an observable over the system's *density matrix*, so that

$$\langle\hat{O}\rangle = \mathrm{Tr}(\rho_Q\hat{O}), \tag{28}$$

where $\hat{O}$ is an operator associated with the observable we would like to measure, and

$$\rho_Q = \mathrm{Tr}_A|\Psi_{QA}\rangle\langle\Psi_{QA}|, \tag{29}$$

is obtained from the quantum wave function Ψ_{QA} (for the combined system QA, since neither Q nor the measurement device A separately have a wave function after the entanglement occurred) by *tracing out* the measurement device. However, technically, we are observing the states of the detector, not the states of the quantum system, so instead we need to obtain

$$\rho_A = \mathrm{Tr}_Q|\Psi_{QA}\rangle\langle\Psi_{QA}|, \tag{30}$$

by averaging over the states of the quantum system (which strictly speaking is *not* being observed) and the expectation value of the measurement is instead

$$\langle\hat{O}\rangle = \mathrm{Tr}(\rho_A\hat{O}). \tag{31}$$

The uncertainty about the quantum system is then assumed to be given by the uncertainty in the measurement device A, and can be calculated simply using the von Neumann entropy [6, 9]:

$$S(\rho_A) = -\mathrm{Tr}\rho_A \log \rho_A. \tag{32}$$

If Q has been measured in A's eigenbasis, then the density matrix ρ_A is diagonal, and von Neumann entropy turns into Shannon entropy, as we expect. Indeed, measuring with respect to the system's eigenbasis is precisely the classical limit: entanglement does not happen under these conditions.

Quantum Information Theory, of course, needs concepts such as conditional entropies and mutual entropies besides the von Neumann entropy. They can be defined in a straightforward manner [14, 15], but their interpretation needs care. For example, we can define a conditional entropy in analogy to Shannon theory as

$$S(A|B) = S(AB) - S(B)$$
$$= -\mathrm{Tr}_{AB}(\rho_{AB}\log\rho_{AB}) + \mathrm{Tr}_B(\rho_B\log\rho_B), \tag{33}$$

where $S(AB)$ is the joint entropy of two systems A and B. But can we write this entropy in terms of a conditional density matrix, just as we were able to write the conditional Shannon entropy in terms of a conditional probability? The answer is yes and no: a definition of conditional von Neumann entropy in terms of a conditional density *operator* $\rho_{A|B}$ exists [15, 16], but this operator is technically not a density matrix (its trace is not equal to one), and the eigenvalues of this matrix are very peculiar: they can *exceed* one (this is of course not possible for probabilities). Indeed, the eigenvalues can exceed one only when the system is entangled. As a consequence, quantum conditional entropies can be *negative* [15]. This negative quantum entropy has an operational meaning in quantum information theory: it quantifies how much additional information must be conveyed in order to transport a quantum state if part of a distributed quantum system is known [17]. If this "partial information" is negative, the sender and receiver can use the states for future communication. In Fig. 1(a), we can see a quantum communication process known as "quantum teleportation" [18], in which the quantum wavefunction of a *qubit* (the quantum analogue of the usual bit, which is a quantum particle that can exist in superpositions of zero and one) is transported from the sender "A" (often termed "Alice") to the receiver "B" (conventionally known as "Bob"). This can be achieved using an entangled pair of particles $e\bar{e}$ (an ebit–anti-ebit pair), where "ebit" stands for entangled bit [19]. This pair carries no information, but each element of the pair carries partial information: in this case the ebit carries one bit, while the anti-ebit carries minus one bit. Bob sends the ebit over to Alice, who performs a joint measurement M of the pair and sends the two classical bits of information back to Bob (see Fig. 1(a)). Armed with the two classical bits, Bob in turn can now perform a unitary operation U on the anti-ebit he has been carrying around, and transform it into the original qubit that Alice had intended to convey. In this manner, Bob has used the negative "partial information" in his anti-ebit to recover the full quantum state, using only classical information. Note that the anti-ebit with negative partial information traveling forwards in time can be seen as an ebit with positive partial information traveling backwards in time [15]. The process of

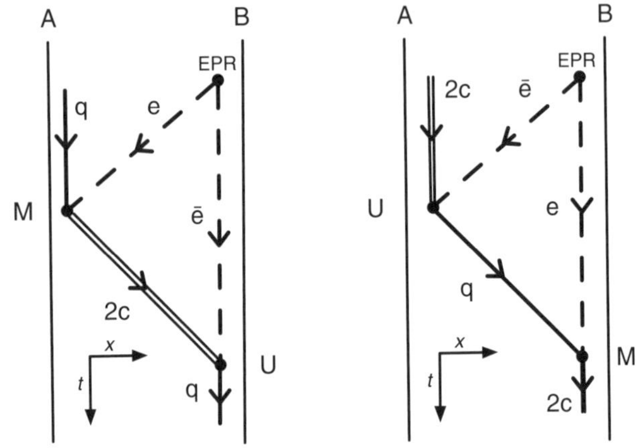

(a) (b)

Fig. 1. Using negative partial information for quantum communication. (a) In these diagrams, time runs from top to bottom, and space is horizontal. The line marked "A" is Alice's space-time trajectory, while the line marked "B" is Bob's. Bob creates an $e\bar{e}$ pair (an Einstein-Podolski-Rosen pair) close to him, and sends the ebit over to Alice. Alice, armed with an arbitrary quantum state q, performs a joint measurement M on both e and q, and sends the two classical bits $2c$ she obtains from this measurement back to Bob (over a classical channel). When Bob receives these two cbits, he performs one out of four unitary transformations U on the anti-ebit he is still carrying, conditionally on the classical information he received. Having done this, he recovers the original quantum state q, which was "teleported" over to him. The partial information in e is one bit, while it is minus one for the anti-ebit. (b) In superdense coding, Alice sends two classical bits of information $2c$ over to Bob, but using only a single qubit in the quantum channel. This process is in a way the "dual" to the teleportation process, as Alice encodes the two classical bits by performing a conditional unitary operation U on the anti-ebit, while it is Bob that performs the measurement M on the ebit he kept and the qubit Alice sent. Figure adapted from [15].

super-dense coding [20] can be explained in a similar manner (see Fig. 1(b)), except here Alice manages to send 2 classical bits by encoding them on the single anti-ebit she received from Bob.

Quantum mutual entropy is perhaps even harder to understand. We can again define it simply in analogy to (15) as [14, 15, 21]

$$S(A : B) = S(A) + S(B) - S(AB), \tag{34}$$

but what does it mean? For starters, this quantum mutual entropy can be twice as large as the entropy of any of the subsystems, so A and B can share more quantum entropy than they even have by themselves! Of course, this is due to the fact, again, that "selves" do not exist anymore after entanglement. Also, in the classical theory, information, that is, shared entropy, could be used to make predictions, and therefore to reduce the uncertainty we have about the system that we share entropy with. But that's not possible in quantum mechanics. If, for example, I measure the spin of a quantum particle that is in an even superposition of its spin-up and spin-down state, my measurement device will show me spin-up half the time, and spin-down half the time, that is, my measurement device has an entropy of one bit. It can also be shown that the shared entropy is *two bits* [15]. But this shared entropy cannot be used to make predictions about the actual spin. Indeed, for the case of the even superposition, I still do not know anything about it [22]! On the other hand, it is possible, armed with my measurement result, to make predictions about the state of *other* detectors measuring the same spin. And even though all these detectors will agree about their result, technically they agree about a random variable (the state of the measurement device), not the actual state of the spin they believe their measurement device to reflect [23]. Indeed, what else could they agree on, since the spin does not have a state? Only the combined system with all the measurement devices that have ever interacted with it, does [24].

Still, the quantum mutual entropy plays a central role in quantum information theory, because it plays a similar role as the classical mutual entropy in the construction of the capacity of an entanglement-assisted channel [25, 26]. In this respect, it is unsurprising that the mutual entropy between two qubits can be as large as 2, as this is the capacity of the superdense coding channel described in Fig. 1(b) [25].

The extension of Shannon's theory into the quantum regime not only throws new light on the measurement problem, but it also helps in navigating the boundary between classical and quantum physics. According to standard lore, quantum systems (meaning systems described by a quantum wave function) "become" classical in the macroscopic limit, that is, if the action unit associated with that system is much larger than $\hbar$. Quantum information theory has thoroughly refuted this notion, since we now know that macroscopic bodies can be entangled just as microscopic ones can [27]. Instead, we realize that quantum systems appear to follow the rules of classical mechanics if parts of their wave function are averaged over [such as in Eq. (29)], that is, if the experimenter is not in total control of all the degrees of freedom that make up the quantum system. Because entanglement, once achieved, is not undone by the distance between entangled parts, almost all systems will seem classical unless expressly prepared, and then protected from interaction with uncontrollable quantum systems. Unprotected quantum systems spread their state over many variables very quickly: a process known as *decoherence* of the quantum state [28].

3.2.1. *Example: quantum thermodynamics*

A simple example that illustrates the use of information theory in (quantum) thermal physics is the Heisenberg dimer model, defined by the Hamiltonian

$$H = J\vec{s}_1 \otimes \vec{s}_2 = \frac{J}{4}\vec{\sigma}\otimes\vec{\sigma}, \tag{35}$$

where $\vec{\sigma} = (\sigma_x, \sigma_y, \sigma_z)$ are the Pauli matrices. The system has three degenerate excited states with energy $J/4$, and a (singlet) ground state with energy $-3J/4$. The thermal density matrix of the two-spin system can be written in the product basis as (here, $\beta = 1/T$ is the inverse temperature)

$$\rho_{12} = \frac{e^{-\beta H}}{Z} = \frac{e^{\frac{\beta J}{4}}}{Z}\begin{pmatrix} e^{-\frac{\beta J}{2}} & 0 & 0 & 0 \\ 0 & \cosh\frac{\beta J}{2} & -\sinh\frac{\beta J}{2} & 0 \\ 0 & -\sinh\frac{\beta J}{2} & \cosh\frac{\beta J}{2} & 0 \\ 0 & 0 & 0 & e^{-\frac{\beta J}{2}} \end{pmatrix}$$

where $Z = \mathrm{Tr}e^{-\beta H} = e^{\frac{3\beta J}{4}} + 3e^{-\frac{\beta J}{4}}$. We can calculate the von Neumann entropy of the joint system as

$$S(\rho_{12}) = -\mathrm{Tr}\rho_{12}\log\rho_{12} = \log Z + \beta E, \tag{36}$$

where E is the energy

$$E = \mathrm{Tr}\rho_{12}H = \frac{3J}{4}\frac{1-e^{\beta J}}{3+e^{\beta J}}. \tag{37}$$

The marginal density matrices for each of the spin subsystems turn out to be

$$\rho_1 = \rho_2 = \begin{pmatrix} 1 & 0 \\ 0 & 1 \end{pmatrix}$$

as can easily be seen from inspecting ρ_{12} above, so that $S(\rho_1) = S(\rho_2) = 1$. Using (34) we can calculate the mutual entropy between the quantum subsystems to find

$$S(1:2) = 2 - \log Z - \beta E, \tag{38}$$

which is formally analogous to the classical result (19), but has very peculiar quantum properties instead. In the infinite temperature limit $\beta \to 0$ we see that $Z \to 4$ while $E \to 0$, so the shared entropy vanishes in that limit as it should: no interactions can be maintained. But it is clear that at any finite temperature, the quantum interaction between the spins creates correlations that can be quantified by the mutual von Neumann entropy between the spins. In particular, in the limit of zero temperature we find

$$\log Z + \beta E \underset{T \to 0}{\longrightarrow} 0, \tag{39}$$

that is, the joint entropy of the spins $S(\rho_{12})$ vanishes and $S(1:2) \to 2$. In that case, the mutual von Neumann entropy is that of a pure Einstein-Podolski-Rosen pair: the singlet solution

$$|\Psi\rangle = \frac{1}{\sqrt{2}} \left(|\uparrow\downarrow\rangle - |\downarrow\uparrow\rangle \right), \tag{40}$$

and exceeds by a factor of two the entropy of any of the spins it is composed of. We recognize the wavefunction of the ground state of the Heisenberg dimer at zero temperature as the entangled $e\bar{e}$ pair that we encountered earlier, and that was so useful in quantum teleportation and superdense coding. We will study its behavior under Lorentz transformations below.

4. Relativistic Theory

Once convinced that information theory is a statistical theory about the relative states of detectors in a physical world, it is clear that we must worry not only about quantum detectors, but about moving ones as well. Einstein's special relativity established an upper limit for the speed at which information can be transmitted, without the need to cast this problem in an information-theoretic language. But in hindsight, it is clear that the impossibility of superluminal signaling could just as well have been the result of an analysis of the information transmission capacity of a communication channel involving detectors moving at constant speed with respect to each other. As a matter of fact, Jarett and Cover calculated the capacity of an "additive white noise Gaussian" (AWNG) channel [4] for information transmission for the case of moving observers, and found [29]

$$C = W\log(1+\alpha\mathrm{SNR}), \tag{41}$$

where W is the bandwidth of the channel, SNR is the signal-to-noise ratio, and $\alpha = \nu'/\nu$ is the Doppler shift. As the relative velocity $v/c \to 1$, $\alpha \to 0$ and the communication capacity vanishes. In the limit $\alpha = 1$, the common capacity formula for the common Gaussian channel with limited bandwidth [4] is recovered. Note that in the limit of an infinite bandwidth channel, Eq. (41) becomes

$$C = \alpha\,\mathrm{SNR}\log_2(e) \text{ bits per second.} \tag{42}$$

Historically, this calculation seems to have been an anomaly: no-one else seems to have worried about an "information theory of moving bodies", not the least because such a theory had, or indeed has, little immediate relevance. Interestingly, the problem that Jarett and Cover addressed with their calculation [29] was the famous "twin-paradox": a thought experiment in special relativity that involves a twin journeying into space at high-speed, only to turn around to find that his identical twin that stayed behind has aged faster. Relativistic information theory gives a nice illustration of the resolution of the paradox, where the U-turn that the traveling twin must undergo creates a switch in reference frames that affects the information transmission capacities between the twins, and accounts for the differential aging.

A standard scenario that would require relativistic information theory thus involves two random variables moving with respect to each other. The question we may ask is whether relative motion is going to affect any shared entropy between the variables. First, it is important to point out that Shannon entropy is a *scalar*, and we therefore do not expect it to transform under Lorentz transformations. This is also intuitively clear if we adopt the "strict" interpretation of entropy as being *unconditional* (and therefore equal to the logarithm of the number of degrees of freedom). On the other hand, probability distributions (and the associated conditional entropies) could conceivably change under Lorentz transformations. How is this possible given the earlier statement that entropy is a scalar?

We can investigate this with a *gedankenexperiment* where the system under consideration is an ideal gas, with particle velocities distributed according to the Maxwell distribution. In order to define entropies, we have to agree about which degrees of freedom we are interested in. Let us say that we only care about the two components of the velocity of particles confined in the $x-y$-plane. Even at rest, the mutual entropy between the particle velocity components $H(v_x : v_y)$ is non-vanishing, due to the finiteness of the magnitude of v. A detailed calculation[c] using continuous variable entropies of the Maxwell distribution shows that, at rest

$$H(v_x : v_y) = \log(\pi/e). \tag{43}$$

The Maxwell velocity distribution, on the other hand, will surely change under Lorentz transformations in, say, the x-direction, because clearly the components are affected differently by the boost. In particular, it can be shown that the mutual entropy between v_x and v_y will *rise* monotonically from $\log(\pi/e)$, and tend to a constant value as the boost-velocity $\beta = v/c \to 1$. But of course, β is just another variable characterizing the moving system, and if this is known precisely, then we ought to be able to recover Eq. (43), and the apparent change in information is due entirely to a reduction in the uncertainty $H(v_x)$. This example shows that in information theory, even if the entire system's entropy does not change under Lorentz transformations, the entropies of subsystems, and therefore also information, can.

While a full theory of relativistic information does not exist, pieces of such a theory can be found when digging through the literature. For example, relativistic thermodynamics is a limiting case of relativistic information theory, simply because as we have seen above, thermodynamical entropy is a limiting case of Shannon entropy. But unlike in the case constructed above, we do not have the freedom to choose our variables in thermodynamics. Hence, the invariance of entropy under Lorentz transformations is assured via Liouville's theorem, because the latter guarantees that the phase-space volume occupied by a system is invariant. Yet, relativistic thermodynamics is an odd theory, not the least because it is intrinsically inconsistent: the concept of equilibrium becomes dubious. In thermodynamics, equilibrium is defined as a state where all relative motion between the subsystems of an ensemble have ceased. Therefore, a joint system where one part moves with a constant velocity with respect to the other cannot be at equilibrium, and relativistic information theory has to be used instead.

---

[c]R.M. Gingrich, unpublished.

One of the few questions of immediate relevance that relativistic thermodynamics has been able to answer is how the temperature of an isolated system will appear from a moving observer. Of course, temperature itself is an equilibrium concept and therefore care must be taken in framing this question [30]. Indeed, both Einstein [31] and Planck [32] tackled the question of how to Lorentz-transform temperature, with different results. The controversy [33, 34] can be resolved by realizing that no such transformation law can in fact exist [35], as the usual temperature (the parameter associated with the Planckian blackbody spectrum) becomes direction-dependent if measured with a detector moving with velocity $\beta = v/c$ and oriented at an angle θ' with respect to the radiation [36, 37]

$$T' = \frac{T\sqrt{1-\beta^2}}{1-\beta\cos\theta'}. \tag{44}$$

In other words, an ensemble that is thermal in the rest frame is non-thermal in a moving frame, and in particular cannot represent a standard heat bath because it will be *non-isotropic*.

5. Relativistic Quantum Theory

While macroscopic quantities like temperature lose their meaning in relativity, microscopic descriptions in terms of probability distributions clearly still make sense. But in a quantum theory, these probability distributions are obtained from quantum measurements specified by local operators, and the space-time relationship between the detectors implementing these operators becomes important. For example, certain measurements on a joint (i.e. composite) system may require communication between parties, while certain others are impossible *even though* they do not require communication [38]. In general, a relativistic theory of quantum information needs to pay close attention to the behavior of the von Neumann entropy under Lorentz transformation, and how such entropies are being reduced by measurement. In this section, I discuss the effect of a Lorentz transformation on the entropy of a single particle, or a pair of entangled particles. For the latter case, I study how quantum entanglement between particles is affected by global Lorentz boosts. This formalism has later been used to study the effect of *local* Lorentz transformations on the von Neumann entropy of a single particle or a pair of entangled particles, and I will summarize those results too.

5.1. *Boosting quantum entropy*

The entropy of a qubit (which we take here for simplicity to be a spin-1/2 particle) with wave function

$$|\Psi\rangle = \frac{1}{\sqrt{|a|^2+|b|^2}}(a|\uparrow\rangle + b|\downarrow\rangle), \tag{45}$$

(a and b are complex numbers), can be written in terms of its density matrix $\rho = |\Psi\rangle\langle\Psi|$ as

$$S(\rho) = -\mathrm{Tr}(\rho \log \rho). \tag{46}$$

A wave function is by definition a completely known state (called a "pure state"), because the wave function is a complete description of a quantum system. As a consequence, (46) vanishes: we have no uncertainty about this quantum system. As we have seen earlier, it is when that wave function interacts with uncontrolled degrees of freedom that mixed states arise. And indeed, just by Lorentz-boosting a qubit, such mixing will arise [39]. The reason is not difficult to understand. The wave function (45), even though I have just stated that it completely describes the system, in fact only completely describes the *spin* degree of freedom! Just as we saw in the earlier discussion about the classical theory of information, there may always be other degrees of freedom that our measurement device (here, a spin-polarization detector) cannot resolve. Because we are dealing with particles, ultimately we have to consider their *momenta*. A more complete description of the qubit state then is

$$|\Psi\rangle = |\sigma\rangle \times |\vec{p}\rangle, \tag{47}$$

where σ stands for the spin-variable, and $\vec{p}$ is the particle's momentum. Note that the momentum wave function $|\vec{p}\rangle$ is in a product state with the spin wave function $|\sigma\rangle$. This means that both spin and momentum have their own state, they are *unmixed*. But as is taught in every first-year quantum mechanics course, such momentum wave functions (plane waves with perfectly well-defined momentum $\vec{p}$) do not actually exist; in reality, they are wave packets with a momentum *distribution* $f(\vec{p})$, which we may take to be Gaussian. If the system is at rest, the momentum wave function does not affect the entropy of (47), because it is a product.

What happens if the particle is boosted? The spin and momentum degrees *do* mix, which we should have expected because Lorentz transformations always imply frame rotations as well as changes in linear velocity.

The product wave function (47) then turns into

$$|\Psi\rangle \longrightarrow \sum_{\sigma} \int f(\vec{p})|\sigma, \vec{p}\rangle d\vec{p}, \tag{48}$$

which is a state where spin-degrees of freedom and momentum degrees of freedom are entangled. But our spin-polarization detector is insensitive to momentum! Then we have no choice but to average over the momentum, which gives rise to a spin density matrix that is mixed

$$\rho_\sigma = \mathrm{Tr}_{\vec{p}}\left(|\Psi\rangle\langle\Psi|\right), \tag{49}$$

and that consequently has positive entropy. Note, however, that the entropy of the joint spin-momentum density matrix remains unchanged, at zero. Note also that if the momentum of the particle was truly perfectly known from the outset, i.e. a plane wave $|\vec{p}\rangle$, mixing would also not take place [40].

While the preceding analysis clearly shows what happens to the quantum entropy of a spin-1/2 particle under Lorentz transformations (a similar analysis can be done for photons [41]), what is most interesting in quantum information theory is the entanglement *between* systems. While some aspects of entanglement are captured by quantum entropies [42] and the spectrum of the conditional density operator [16], quantifying entanglement is a surprisingly hard problem, currently without a perfect solution. However, some good measures exist, in particular for the entanglement between two-level systems (qubits) and three-or-fewer level systems [43].

5.2. *Boosting quantum entanglement*

If we wish to understand what happens to the entanglement between two massive spin-1/2 particles, say, we have to keep track of four variables, the spin states $|\sigma\rangle$ and $|\lambda\rangle$ and the momentum states $|\vec{p}\rangle$ and $|\vec{q}\rangle$. A Lorentz transformation on the joint state of this two-particle system will mix spins and momenta just as in the previous example. Let us try to find out how this affects entanglement.

A good measure for the entanglement of mixed states, i.e. states that are not pure such as (47), is the so-called *concurrence*, introduced by Wootters [44]. This concurrence $C(\rho_{AB})$ can be calculated for a density matrix ρ_{AB} that describes two subsystems A and B of a larger system, and quantifies the entanglement *between* A and B. For our purposes, we will be interested in the entanglement between the spins σ and λ of our pair. The concurrence is one if two degrees of freedom are perfectly entangled, and vanishes if no entanglement is present.

In order to do this calculation we first have to specify our initial state. We take this to be a state with spin and momentum wave function in a product, but where the spin-degrees of freedom are perfectly entangled in a so-called Bell state:

$$|\sigma,\lambda\rangle = \frac{1}{\sqrt{2}}\left(|\uparrow,\downarrow\rangle - |\downarrow,\uparrow\rangle\right). \tag{50}$$

The concurrence of this state can be calculated to be maximal: $C(\rho_{\sigma\lambda}) = 1$. We now apply a Lorentz boost to this joint state, i.e. we move our spin-polarization detector with speed $\beta = v/c$ with respect to this pair (or, equivalently, we move the pair with respect to the detector). If the momentum degrees of freedom of the particles at the outset are Gaussian distributions unentangled with each other and the spins, the Lorentz boost will entangle them, and the concurrence will drop [45]. How much it drops depends on the ratio between the spread of the momentum distribution σ_r (not to be confused with the spin σ) and the particle's mass m. In Fig. 2 below, the concurrence is displayed for two different such ratios, as a function of the *rapidity* ξ. The rapidity ξ here is just a transformed velocity: $\xi = \sinh\beta$, such that $\xi \to \infty$ as $\beta \to 1$. We can see that if the ratio σ_r/m is not too large, the concurrence will drop but not disappear altogether. But

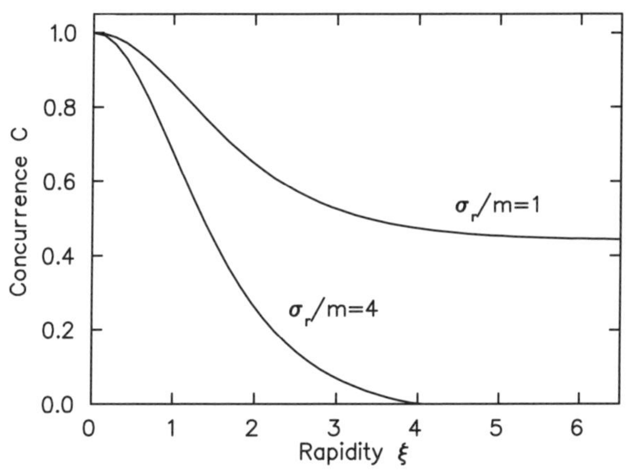

Fig. 2. Spin-concurrence as a function of rapidity, for an initial Bell state with momenta in a product Gaussian. Data is shown for $\sigma_r/m = 1$ and $\sigma_r/m = 4$ (from Ref. [45]).

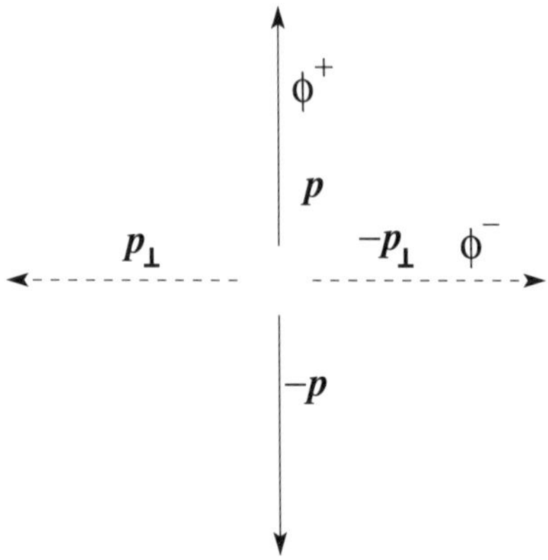

Fig. 3. Superposition of Bell-states Φ^+ and Φ^- at right angles, with the particle pair moving in opposite directions.

if the momentum spread is large compare to the mass, all entanglement can be lost.

Let us consider instead a state that is initially unentangled in spins, but fully entangled in momenta. I depict such a wave function in Fig. 3, where a pair is in a superposition of two states, one moving in opposite directions with momentum $\vec{p}_\perp$ in a relative spin state Φ^- (this is one of the four Bell spin-entangled states, Eq. (50)), and one moving in a plane in opposite orthogonal directions with momentum p, in a relative spin-state Φ^+ (which is Eq. (50) but with a plus sign between the superpositions). It can be shown that if observed at rest, the spins are indeed unentangled. But when boosted to rapidity ξ, the concurrence actually *increases* [45], as for this state (choosing $m = 1$)

$$C(\rho_{AB}) = \frac{p^2(\cosh^2(\xi) - 1)}{(\sqrt{1 + p^2}\cosh(\xi) + 1)^2}. \tag{51}$$

Thus, Lorentz boosts can, under the right circumstances, create entanglement where there had been none before.

A similar analysis can be performed for pairs of entangled photons, even though the physics is quite different [46]. First of all, photons are massless and their quantum degree of freedom is the photon polarization. The masslessness of the photons makes the analysis a bit tricky, because issues of gauge invariance enter into the picture, and as all particles move with constant velocity (light speed), there cannot be a spread in momentum as in the massive case. Nevertheless, Lorentz transformation laws acting on polarization vectors can be identified, and an analysis similar to the one described above can be carried through. The difference is that the entangling effect of the Lorentz boost is now entirely due to the spread in momentum *direction* between the two entangled photon beams. This implies first of all that fully-entangled photon polarizations cannot exist, even at rest, and second that existing entanglement can either be decreased or increased, depending on the angle with which the pair is boosted (with respect to the angle set by the entangled pair), and the rapidity [46].

5.3. *Entanglement in accelerated frames*

The logical extension of the work just described is to allow for *local* Lorentz transformations on quantum particles, that is, to study particles on relativistic (accelerating) orbits or in classical gravitational fields. Alsing and Milburn, for example, studied the quantum teleportation channel I discussed earlier [47]. Because quantum teleportation relies on an entangled pair of particles, the fidelity of quantum teleportation (how well Bob's version of Alice's quantum state agrees with the original) would suffer if acceleration of either Bob or Alice leads to a deterioration of entanglement. This is precisely what happens, but the origin of the deterioration of entanglement is different here: it is not due to the mixing of spin and momentum degrees of freedom, but rather due to the emergence of "Unruh radiation" in the rest frame of the accelerated observer [47, 48]. Unruh radiation is a peculiar phenomenon that is due to the appearance of a sort of "event horizon" for accelerated observers: there are regions of spacetime that are causally disconnected from an accelerated observer, and this disconnected region affects the vacuum fluctuations that occur anywhere in space [49–51]. In this sense, the Unruh radiation is analogous to Hawking radiation, which I will discuss in more detail in the following section. Unruh radiation produces thermal noise in the communication channel, which leads to the breakdown of the fidelity of quantum teleportation. Because this reasoning applies to all quantum communication that relies on the assistance of entanglement, we can conclude that generally the capacity of entanglement-assisted channels would be reduced between accelerated observers [52]. A similar conclusion holds for entangled particles near strong gravitational fields [53]. In that case, it is indeed the Hawking radiation that leads to the deterioration of entanglement between Einstein-Podolski-Rosen pairs.

6. Information in Curved Space Time

While there are clearly many other questions that can conceivably be posed (and hopefully answered) within the relatively new field of relativistic quantum information theory [54], I would like to close this review with some speculations about quantum information theory in curved space time.

That something interesting might happen to entropies in curved space time has been suspected ever since the discovery of Hawking radiation [55] that gave rise to the black hole information paradox [56]. The paradox has two parts and can be summarized as follows: According to standard theory, a non-rotating and uncharged black hole can be described by an entropy that is determined entirely in terms of its mass M (in units where $\hbar = G = 1$):

$$S_{BH} = 4\pi M^2. \tag{52}$$

Presumably, a state that is fully known (that is, one that is correlated with another system that an observer has in its possession) can be absorbed by the black hole. Once that state disappears behind the event horizon, the correlation between that state and its description in the observer's hands seems to disappear: the information cannot be retrieved any longer. Even worse, after a long time, the black hole will have evaporated entirely into thermal (Hawking) radiation, and the information is not only irretrievable, it must have been destroyed. A more technical discussion would argue that black holes appear to have the capability to turn pure states into mixed stated *without* disregarding (tracing over) parts of the wave function. Such a state of affairs is not only paradoxical, but it is in fact completely incompatible not only with the standard model of physics, but with the basic principle of probability conservation.

The second part of the problem has to do with the entropy balance between the black hole and the radiation it emits. When the black hole evaporates via Hawking radiation, the emitted radiation is thermal, and carries entropy

$$S_{\rm rad} \sim T_H^3 \tag{53}$$

with black hole temperature $T_H = (8\pi M)^{-1}$. But the black hole's entropy must also change at the same time, and this is determined by the amount of energy that had to be spent in order to create the virtual particle pairs that gave rise to the radiation. Because mass and temperature of the black hole are inversely related, the entropy decrease of the black hole and the entropy of the emitted radiation *cannot match*. Indeed, we roughly find that

$$dS_{\mathrm{rad}} \approx 4/3 \; dS_{BH}. \tag{54}$$

Now it should be pointed out that the preceding results were obtained within equilibrium thermodynamics in curved space time. But since black holes have negative heat capacity [57], they can *never* be at equilibrium, and the assumptions of that theory are strongly violated. As the concept of information itself is a non-equilibrium one, we should not be surprised if paradoxical results are obtained if equilibrium concepts are used to describe such a case. Still, a resolution of the microscopic dynamics in black hole evaporation is needed. One possible approach is to use quantum information theory to characterize the relative states of the black hole, the stimulated radiation emitted during the formation of the black hole, and the Hawking radiation (spontaneous emission of radiation) created in the subsequent evaporation [23, 58]. As we have lost track of the stimulated radiation, we must always average over it ("trace" it out), which (along with tracing out the causally disconnected region that lies beyond the Schwarzschild radius) creates the positive black hole entropy. In the flat space-time treatment of Ref. [23], the entropy balance between the black hole and the Hawking radiation can be maintained because entanglement is spread between the stimulated radiation, Hawking radiation, and the black hole. While all three are strongly entangled, tracing over the stimulated radiation produces a state of *no* correlations between Hawking radiation and black hole, implying that the Hawking radiation appears purely thermal. But of course, the joint system is still highly entangled, but in order to discover this entanglement we would have to have access to the lost radiation emitted during the formation process. Still, this treatment is unsatisfying because it does not resolve the ultimate paradox: the unitary description only works up until the black hole has shrunk to a particular small size. At that point it appears to break down.

One reason for this breakdown might lie in the inappropriate treatment of quantum entropy in curved space time (the preceding formalism ignored curvature). A more thorough analysis must take into account the causal structure of space time. For example, not all quantum measurements are realizable [38], because only those variables can be simultaneously measured whose separation is space-like. In physics, we do have a theory that correctly describes how different observables interact in a manner compatible with the causal structure of space-time, namely quantum field theory. In order to consistently define quantum entropies then, we must define them within quantum field theory in curved space-time.

The first steps toward such a theory involve defining quantum fields over a manifold separated into an accessible and an inaccessible region. This division will occur along a world-line, and we shall say that the "inside" variables are accessible to me as an observer, while the outside ones are not. Note that the inaccessibility can be due either to causality, or due to an event horizon. Both cases can be treated within the same formalism. States in the inaccessible region have to be averaged over, since states that differ only in the outside region are unresolvable. Let me denote the inside region by R, while the entire state is defined on E. We can now define a set of commuting variables X that can be divided into $X_{\rm in}$ and $X_{\rm out}$. By taking matrix elements of this density matrix of the entire system

$$\rho = |E\rangle\langle E| \tag{55}$$

with the complete set of variables $(X_{\rm in}, X_{\rm out})$, we can construct the inside density matrix (defined on R) as

$$\rho_{\rm in} = {\rm Tr}_{X_{\rm out}}(\rho_{X_{\rm in}X_{\rm out}}). \tag{56}$$

which allows me to define the *geometric entropy* [59] of a state E for an observer restricted to R

$$S_{\rm geom} = -{\rm Tr}(\rho_{\rm in}\log\rho_{\rm in}). \tag{57}$$

Here, the trace is performed using the inside variables only.

This, however, is just the beginning. As with most quantities in quantum field theory, this expression is divergent and needs to be renormalized. Rather than being an inconvenience, this is precisely what we should have expected: after all, we began this review by insisting that entropies only make sense when discussed in terms of the possible measurements that can be made to this system. This is, of course, precisely the role of renormalization in quantum field theory. Quantum entropies can be renormalized via a number of methods, either using Hawking's zeta function regularization procedure [60] or by the "replica trick", writing

$$S_{\rm geom} = -\left[\frac{d}{dn}{\rm Tr}(\rho_{\rm in}^n)\right]_{n=1}, \tag{58}$$

and then writing $dS(n)$ in terms of the expectation value of the stress tensor. A thorough application of this program should reveal components of the entropy due entirely to the curvature of space-time, and which vanish in the flat-space limit. Furthermore, the geometric entropy can be used to write equations relating the entropy of the inside and the outside space-time regions, as

$$S(E) = S(\rho_{\text{in,out}}) = S(\rho_{\text{in}}) + S(\rho_{\text{out}}|\rho_{\text{in}}). \qquad (59)$$

If $S(\rho_{in})$ is the entropy of the black hole radiation (together with the stimulated radiation), then $S(\rho_{\text{out}}|\rho_{\text{in}})$ is the conditional black hole entropy *given* the radiation field, a most interesting quantity in black hole physics.

7. Summary

Entropy and information are statistical quantities describing an observer's capability to predict the outcome of the measurement of a physical system. Once couched in those terms, information theory can be examined in all physically relevant limits, such as quantum, relativistic, and gravitational. Information theory is a non-equilibrium theory of statistical processes, and should be used under those circumstances (such as measurement, non-equilibrium phase transitions, etc.) where an equilibrium approach is meaningless. Because an observer's capability to make predictions (quantified by entropy) is not a characteristic of the object the predictions apply to, it does not have to follow the same physical laws (such as reversibility) as that befitting the objects. Thus, the arrow of time implied by the loss of information under standard time-evolution is even less mysterious than the second law of thermodynamics, which is just a consequence of the former.

In time, a fully relativistic theory of quantum information, defined on curved space-time, should allow us to tackle a number of problems in cosmology and other areas that have as yet resisted a consistent treatment. These developments, I have no doubt, would make Shannon proud.

Acknowledgments

I am grateful to N.J. Cerf for years of very fruitful collaboration in quantum information theory, as well as to R.M. Gingrich and A.J. Bergou for their joint efforts in the relativistic theory. I would also like to acknowledge crucial discussions on entropy, information, and black holes, with P. Cheeseman, J.P. Dowling, and U. Yurtsever. This work was carried out in part with support from the Army Research Office. I would like to acknowledge support from Michigan State University's BEACON Center for the Study of Evolution in Action, where part of this review was written.

References

[1] C. Adami, Three weeks with Hans Bethe. In eds. G.E. Brown and C.-H. Lee, *Hans Bethe and His Physics*, pp. 45–111. World Scientific, Singapore, (2006).

[2] C. Shannon, 'A mathematical theory of communication', *Bell Syst. Tech. J.* **27**, 379–423, 623–656 (1948).

[3] R. Landauer, 'Information is physical', *Physics Today* **44**, 23–29 (1991).

[4] T.M. Cover and J.A. Thomas, *Elements of Information Theory.* (John Wiley, New York, NY, 1991).

[5] G.P. Basharin, 'On a statistical estimate for the entropy of a sequence of independent random variables', *Theory Probability Applic.* **4**, 333–337 (1959).

[6] J. von Neumann, 'Thermodynamik quantenmechanischer Gesamtheiten', *Gött. Nach.* **1**, 272–291 (1927).

[7] M. Tribus and E. McIrvine, 'Energy and information', *Scientific American* **224/9**, 178–184 (1971).

[8] W. Zurek, 'Pointer basis of quantum apparatus': Into what mixture does the wave packet collapse?, *Physical Review D.* **24**, 1516–1525 (1981).

[9] J. von Neumann, *Mathematische Grundlagen der Quantenmechanik.* (Julius Springer, Berlin, 1932).

[10] A.M. Turing, 'On computable numbers, with an application to the Entscheidungsproblem', *Proceedings of the London Mathematical Society* **42**, 230–265 (1937).

[11] K. Gödel, 'Über formal unentscheidbare Sätze der Principia Mathematica und verwandter Systeme', *Monatshefte Math. Physik.* **38**, 173–193 (1931).

[12] G. Chaitin, *The Limits of Mathematics.* (Springer, Singapore, 1997).

[13] C.S. Calude, *Information and Randomness.* (Springer Verlag, Berlin, Heidelberg, New York, 1998), 2nd edition.

[14] A. Wehrl, 'General properties of entropy', *Reviews of Modern Physics* **50**, 221–260 (1978).

[15] N.J. Cerf and C. Adami, 'Negative entropy and information in quantum mechanics', *Physical Review Letters* **79**, 5194–5197 (1997).

[16] N.J. Cerf and C. Adami, 'Quantum extension of conditional probability', *Physical Review A.* **60**, 893–897 (1999).

[17] M. Horodecki, J. Oppenheim and A. Winter, 'Partial quantum information', *Nature.* **436**, 673–676 (2005).

[18] C.H. Bennett, G. Brassard, C. Crepeau, R. Jozsa, A. Peres and W.K. Wootters, 'Teleporting an unknown quantum state via dual classical and Einstein-Podolski-Rosen channels', *Physical Review Letters* **70**, 1895–1899 (1993).

[19] C.H. Bennett, H.J. Bernstein, S. Popescu and B. Schumacher, 'Concentrating partial entanglement by local operations', *Physical Review A.* **53**, 2046–2052 (1996).

[20] C. Bennett and S. Wiesner, 'Communication via one- and two-particle operators on Einstein-Podolsky-Rosen states', *Phys. Rev. Lett.* **69**, 2881–2884 (1992).

[21] M. Ohya, 'On compound state and mutual information in quantum information-theory', *IEEE Transactions On Information Theory* **29**, 770–774 (1983).
[22] N.J. Cerf and C. Adami, 'Information theory of quantum entanglement and measurement', *Physica D.* **120**, 62–81 (1998).
[23] C. Adami and N.J. Cerf, 'What information theory can tell us about quantum reality', *Quantum Computing and Quantum Communications (Lecture Notes in Physics).* **1509**, 258–268 (1999).
[24] C. Adami, 'Quantum mechanics of consecutive measurements'. arxiv.org eprint arXiv:0911.1142v2, (2010).
[25] C. Adami and N.J. Cerf, 'von Neumann capacity of noisy quantum channels', *Phys. Rev. A.* **56**, 3470–3483 (1997).
[26] C.H. Bennett, P.W. Shor, J.A. Smolin and A.V. Thapliyal, 'Entanglement-assisted capacity of a quantum channel and the reverse shannon theorem', *IEEE Transactions On Information Theory.* **48**, 2637–2655 (2002).
[27] B. Julsgaard, A. Kozhekin and E.S. Polzik, 'Experimental long-lived entanglement of two macroscopic objects', *Nature.* **413**, 400–403 (2001).
[28] W. Zurek, 'Decoherence, einselection and the quantum origins of the classical', *Reviews of Modern Physics.* **75**, 715–775 (2003).
[29] K. Jarett and T.M. Cover, 'Asymmetries in relativistic information-flow', *IEEE Transactions On Information Theory* **27**, 152–159 (1981).
[30] R. Aldrovandi and J. Gariel, 'On the riddle of the moving thermometer', *Phys. Lett. A.* **170**, 5 (1992).
[31] A. Einstein, 'Über das Relativitätsprinzip und die aus demselben gezogenen Folgerungen', *Jahrb. f. Rad. und Elekt.* **4**, 411 (1907).
[32] M. Planck, 'Zur Dynamik bewegter Systeme', *Ann. d. Phys.* **26**, 1 (1908).
[33] H. Ott, 'Lorentz-Transformation der Wärme und der Temperatur', *Zeitschrift f. Physik.* **175**, 70 (1963).
[34] H. Arzeliès, 'Transformation relativiste de la temperature et de quelques autres grandeurs thermodynamiques', *Nuovo Cimento.* **35**, 792 (1965).
[35] P.T. Landsberg and G.E.A. Matsas, Laying the ghost of the relativistic temperature transformation, *Physics Letters A.* **223**, 401–403 (1996).
[36] W. Pauli, 'Die Relativitätstheorie'. In *Enzyklopädie der Mathematischen Wissenschaften*, vol. 5/2, pp. 539–776. Teubner, Leipzig, (1921).
[37] P.J.B. Peebles and D. T. Wilkinson, 'Comment on the anisotropy of the primeval fireball', *Physical Review.* **174**, 2168 (1968).
[38] D. Beckman, D. Gottesman, M.A. Nielsen and J. Preskill, 'Causal and localizable quantum operations', *Physical Review A.* **64**, (2001).
[39] A. Peres, P.F. Scudo and D.R. Terno, 'Quantum entropy and special relativity', *Physical Review Letters.* **88**, (2002).
[40] P.M. Alsing and G.J. Milburn, Lorentz invariance of entanglement, *Quant. Information and Comupting.* **2**, 487–512 (2002).
[41] A. Peres and D.R. Terno, 'Relativistic Doppler effect in quantum communication', *Journal of Modern Optics.* **50**, 1165–1173 (2003).
[42] C.H. Bennett, D.P. DiVincenzo, J.A. Smolin and W.K. Wootters, Mixed-state entanglement and quantum error correction, *Physical Review A.* **54**, 3824–3851 (1996).

[43] M. Plenio and S. Virmani, 'An introduction to entanglement measures', *Quant. Inf. Comp.* **1**, 1–51 (2007).
[44] W.K. Wootters, 'Entanglement of formation of an arbitrary state of two qubits', *Physical Review Letters* **80**, 2245–2248 (1998).
[45] R.M. Gingrich and C. Adami, 'Quantum entanglement of moving bodies', *Physical Review Letters* **89**, 270402 (2002).
[46] R.M. Gingrich, A.J. Bergou and C. Adami, 'Entangled light in moving frames', *Physical Review A.* **68**, 042102 (2003).
[47] P.M. Alsing and G.J. Milburn, 'Teleportation with a uniformly accelerated partner', *Physical Review Letters* **91**, 180404 (2003).
[48] I. Fuentes-Schuller and R. Mann, 'Alice falls into a black hole: Entanglement in noninertial frames', *Physical Review Letters* **95**, 120404 (2005). ISSN 0031-9007.
[49] S. Fulling, 'Nonuniqueness of canonical field quantization in Riemannian space-time', *Physical Review D.* **7**, 2850–2862 (1973).
[50] P.C.W. Davies, 'Scalar production in Schwarzschild and Rindler metrics', *J. Phys. A.* **8**, 609 (1975).
[51] W. Unruh, 'Notes on black-hole evaporation', *Physical Review D.* **14**, 870–892 (1976).
[52] P.M. Alsing, I. Fuentes-Schuller, R.B. Mann and T.E. Tessier, 'Entanglement of Dirac fields in noninertial frames', *Physical Review A.* **74**, 032326 (2006).
[53] H. Terashima and M. Ueda, 'Einstein-Podolsky-Rosen correlation in a gravitational field', *Physical Review A.* **69**, 032113 (2004).
[54] A. Peres and D.R. Terno, 'Quantum information and relativity theory', *Reviews of Modern Physics.* **76**, 93–123 (2004).
[55] S.W. Hawking, 'Particle creation by black holes', *Commun. Math. Phys.* **43**, 199–220 (1975).
[56] J. Preskill. 'Do black holes destroy information'? In eds. S. Kalara and D. Nanopoulos, *International Symposium on Black Holes, Membranes, Wormholes and Superstrings*, p. 22, Singapore, (1993). World Scientific.
[57] P.T. Landsberg, 'Thermodynamics inequalities with special reference to negative heat capacities and black holes', *Nuclear Physics B — Proceedings Supplements* **5**, 316–321 (1988).
[58] C. Adami and G. ver Steeg. 'Black holes conserve information in curved-space quantum field theory'. arXiv:gr-qc/0407090, (2004).
[59] C. Holzhey, F. Larsen and F. Wilczek, 'Geometric and renormalized entropy in conformal field theory', *Nucl. Phys. B.* **424**, 443 (1994).
[60] S. Hawking, Zeta function renormalization of path integrals in curved space time, *Commun. Math. Phys.* **55**, 133 (1977).

Chapter 3

Hadron Production in Ultra-relativistic Nuclear Collisions and the QCD Phase Diagram: an Update

P. Braun-Munzinger* and J. Stachel†

We summarize our current understanding of the connection between the QCD phase line and the chemical freeze-out curve as deduced from thermal analyses of yields of particles produced in central collisions between relativistic nuclei.

Quantum chromodynamics (QCD), the theory of strong interactions, contains, as a key prediction, a new phase of matter, the quark-gluon plasma (QGP). This partonic QGP was the main ingredient (in addition to leptons and neutrinos) of the matter in the early universe until about $10\,\mu$s after the big bang. Around that time the QGP was transformed into hadronic matter through the QCD phase transition. In general, strongly interacting matter in equilibrium is characterized by two quantities, the temperature T and the baryon chemical potential μ_B or equivalently, the (net) baryon density n_B.

How does this phase transition come about? Due to asymptotic freedom, the strong coupling constant α_s runs, i.e. diminishes with increasing energy scale, implying that interactions among strongly interacting particles will get weaker as T or μ_B (or both) are increased. One may ask at what

*GSI Helmholtzzentrum für Schwerionenforschung, D-64291 Darmstadt, Germany; ExtreMe Matter Institute EMMI, GSI, D-64291 Darmstadt, Germany; Technische Universität Darmstadt, D-64289 Darmstadt, Germany; Frankfurt Institute for Advanced Studies, J.W. Goethe Universität, D-60438 Frankfurt, Germany.
†Physikalisches Institut der Universität Heidelberg, D-69120 Heidelberg, Germany.

energy scale will there be new physics such as deconfinement for strongly interacting matter in equilibrium? The only two scales relevant appear to be $\Lambda_{\mathrm{QCD}} \approx 200\,\mathrm{MeV}$ along the temperature axis and the nucleon mass m_N along the μ_B axis. From phenomenological considerations and based on ever more accurate solutions of QCD on a discrete space-time lattice ("lattice-QCD") convincing arguments were put forward to demonstrate that strongly interacting matter undergoes indeed a (phase) transition from a dense hadronic medium where all constituents (hadrons) are confined to a deconfined plasma of interacting quarks and gluons. This implies the existence of a line in the $(T - \mu_B)$ plane, the QCD phase boundary, anchored by parametrically critical parameters $T_c(\mu_B = 0) \approx \Lambda_{\mathrm{QCD}}$ and $(\mu_B)_c(T = 0) \approx m_N$. For recent detailed reviews on the QCD phase diagram, where many of the above short arguments are exposed in detail and the state of theory and relevant experiments is summarized, see Refs. [1,2].

This article is an attempt to update a paper on the subject which we wrote 15 years ago [3] in honor of Gerry Brown's 70th birthday. While we have (with one exception [4]) never published together with Gerry, his constant probing and provocative questions on the QGP and related areas have, during nearly 2 decades of close scientific interactions, deeply influenced our thinking on the subject. The experimental basis for making a connection between data from ultra-relativistic nuclear collisions and the QCD phase boundary is dramatically improved after 15 years of intense research on the subject. Given all this we consider it an honor and an opportunity to be able to summarize our most recent thinking on the subject as part of a Festschrift on the occasion of Gerry's 85th birthday.

Strongly interacting matter at high temperature and baryon density is produced in collisions between atomic nuclei at ultra-relativistic energies. Depending on the center-of-mass energy reached in the collisions the temperature and baryon density can be "tuned" systematically, as discussed in detail below. A sequence of such experiments conducted over the past 25 years over a wide range of energies has provided convincing evidence for a new state of matter, see e.g. Refs. [5–12]. In November 2010, first results from Pb–Pb collisions at the Large Hadron Collider at CERN have provided a first glimpse into fireballs with ultra-high temperatures (approaching $1\,\mathrm{GeV}$) and opened a new era in quark matter research [13–18].

The range of temperatures and densities reached in these experiments is clearly of the order of or (for RHIC and LHC experiments) significantly exceeding the typical values discussed above. For a review and summary of the evidence, see Ref. [19]. It has been much harder to assess the degree of equilibration reached in the collisions and to provide direct evidence on the QCD phase boundary. Below we will argue that the most direct information on equilibration and the QCD phase boundary available to-date is obtained from analyses of the multiplicities of hadrons produced in central collisions between ultra-relativistic nuclei.

These data have been analyzed independently by various groups, using hadron resonance gas models with chemical freeze-out as a key ingredient. In this resonance gas approach, described in detail in Ref. [20], it is assumed that the final hadron yields result from the decay of fully equilibrated hadronic matter, comprising the full QCD hadron mass spectrum — and thereby implicitely containing all of QCD. The measured hadrons, those not decaying further by the strong interaction, are then obtained by summing up all strong decay products of the hadron mass spectrum with their corresponding thermal weight. By assuming that the measured hadrons are generated at a common surface at which all particles simultaneously decouple, values of the baryon chemical potential, μ_B, and temperature, T, on this surface, the chemical freeze-out surface, are extracted. Fitting these two parameters, μ_b and T, together with the volume parameter gives values for the particle abundances which are in very good agreement with experiment [21–34], for an overall summary see Ref. [20].

These results clearly demonstrate that, at chemical freeze-out, a very high degree of equilibration is reached in central nucleus-nucleus collisions, implying that matter in the proper sense with thermodynamic properties has been formed at the latest at this stage of the collision. We note that the system most likely is in or very close to equilibrium much earlier in the collision, as demonstrated by, among other things, hydrodynamic analysis of azimuthal anisotropies ("elliptic flow"). For a brief summary see [12, 19].

The resulting values of μ_B and T are shown in Fig. 1 as functions of center-of-mass energy per nucleon pair. We note first a very important outcome of these investigations: near 10 GeV center-of-mass energy, the temperature saturates with increasing beam energy, reaching an asymptotic value of about 160 MeV, while the baryon chemical potential decreases smoothly. This temperature saturation concept can be tested thoroughly when the newly taken data at the much higher energy of $\sqrt{s_{NN}} = 2.76\,\text{TeV}$ from the LHC are analyzed.

All these data points have been obtained by analyzing hadron yields, implying the existence of a limiting temperature to which a hadron "gas" can be heated, as has been conjectured more than 40 years ago by Hagedorn [35]. The constancy of T above $\sqrt{s_{NN}} = 10\,\text{GeV}$ indicates that a

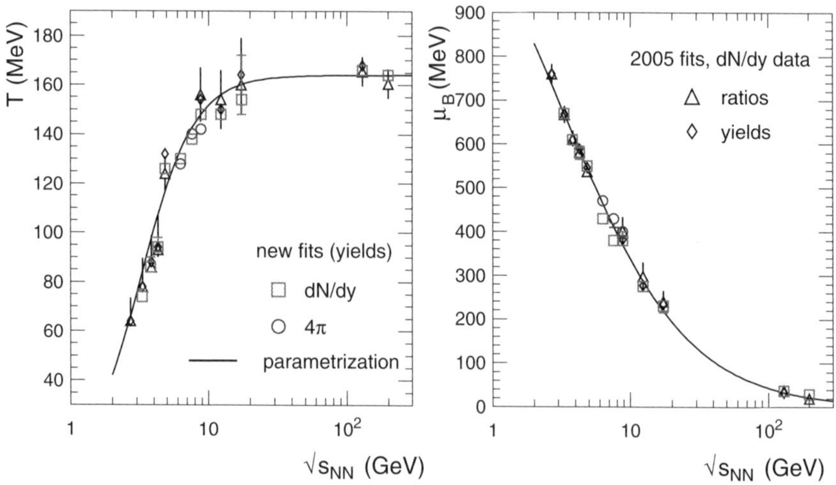

Fig. 1. The temperature and baryon chemical potential of Statistical Model fits to hadro-chemical abundances as a function of center-of-mass energy per nucleon pair for collisions of heavy nuclei (Figure taken from Refs. [33, 34]).

boundary has been reached. Is this the QCD phase boundary? Could there be hadronic matter hotter than is observed here? Since the data analyzed here comprise the full hadronic spectrum, this would imply the existence of a hot and dense hadronic medium between the QCD phase boundary and the chemical freeze-out curve that leaves no trace in hadronic observables. As we will discuss below, this would contradict the existence of chemical freeze-out outlined above.

Plotting these temperature-chemical potential pairs for all available energies results in a phase diagram-like picture as is illustrated in Fig. 2. In the μ_B region from 800 to 400 MeV, as T increases from 50 to 150 MeV, the experimental points rise approximately linearly. In contrast, below $\mu_B \simeq 400$ MeV, the temperature is approximately constant, $T \simeq 160$ MeV. The highest collision energies studied to date at RHIC are those for which $\mu_B \simeq 25$ MeV. Soon an experimental point at $\mu_B \simeq 1$ MeV will be available from the recent campaign at the LHC. Also shown on this plot are lines of fixed energy per particle and fixed entropy density per T^3 as well as a line of hadron percolation [32, 38, 39].

These experimental results can be compared to the phase boundary computed on the lattice [41, 42]. Numerical simulations in lattice QCD can be performed at non-zero temperature only for small values of μ_B, typically

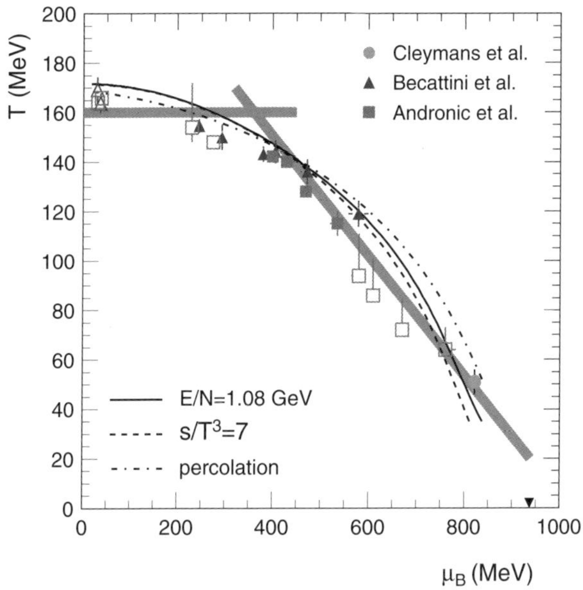

Fig. 2. The decoupling temperatures and chemical potentials extracted by Statistical Model fits to experimental data. The freeze-out points are from Refs. [28,33,36,37]. The open points are obtained from fits to data at mid-rapidity, the full-points refer to 4π data. The inverse triangle at $T = 0$ indicates the position of normal nuclear matter at $T = 0$. The lines are different model calculations to provide a phenomenological understanding of the freeze-out curve [32,38,39]. The shaded lines are drawn to indicate different regimes in this diagram. Figure taken from Ref. [40].

$\mu_B < T$, without running into the "sign" problem. At $\mu_B = 0$, these simulations indicate that there is no true phase transition from hadronic matter to a quark-gluon plasma, but rather a very rapid rise in the energy density at a temperature T_c of 155–175 MeV within current systematic errors, in very close agreement with the chemical freeze-out temperatures determined for small values of μ_B.

Further, studies using the lattice technique imply that T_c decreases very little as μ_B increases, at least for moderate values of μ_B. A very recent study [43] provides a quantitative estimate of the curvature of the phase boundary near $\mu_B = 0$. By comparison to the curvature of the line of fixed energy/particle [39] which has been constructed to get a phenomenological understanding of the complete chemical freeze-out curve, these authors argue that there is a significant difference between the data on chemical freeze-out and the lattice predictions, implying in their view the existence of a dense hadronic phase between chemical freeze-out and the QCD phase boundary.

A precise determination of the curvature of the chemical freeze-out curve can, however, only be obtained by analyzing the data themselves, not a curve such as the line of fixed energy/particle or any of the other lines shown in Fig. 2. We have recently investigated [44] the curvature of the chemical freeze-out curve in the region $\mu_B < 250\,\text{MeV}$. The results are presented in Fig. 3. The chemical freeze-out points are taken from [40]. In addition, we show in Fig. 3 also fit results from the STAR collaboration [45], where only pions, kaons and protons are analyzed to determine the freeze-out curve. This leads to a slightly lower chemical freeze-out temperature, but we believe that its energy (or μ_B) dependence contains the correct information. Independent of which chemical freeze-out parameters are used, the resulting description using the parabolic fit function as in Ref. [43] (long dashes) and the same function with twice the slope parameter (short dashes) are in agreement with the data, within the (still significant systematic) uncertainties, as shown in Fig. 3. For reference we include, in Fig. 3, the

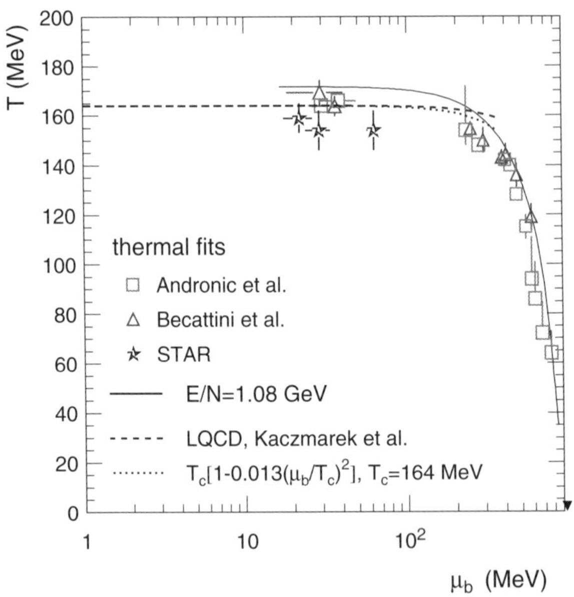

Fig. 3. Comparison of the experimental μ_B dependence of the chemical freeze-out temperature with recent lattice predictions [43]. The STAR chemical freeze-out values are from Ref. [45], the other values are from Ref. [40]. For more discussion see text.

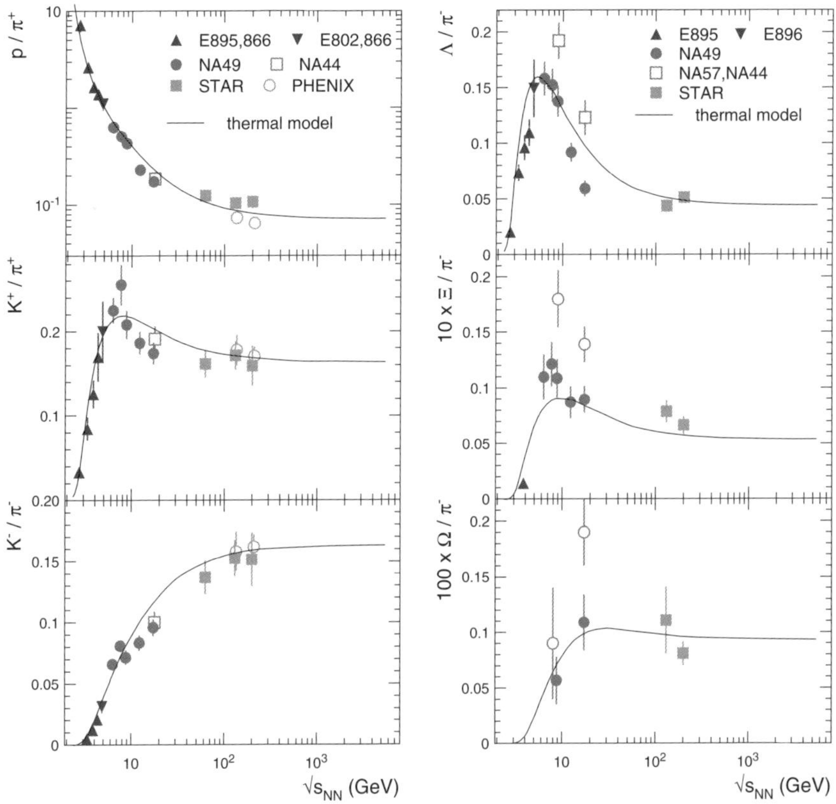

Fig. 4. Energy dependence of hadron yields relative to pions. The points are experimental data from various experiments. Lines are results of the Statistical Model calculations. The Figure is taken from Refs. [33, 34].

line of constant E/N [39]. From the available data one cannot construct evidence for a dense hadronic phase between chemical freeze-out and the phase boundary.

With the parametrizations of T and μ_B from Fig. 1 one can compute the energy dependence of the production yields of various hadrons relative to pions, shown in Fig. 4. Important for our purposes is the observation that there are peaks in the abundances of strange to non-strange particles at center-of-mass energies near 10 GeV, i.e. where the temperature reaches its limiting value. In particular, the K^+/π^+ and Λ/π ratios exhibit rather pronounced maxima there. We further note that in the region near 10 GeV, there is also a minimum in the chemical freeze-out volume [31, 46] obtained

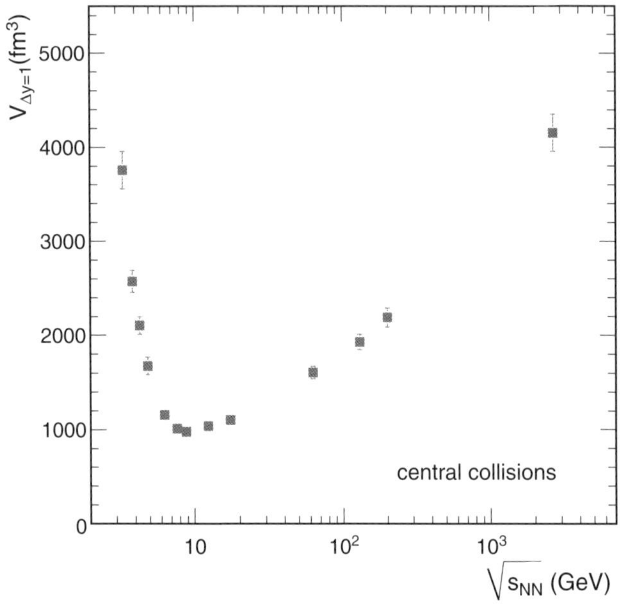

Fig. 5. Energy dependence of the volume for central nucleus-nucleus collisions. The chemical freeze-out volume dV/dy for one unit of rapidity is taken from Ref. [31]. The LHC point is from analysis of the recent ALICE data Ref. [13].

from the Statistical Model fit to particle yields [31, 34], as well as in the thermal freeze-out volume obtained from the Hanbury-Brown and Twiss (HBT) radii of the fireball [47]. The energy dependence of the chemical freeze-out volume is shown in Fig. 5. Included in this figure is the most recent point at LHC energy from ALICE [13] obtained by analyzing the pseudo-rapidity density of charged particles for central Pb–Pb collisions.

These experimental observations have long resisted interpretation in terms of a transition between hadronic matter and a quark-gluon plasma.[a] The general structures observed in the data are well reproduced only by the most recent model calculations [33]. There, it is shown that these structures arise due to the interplay between the limit in hadronic temperature (see Fig. 1), in our interpretation due to the QCD phase transition, and the rapid decrease of μ_B with increasing energy. The minimum near a center-of-mass energy of 10 GeV and the rise towards higher energies in the thermal

[a]We note the interpretation given in Ref. [48], obtained within a schematic 1st order phase transition model.

freeze-out volume [47] was explained as due to the increasing meson to baryon ratio in this region combined with a meson-meson cross section which is significantly smaller than the pion-nucleon cross section. The chemical freeze-out volume depicted in Fig. 5 behaves similarily, with a minimum at an energy close to where the chemical freeze-out temperature saturates. At the highest (LHC) energy the chemical freeze-out volume reaches, for Pb–Pb central collisions, a value of about 4 times, the thermal freeze-out volume about 7 times that of a Pb nucleus. The strong volume increase between RHIC and LHC energy, where neither T nor μ_B are expected to change much, provides a strong argument that chemical freeze-out takes place at a fixed density. These findings further strengthen the connection between hadron gas and quark-gluon plasma.

The same chemical freeze-out mechanism also governs the production of light nuclei and their anti-particles and even complex and exotic objects such as light hypernuclei and their anti-particles [49–51]. This comes at first glance as a surprize to many, as the binding energy of these objects is typically 1–2 orders of magnitude smaller than the chemical freeze-out temperature of $T \approx 160\,\mathrm{MeV}$. Can it be that such loosely bound objects are messengers of the phase transition between hadronic matter and the quark-gluon plasma? Entropy conservation comes to the rescue! After the phase transition the hot fireball expands adiabatically and the overall entropy as well as the entropy/baryon are conserved quantities. As was realized already more than 30 years ago [52] the assembly yield of complex nuclei from a gas of hot nucleons is a measure of the entropy/baryon, and hence can be used to diagnose the hot and dense phase of the collision. The answer to the above question is therefore strongly affirmative: the yield of composite objects such as light nuclei and hyper-nuclei (and their anti-particles) produced in central nucleus-nucleus collisions at ultra-rlativistic energy is a direct measure of the entropy/baryon at chemical freeze-out, and hence, of the QCD phase transition.

At RHIC energies, chemical freeze-out was shown [53] to take place very close (within less than about $10\,\mathrm{MeV}$) to the phase boundary, driven by multi-particle collisions in the high density regime of, and by the the rapid density change across the phase transition. Further it is argued that freeze-out ends when the system is fully hadronized, i.e. at comparatively low density in the hadronic phase. Without a phase transition, freeze-out in a purely hadronic medium takes place over a considerable time and temperature range, as has been recently demonstrated from very general considerations [54]. This would necessarily lead to different freeze-out parameters for each hadron species due to widely different hadronic cross sections. This is not observed. Conversely, the rapid change of density with temperature near the phase transition makes the system insensitive to the different cross sections. We conclude that the observed (nearly) simultaneous chemical freeze-out of all hadrons is inconsistent with the hypothesis of a dense hadronic phase between the chemical freeze-out line and the phase boundary, at least for relatively small values of $\mu_B < 250\,\text{MeV}$.

We believe, however, that the above argument is generic [53]: to ensure simultaneous freeze-out (within a very small interval in temperature and chemical potential) of all hadrons, the freeze-out curve has to be very close to a line with a rapid density change, significantly faster than can be obtained within the scenario of an expanding hadronic medium. An immediate consequence of this would be that the chemical freeze-out curve delineates phase boundaries, not only for small values of μ_B but everywhere.

But what provides the phase boundary for large values of μ_B, where the deconfinement transition seems far away, at least if one follows the guidance from lattice QCD calculations? In line with Gerry Brown's daring attitude concerning new scientific directions we have recently speculated [40] that the transition from hadronic to quarkyonic matter might provide the missing link. Another, related possibility might be that, at high baryon density, the lines corresponding to the chiral and deconfinement phase transition split, leading to a region of chirally symmetric but still confined matter in the QCD phase diagram.

Whatever the case may be, it is clear that the phase diagram at high baryon density is difficult to explore, both experimentally and theoretically. Speculations have been put forward about the possible existence of a critical endpoint [55] or a triple point [40] in the QCD phase diagram. However, despite significant efforts, no clear signals have been discovered to-date [56, 57] of either critical points or, more generally, of dense phases beyond chemical freeze-out. This research on the high baryon density region of the phase diagram remains a challenge for the future and will have to be largely driven by the running campaign at RHIC [57] and the planned new experiments at FAIR [58] and DUBNA [59].

We have provided strong evidence that, for not too large values of the baryo-chemical potential $\mu_B < 250\,\text{MeV}$, the experimentally observed chemical freeze-out curve closely coincides with the QCD phase boundary between the hadronic world and the quark-gluon plasma. This implies a direct connection between results from experiments with ultra-relativistic nuclear collisions and a fundamental prediction of QCD concerning the phase structure of strongly interacting matter. For large values of μ_B, i.e. large net baryon densities, we can presently only speculate, but in Gerry Brown fashion have put forward an argument that also there the chemical freeze-out curve is driven by a phase transition.

Acknowledgments

The authors would like to thank Anton Andronic for many important contributions on the subject discussed.

References

[1] P. Braun-Munzinger and J. Wambach, *Rev. Mod. Phys.* **81**, 1031 (2009).
[2] K. Fukushima and T. Hatsuda, *Rept. Prog. Phys.* **74**, 014001 (2011); arXiv:1005.4814 [hep-ph].
[3] P. Braun-Munzinger and J. Stachel, *Nucl. Phys. A* **606**, 320 (1996); arXiv:nucl-th/9606017.
[4] G.E. Brown, J. Stachel and G. Welke, *Phys. Lett. B* **253**, 19 (1991).
[5] H.J. Specht, *Nucl. Phys. A* **698**, 341c (2002).
[6] U.W. Heinz and M. Jacob, arXiv:nucl-th/0002042.
[7] J. Stachel, *Nucl. Phys. A* **654**, 119c (1999).
[8] I. Arsene *et al.*, [BRAHMS Collaboration], *Nucl. Phys. A* **757**, 1 (2005).
[9] B.B. Back *et al.*, [PHOBOS Collaboration], *Nucl. Phys. A* **757**, 28 (2005).
[10] J. Adams *et al.*, [STAR Collaboration], *Nucl. Phys. A* **757**, 102 (2005).
[11] K. Adcox *et al.*, [PHENIX Collaboration], *Nucl. Phys. A* **757**, 184 (2005).
[12] P. Braun-Munzinger and J. Stachel, *Nature* **448**, 302 (2007).
[13] K. Aamodt *et al.*, [ALICE Collaboration], *Phys. Rev. Lett.*, in print, arXiv:1011.3916 [nucl-ex].
[14] K. Aamodt *et al.*, [ALICE Collaboration], *Phys. Rev. Lett.*, in print, arXiv:1011.3914 [nucl-ex].
[15] K. Aamodt *et al.*, [ALICE Collaboration], *Phys. Lett. B*, in print, arXiv:1012.1004 [nucl-ex].
[16] K. Aamodt *et al.*, [ALICE Collaboration], arXiv:1012.1657 [nucl-ex].
[17] K. Aamodt *et al.*, [ALICE Collaboration], arXiv:1012.4035 [nucl-ex].
[18] G. Aad *et al.*, [ATLAS Collaboration], *Phys. Rev. Lett.* **105**, 252203 (2010).
[19] M. Gyulassy and L. McLerran, *Nucl. Phys. A* **750**, 30 (2005) [arXiv:nucl-th/0405013].
[20] P. Braun-Munzinger, K. Redlich and J. Stachel, in Quark-Gluon Plasma 3, Eds. R.C. Hwa and X.N. Wang, (World Scientific Publishing, 2004) 491; nucl-th/0304013.
[21] P. Braun-Munzinger, J. Stachel, J.P. Wessels and N. Xu, *Phys. Lett. B* **344**, 43 (1995).
[22] P. Braun-Munzinger, J. Stachel, J.P. Wessels and N. Xu, *Phys. Lett. B* **365**, 1 (1996).

[23] J. Cleymans and K. Redlich, *Phys. Rev. C* **60**, 054908 (1999).
[24] U.W. Heinz, *Nucl. Phys. A* **661**, 140 (1999).
[25] J. Letessier and J. Rafelski, *Int. J. Mod. Phys. E* **9**, 107 (2000).
[26] P. Braun-Munzinger, D. Magestro, K. Redlich and J. Stachel, *Phys. Lett. B* **518**, 41 (2001).
[27] W. Florkowski, W. Broniowski and M. Michalec, *Acta Phys. Polon. B* **33**, 761 (2002).
[28] F. Becattini, J. Manninen and M. Gaździcki, *Phys. Rev. C* **73**, 044905 (2006).
[29] F. Becattini *et al.*, *Phys. Rev. C* **64**, 024901 (2001).
[30] J. Cleymans, H. Oeschler, K. Redlich and S. Wheaton, *Eur. Phys. J. A* **29**, 119 (2006).
[31] A. Andronic, P. Braun-Munzinger and J. Stachel, *Nucl. Phys. A* **772**, 167 (2006).
[32] J. Cleymans, H. Oeschler, K. Redlich and S. Wheaton, *Phys. Rev. C* **73**, 034905 (2006).
[33] A. Andronic, P. Braun-Munzinger and J. Stachel, *Phys. Lett. B* **673**, 142 (2009); Erratum, *ibid.* B **678**, 516 (2009).
[34] A. Andronic, P. Braun-Munzinger and J. Stachel, *Acta Phys. Polon. B* **40**, 1005 (2009).
[35] R. Hagedorn, *Nuovo Cim. Suppl.* **3**, 147 (1965); *Nuovo Cim. A* **56**, 1027 (1968).
[36] J. Manninen and F. Becattini, *Phys. Rev. C* **78**, 054901 (2008).
[37] J. Cleymans *et al.*, *Phys. Rev. C* **59**, 1663 (1999).
[38] V. Magas and H. Satz, *Eur. Phys. J. C* **32**, 115 (2003).
[39] J. Cleymans and K. Redlich, *Phys. Rev. Lett.* **81**, 5284 (1998).
[40] A. Andronic *et al.*, *Nucl. Phys. A* **837**, 65 (2010); arXiv:0911.4806 [hep-ph].
[41] C. DeTar and U.M. Heller, *Eur. Phys. J. A* **41**, 405 (2009).
[42] Y. Aoki, Z. Fodor, S.D. Katz and K.K. Szabo, *Phys. Lett. B* **643**, 46 (2006); M. Cheng *et al.*, *Phys. Rev. D* **77**, 014511 (2008); A. Bazavov *et al.*, *Phys. Rev. D* **80**, 014504 (2009).
[43] F. Karsch *et al.*, arXiv:1011.3130 [hep-lat].
[44] A. Andronic, P. Braun-Munzinger and J. Stachel, Manuscript in preparation.
[45] B.I. Abelev *et al.*, [STAR Collaboration], *Phys. Rev.* **C79**, 034909 (2009).
[46] V.D. Toneev and A.S. Parvan, *J. Phys. G* **31**, 583 (2005).
[47] D. Adamova *et al.*, [CERES Collaboration], *Phys. Rev. Lett.* **90**, 022301 (2003).
[48] M. Gazdzicki and M.I. Gorenstein, *Acta Phys. Polon. B* **30**, 2705 (1999).
[49] P. Braun-Munzinger and J. Stachel, *J. Phys. G* **21**, L17 (1995).
[50] P. Braun-Munzinger and J. Stachel, *J. Phys. G* **28**, 1971 (2002), nucl-th/0112051.
[51] A. Andronic, P. Braun-Munzinger, J. Stachel and H. Stöcker, *Phys. Lett. B* submitted Oct. 15, 2010, arXiv:1010.2995 [nucl-th].
[52] P.J. Siemens and J.I. Kapusta, *Phys. Rev. Lett.* **43**, 1486 (1979).
[53] P. Braun-Munzinger, J. Stachel and C. Wetterich, *Phys. Lett. B* **596**, 61 (2004).
[54] J. Knoll, *Nucl. Phys. A* **821**, 235 (2009).

[55] M.A. Stephanov, K. Rajagopal and E.V. Shuryak, *Phys. Rev. Lett.* **81**, 4816 (1998); *Phys. Rev. D* **60**, 114028 (1999).
[56] The CBM Physics Book: Compressed Baryonic Matter in Laboratory Experiments, B. Friman, C. Hoehne, J. Knoll, S. Leupold, J. Randrup, R. Rapp, P. Senger, editors, Lecture Notes In Physics, (Springer, Heidelberg, 2010).
[57] M.M. Aggarwal *et al.*, [STAR Collaboration], *Phys. Rev. Lett.* **105**, 022302 (2010); arXiv:1004.4959 [nucl-ex].
[58] J. Steinheimer, H. Stöcker, I. Augustin, A. Andronic, T. Saito and P. Senger, *Prog. Part. Nucl. Phys.* **62**, 313 (2009).
[59] A.N. Sissakian and A.S. Sorin [NICA Collaboration], *J. Phys. G* **36**, 064069 (2009).

Chapter 4

The Nuclear Shell Model for Nuclei in the Region of ^{208}Pb

B. Alex Brown*

In "Bird of Passage", Rudolf Peierls summarizes the situation of nuclear theory in the 1940's: "The success of the concept of the compound nucleus, consisting of strongly interacting nucleons, which Niels Bohr introduced in 1936, had mesmerized everybody into believing that there was no chance of describing nuclei in terms of the more or less independent motion of individual nucleons, the so-called shell model. Therefore the success of this model in 1949 came as a great surprise. Some of us, notably Brian Flowers and later Gerry Brown, studied various applications of the model." The many applications that Gerry developed laid the foundations for the development of nuclear theory.

The shell model has become the standard model of nuclear structure. It provides a framework for relating many observables with a few fundamental parameters in a truncated model space. Based on the work started by Gerry we have the theoretical tools to understand the origin of the parameters in terms of the nucleon-nucleon interaction. A more complete understanding of the shell model is starting to emerge from ab-initio methods based upon realistic two-nucleon and three-nucleon interactions applied to light nuclei and nuclear matter.

The key for the success of the nuclear shell model is that there are energy gaps in the independent-particle spectrum that are large enough so that the structure of some "doubly-magic" nuclei are dominated by the closed-shell configuration for both protons and neutrons. Thus, we can use perturbation theory to improve the wavefunction for the closed-shell nuclei (such as in the coupled-cluster methods), or to calculate the spectra of nuclei

*Department of Physics and Astronomy, and National Superconducting Cyclotron Laboratory, Michigan State University, East Lansing, Michigan 48824-1321, USA.

that are obtained by adding valence nucleons to a closed-shell. Perhaps the best example is provided by ^{208}Pb where the 82 protons and 126 neutrons provide the closed shells. ^{208}Pb is near stability, and many hundreds of experiments have provide rather complete information about the structure in this mass region. Almost all of these data can be understood in terms of the nuclear shell model. The next best example is ^{132}Sn. Since ^{132}Sn lies further from stability, the experimental information is not as complete, but is currently being filled in with rare-isotope beam experiments. Comparison between ^{132}Sn and ^{208}Pb provides information on the stability and evolution of shell-model properties towards the nuclear drip lines.

One of key contributions to the development of the shell model were the papers by Gerry Brown and Tom Kuo on "Structure of finite nuclei and the free nucleon-nucleon interactions: General discussion of the effective force" [1] and "Structure of finite nuclei and the free nucleon-nucleon interactions: An application to ^{18}O and ^{18}F" [2]. It was shown that some of the states in these light nuclei could be understood in an sd-shell model space from the known nucleon-nucleon interaction. In addition to the first-order G-matrix interaction, it was crucial to include second-order core-polarization [3] to take into account core breaking. Shortly after this Herling and Kuo successfully applied this method to the spectra for two-nucleons outside of the ^{208}Pb core [4].

As a graduate student at Stony Brook I carried out experiments under the guidance of Dave Fossan. But at Stony Brook the interactions between theory and experiment were very strong. Everyone was invited to the theory "bull sessions" where we heard from the theory faculty and a constant stream of visitors talking on their new results with many discussions often led by Gerry. I had gathered enough experience in nuclear theory to continue as a Research Associate with Akito Arima at the University of Tokyo in 1974 when he returned to Japan after his stay in Gerry's group at Stony Brook.

In the remaining part of this contribution I want to discuss some results for the applications of effective interaction theory with configuration interactions to nuclei above ^{208}Pb that illustrate the essential aspects of the theory and the sensitivity of the results to various assumptions. With some phenomenological adjustments to the strength of the second-order contributions or to the monopole parts, the Herling and Kuo interaction has provided a basis for understanding for almost 40 years [5–7]. More recently the Naples group has used modern interactions and methods for the short-range correlations and have found excellent agreement with the spectra for two valence nucleons [8, 9]. For the single-particle part of the Hamiltonian

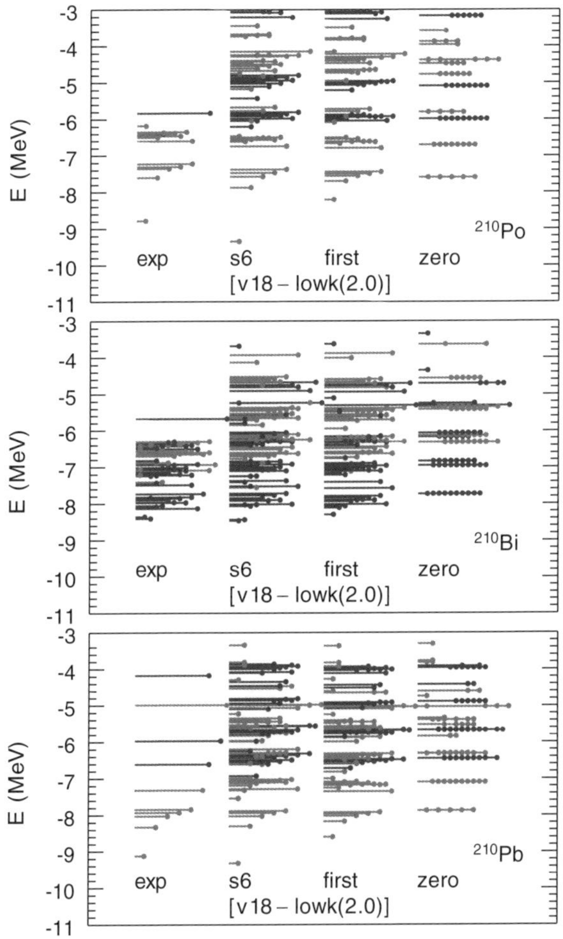

Fig. 1. Energies for states in nuclei with $A = 210$ relative to that of ^{208}Pb. The length of the line indicates the spin J. The lowest few levels for ^{210}Po and ^{210}Pb have $J^{\pi} = 0^{+}$, 2^{+}, 4^{+}, 6^{+} and 8^{+}. For each nucleus the experiment on the left is compared to three calculations based upon the Argonne V18 interaction and V_{lowk} with $\Lambda = 2.0\,\mathrm{fm}^{-1}$. From right to left, zeroth-order with no interaction, first order in the interaction and first plus second-order up to $6\hbar\omega$.

I use the experimental single-particle energies shown in Fig. 1 of [6] that are similar to those used by Herling and Kuo. The computer codes for the two-body effective interaction are those by Morten Hjorth-Jensen [10]. The code NuShell [11] was used for the configuration mixing calculations.

Figure 1 compares the results for two-particle with three levels of the truncation in the effective interaction. The results on the right-hand side obtained with no residual interaction illustrates the essential role played by the single-particle energy level spacing. The comparison of first-order with the first plus second order carried out up to 6 $\hbar\omega$ in the energy denominator (s6) show the essential role played by the second-order core-polarization contribution. The largest second-order contribution is for the "pairing" energy of the 0^+ state in ^{210}Po and ^{210}Pb. The bubble diagram is most important and for this diagram a large contribution comes from the tensor interaction between the valence protons in ^{210}Po and the core neutrons and between the valence neutrons in ^{210}Pb the core protons.

Figure 2 compares the results obtained with three versions of the interaction fitted to nucleon-nucleon scattering data; Argonne V18 [12], CD-Bonn [13] and N^{3}LO [14]. They all give essentially the same result indicated that the the uncertainty coming from the nucleon-nucleon interaction is small compared to other sources of change.

Figure 3 compares the results obtained with three ways of dealing with the short-range (high-momentum) part of the potential. Kuo and Brown used the Brueckner G-matrix method. More recently the V_{lowk} method was developed by Kuo and Bogner [15]. Results obtained with two value of the momentum cut-off are shown. The higher cut-off of 2.5 fm^{-1} gives results similar to the G matrix. It is argued that that results obtained for a given Λ value should be modified by a Λ dependent three-nucleon interaction. One observes that with second-order in the interaction the agreement with experimental spread in energies is better with $\Lambda = 2.5\,\text{fm}^{-1}$ which indicates that three-body contribution to the energy spread may be smaller for this case.

The agreement with the two-particle spectra is good. But one can see some systematic problems. For example, the lowest theoretical levels with $J^\pi = 4^+, 6^+$ and 8^+ are always about 100 keV lower than experiment. The disagreement becomes larger when many nucleons are added to the core. The valence interactions based upon renormalized two-body interactions tend to over-bind relative to experiment (see Fig. 36 in [8]).

One thing missing from these calculations is the three-body interaction. The role of the three-body interaction is illustrated schematically in Fig. 4 with Wick's theorem applied to a closed shell. The three-body interaction (last line in the figure) modifies the single-particle energies. But when we use experimental single-particle energies this is taken into account. The main part that is missing is the bottom diagram in the solid line box due to

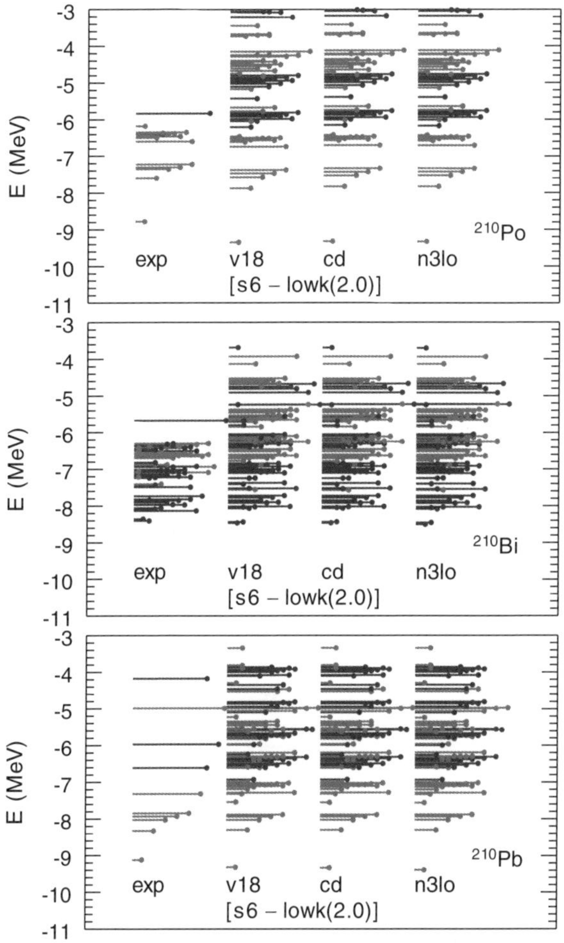

Fig. 2. See caption to Fig. 1. For each nucleus the experiment on the left is compared to three calculations based upon the V_{lowk} method with first plus second order up to $6\hbar\omega$. From right to left, results obtained with the N^{3}LO interaction, the CD-Bonn interaction and the Argonne V18 interaction.

the three-body interaction of two valence nucleons with the core nucleons. The nature of the three-body interaction is still relative uncertain. But it is contained implicitly in the energy-density functionals such as Skyrme that have parameters fitted to nuclear binding energies.

Thus we have recently proposed to modify the valence interaction by the following procedure [16]. First calculate the single-particle energy

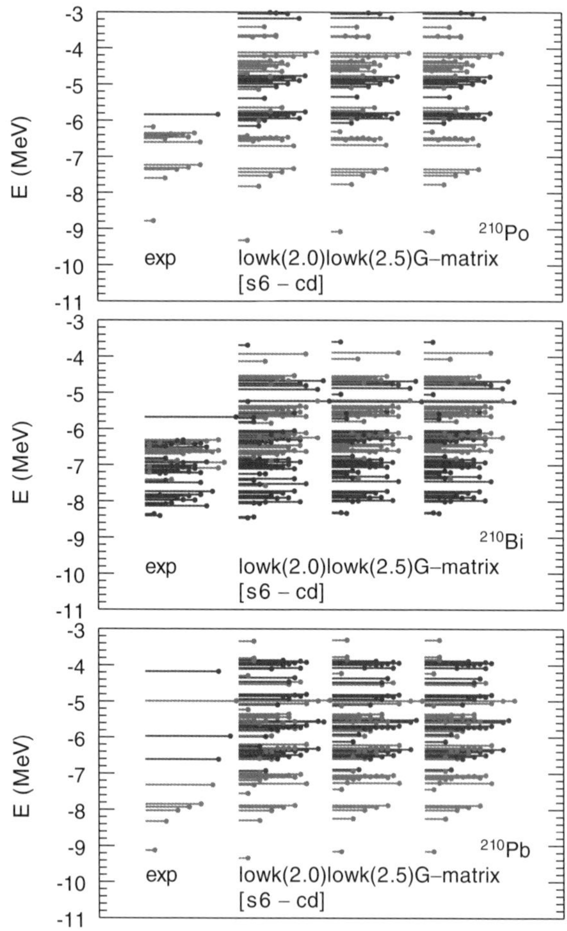

Fig. 3. See caption to Fig. 1. For each nucleus the experiment on the left is compared to three calculations based upon the CD Bonn potential with first plus second order up to $6\hbar\omega$. From right to left, first order matrix elements obtained with the G-matrix procedure, with the V_{lowk} procedure with $\Lambda = 2.5\,\mathrm{fm}^{-1}$ and $\Lambda = 2.0\,\mathrm{fm}^{-1}$.

from EDF with

$$e_a = E(^{208}\mathrm{Pb} + a) - E(^{208}\mathrm{Pb}). \tag{1}$$

Then calculate an average (monopole) interaction for all possible sets of two-valence nucleons with

$$\bar{V}_{ab,\mathrm{EDF}} = E(^{208}\mathrm{Pb} + a + b) - E(^{208}\mathrm{Pb}) - e_a - e_b. \tag{2}$$

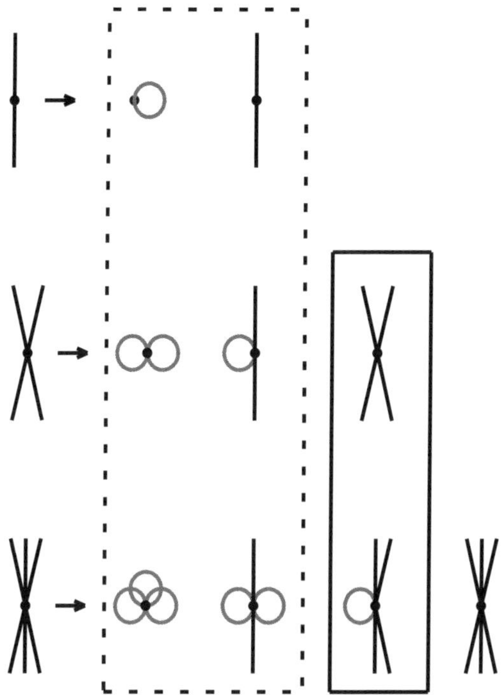

Fig. 4. Schematic diagram for the terms in the Hamiltonian obtained from Wick's theorem for a closed shell. The circles represent the summation over the orbitals in the closed shell. The black lines represent the valence particles and/or holes.

where $E(^{208}\text{Pb} + a + b)$ is the single-reference EDF energy of the configuration for a closed shell plus two nucleons constrained to be in orbitals a and b restricted to spherical symmetry. The EDF-monopole contains both terms in the solid-line box in Fig. 2, and also contains higher-order contributions implicit in the EDF functional including possible quadratic terms in the mass dependence of the single-particle energies. Whereas, the N^{3}LO monopole only contains the valence two-body interaction corrected to second order. We modify the monopole part of the valence two-body interaction (based for example on N^{3}LO) to reproduce the results of Eq. (2). With this modification, the CI calculations closely reproduce the EDF calculations for single-Slater determinants, even when relatively many valence nucleons are added. Thus, the CI calculations are constrained to reproduce the trends of closed-shell energies and effective single-particle energies obtained with the EDF. The final step is to modify all of the

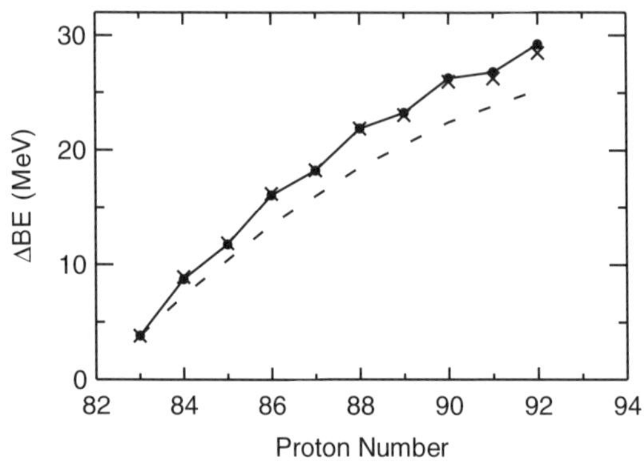

Fig. 5. Binding energies relative to ^{208}Pb. Experiment are the points connected by a line. The results of the CI(N^{3}LO+EDF) are shown by crosses. The results of the single-reference EDF calculation constrained to spherical symmetry are shown by the dashed line.

valence TBME by

$$\langle abJ \mid V \mid abJ \rangle_{\text{eff}} = \langle abJ \mid V \mid abJ \rangle_{\text{N}^3\text{LO}} - \bar{V}_{ab,\text{N}^3\text{LO}} + \bar{V}_{ab,\text{EDF}}. \tag{3}$$

The CI calculations were carried out up to ^{218}U ($Z = 92$) where the M-scheme dimension is about 1.5 million. The results for the ground state energies are compared to experiment in Fig. 5. The EDF calculation is based upon the spherical $(0h_{9/2})^n$ configuration with $n = 1$ to 10. The difference between EDF and CI can be regarded as the correlation energy in the nuclear ground state, in this case mainly due to the "pairing" interaction. The correlation results in wavefunctions that are highly mixed in the valence proton basis. For example the CI ground state of ^{218}U contains only 4.7% of the $(0h_{9/2})^{10}$ component. Up to $Z = 88$ the difference between experiment and theory for the binding energy relative to ^{208}Pb is on the order of 100 keV, and after this it gradually increases to about 700 keV for ^{218}U. Without the EDF corrections in Eq. (3), ^{218}U would be over-bound by about 5 MeV.

Thus with a combination of EDF and second-order effective interaction theory we have an "ab-initio" calculation for energies and wavefunctions for nuclei with valence nucleons outside of a ^{208}Pb core. To make this a truly ab-initio model we need to be able to derive the Skyrme parameters from the same two and three nucleon interactions that are used to describe light nuclei and nuclear matter within the framework of a closed-shell configuration for ^{208}Pb.

Acknowledgments

This work is partly supported by NSF Grant PHY-0758099 and the DOE UNEDF-SciDAC grant DE-FC02-09ER41585.

References

[1] G.E. Brown and T.T.S. Kuo, *Nucl. Phys.* **A92**, 481 (1967).
[2] T.T.S. Kuo and G.E. Brown, *Nucl. Phys.* **A85**, 40 (1966).
[3] G.F. Bertsch, *Nucl. Phys.* **74**, 234 (1965).
[4] G.H. Herling and T.T.S. Kuo, *Nucl. Phys.* **181**, 113 (1972).
[5] J.B. McGrory and T.T.S. Kuo, *Nucl. Phys.* **A247**, 283 (1975).
[6] E.K. Warburton and B.A. Brown, *Phys. Rev. C* **43**, 602 (1991).
[7] E. Caurier, M. Rejmund and H. Grawe, *Phys. Rev. C* **67**, 054310 (2003).
[8] L. Coraggio, A. Covello, A. Gargano, N. Itaco and T.T.S. Kuo, *Prog. in Part. and Nucl. Phys.* **62**, 135 (2009).
[9] L. Coraggio, A. Covello, A. Gargano and N. Itaco, *Phys. Rev. C* **76**, 061313(R) (2007).
[10] http://www.fys.uio.no/compphys/cp/software.html.
[11] Nushell@MSU, B.A. Brown and W.D.M. Rae, MSU-NSCL report (2007).
[12] R.B. Wiringa, V.G.J. Stoks and R. Schiavilla, *Phys. Rev. C* **51**, 38 (1995).
[13] R. Machleidt, *Phys. Rev. C* **63**, 024001 (2001).
[14] D.R. Entem and R. Machleidt, *Phys. Rev. C* **68**, 041001 (2003).
[15] S.K. Bogner, R.J. Furnstahl and A. Schwenk, *Prog. in Part. and Nucl. Phys.* **65**, 94 (2010).
[16] B.A. Brown, A. Signoracci, M. Hjorth-Jensen, *Phys. Lett. B* **695**, 507 (2011).

Chapter 5

Nuclear Medium Effects from Hadronic Atoms

E. Friedman*,† and A. Gal*,‡

The state of the art in the study of π^-, K^- and Σ^- atoms, along with the in-medium nuclear interactions deduced for these hadrons, is reviewed. A special emphasis is placed on recent developments in $\bar{K}$–nuclear physics, where a strongly attractive density dependent K^-–nuclear potential of order 150–200 MeV in nuclear matter emerges by fitting K^-–atom data. This has interesting repercussions on $\bar{K}$ quasibound *nuclear* states, on the composition of strange hadronic matter and on $\bar{K}$ condensation in self bound hadronic systems.

1. Introduction

Hadronic atoms have played an important role in elucidating in-medium properties of hadron-nucleon (hN) interactions near threshold [1, 2]. Hadronic atom data consist primarily of strong interaction level shifts, widths and yields, derived from h^--atom X-ray transitions. These data are analyzed in terms of optical potentials $V^h_{\rm opt} = t_{hN}(\rho)\rho$ which are functionals of the nuclear density $\rho(r)$, capable of handling large data sets across the periodic table in order to identify characteristic entities that may provide a link between experiment and microscopic approaches. Here, $t_{hN}(\rho)$ is a density dependent in-medium hN t matrix at threshold, satisfying the low-density limit $t_{hN}(\rho) \to t^{\rm free}_{hN}$, with the free-space t matrix $t^{\rm free}_{hN}$, as $\rho \to 0$. A schematic summary of lessons gained by analyzing the available hadronic atom data in terms of optical potentials is given in Table 1, where the number of data points included in these analyses is shown in parentheses

*Racah Institute of Physics, The Hebrew University, Jerusalem 91904, Israel.
†E-mail: elifried@vms.huji.ac.il, ‡E-mail: avragal@vms.huji.ac.il

Table 1. Hadronic atom scenarios which except for $\bar{p}$ atoms are discussed in the present review.

Hadron	Re $V_{\rm opt}$	Im $V_{\rm opt}$	Comments
π^-	Repulsive in bulk	Moderate	Excellent data (100)
	Attractive on surface		Well understood
Σ^-	Repulsive in bulk	Moderate	Limited data (23)
	Attractive outside		Poorly understood
K^-	Attractive	Absorptive	Good data (65)
	Deep or shallow?		*Open problems*
$\bar{p}$	Model dependent	Very absorptive	Excellent data (90)

in the last column. A more exhaustive discussion of π^-, Σ^- and K^- atoms follows below.

2. Partial Restoration of Chiral Symmetry from Pionic Atoms

Density independent optical potential global fits to pionic atom data across the periodic table reveal an anomalous s-wave repulsion [1, 2], a major component of which is due to a too repulsive isovector πN amplitude b_1 with respect to the free-space value $b_1^{\rm free}$. This is demonstrated on the l.h.s. of Fig. 1 which shows results of global fits to pionic atom data as a function of a parameter γ, related to the difference between neutron and proton rms radii:

$$r_n - r_p = \gamma \frac{N-Z}{A} + \delta. \tag{1}$$

Applying a finite-range folding with rms radius 0.9 fm to the p-wave part of $V^{\pi}_{\rm opt}$, the resulting χ^2 minimum for a "skin" neutron distribution is obtained at $\gamma = 1.1 \pm 0.1$ fm.[a] The r.h.s. of the figure shows results of global fits with empirical energy dependence imposed on the s-wave amplitudes b_0 and b_1 [4], and more importantly with a DD renormalization of b_1:

$$b_1(\rho) = \frac{b_1}{1 - \frac{\sigma\rho}{m_\pi^2 f_\pi^2}}, \tag{2}$$

where $f_\pi = 92.4$ MeV is the pion weak decay constant and $\sigma \approx 50$ MeV is the πN σ term. Equation (2) was derived by Weise [5] considering an

[a] For a recent discussion of the role of neutron distributions in hadronic atoms, see Ref. [3].

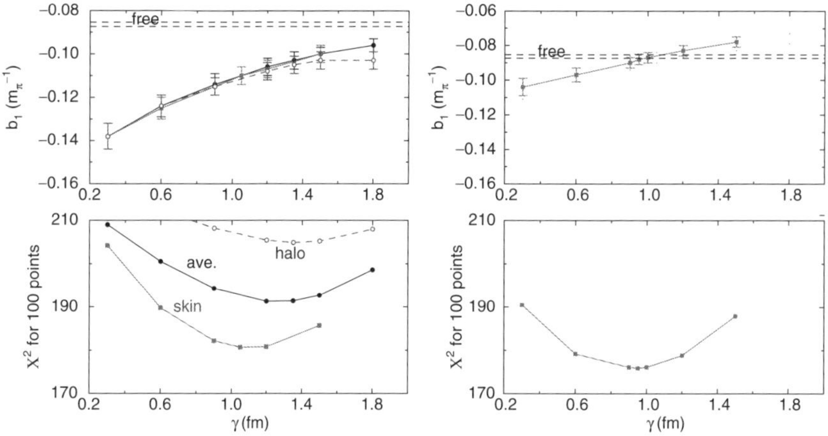

Fig. 1. Fits to 100 π^--atom data points from Ne to U as a function of γ. Left: density-independent (DI) b_1. Right: density-dependent (DD) b_1. Figure adapted from Ref. [2].

in-medium extension of the Tomozawa-Weinberg (TW) LO chiral limit for b_1 in terms of f_π [6] which is then related to the quark condensate $\langle \bar{q}q \rangle$:

$$b_1(\rho) = -\frac{\mu_{\pi N}}{8\pi f_\pi^2(\rho)}, \quad \frac{f_\pi^2(\rho)}{f_\pi^2} = \frac{\langle \bar{q}q \rangle_\rho}{\langle \bar{q}q \rangle_0} \simeq 1 - \frac{\sigma \rho}{m_\pi^2 f_\pi^2}. \tag{3}$$

The figure makes it evident that the magnitude of b_1 on the r.h.s., following the DD renormalization, is systematically smaller than that on the l.h.s., and at the χ^2 minimum it agrees perfectly with $b_1^{\rm free}$.

A similar conclusion was reached in Refs. [7, 8] from measurements of $1s$ "deeply bound" pionic atoms of Sn isotopes. The advantage of using $1s$ "deeply bound" levels is that the p-wave πN interaction plays there a secondary role, but this merit is more than compensated by the considearably increased errors associated with smaller data sets, as shown in Table 2. The uncertainty listed in the fourth column makes it clear that the "deeply bound" atoms alone do not give conclusive evidence for the need to renormalize b_1. In fact, Suzuki *et al.* [8] considered "deeply bound" $1s$ levels in three Sn isotopes together with "normal" $1s$ levels in ^{16}O, ^{20}Ne and ^{28}Si, 12 data points in total yielding $b_1 = -0.1149 \pm 0.0074\, m_\pi^{-1}$. However, this small uncertainty excludes the uncertainty from the p-wave πN potential which was held fixed in their analysis. A more realistic uncertainty for this type of deduction is given in column 3 of the table.

Table 2. Values of b_1 derived from density-independent fits to pionic atom data sets listed in Ref. [9]. For comparison, $b_1^{\rm free} = -0.0864 \pm 0.0010\, m_\pi^{-1}$ [10]. "Deep" refers to $1s$ π^- "deeply bound" states in ^{205}Pb [11] and ^{115,119,123}Sn [8] observed in (d, ^{3}He) experiments at GSI.

Data	Global ^{20}Ne to ^{238}U	Light $N = Z$ + "deep" $1s$ only	"Deep" $1s$ only
Points	100	20	8
$b_1 (m_\pi^{-1})$	-0.104 ± 0.006	-0.104 ± 0.013	-0.130 ± 0.036

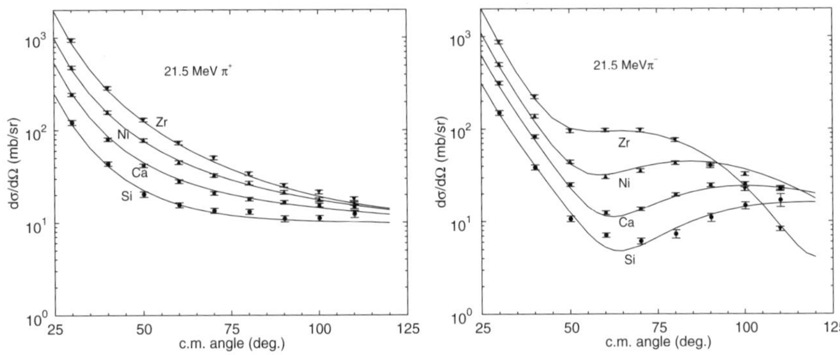

Fig. 2. Low energy $\pi^\pm$ nucleus elastic scattering angular distributions [12] reproduced with $b_1(\rho)$ ansatz, Eq. (2).

The renormalization of b_1 derived from pionic atoms is consistent with that shown recently to be also required in low energy π-nucleus scattering, as demonstrated in Fig. 2 for 21.5 MeV $\pi^\pm$ scattered off isotopes of Si, Ca, Ni and Zr at PSI [12].

3. Σ Nuclear Repulsion from Σ^- Atoms

A vast body of $(K^-, \pi^\pm)$ spectra indicate a repulsive and moderately absorptive Σ nuclear potential V^Σ, with a substantial isospin dependence [13, 14]. These data, including recent (π^-, K^+) spectra [15] and related DWIA analyses [16], provide credible evidence that Σ hyperons generally do not bind in nuclei. A repulsive component of a DD Σ nuclear potential was already deduced in the mid 1990s from Σ^- atom data [17,18], as shown in Fig. 3. In fact, $V_{\rm R}^\Sigma$ is attractive at low densities outside the nucleus, as enforced by the observed "attractive" Σ^- atomic level shifts, changing

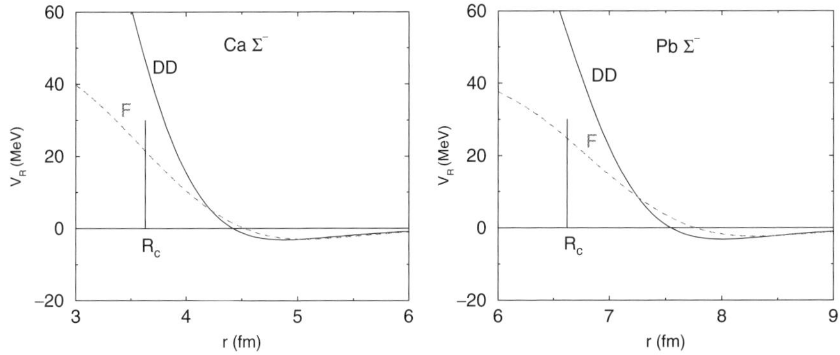

Fig. 3. V_R^{Σ} from fits to Σ^- atoms [2]. The transition from attraction to repulsion occurs well outside of R_c, the half-density radius of the charge distribution in Ca (l.h.s.) and Pb (r.h.s.).

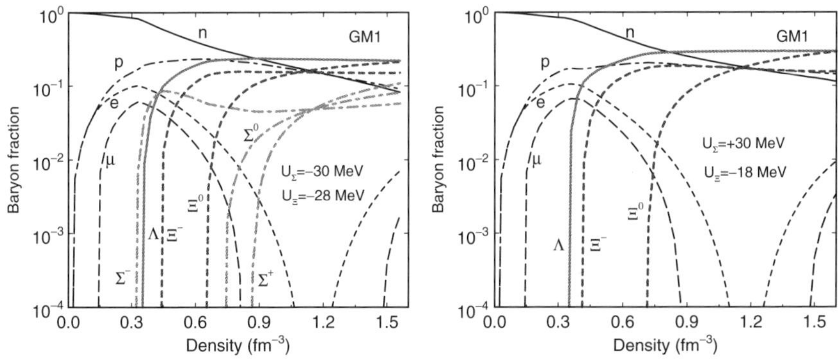

Fig. 4. RMF calculations of baryon & lepton fractions in neutron star matter for different scenarios of hyperon nuclear potential depths U_Y. Figure adapted from Ref. [19].

into repulsion on approach of the nuclear radius. The precise magnitude and shape of V_R^{Σ} within the nucleus, however, are model dependent as demonstrated by the difference between potentials DD and F (defined in the Appendix). This repulsion bears interesting consequences for the balance of strangeness in the inner crust of neutron stars, primarily by delaying to higher densities, or even aborting the appearance of Σ^- hyperons, as shown in Fig. 4.

The G-matrices constructed from Nijmegen soft-core potential models have progressed throughout the years to produce Σ repulsion in symmetric

Table 3. Isoscalar and isovector Σ nucleus potentials, Eq. (4) in MeV, calculated in Nijmegen soft-core potential models [20] at $k_F = 1.35\,\mathrm{fm}^{-1}$, corresponding to nuclear-matter density.

	97f	04d	06d	08a	08b	Phenom.	Refs.
V_0^Σ	-13.9	-26.0	-1.2	$+13.4$	$+20.3$	$+30 \pm 20$	[2, 16]
V_1^Σ	-30.4	$+30.4$	$+52.6$	$+64.5$	$+85.2$	$\approx +80$	[21]

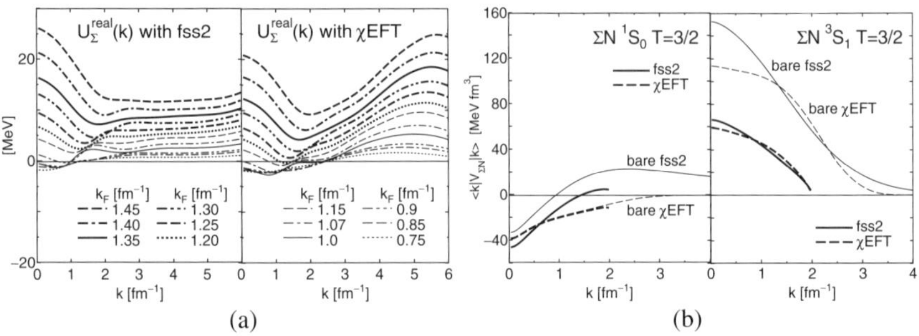

Fig. 5. Left: Isoscalar Σ nuclear potentials calculated in two models, Refs. [22,23]. Right: $(\Sigma N)_{T=3/2}$ potentials calculated in these models. Figure adapted from Ref. [24].

nuclear matter, as demonstrated in Table 3 using the parametrization

$$V_R^\Sigma = V_0^\Sigma + \frac{1}{A} V_1^\Sigma \mathbf{T}_A \cdot \mathbf{t}_\Sigma. \tag{4}$$

In the latest Nijmegen ESC08 model [20], this repulsion is dominated by repulsion in the $T = 3/2$, ${}^3S_1 - {}^3D_1$ ΣN channel where a strong short distance Pauli exclusion repulsion for quarks arises in SU(6) quark-model RGM [22] and in chiral EFT [23] calculations, as seen on the r.h.s. of Fig. 5. These model calculations also lead to Σ nuclear repulsion, shown in momentum space on the l.h.s. of the figure. A strong repulsion appears also in a recent SU(3) chiral perturbation calculation [25] which yields $V_0^\Sigma \approx 60\,\mathrm{MeV}$. Phenomenologically $V_0^\Sigma > 0$ and $V_1^\Sigma > 0$, as listed in the table, and the resulting Σ-nuclear potential V_R^Σ is repulsive.[b]

[b]In the case of ${}^4_\Sigma$He, the only known quasibound Σ hypernucleus [26, 27], the isovector term provides substantial attraction owing to the small value of A towards binding the $T = 1/2$ hypernuclear configuration, while the isoscalar repulsion reduces the quasibound level width [28].

4. $\bar{K}$-Nucleus Potentials from K^- Atoms

The gross features of low-energy $\bar{K}N$ physics are encapsulated in the leading-order Tomozawa-Weinberg (TW) vector term of the chiral effective Lagrangian [6]. The Born approximation for the $\bar{K}$-nuclear potential $V_{\rm TW}^{\bar{K}}$ due to this TW interaction term yields a sizable attraction:

$$V_{\rm TW}^{\bar{K}} = -\frac{3}{8f_\pi^2}\,\rho \sim -55\,\frac{\rho}{\rho_0} \quad ({\rm MeV}) \tag{5}$$

for $\rho_0 = 0.16\,{\rm fm}^{-3}$. Iterating the TW term plus the less significant NLO terms, within an *in-medium* coupled-channel approach constrained by the $\bar{K}N-\pi\Sigma-\pi\Lambda$ data near the $\bar{K}N$ threshold, roughly doubles this $\bar{K}$-nucleus attraction [29]. A major uncertainty in these chirally based studies arises from fitting the $\Lambda(1405)$ resonance by the imaginary part of the $(\pi\Sigma)_{I=0}$ amplitude calculated within the same coupled channel chiral scheme. Yet, irrespective of this uncertainty, the $\Lambda(1405)$ which may be viewed as a K^-p quasibound state quickly dissolves in the nuclear medium at low density, so that the repulsive free-space scattering length a_{K^-p}, as function of ρ, becomes *attractive* well below ρ_0. Adding the weakly density dependent $I=1$ attractive scattering length a_{K^-n}, the resulting in-medium $\bar{K}N$ isoscalar scattering length $b_0(\rho) = \frac{1}{2}(a_{K^-p}(\rho) + a_{K^-n}(\rho))$ translates into a strongly attractive $V^{\bar{K}}$ [30, 31]:

$$V_R^{\bar{K}}(\rho) \sim -\frac{2\pi}{\mu_{KN}}{\rm Re}\,b_0(\rho_0)\,\rho_0\,\frac{\rho}{\rho_0} \approx -110\,\frac{\rho}{\rho_0} \quad ({\rm MeV}). \tag{6}$$

Shallower potentials, $V_R^{\bar{K}}(\rho_0) \sim -(40-60)$ MeV, were obtained by imposing a Watson-like self-consistency requirement [30, 32]. However, it was shown recently that stronger attraction, $V_R^{\bar{K}}(\rho_0) \sim -(80-100)$ MeV, arises in similar chiral approaches [33] by considering the energy dependence of the in-medium $\bar{K}N$ scattering amplitude below threshold [34].

Comprehensive fits to the strong-interaction shifts and widths of K^--atom levels, begun in the mid 1990s [35], have yielded DD deeply attractive and strongly absorptive optical potentials with nuclear-matter depth $-V_R^{\bar{K}}(\rho_0) \sim (150-200)$ MeV at threshold [35]. The l.h.s. of Fig. 6 illustrates for ^{58}Ni the real part of $\bar{K}$-nucleus potentials obtained from a global fit to the data in several models and, in parentheses, the corresponding values of χ^2 for 65 K^--atom data points. A model-independent Fourier-Bessel (FB) fit [36] is also shown, within an error band. Just three terms in the FB series, added to a $t\rho$ potential, suffice to achieve a χ^2 as low as 84 and to make the potential extremely deep, in agreement with the density-dependent best-fit

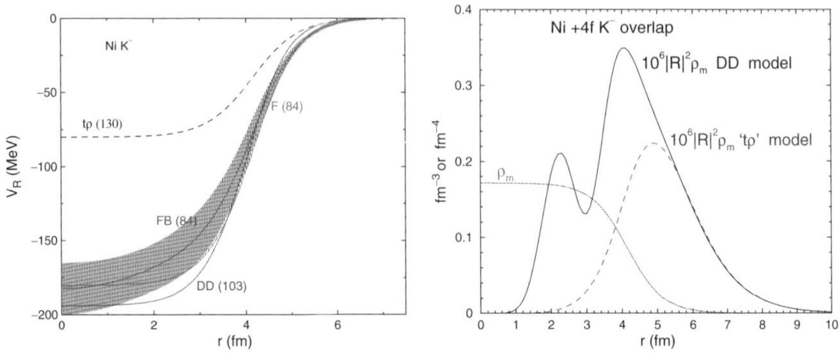

Fig. 6. Left: $V_R^{\bar{K}}$ in Ni from global fits to 65 K^- atom data points, with χ^2 values in parentheses [2]. The shaded area approximates the uncertainty in the Fourier-Bessel fit. Right: Overlap of K^- atomic $4f$ radial wavefunctions R squared with Ni matter density ρ_m.

potentials DD and F. In particular, potential F provides by far the best fit ever reported for any global K^--atom data fit [37], and the lowest χ^2 value as reached by the FB method.

Shown on the r.h.s. of Fig. 6 are overlaps of the $4f$ atomic radial wavefunction squared with the matter density ρ_m in ^{58}Ni for two of the models exhibited on the l.h.s. of the figure. The $4f$ atomic orbit is the last circular K^- atomic orbit from which the K^- meson undergoes nuclear absorption. The figure demonstrates that, whereas this overlap for the shallower $t\rho$ potential peaks at nuclear density of order 10% of ρ_0, it peaks at about 60% of ρ_0 for the deeper DD potential and has a secondary peak well inside the nucleus. The double-peak structure indicates the existence of a K^- strong-interaction $\ell = 3$ quasibound state for the DD potential. It is clear that whereas within the $t\rho$ potential there is no sensitivity to the interior of the nucleus, the opposite holds for the density dependent F potential which accesses regions of full nuclear density. This owes partly to the smaller imaginary part of F.

Given the repercussions of deeply attractive potentials on the equation of state of dense matter, it is important to explore the stability of these best-fit solutions to variations in the data selection and the fitting procedure. The most obvious question to ask is whether the resulting best-fit potentials depend strongly on the size, composition and accuracy of the data set studied. Regarding size and composition, following an earlier discussion [38] it has been observed recently [39] that the DD deep potentials and the

Table 4. Full and reduced K^- atom data set fits. The reduced set consists of $2p, 3d, 4f, 5g, 7i$ shifts, widths and yields in C, Si, Ni, Sn and Pb targets, respectively [39].

	Shallow potential			Deep potential		
N	Re $b(\rho_0)$	Im $b(\rho_0)$	χ^2	Re $b(\rho_0)$	Im $b(\rho_0)$	χ^2
65	0.62 ± 0.05	0.93 ± 0.04	130	1.44 ± 0.03	0.59 ± 0.03	84
15	0.78 ± 0.13	0.92 ± 0.11	44	1.47 ± 0.05	0.55 ± 0.06	26

DI relatively shallow potentials, as well as the superiority of DD to DI in terms of quality of fit, persist upon decreasing the size of the data set. This is demonstrated in Table 4 upon reducing the 65 data point global set down to 15 data points from five targets spread over the entire periodic table (C, Si, Ni, Sn, Pb). Similar results hold for any four out of these five targets. It makes sense then to repeat some of the 30–40 years old K^- atom measurements, making use of modern techniques, in order to acquire a minimum size canonical set of data with reduced statistical errors and with common systematics. For the specific set proposed in Ref. [39], the (lower level) widths are directly measurable yet not excessively large to make it difficult to observe the feeding X-ray transition above the background. Similarly, the relative yields of the upper to lower level transitions are of the order of 10% and higher. Fitting to such a data set with improved accuracy could resolve the issue of deep vs. shallow potentials and determine how deep is "deep".

A fairly new and independent evidence in favor of extremely deep $\bar{K}$-nucleus potentials is provided by (K^-, n) and (K^-, p) spectra taken at KEK on ^{12}C [40] and very recently also on ^{16}O [41] at $p_{K^-} = 1\,\text{GeV/c}$. The ^{12}C spectra are shown on the l.h.s. of Fig. 7, where the solid lines represent calculations (outlined in Ref. [43]) using potential depths in the range 160–190 MeV. The dashed lines correspond to using relatively shallow potentials of depth about 60 MeV which may be considered excluded by these data. However, Magas *et al.* [44] have recently expressed concerns about protons of reactions other than those *directly* emanating in the (K^-, p) reaction and which could explain part of the bound-state region of the measured spectrum without invoking a very deep $\bar{K}$-nuclear potential. A sufficientlly deep potential would allow quasibound states bound by over 100 MeV, for which the major $\bar{K}N \to \pi\Sigma$ decay channel is blocked, resulting in relatively narrow $\bar{K}$-nuclear states. Of course, a fairly sizable

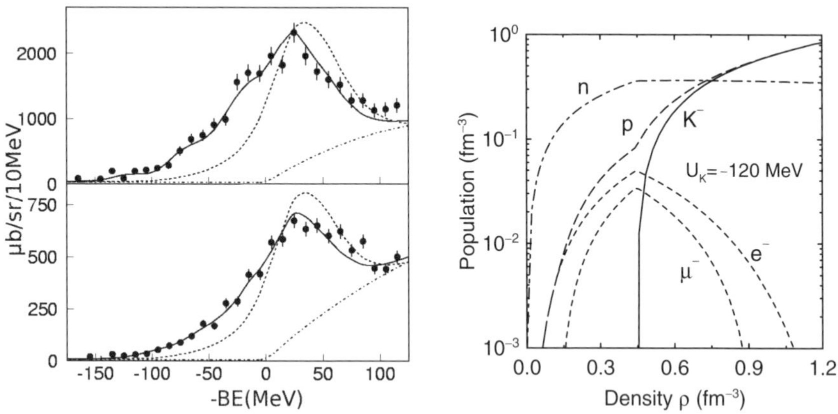

Fig. 7. Left: KEK-PS E548 missing mass (K^-, n) (upper) & (K^-, p) (lower) spectra on ^{12}C at $p_{K^-} = 1\,\mathrm{GeV/c}$ [40]. Right: calculated neutron-star population as a function of density. The neutron density stays nearly constant once kaons condense [42].

extrapolation is involved in this case using an energy-independent potential determined largely near threshold. Furthermore, the best-fit $V_I^{\bar{K}}$ imaginary depths of 40–50 MeV imply that $\bar{K}$-nuclear quasibound states are broad, as studied in Refs. [37, 47].

A robust consequence of the sizable $\bar{K}$-nucleus attraction is that K^- condensation, when hyperon degrees of freedom are ignored, could occur in neutron star matter at about 3 times nuclear matter density, as shown on the r.h.s. of Fig. 7. Comparing it with Fig. 4 for neutron stars, but where strangeness materialized through hyperons, one may ask whether $\bar{K}$ mesons condense also in the presence of hyperons. This question was posed within RMF calculations of neutron star matter long ago and answered negatively [45, 46], but only recently it was posed for strange hadronic matter in Ref. [48] by calculating multi-$\bar{K}$ nuclear configurations. Figure 8 demonstrates a remarkable saturation of K^- separation energies B_{K^-} calculated in multi-K^- nuclei, independently of the applied RMF model as shown on the l.h.s. for three different nuclear RMF schemes. The r.h.s. of the figure demonstrates that this saturation persists already in the most straightforward $\sigma + \omega$ model, primarily owing to the repulsion induced by the vector ω field between like $\bar{K}$ mesons. The additional vector fields ρ and ϕ only add repulsion, thus strengthening the saturation. The effect of $V_I^{\bar{K}}$ is noticeable only below $B_{\bar{K}} \approx 100\,\mathrm{MeV}$, as seen by the departure of the lowest green line with respect to the lowest red line.

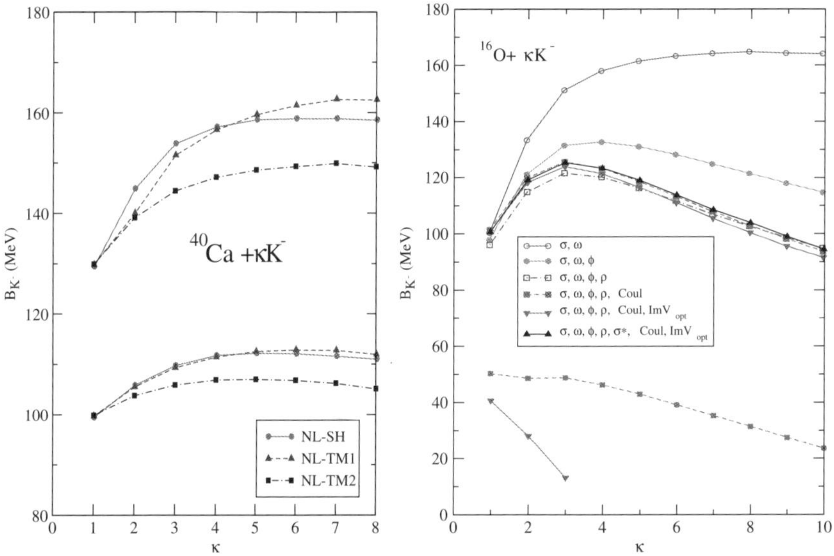

Fig. 8. Saturation of K^- separation energy $B_{\bar{K}}$ in RMF calculations of multi K^- nuclei as a function of the number κ of K^- mesons [48]. Left: RMF model dependence. Right: dependence on RMF ingredients.

The saturation values of B_{K^-} do not allow conversion of hyperons to $\bar{K}$ mesons through the strong decays $\Lambda \to p + K^-$ or $\Xi^- \to \Lambda + K^-$ in multi-strange hypernuclei, which therefore remain the lowest-energy configuration for multi-strange systems [49]. This provides a powerful argument against $\bar{K}$ condensation in the laboratory, under strong-interaction equilibrium conditions [48]. It does not apply to kaon condensation in neutron stars, where equilibrium configurations are determined by weak-interaction conditions. This work has been recently generalized to multi-K^- *hypernuclei* [50].

Appendix: Density Dependent Optical Potentials

Here we specify the functional form of two density dependent optical potentials used in studies of hadronic atoms. For a recent application to K^- atoms, see Table 1 of Ref. [37].

- The DD form is based on modifying the effective scattering length b_0 (e.g. Eq. (6)):

$$V^h(r) \sim -\frac{2\pi}{\mu_{hN}}\, b_0\, \rho(r) \Rightarrow b_0 \to b_0 + B_0 \left\{\frac{\rho(r)}{\rho_0}\right\}^{\alpha}, \quad \alpha > 0, \qquad \text{(A.1)}$$

where $\rho_0 = 0.16\,\text{fm}^{-3}$ is a central nuclear density. It is possible then to respect the "low density limit" by keeping b_0 fixed, $b_0 = b_0^{\text{free}}$, while varying the parameters B_0 and α.

- The F form is based on modifying b_0 as follows:

$$b_0 \to B_0\, F(r) + b_0\, [1 - F(r)]. \tag{A.2}$$

The density-like function $F(r)$ is defined as

$$F(r) = \frac{1}{e^x + 1}, \quad x = \frac{r - R_x}{a_x}. \tag{A.3}$$

Clearly, $F(r) \to 1$ for $r \ll R_x$ which defines an internal region and similarly $[1 - F(r)] \to 1$ for $r \gg R_x$ which defines an external region. Thus R_x forms an approximate border between internal and external regions, and *if* R_x is close to the nuclear surface then the two regions do correspond to the high-density and low-density regions of nuclei, respectively. In global fits across the periodic table, R_x is parametrized as $R_x = R_{x0}A^{1/3}+\delta_x$ and the parameters B_0, R_{x0} and δ_x are varied upon in the least-squares fit, while gridding on values of a_x around 0.5 fm. The parameter b_0 may be held fixed at its free hN value, but the results often depend very little on its precise value.

Acknowledgments

On the occasion of Gerry Brown's 85th birthday Festschrift, we dedicate this mini review to him who commissioned our two past reviews [1,2] on similar subjects. This work was supported in part by the SPHERE collaboration within the HadronPhysics2 Project No. 227431 of the EU initiative FP7.

References

[1] C.J. Batty, E. Friedman and A. Gal, *Phys. Rept.* **287**, 385 (1997).
[2] E. Friedman and A. Gal, *Phys. Rept.* **452**, 89 (2007), and references therein.
[3] E. Friedman, *Hyp. Int.* **193**, 33 (2009), and references therein.
[4] E. Friedman and A. Gal, *Phys. Lett. B* **578**, 85 (2004).
[5] W. Weise, *Nucl. Phys. A* **690**, 98c (2001).
[6] Y. Tomozawa, *Nuovo Cimento A* **46**, 707 (1966), S. Weinberg, *Phys. Rev. Lett.* **17**, 616 (1966).
[7] E.E. Kolomeitsev, N. Kaiser and W. Weise, *Phys. Rev. Lett.* **90**, 092501 (2003).
[8] K. Suzuki *et al.*, *Phys. Rev. Lett.* **92**, 072302 (2004).

[9] E. Friedman and A. Gal, *Nucl. Phys. A* **724**, 143 (2003).
[10] J. Marton, *Nucl. Phys. A* **790**, 328c (2007).
[11] H. Geissel *et al.*, *Phys. Rev. Lett.* **88**, 122301 (2002).
[12] E. Friedman *et al.*, *Phys. Rev. Lett.* **93**, 122302 (2004), *Phys. Rev. C* **72**, 034609 (2005).
[13] C.B. Dover, D.J. Millener and A. Gal, *Phys. Rept.* **184**, 1 (1989), and references therein.
[14] S. Bart *et al.* [BNL E887], *Phys. Rev. Lett.* **83**, 5238 (1999).
[15] H. Noumi *et al.* [KEK E438], *Phys. Rev. Lett.* **89**, 072301 (2002), *Phys. Rev. Lett.* **90**, 049902(E) (2003), P.K. Saha *et al.*, *Phys. Rev. C* **70**, 044613 (2004).
[16] M. Kohno, Y. Fujiwara, Y. Watanabe, K. Ogata and M. Kawai, *Prog. Theor. Phys.* **112**, 895 (2004), *Phys. Rev. C* **74**, 064613 (2006), T. Harada and Y. Hirabayashi, *Nucl. Phys. A* **759**, 143 (2005), **767**, 206 (2006).
[17] C.J. Batty, E. Friedman and A. Gal, *Phys. Lett. B* **335**, 273 (1994), *Prog. Theor. Phys. Suppl.* **117**, 227 (1994).
[18] J. Mareš, E. Friedman, A. Gal and B.K. Jennings, *Nucl. Phys. A* **594**, 311 (1995).
[19] J. Schaffner-Bielich, *Nucl. Phys. A* **835**, 279 (2010), and references therein.
[20] A. Rijken Th, M.M. Nagels and Y. Yamamoto, *Nucl. Phys. A* **835**, 160 (2010), and references therein.
[21] C.B. Dover, A. Gal and D.J. Millener, *Phys. Lett. B* **138**, 337 (1984).
[22] Y. Fujiwara, Y. Suzuki and C. Nakamoto, *Prog. Part. Nucl. Phys.* **58**, 439 (2007), and references therein.
[23] H. Polinder, J. Haidenbauer and U.G. Meißner, *Nucl. Phys. A* **779**, 244 (2006).
[24] M. Kohno, *Phys. Rev. C* **81**, 014003 (2010).
[25] N. Kaiser, *Phys. Rev. C* **71**, 068201 (2005).
[26] R.S. Hayano *et al.*, *Phys. Lett. B* **231**, 355 (1989).
[27] T. Nagae *et al.* [BNL E905], *Phys. Rev. Lett.* **80**, 1605 (1998).
[28] T. Harada, *Phys. Rev. Lett.* **81**, 5287 (1998).
[29] B. Borasoy, R. Nißler and W. Weise, *Eur. Phys. J. A* **25**, 79 (2005).
[30] A. Cieplý, E. Friedman, A. Gal and J. Mareš, *Nucl. Phys. A* **696**, 173 (2001).
[31] W. Weise and R. Härtle, *Nucl. Phys. A* **804**, 173 (2008).
[32] A. Ramos and E. Oset, *Nucl. Phys. A* **671**, 481 (2000).
[33] A. Cieplý and J. Smejkal, *Eur. Phys. J. A* **43**, 191 (2010).
[34] A. Cieplý, E. Friedman, A. Gal, D. Gazda and J. Mareš, arXiv: 1102.4515, A. Cieplý, E. Friedman, A. Gal and V. Krejčiřík, *Phys. Lett. B* **698**, 226 (2011).
[35] E. Friedman, A. Gal and C.J. Batty, *Phys. Lett. B* **308**, 6 (1993), *Nucl. Phys. A* **579**, 518 (1994).
[36] N. Barnea and E. Friedman, *Phys. Rev. C* **75** 022202(R) (2007).
[37] J. Mareš, E. Friedman and A. Gal, *Nucl. Phys. A* **770**, 84 (2006).
[38] E. Friedman, A. Gal, J. Mareš and A. Cieplý, *Phys. Rev. C* **60**, 024314 (1999).
[39] E. Friedman, *Int. J. Mod. Phys. A* **26**, 468 (2011).
[40] T. Kishimoto *et al.* [KEK E548], *Prog. Theor. Phys.* **118**, 181 (2007).
[41] T. Kishimoto, *Nucl. Phys. A* **827**, 321c (2009).

[42] N.K. Glendenning and J. Schaffner-Bielich, *Phys. Rev. C* **60**, 025803 (1999).
[43] J. Yamagata, H. Nagahiro and S. Hirenzaki, *Phys. Rev. C* **74**, 014604 (2006).
[44] V.K. Magas, J. Yamagata-Sekihara, S. Hirenzaki, E. Oset and A. Ramos, *Phys. Rev. C* **81**, 024609 (2010).
[45] R. Knorren, M. Prakash and P.J. Ellis, *Phys. Rev. C* **52**, 3470 (1995).
[46] J. Schaffner and I.N. Mishustin, *Phys. Rev. C* **53**, 1416 (1996).
[47] D. Gazda, E. Friedman, A. Gal and J. Mareš, *Phys. Rev. C* **76**, 055204 (2007).
[48] D. Gazda, E. Friedman, A. Gal and J. Mareš, *Phys. Rev. C* **77**, 045206 (2008).
[49] J. Schaffner-Bielich and A. Gal, *Phys. Rev. C* **62**, 034311 (2000), and references therein.
[50] D. Gazda, E. Friedman, A. Gal and J. Mareš, *Phys. Rev. C* **80**, 035205 (2009).

Chapter 6

Three-Body Interactions in Fermi Systems

B. Friman* and A. Schwenk†

We show that the contributions of three-quasiparticle interactions to normal Fermi systems at low energies and temperatures are suppressed by n_q/n compared to two-body interactions, where n_q is the density of excited or added quasiparticles and n is the ground-state density. For finite Fermi systems, three-quasiparticle contributions are suppressed by the corresponding ratio of particle numbers N_q/N. This is illustrated for polarons in strongly interacting spin-polarized Fermi gases and for valence neutrons in neutron-rich calcium isotopes.

1. Introduction

Three-nucleon (3N) forces and advancing microscopic many-body methods are a frontier in the physics of nuclei and nucleonic matter in stars. New facets of 3N forces are revealed in neutron-rich nuclei, such as their role in determining the location of the neutron dripline [1,2] and in elucidating the doubly-magic nature of ^{48}Ca [3]. Three- and higher-body forces are also the dominant uncertainty in constraining the properties of neutron-rich matter at nuclear densities and thus the structure of neutron stars [4, 5]. At the same time, 3N forces are at the center of developing systematic interactions based on effective field theory (EFT) [6], thus linking the nuclear forces frontier with the experimental exploration of neutron-rich nuclei.

For light nuclei, ab-initio methods have unambiguously established the quantitative role of 3N forces for ground-state properties, excitations and

*GSI Helmholtzzentrum für Schwerionenforschung GmbH, 64291 Darmstadt, Germany. E-mail: b.friman@gsi.de
†ExtreMe Matter Institute EMMI, GSI Helmholtzzentrum für Schwerionenforschung GmbH, 64291 Darmstadt, Germany; Institut für Kernphysik, Technische Universität Darmstadt, 64289 Darmstadt, Germany. E-mail: schwenk@physik.tu-darmstadt.de

reactions [7, 8]. These results are consistent with estimates of 3N force contributions $\langle V_{3N} \rangle \sim 0.1\text{--}1\,\text{MeV}$ per triplet in few-nucleon systems [9]. However, this scaling with the number of triplets, which would suggest that 3N forces become more important than two-nucleon (NN) interactions with increasing particle number (independent of the density), has to break down for larger nuclei. This is because nuclear forces have a finite range R (pion-exchange or shorter), so that in larger systems the interactions of three particles are restricted to volumes $R^3 < L^3$, where L is the size of the nucleus. As a result, in nuclear matter 3N forces scale with the density n (and not the number of triplets to pairs) compared to two-nucleon interactions $\langle V_{3N} \rangle \sim n\, R^3 \, \langle V_{NN} \rangle$ (compare, e.g. the results in Ref. [10]).

For larger nuclei, three-body interactions can be classified according to a finite-density reference state. In Fermi liquid theory, this corresponds to the interacting ground state, which is often taken to be a core nucleus. While 3N force contributions to the energy of the core nucleus are important, at the level of accuracy of present calculations for medium-mass nuclei, there is no evidence for residual three-body interactions between *valence* nucleons. For example, a recent analysis concludes [11]: "So far, no evidence was found for the effects of three-body interactions on states of valence nucleons. In case where rather pure shell-model configurations were observed, states and energies were well determined by effective two-body interactions". Clearly this must depend on the number of particles in the core, because in light nuclei with only valence particles, three-body interactions are significant. In the framework of Fermi liquid theory, this suggests that three-quasiparticle interactions are small at low energies, both for excitations and when valence nucleons are added.

In this paper, we discuss the impact of three-quasiparticle interactions in normal Fermi systems. These questions have been touched on briefly in the literature [12, 13]. For example, Brandow writes: "The weakness of the apparent three-body effects is the essential content of the statement that a Fermi liquid may be viewed as a low-density gas of weakly interacting quasiparticles". Considering the developments for 3N forces, it is however important to revisit these issues. After an introduction to Fermi liquid theory, we show in Sec. 3 that the contributions of three-quasiparticle interactions to normal Fermi systems at low energies and temperatures are suppressed by n_q/n compared to two-body interactions, where n_q is the density of excited or added quasiparticles and n is the ground-state density, or by the corresponding ratio of particle numbers N_q/N. This explains why the shell model with effective two-body interactions works so well. It also enables us to estimate at which level residual three-body interactions are expected to contribute. These results demonstrate that there is a change from few-body systems and light nuclei to normal Fermi systems. For larger nuclei and nucleonic matter, the contributions from residual three-body interactions are small when the system is weakly excited (including excitations where valence nucleons are added), even if 3N forces are significant for the interacting ground state (the core nucleus).

We dedicate this paper to Gerry Brown on the occasion of his 85th birthday. Gerry is one of the pioneers in the theory and microscopic understanding of Fermi systems. We were extremely fortunate to start on many-body problems and Fermi liquid theory with him as a teacher and mentor when we were students in Stony Brook. Using Gerry's words in "Fly with eagles" [14], he has been an eagle for generations of nuclear theorists and certainly for both of us.

2. Fermi Liquid Theory as an Effective Theory

Much of our understanding of strongly interacting Fermi systems at low energies and temperatures goes back to the seminal work of Landau in the late fifties [15–17]. Landau was able to express macroscopic observables in terms of microscopic properties of the elementary excitations, the so-called quasiparticles, and their residual interactions. In order to illustrate Landau's arguments here, we consider a uniform system of non-relativistic spin-1/2 fermions at zero temperature.

Landau assumed that the low-energy, elementary excitations of the interacting system can be described by effective degrees of freedom, the quasiparticles. Due to translational invariance, the states of the uniform system are eigenstates of the momentum operator. The quasiparticles are much like single-particle states in the sense that for each momentum there is a well-defined quasiparticle energy.[a] Landau assumed that there is a one-to-one correspondence between the quasiparticles and the single-particle states of a free Fermi gas. For a superfluid system, this one-to-one correspondence does not exist, and Landau's theory must be suitably modified, as discussed by Larkin and Migdal [18] and Leggett [19].

[a]In general, a quasiparticle state is not an energy eigenstate, but rather a resonance with a non-zero width. For quasiparticles close to the Fermi surface, the width is small and the corresponding life-time is large; hence the quasiparticle concept is useful for time scales short compared to the quasiparticle life-time.

The one-to-one correspondence starts from a free Fermi gas consisting of N particles, where the ground state is given by a filled Fermi sphere in momentum space. The particle number density n and the ground-state energy E_0 are given by (with $\hbar = c = 1$)

$$n = \frac{1}{V} \sum_{\mathbf{p}\sigma} n^0_{\mathbf{p}\sigma} = \frac{k_{\mathrm{F}}^3}{3\pi^2} \quad \text{and} \quad E_0 = \sum_{\mathbf{p}\sigma} \frac{\mathbf{p}^2}{2m} n^0_{\mathbf{p}\sigma} = \frac{3}{5} \frac{k_{\mathrm{F}}^2}{2m} N \,, \tag{1}$$

where k_{F} denotes the Fermi momentum, V the volume, and $n^0_{\mathbf{p}\sigma} = \theta(k_{\mathrm{F}} - |\mathbf{p}|)$ is the Fermi-Dirac distribution function at zero temperature for particles with momentum $\mathbf{p}$, spin projection σ, and mass m. By adding particles or holes, the distribution function is changed by $\delta n_{\mathbf{p}\sigma} = n_{\mathbf{p}\sigma} - n^0_{\mathbf{p}\sigma}$, and the total energy of the system by

$$\delta E = E - E_0 = \sum_{\mathbf{p}\sigma} \frac{\mathbf{p}^2}{2m} \, \delta n_{\mathbf{p}\sigma} \,. \tag{2}$$

When a particle is added in the state $\mathbf{p}\sigma$, one has $\delta n_{\mathbf{p}\sigma} = 1$ and when a particle is removed (a hole is added) $\delta n_{\mathbf{p}\sigma} = -1$.

In the interacting system the corresponding state is one with a quasiparticle added or removed, and the change in energy is given by

$$\delta E = \sum_{\mathbf{p}\sigma} \varepsilon_{\mathbf{p}\sigma} \, \delta n_{\mathbf{p}\sigma} \,, \tag{3}$$

where $\varepsilon_{\mathbf{p}\sigma} = \delta E / \delta n_{\mathbf{p}\sigma}$ denotes the quasiparticle energy. When two or more quasiparticles are added to the system, an additional term takes into account the interaction between the quasiparticles:

$$\delta E = \sum_{\mathbf{p}\sigma} \varepsilon^0_{\mathbf{p}\sigma} \, \delta n_{\mathbf{p}\sigma} + \frac{1}{2V} \sum_{\mathbf{p}_1\sigma_1, \mathbf{p}_2\sigma_2} f_{\mathbf{p}_1\sigma_1\mathbf{p}_2\sigma_2} \, \delta n_{\mathbf{p}_1\sigma_1} \, \delta n_{\mathbf{p}_2\sigma_2} \,. \tag{4}$$

Here $\varepsilon^0_{\mathbf{p}\sigma}$ is the quasiparticle energy in the ground state. In Sec. 3, we will show that the expansion in δn is general and does not require weak interactions. The small expansion parameter in Fermi liquid theory is the density of quasiparticles, or equivalently the excitation energy, and not the strength of the interaction. This allows a systematic treatment of strongly interacting systems at low temperatures.

In normal Fermi systems, the quasiparticle concept makes sense only for states close to the Fermi surface, where the quasiparticle life-time $\tau_{\mathbf{p}} \sim (p - k_{\mathrm{F}})^2$ is long [20]. In particular, states deep in the Fermi sea, which are occupied in the ground-state distribution, do not correspond to well-defined quasiparticles. Accordingly, we refer to the interacting ground state that corresponds to a filled Fermi sea in the non-interacting system as a state with no quasiparticles. In a weakly excited state the quasiparticle distribution $\delta n_{\mathbf{p}\sigma}$ is generally non-zero only for states close to the Fermi surface.

The second term in Eq. (4), the quasiparticle interaction $f_{\mathbf{p}_1\sigma_1\mathbf{p}_2\sigma_2}$, has no correspondence in the non-interacting Fermi gas. In an excited state with more than one quasiparticle, the quasiparticle energy is modified according to

$$\varepsilon_{\mathbf{p}\sigma} = \frac{\delta E}{\delta n_{\mathbf{p}\sigma}} = \varepsilon^0_{\mathbf{p}\sigma} + \frac{1}{V} \sum_{\mathbf{p}_2\sigma_2} f_{\mathbf{p}\sigma\mathbf{p}_2\sigma_2}\, \delta n_{\mathbf{p}_2\sigma_2} \,, \tag{5}$$

where the changes are effectively proportional to the quasiparticle density.

The quasiparticle interaction can be understood microscopically from the second variation of the energy with respect to the quasiparticle distribution,

$$f_{\mathbf{p}_1\sigma_1\mathbf{p}_2\sigma_2} = V\, \frac{\delta^2 E}{\delta n_{\mathbf{p}_1\sigma_1} \delta n_{\mathbf{p}_2\sigma_2}} = V\, \frac{\delta \varepsilon_{\mathbf{p}_1\sigma_1}}{\delta n_{\mathbf{p}_2\sigma_2}} \,. \tag{6}$$

Diagrammatically, this variation corresponds to cutting one of the fermion lines in a given energy diagram and labeling the incoming and outgoing fermion by $\mathbf{p}_1\sigma_1$, followed by a second variation leading to $\mathbf{p}_2\sigma_2$. For the uniform system, the resulting contributions to $f_{\mathbf{p}_1\sigma_1\mathbf{p}_2\sigma_2}$ are quasiparticle reducible in the particle-particle and in the exchange particle-hole (induced interaction) channels, but irreducible in the direct particle-hole (zero sound) channel [21–23]. The zero-sound-channel reducible diagrams are generated by the particle-hole scattering equation [17]. With Babu, one of Gerry's seminal contributions to Fermi liquid theory was to derive an integral equation that self-consistently takes into account induced interactions due to the polarization of the medium [23]. The Babu-Brown induced interaction is still one of the few non-perturbative approaches for calculating Fermi liquid parameters that have been implemented in practice.

Landau's theory of normal Fermi liquids is an effective low-energy theory in the modern sense [24, 25]. The effective theory incorporates the symmetries of the system and the low-energy couplings can be fixed by experiment or calculated microscopically based on the underlying theory. In an isotropic and spin-saturated system, such as liquid ^{3}He, the quasiparticle interaction can be decomposed as

$$f_{\mathbf{p}_1\sigma_1\mathbf{p}_2\sigma_2} = f^s_{\mathbf{p}_1\mathbf{p}_2} + f^a_{\mathbf{p}_1\mathbf{p}_2}\, \boldsymbol{\sigma}_1 \cdot \boldsymbol{\sigma}_2 \,, \tag{7}$$

which is the most general form consistent with $SU(2)$ spin symmetry.[b] For nuclear systems, the quasiparticle interaction includes additional terms that take into account the isospin dependence and non-central tensor contributions [26–28]. However, for our discussion here, the spin and isospin dependence is not important.

For the uniform system, Eq. (6) yields the quasiparticle interaction only for forward scattering (low momentum transfers). In the particle-hole channel, this corresponds to the long-wavelength limit. This restriction, which is consistent with considering low excitation energies, constrains the momenta $\mathbf{p}_1$ and $\mathbf{p}_2$ to be close to the Fermi surface, $|\mathbf{p}_1| = |\mathbf{p}_2| = k_{\rm F}$. The quasiparticle interaction then depends only on the angle between $\mathbf{p}_1$ and $\mathbf{p}_2$. It is convenient to expand this dependence on Legendre polynomials

$$f^{s/a}_{\mathbf{p}_1\mathbf{p}_2} = f^{s/a}(\cos\theta_{\mathbf{p}_1\mathbf{p}_2}) = \sum_l f^{s/a}_l \, P_l(\cos\theta_{\mathbf{p}_1\mathbf{p}_2}) \,, \tag{8}$$

and to define the dimensionless Landau Parameters $F^{s/a}_l$ by

$$F^{s/a}_l = N(0)\, f^{s/a}_l \,, \tag{9}$$

where $N(0) = \frac{1}{V}\sum_{\mathbf{p}\sigma} \delta(\varepsilon_{\mathbf{p}\sigma} - \mu) = m^* k_{\rm F}/\pi^2$ denotes the quasiparticle density of states at the Fermi surface.

The Landau parameters can be directly related to macroscopic properties of the system. Here we mention only the specific heat c_V, which at low temperature is determined by the effective mass given by F^s_1,

$$\frac{m^*}{m} = 1 + \frac{F^s_1}{3} \,, \tag{10}$$

$$c_V = \frac{m^* k_{\rm F}}{3}\, k_B^2 T \,, \tag{11}$$

and the incompressibility κ, which is related to F^s_0,

$$\kappa = -\frac{9V}{n}\frac{\partial P}{\partial V} = \frac{3k_{\rm F}^2}{m^*}\,(1 + F^s_0) \,. \tag{12}$$

Fermi liquid theory has been very successful in describing low-temperature Fermi liquids, in particular liquid ^{3}He [20]. The first applications to nuclear

[b] In nuclear physics the notation $f_{\mathbf{p}_1\mathbf{p}_2} = f^s_{\mathbf{p}_1\mathbf{p}_2}$ and $g_{\mathbf{p}_1\mathbf{p}_2} = f^a_{\mathbf{p}_1\mathbf{p}_2}$ is generally used.

systems were pioneered by Migdal [26] and first microscopic calculations for nuclei and nuclear matter by Gerry and collaborators (for a review, see Ref. [27]). Recently, advances using renormalization group (RG) methods for nuclear forces [29] and Fermi systems [24] have lead to the development of a non-perturbative RG approach for nucleonic matter [30], to a first complete study of the spin structure of induced interactions [28], and to new calculations of Fermi liquid parameters [31, 32].

3. Three-Quasiparticle Interactions

In Sec. 2, we introduced Fermi liquid theory as an expansion in the density of quasiparticles $\delta n/V$. In applications of Fermi liquid theory to date, even for liquid ^{3}He, which is a very dense and strongly interacting system, this expansion is truncated after the second-order $(\delta n)^2$ term, including only pairwise interactions of quasiparticles. However, for a strongly interacting system, there is a priori no reason that three-body (or higher-body) interactions between quasiparticles are small. In this section, we discuss the convergence of this expansion. Three-quasiparticle interactions arise from iterated two-body forces, leading to three- and higher-body clusters in the linked-cluster expansion, or through many-body forces. While three-body forces play an important role in nuclear physics [6], little is known about them in other Fermi liquids. Nevertheless, in strongly interacting systems, the contributions of many-body clusters can in general be significant, leading to potentially important $(\delta n)^3$ terms in the Fermi liquid expansion, also in the absence of three-body forces:

$$\delta E = \sum_1 \varepsilon_1^0 \, \delta n_1 + \frac{1}{2V} \sum_{1,2} f_{1,2}^{(2)} \, \delta n_1 \, \delta n_2 + \frac{1}{6V^2} \sum_{1,2,3} f_{1,2,3}^{(3)} \, \delta n_1 \, \delta n_2 \, \delta n_3 \, . \tag{13}$$

Here $f^{(n)}_{1,\ldots,n}$ denotes the n-quasiparticle interaction (the Landau interaction is $f \equiv f^{(2)}$) and we have introduced the short-hand notation $n \equiv \mathbf{p}_n, \sigma_n$.

In order to better understand the expansion, Eq. (13), around the interacting ground state with N fermions, consider exciting or adding N_q quasiparticles with $N_q \ll N$. The microscopic contributions from many-body clusters or from many-body forces can be grouped into diagrams containing zero, one, two, three, or more quasiparticle lines. The terms with zero quasiparticle lines contribute to the interacting ground state for $\delta n = 0$, whereas the terms with one, two, and three quasiparticle lines contribute to

ε_1^0, $f_{1,2}^{(2)}$, and $f_{1,2,3}^{(3)}$, respectively (these also depend on the ground-state density due to the N fermion lines). The terms with more than three quasiparticle lines would contribute to higher-quasiparticle interactions. Because a quasiparticle line replaces a line with N fermions when going from ε_1^0 to $f_{1,2}^{(2)}$, and from $f_{1,2}^{(2)}$ to $f_{1,2,3}^{(3)}$, it is intuitively clear that the contributions due to three-quasiparticle interactions are suppressed by N_q/N compared to two-quasiparticle interactions, and that the Fermi liquid expansion is effectively an expansion in N_q/N or n_q/n [33]. This will be discussed in detail and illustrated with examples in the following sections.

3.1. *General considerations*

Fermi liquid theory applies to normal Fermi systems at low energies and temperatures, or equivalently at low quasiparticle densities. We first consider excitations that conserve the net number of quasiparticles, $\delta N = \sum_{\mathbf{p}\sigma} \delta n_{\mathbf{p}\sigma} = 0$, so that the number of quasiparticles equals the number of quasiholes. This corresponds to the lowest energy excitations of normal Fermi liquids. We denote their energy scale by Δ. Excitations with one valence particle or quasiparticle added start from energies of order the chemical potential μ. In the case of $\delta N = 0$, the contributions of two-quasiparticle interactions are of the same order as the first-order δn term, but three-quasiparticle interactions are suppressed by Δ/μ [13]. This is the reason that Fermi liquid theory with only two-body Landau parameters is so successful in describing even strongly interacting and dense Fermi liquids. This counting is best seen from the variation of the free energy $F = E - \mu N$,

$$\begin{aligned} \delta F &= \delta(E - \mu N) \\ &= \sum_1 (\varepsilon_1^0 - \mu)\,\delta n_1 + \frac{1}{2V}\sum_{1,2} f_{1,2}^{(2)}\,\delta n_1\,\delta n_2 + \frac{1}{6V^2}\sum_{1,2,3} f_{1,2,3}^{(3)}\,\delta n_1\,\delta n_2\,\delta n_3\,, \end{aligned} \tag{14}$$

which for $\delta N = 0$ is equivalent to δE of Eq. (13). The quasiparticle distribution is $|\delta n_{\mathbf{p}\sigma}| \sim 1$ within a shell around the Fermi surface $|\varepsilon_{\mathbf{p}\sigma}^0 - \mu| \sim \Delta$. The first-order δn term is therefore proportional to Δ times the number of quasiparticles $\sum_{\mathbf{p}\sigma} |\delta n_{\mathbf{p}\sigma}| = N_q \sim N(\Delta/\mu)$,

$$\sum_1 (\varepsilon_1^0 - \mu)\,\delta n_1 \sim \frac{N\Delta^2}{\mu}\,. \tag{15}$$

Correspondingly, the contribution of two-quasiparticle interactions yields

$$\frac{1}{2V}\sum_{1,2} f^{(2)}_{1,2}\,\delta n_1\,\delta n_2 \sim \frac{1}{V}\,\langle f^{(2)}\rangle \left(\frac{N\Delta}{\mu}\right)^2 \sim \langle F^{(2)}\rangle\,\frac{N\Delta^2}{\mu}\,, \qquad (16)$$

where $\langle F^{(2)}\rangle = n\,\langle f^{(2)}\rangle/\mu$ is an average dimensionless coupling on the order of the Landau parameters. Even in the strongly interacting, scale-invariant case (see Sec. 3.2) $\langle f^{(2)}\rangle \sim 1/k_{\rm F}$; hence $\langle F^{(2)}\rangle \sim 1$ and the contribution of two-quasiparticle interactions is of the same order as the first-order term. However, the three-quasiparticle contribution is of order

$$\frac{1}{6V^2}\sum_{1,2,3} f^{(3)}_{1,2,3}\,\delta n_1\,\delta n_2\,\delta n_3 \sim \frac{n^2}{\mu}\,\langle f^{(3)}\rangle\,\frac{N\Delta^3}{\mu^2} \sim \langle F^{(3)}\rangle\,\frac{N\Delta^3}{\mu^2}\,. \qquad (17)$$

Therefore at low excitation energies this is suppressed by Δ/μ, compared to two-quasiparticle interactions, even if the dimensionless three-quasiparticle interaction $\langle F^{(3)}\rangle = n^2\langle f^{(3)}\rangle/\mu$ is strong (of order 1). Similarly, higher n-body interactions are suppressed by $(\Delta/\mu)^{n-2}$. Normal Fermi systems at low energies are weakly coupled in this sense. The small parameter is the ratio of the excitation energy per particle to the chemical potential. These considerations hold for *all* normal Fermi systems where the underlying interparticle interactions are finite range.

The Fermi liquid expansion in Δ/μ is equivalent to an expansion in $N_q/N \sim \Delta/\mu$, the ratio of the number of quasiparticles and quasiholes N_q to the number of particles N in the interacting ground state, or an expansion in the density of excited quasiparticles over the ground-state density, n_q/n.

We conclude this section with a discussion of the expansion for the energy δE given by Eq. (13), for the case where N_q quasiparticles or valence particles are added to a Fermi-liquid ground state. In this case, $\delta N \neq 0$ and the first-order term is

$$\sum_{1} \varepsilon^0_1\,\delta n_1 \sim \mu N_q \sim \mu\,\frac{N\Delta}{\mu} \sim N\Delta\,, \qquad (18)$$

while the contribution of two-quasiparticle interactions is suppressed by $N_q/N \sim \Delta/\mu$ and that of three-quasiparticle interactions by $(N_q/N)^2$. Therefore, either for $\delta N = 0$ or $\delta N \neq 0$, the contributions of three-quasiparticle interactions are suppressed for normal Fermi systems at low excitation energies. In the following sections, we will illustrate this for polarons in strongly interacting spin-polarized Fermi gases and for valence neutrons in neutron-rich calcium isotopes.

3.2. *Strongly interacting spin-polarized Fermi gases*

Experiments with spin-polarized Fermi gases [34–38] enable a unique exploration of strongly interacting Fermi systems and universal properties. We consider a system with two spin states and large S-wave scattering length interactions. In the limit of extreme population imbalance, the physics is governed by a single spin-down fermion interacting strongly with the spin-up Fermi sea. This spin-down fermion forms a quasiparticle, the so-called Fermi polaron [39], with energy E_p and effective mass m_p^*. Polarons have been directly observed in cold atomic gases using rf spectroscopy [37].

For large scattering lengths, $1/a_{\rm s} = 0$, the polaron energy is universal (it depends only on the density of spin-up fermions): $E_p = \eta\, \varepsilon_{{\rm F}\uparrow}$, where $\varepsilon_{{\rm F}\uparrow} = (6\pi^2 n_\uparrow)^{2/3}/(2m)$ is the spin-up Fermi energy [40]. The polaron binding $\eta = -0.615$ and effective mass $m_p^*/m = 1.20(2)$ have been determined using Monte-Carlo methods [41] and are in excellent agreement with experiment [38]. The polaron energy constrains the energy gain for large asymmetries in the Fermi liquid phase of spin-polarized Fermi gases. This determines the critical polarization for superfluidity and sets limits on the phase diagram and the existence of partially-polarized phases [36, 40, 42].

We can expand the strongly interacting spin-polarized Fermi gas around the fully polarized system with $N_\uparrow$ particles by adding $N_q = N_\downarrow \ll N_\uparrow$ polarons. Following Eq. (13), the change in the energy density is given by [43]

$$\frac{\delta E_\downarrow}{V} = \varepsilon_1^0\, n_\downarrow + \frac{1}{2}\, f^{(2)}\, n_\downarrow^2 + \frac{1}{6}\, f^{(3)}\, n_\downarrow^3\,, \tag{19}$$

where ε_1^0 is the average quasiparticle energy with contributions from both the polaron binding and the kinetic energy (with the single-polaron effective mass),

$$\varepsilon_1^0 = \eta\, \varepsilon_{{\rm F}\uparrow} + \frac{3}{5}\, \frac{(6\pi^2 n_\downarrow)^{2/3}}{2m_p^*}\,. \tag{20}$$

The average two-quasiparticle interaction $f^{(2)}$ is due to induced interactions mediated by the spin-up Fermi sea [44] and scales with the Fermi energy and the density of spin-up fermions,

$$f^{(2)} = \frac{\varepsilon_{{\rm F}\uparrow}}{n_\uparrow}\, F^{(2)}\,, \tag{21}$$

and correspondingly for the three-quasiparticle interaction,

$$f^{(3)} = \frac{\varepsilon_{\mathrm{F}\uparrow}}{n_\uparrow^2} F^{(3)} . \tag{22}$$

For large scattering lengths, $1/a_{\mathrm{s}} = 0$, the only scale is set by the spin-up density, and therefore the average $F^{(2)}$ and $F^{(3)}$ are dimensionless constants. In general, the two- and three-quasiparticle interactions also depend on the angles between the quasiparticles close to the Fermi surface (in particular, the effective mass at finite polaron density is given by an appropriately defined $l = 1$ Landau parameter [45]), but for the general estimates here, we can consider an average interaction relevant for the energy contribution. If additional scales are significant, such as the effective range or other ranges R, $F^{(2)}$ and $F^{(3)}$ will depend on $R^3 n_\uparrow$. Monte-Carlo calculations of $F^{(2)}$ give $F^{(2)} = 6B/5 \approx 0.17$ ($B \approx 0.14$ in Ref. [46]), which is small (compared to the normal symmetric phase) due to the Pauli principle for spin-down fermions. These scaling results for large scattering lengths demonstrate nicely the suppression of three-quasiparticle contributions by $(F^{(3)}/F^{(2)})(n_\downarrow/n_\uparrow)$, in line with the results of the previous section.

3.3. *Neutron-rich nuclei*

Next we illustrate the suppression of three-quasiparticle terms for finite Fermi systems. We consider valence neutrons in neutron-rich calcium isotopes, where the interacting ground state is taken to be the ^{40}Ca core.[c] This is also an interesting system, because recent calculations (with empirical single-particle or quasiparticle energies) have shown that the dominant contributions from chiral 3N forces are due to interactions between two valence neutrons and one core nucleon [3]. This corresponds to the normal-ordered two-body part of 3N forces, which is enhanced by the number of core nucleons. In the language of Fermi liquid theory, these are 3N force contributions to the two-quasiparticle interaction.

As shown in Fig. 1, the ground-state energies of neutron-rich calcium isotopes are well reproduced with effective two-body interactions in the shell

[c]The properties of medium-mass (and heavier) nuclei are often also calculated in energy-density functional approaches, where particle-particle (pair) correlations are included by generalizing the ground state to a (particle-number-projected) superfluid ground state. Figure 1 shows that pairing effects responsible for the odd-even mass staggering are relatively weak in nuclei.

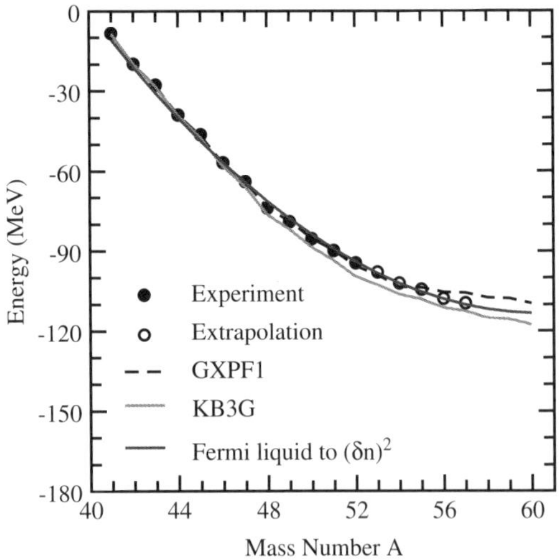

Fig. 1. Ground-state energies of calcium isotopes relative to ^{40}Ca as a function of mass number A, taken from the AME2003 atomic mass evaluation [49] based on experimental energies to ^{52}Ca and an extrapolation from ^{53}Ca to ^{58}Ca (the heaviest neutron-rich calcium isotope known to exist [50]). In addition, we show the energies obtained from shell-model calculations with phenomenological two-body interactions KB3G [47] and GXPF1 [48], and the fit using the Fermi liquid expansion, Eq. (23), to second order.

model [47, 48]. The differences between the phenomenological interactions are however amplified with increasing neutron number.

For a finite system, the Fermi liquid expansion, Eq. (13), is given by:

$$\delta E \sim \mu \left[N_q + \langle F^{(2)} \rangle \, \frac{N_q(N_q-1)}{2A} + \langle F^{(3)} \rangle \, \frac{N_q(N_q-1)(N_q-2)}{6A^2} \right] , \qquad (23)$$

where we have used $\varepsilon_1^0 \sim \mu$ and the dimensional scaling of $f_{12}^{(2)} \sim \mu/n$ and $f_{123}^{(3)} \sim \mu/n^2$ [see Eqs. (16) and (17)]. In this example, N_q is the number of valence neutrons and $A = 40$ the number of core nucleons. A fit to the ground-state energies of the AME2003 atomic mass evaluation [49] including only two-quasiparticle interactions yields $\mu = -10.8\,\text{MeV}$ and $\langle F^{(2)} \rangle = -2.0$. Overall a very good description of the energies is obtained, and the fit from ^{41}Ca to ^{58}Ca is not sensitive to three-quasiparticle interactions. The value for μ is consistent with typical single-particle energies in ^{41}Ca and $\langle F^{(2)} \rangle \sim 1$ is expected for nuclear interactions. We could improve the description further by accounting for the dependence of the two-body interaction on the single-particle orbitals (as in the shell model), instead of using an average $\langle F^{(2)} \rangle$.

Moreover, the Fermi liquid expansion provides an estimate of the contribution from three-quasiparticle interactions. With $\langle F^{(3)} \rangle \sim 1$, which is likely to be an overestimate, because three-neutron interactions are suppressed by the Pauli principle, the corresponding energy contributions to ^{52}Ca and ^{58}Ca ($N_q = 12$ and 18) are $\delta E \sim 1.5$ and 5.5 MeV. This is a factor 2 smaller than the spread of the shell-model results in Fig. 1. However, due to the uncertainty in the strength of $\langle F^{(3)} \rangle$, microscopic calculations or more global shell-model analyses of $\langle F^{(3)} \rangle$ are important to improve this estimate. Finally, the convergence of the Fermi liquid expansion is improved and the suppression of three-quasiparticle contributions is even more effective in heavier nuclei.

4. Concluding Remarks

We have shown that for normal Fermi systems at low excitation energies the contributions of three-quasiparticle interactions are suppressed by the ratio of the quasiparticle density to the ground-state density, or equivalently by the ratio of the excitation energy over the chemical potential. This holds for excitations that conserve the number of particles (excited states of the interacting ground state) as well as for excitations that add or remove valence particles. This suppression is general and applies to strongly interacting systems even with strong, but finite-range three-body forces. However, this does not imply that the contributions from 3N forces to the interacting ground-state energy (the energy of the core nucleus in the context of shell-model calculations), to quasiparticle energies (single-particle energies), or to two-quasiparticle interactions (effective two-body interactions) are small. The argument only applies to the effects of residual three-body interactions at low energies.

The Fermi liquid suppression of three-quasiparticle interactions can be tested in large-scale shell model calculations, and with advances in ab-initio methods for larger nuclei, in no-core shell model calculations with a core [51], in coupled-cluster theory [52] and with nuclear lattice simulations [53]. For interparticle interactions where a finite-density reference state is close to the interacting ground state, the Fermi liquid expansion also implies that normal-ordered three-body interactions are small. This can explain why, for low-momentum interactions [29], calculations of nuclei [54,55] and nucleonic matter [10,56] at the normal-ordered two-body level are so successful.

Acknowledgments

We are deeply grateful to Gerry for his very personal and continuous support and for the scientific inspiration in physics, through numerous discussions, and in the unique atmosphere he created in Stony Brook. Without his tremendous investment in people, we would not be where we are today.

We also thank R. J. Furnstahl and C. J. Pethick for useful discussions on these topics. This work was supported in part by the Helmholtz Alliance Program of the Helmholtz Association, contract HA216/EMMI "Extremes of Density and Temperature: Cosmic Matter in the Laboratory" and the DFG through grant SFB 634.

References

[1] T. Otsuka, T. Suzuki, J.D. Holt, A. Schwenk and Y. Akaishi, *Phys. Rev. Lett.* **105**, 032501 (2010).
[2] G. Hagen, T. Papenbrock, D.J. Dean, M. Hjorth-Jensen and B. Velamur Asokan, *Phys. Rev. C* **80**, 021306 (2009).
[3] J.D. Holt, T. Otsuka, A. Schwenk and T. Suzuki, arXiv:1009.5984.
[4] K. Hebeler, J.M. Lattimer, C.J. Pethick and A. Schwenk, *Phys. Rev. Lett.* **105**, 161102 (2010).
[5] S. Gandolfi, J. Carlson and S. Reddy, arXiv:1101.1921.
[6] E. Epelbaum, H.-W. Hammer and U.-G. Meißner, *Rev. Mod. Phys.* **81**, 1773 (2009).
[7] S.C. Pieper, *Riv. Nuovo Cim.* **031**, 709 (2008).
[8] P. Navrátil, S. Quaglioni, I. Stetcu and B.R. Barrett, *J. Phys. G* **36**, 083101 (2009).
[9] J.L. Friar, "Nuclear Scales" in *Nuclear Physics with Effective Field Theory*, Eds. R. Seki, U. van Kolck and M.J. Savage (World Scientific, Singapore, 1998); nucl-th/9804010.
[10] S.K. Bogner, A. Schwenk, R.J. Furnstahl and A. Nogga, *Nucl. Phys. A* **763**, 59 (2005); K. Hebeler, S.K. Bogner, R.J. Furnstahl, A. Nogga and A. Schwenk, *Phys. Rev. C* **83**, 031301(R) (2011).
[11] P. van Isacker and I. Talmi, *Europhys. Lett.* **90**, 32001 (2010).
[12] B.H. Brandow, *Rev. Mod. Phys.* **39**, 771 (1967).
[13] C.J. Pethick, "Selected Topics in the Theory of Normal Fermi Liquids", in *Lectures in Theoretical Physics*, Eds. K.T. Mahantappa and W.E. Brittin (Gordon and Breach, New York, 1969), p. 187.
[14] G.E. Brown, *Annu. Rev. Nucl. Part. Sci.* **51**, 1 (2001).
[15] L.D. Landau, *Sov. Phys. JETP* **3**, 920 (1957).
[16] L.D. Landau, *Sov. Phys. JETP* **5**, 101 (1957).
[17] L.D. Landau, *Sov. Phys. JETP* **8**, 70 (1959).
[18] A. Larkin and A.B. Migdal, *Sov. Phys. JETP* **17**, 1146 (1963).
[19] A.J. Leggett, *Phys. Rev. A* **140**, 1869 (1965); *ibid.* **147**, 119 (1966).

[20] G. Baym and C.J. Pethick, *Landau Fermi Liquid Theory: Concepts and Applications* (Wiley, New York, 1991).
[21] A.A. Abrikosov and I.M. Khalatnikov, *Rept. Prog. Phys.* **22**, 329 (1959).
[22] G.E. Brown, *Rev. Mod. Phys.* **43**, 1 (1971).
[23] S. Babu and G.E. *Brown, Ann. Phys.* **78** (1973) 1.
[24] R. Shankar, *Rev. Mod. Phys.* **66**, 129 (1994).
[25] J. Polchinski, "Effective Field Theory and the Fermi Surface", in *Proceedings of the 1992 Theoretical Advanced Studies Institute in Elementary Particle Physics*, Eds. J. Harvey and J. Polchinski (World Scientific, Singapore, 1993), hep-th/9210046.
[26] A.B. Migdal, *Theory of Finite Fermi Systems and Applications to Atomic Nuclei* (Interscience, New York, 1967).
[27] S.-O. Bäckman, G.E. Brown and J. Niskanen, *Phys. Rept.* **124**, 1 (1985).
[28] A. Schwenk and B. Friman, *Phys. Rev. Lett.* **92**, 082501 (2004).
[29] S.K. Bogner, R.J. Furnstahl and A. Schwenk, *Prog. Part. Nucl. Phys.* **65**, 94 (2010).
[30] A. Schwenk, B. Friman and G.E. Brown, *Nucl. Phys. A* **713**, 191 (2003).
[31] A. Schwenk, G.E. Brown and B. Friman, *Nucl. Phys. A* **703**, 745 (2002).
[32] N. Kaiser, *Nucl. Phys. A* **768**, 99 (2006).
[33] D. Pines and P. Nozières, *The Theory of Quantum Liquids* (Volume 1, Advanced Book Classics, Westview Press, 1999).
[34] M.W. Zwierlein, A. Schirotzek, C.H. Schunck and W. Ketterle, *Science* **311**, 492 (2006); Y. Shin, M.W. Zwierlein, C.H. Schunck, A. Schirotzek and W. Ketterle, *Phys. Rev. Lett.* **97**, 030401 (2006).
[35] G.B. Partridge, W. Li, R.I. Kamar, Y.-a. Liao and R.G. Hulet, *Science* **311**, 503 (2006).
[36] Y.-i. Shin, C.H. Schunck, A. Schirotzek and W. Ketterle, *Nature* **451**, 689 (2008).
[37] A. Schirotzek, C.-H. Wu, A. Sommer and M. Zwierlein, *Phys. Rev. Lett.* **102**, 230402 (2009).
[38] S. Nascimbène, N. Navon, K.J. Jiang, F. Chevy and C. Salomon, *Nature* **463**, 1057 (2010).
[39] F. Chevy and C. Mora, *Rep. Prog. Phys.* **73**, 112401 (2010).
[40] F. Chevy, *Phys. Rev. A* **74**, 063628 (2006).
[41] N. Prokof'ev and B. Svistunov, *Phys. Rev. B* **77**, 020408(R) (2008); *ibid.* **77**, 125101 (2008).
[42] A. Bulgac and M.M. Forbes, *Phys. Rev. A* **75**, 031605(R) (2007).
[43] C. Lobo, A. Recati, S. Giorgini and S. Stringari, *Phys. Rev. Lett.* **97**, 200403 (2006).
[44] C. Mora and F. Chevy, *Phys. Rev. Lett.* **104**, 230402 (2010); Z. Yu, S. Zöllner and C.J. Pethick, *Phys. Rev. Lett.* **105**, 188901 (2010).
[45] O. Sjöberg, *Nucl. Phys. A* **265**, 511 (1976).
[46] S. Pilati and S. Giorgini, *Phys. Rev. Lett.* **100**, 030401 (2008).
[47] A. Poves, J. Sánchez-Solano, E. Caurier and F. Nowacki, *Nucl. Phys. A* **694**, 157 (2001).

[48] M. Honma, T. Otsuka, B.A. Brown and T. Mizusaki, *Phys. Rev. C* **69**, 034335 (2004).
[49] G. Audi, A.H. Wapstra and C. Thibault, *Nucl. Phys. A* **729**, 337 (2003).
[50] M. Langevin *et al.*, *Phys. Lett. B* **130**, 251 (1983); M. Bernas *et al.*, *Phys. Lett. B* **415**, 111 (1997); O.B. Tarasov *et al.*, *Phys. Rev. C* **80**, 034609 (2009).
[51] A.F. Lisetskiy, B.R. Barrett, M.K.G. Kruse, P. Navrátil, I. Stetcu and J.P. Vary, *Phys. Rev. C* **78**, 044302 (2008).
[52] G. Hagen, T. Papenbrock, D.J. Dean and M. Hjorth-Jensen, *Phys. Rev. C* **82**, 034330 (2010).
[53] E. Epelbaum, H. Krebs, D. Lee and U.-G. Meißner, *Eur. Phys. J. A* **45**, 335 (2010).
[54] G. Hagen, T. Papenbrock, D.J. Dean, A. Schwenk, A. Nogga, M. Włoch and P. Piecuch, *Phys. Rev. C* **76**, 034302 (2007).
[55] K. Tsukiyama, S.K. Bogner and A. Schwenk, arXiv:1006.3639.
[56] K. Hebeler and A. Schwenk, *Phys. Rev. C* **82**, 014314 (2010).

Chapter 7

Meson Assisted Strange Dibaryons

A. Gal*

The state of the art in dibaryons with strangeness is reviewed, including the K^-pp dibaryon which signals the onset of $\bar{K}$–nuclear binding. A new type of strange dibaryons is highlighted, where the primary binding mechanism is provided by strong p-wave pion interactions, as demonstrated by a quasibound $(I = \frac{3}{2}, J^P = 2^+)$ πYN dibaryon calculation.

1. Introduction

The quark model (QM) has been very successful in reproducing SU(3) flavor octet (8) and decuplet (10) baryon masses, and to a somewhat lesser extent also higher mass baryon resonances, in terms of three-quark ($3q$) configurations. It is remarkable then that subsequent predictions of $6q$ configurations of bound or quasibound dibaryons have not been found to be realized in Nature, and that todate there is not even one single unambiguously established dibaryon. This statement is challenged by recent indications from $np \to d\pi\pi$ reactions at CELSIUS-WASA of a resonance structure at $M_R \approx 2.36\,\text{GeV}$ and $\Gamma_R \approx 80\,\text{MeV}$ that could be interpreted as a $\Delta\Delta$ dibaryon bound by about 100 MeV, but still about 200 MeV above the $d\pi\pi$ threshold [1]. We note that quark cluster calculations for $L = 0$ $6q$ configurations [2] come up with only a weakly bound $(I, J^P) = (0, 3^+)$ $\Delta\Delta$ dibaryon which in terms of a NN configuration corresponds to a high-lying $^3D_3$ pn resonance.

In the strange sector, Jaffe's dibaryon H with strangeness $\mathcal{S} = -2$ and quantum numbers $(I, J^P) = (0, 0^+)$ which was predicted as a genuinely bound state well below the $\Lambda\Lambda$ threshold, perhaps the most cited ever prediction made for any dibaryon [3], has not been confirmed experimentally

*Racah Institute of Physics, The Hebrew University, 91904 Jerusalem, Israel.

in spite of several comprehensive searches [4]. Another equally ambitious early prediction was made by Goldman *et al.* [5], also using a variant of the MIT bag model, for $\mathcal{S} = -3$ dibaryons with (I, J^P) values $(\frac{1}{2}, 1^+)$ and $(\frac{1}{2}, 2^+)$, dominated by ΩN structure and lying below the $\Xi\Lambda$ threshold. More realistic quark cluster calculations by Oka *et al.* [6], applying resonating group methods, did not confirm Jaffe's deeply bound H, placing it just below the ΞN threshold as a resonance about 26 MeV above the $\Lambda\Lambda$ threshold. The underlying binding mechanism common to all of these orbital angular momentum $L = 0$ configurations is the color-magnetic gluon exchange interaction between quarks, a feature emphasized by Oka [7] who systematically studied $L = 0$ dibaryon configurations that could benefit from short-range attraction. Following earlier quark cluster calculations [2, 6], these calculations resulted in no strange dibaryon bound states, and for the ΩN-dominated $\mathcal{S} = -3$ bound-state configurations predicted in Ref. [5], in particular, only a $(I, J^P) = (\frac{1}{2}, 2^+)$ quasibound state resulted.

For strangeness $\mathcal{S} = -1$, the focus of this review, old $K^- d \to \pi^- \Lambda p$ data [8] suggested resonant Λp structures at the ΣN threshold and 10 MeV above it. However, a $(I, J^P) = (\frac{1}{2}, 1^+)$ ΣN quasibound state is not necessarily required in order to reproduce the general shape of the Λp spectrum, as shown by multichannel Faddeev calculations [9,10]. A cusp-like structure at the ΣN threshold region could arise from the particularly strong one pion exchange tensor interaction in the $\Lambda N - \Sigma N$ coupled-channel $^3S_1 - {}^3D_1$ partial waves. Several low-lying $L = 1$ ΛN resonances were predicted in singlet and triplet configurations in a QM study by Mulders *et al.* [11], but negative results, particularly for the singlet resonance, were reported in dedicated K^- initiated experiments [12, 13] near the ΣN threshold. YN invariant mass spectra taken in recent $pp \to K^+ X$ experiments at COSY [14], shown in Fig. 1, give evidence for a cusp behavior at the ΣN threshold, but no further imminent structure below or above. The portion of the spectrum below the kinematical range spanned in Fig. 1 is shown in detail in Fig. 2, countering earlier evidence from Saturne [16] for a dibaryon signal at $M_X = 2097$ MeV.

In the present paper, following a brief summary of dibaryon theoretical expectations, I introduce the notion of pion assisted dibaryons, $\pi BB'$. The idea is to enhance the binding of a $L = 0$ BB' configuration through the strong attraction provided by p-wave πB resonances. For the πNN system this scenario requires quantum numbers $(I = 2, J^P = 2^+)$ to maximize the attraction from each one of the $(3, 3)$ resonating πN subsystems, but then the NN subsystem must have $(I_{NN} = 1, J^P = 1^+)$ which is Pauli forbidden.

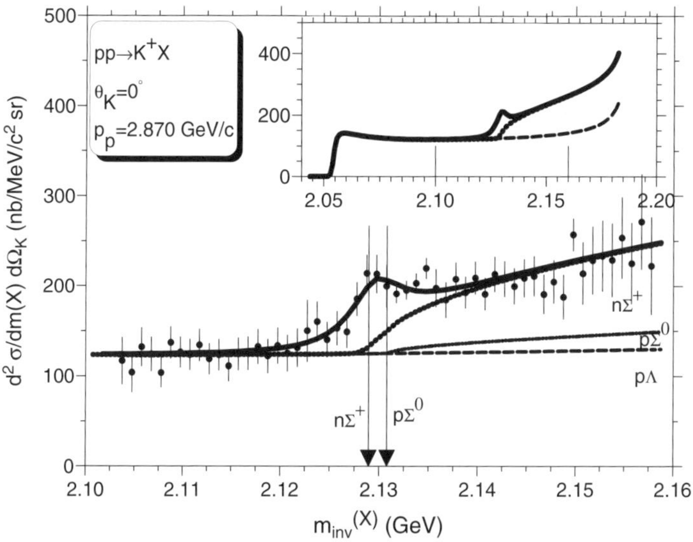

Fig. 1. YN invariant mass spectrum taken by the HIRES Colaboration at COSY [14] around the ΣN threshold. The solid curve is a combined fit to $pp \to K^+(\Lambda p, \Sigma^0 p, \Sigma^+ n)$, including a BW fit to the threshold cusp.

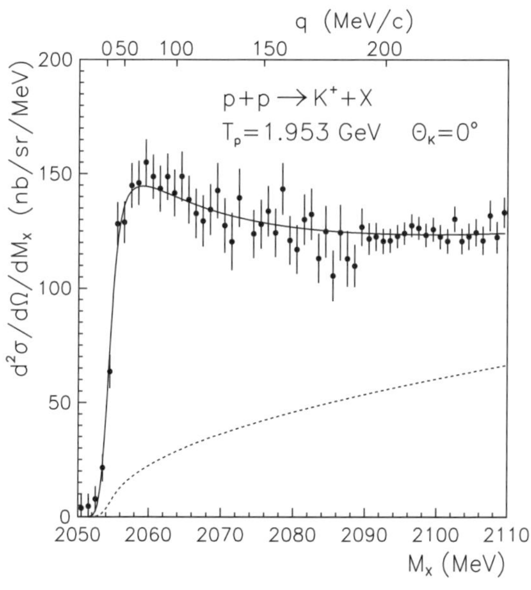

Fig. 2. ΛN invariant mass spectrum below the ΣN threshold taken by the HIRES Collaboration at COSY [15].

Hence this mechanism fails for $\mathcal{S}=0$, although $(I=2, J^P=2^+)$ is allowed for a πNN configuration with $(I_{NN}=1, {}^{2s+1}L={}^3P)$ and $\ell_\pi=0,2,4$, and also for a ΔN configuration, particularly in $L=0$. Dibaryons of this sort that cannot be composed of two nucleons were discussed long ago in the QM and termed "extraneous dibaryons" [17]. The lowest ones predicted are $(I=0, J^P=0^-,2^-)$ at $M\approx 2120\,\text{MeV}$, with the 0^- subsequently assigned to a πNN resonance suggested by observed anomalies in (π^+,π^-) reactions on nuclei at $T_\pi=50$ MeV $(M_{\pi NN}\sim 2065\,\text{MeV})$ [18]. Such a state can be realized for two nucleons in $(I_{NN}=1, {}^{2s+1}L_J={}^1S_0)$ plus an s-wave pion. There has been no solid experimental support for this dibaryon candidate ever since.

The limitations imposed by the Pauli principle in the $\mathcal{S}=0$ sector have no counterpart in the $\mathcal{S}=-1$ sector, where a $\pi\Lambda N$ stretched configuration $(I=\frac{3}{2}, J^P=2^+)$ is allowed. A possible quasibound state in this configuration has been recently studied [19, 20] by solving three-body Faddeev equations with ${}^3S_1-{}^3D_1$, $\Lambda N-\Sigma N$ coupled channels chiral QM local interactions, and coupled πY $(Y\equiv\Lambda,\Sigma)$ and πN separable p-wave interactions fitted to the position and decay parameters of the $\Sigma(1385)$ and $\Delta(1232)$ resonances, respectively. The results exhibit strong sensitivity to the p-wave πY interaction, the least phenomenologically constrained interaction in this calculation, with a $\pi\Lambda N$ quasibound state persisting over a wide range of acceptable parametrizations.

Finally, we briefly review the ongoing study, both theoretically and experimentally, of a more familiar example of meson assisted dibaryons: a K^-pp quasibound state in the $\mathcal{S}=-1$ sector, driven by the $\Lambda(1405)$ which within $\bar{K}N-\pi\Sigma$ coupled channels appears as a s-wave K^-p quasibound state (QBS). For a recent overview of K^-pp and its implications to $\bar{K}$–nuclear QBS phenomenology, see Ref. [21].

2. Deuteron-Like Dibaryon Candidates

Here we review BB' deuteron-like dibaryon candidate configurations made out of $8_{\rm f}$ baryons in which no explicit quark degrees of freedom are considered. A well known example is the $\mathcal{S}=0$ $(I=0, J^P=1^+)$ NN weakly bound deuteron. It is established experimentally that for $\mathcal{S}=-1$, the ΛN and ΣN interactions are too weak to provide binding. The associated scattering data have been used in several YN potential fits, respecting $\text{SU}(3)_{\rm f}$ within well defined symmetry breaking schemes, to make predictions for stranger BB' systems. There is almost general consensus,

Table 1. $\mathcal{S} = -2, -3, -4$ deuteron-like $L = 0$ dibaryon candidates from the Nijmegen meson exchange model NSC97 [22] and from EFT predictions [23]. Plus means yes, minus means no.

Strangeness	$\mathcal{S} = -2$		$\mathcal{S} = -3$		$\mathcal{S} = -4$
BB'	$\Sigma\Sigma$	$\Lambda\Xi$	$\Sigma\Xi$		$\Xi\Xi$
$(I, {}^{2s+1}[L=0]_j)$	$(2, {}^1S_0)$	$(\frac{1}{2}, {}^1S_0)$	$(\frac{3}{2}, {}^1S_0)$	$(\frac{3}{2}, {}^3S_1)$	$(1, {}^1S_0)$
NSC97	+	−	+	+	+
EFT	−	+	+	−	+

supported also by scarce $\Lambda\Lambda$ data, that there are no bound states for $\mathcal{S} = -2$, except for a $\Sigma\Sigma$ ($I = 2, J^P = 0^+$) quasibound state. However, for $\mathcal{S} = -3, -4$ there are several bound-state candidates, as listed in Table 1. The bound states listed in the table are due to two methodologies:

- The latest Nijmegen extended-soft-core (ESC) meson-exchange model, with broken $SU(3)_f$ for $\mathcal{S} = 0, -1, -2$, claims no quasibound states (QBS) [24] while providing no predictions yet for stranger systems. Earlier soft-core versions, e.g. NSC97 [22], found QBS for $\mathcal{S} = -3, -4$ as listed in Table 1. The three ${}^1S_0$ bound states in this model are in one-to-one correspondence with BB' states assigned in $SU(3)_f$ to the 27_f representation which includes the NN ${}^1S_0$ virtual state. Similarly, the ${}^3S_1$ bound state in this model, as listed in the table, is also the only BB' state which together with the deuteron is assigned in $SU(3)_f$ to the $\overline{10}_f$ representation.
- The Bonn-Jülich chiral effective field theory (EFT) model, applied in lowest order to $\mathcal{S} = -1, -2, -3, -4$ with low energy constants (LEC) constrained by $SU(3)_f$ and fitted to low-energy YN data [25] finds no QBS for $\mathcal{S} = -1$ and also for $\mathcal{S} = -2$ [26]. The $\mathcal{S} = -3, -4$ sectors require only the five LECs determined in the YN sector fit, independently of the sixth LEC required in the $\mathcal{S} = -2$ sector (this LEC is consistent with zero). This is how one gets predictions [23] for the $\mathcal{S} = -3, -4$ dibaryon candidates listed in Table 1. The predicted binding energies are all in the several MeV range, in agreement with what one expects for deuteron-like dibaryons where pion dynamics is dominant. The model dependence of these predictions is assessed within the model by varying a cutoff momentum in the range 550−700 MeV/c. Additional model dependence is likely to arise in next-to-leading-order evaluations. While the predictions of the EFT model differ for some

states and agree for other ones with those of the NSC97 model, both these models predict strong attraction for all the configurations listed in the table.

3. Six-Quark Dibaryon Configurations

Early discussions of $6q$ dibaryons were based on symmetry considerations related to the assumed dominance of a color-magnetic (CM) gluon exchange interaction

$$V_{\rm CM} = \sum_{i<j} -(\lambda_i \cdot \lambda_j)(s_i \cdot s_j)v(r_{ij}), \tag{1}$$

where λ_i and s_i are the color and spin operators of the i-th quark and $v(r_{ij})$ is a flavor-conserving short-ranged spatial interaction between quarks i, j. For $L = 0$ orbitally symmetric color-singlet n-quark cluster, the matrix element of $v(r_{ij})$ is independent of the particular i, j pair and is denoted $\mathcal{M}_0$, allowing for a closed form summation over i and j in Eq. (1) with the result:

$$\langle V_{\rm CM}\rangle = \left[-\frac{n(10-n)}{4} + \Delta\mathcal{P}_{\rm f} + \frac{s(s+1)}{3}\right]\mathcal{M}_0, \tag{2}$$

where $\mathcal{P}_{\rm f}$ sums over ± 1 for any symmetric/antisymmetric flavor pair, $\Delta\mathcal{P}_{\rm f}$ means with respect to the SU(3)$_{\rm f}$ 1 antisymmetric representation of n quarks, $n = 3$ for baryons and $n = 6$ for dibaryons, s is the total Pauli spin, and where $\mathcal{M}_0 \sim 75\,$MeV from the $\Delta - N$ mass difference. The leading $\mathcal{S} = 0, -1, -2, -3$ dibaryon candidates are listed in Table 2 following Ref. [7], where $\Delta\langle V_{\rm CM}\rangle$ stands for the CM interaction gain $\langle V_{\rm CM}\rangle_{6q} - \langle V_{\rm CM}\rangle_B - \langle V_{\rm CM}\rangle_{B'}$ in the $6q$ dibaryon configuration with respect to the sum of CM contributions from the separate B and B' $3q$ baryons that define the lowest BB' threshold. The table shows clearly the prominence of the $\mathcal{S} = -2\, H$ dibaryon.

More realistic $6q$ calculations, e.g. Refs. [7, 27], employ quark-cluster models (QCM) that break SU(3)$_{\rm f}$ and account for full quark antisymmetrization, also making contact via resonating group methods (RGM) with related BB' coupled channels and thresholds. Input interactions are shown in Fig. 3, taken from Ref. [28], where the left-hand side diagram corresponds to a gluon-exchange mediated BB' quark-exchange interaction and the right-hand side diagrams correspond to instanton induced interactions [29], a $3q$ as well as a nonstrange $2q$ version. Recent RGM quark calculations

Table 2. Leading $6q$ $L = 0$ dibaryon candidates [7], their BB' structure and the CM interaction gain with respect of the lowest BB' threshold calculated by means of Eq. (2). Asterisks are used for the 10_f baryons $\Sigma^* \equiv \Sigma(1385)$ and $\Xi^* \equiv \Xi(1530)$. The symbol [i,j,k] stands for the Young tablaux of the $SU(3)_f$ representation, with i arrays in the first row, j arrays in the second row and k arrays in the third row, from which $\mathcal{P}_f$ is evaluated.

$-\mathcal{S}$	$SU(3)_f$	I	J^π	BB' structure	$\frac{\Delta\langle V_{CM}\rangle}{M_0}$
0	[3,3,0] $\overline{10}$	0	3^+	$\Delta\Delta$	0
1	[3,2,1] 8	1/2	2^+	$\frac{1}{\sqrt{5}}(N\Xi^* + 2\Delta\Sigma)$	−1
2	[2,2,2] 1	0	0^+	$\frac{1}{\sqrt{8}}(\Lambda\Lambda + 2N\Xi - \sqrt{3}\Sigma\Sigma)$	−2
3	[3,2,1] 8	1/2	2^+	$\frac{1}{\sqrt{5}}(\sqrt{2}N\Omega - \Lambda\Xi^* + \Sigma^*\Xi - \Sigma\Xi^*)$	−1

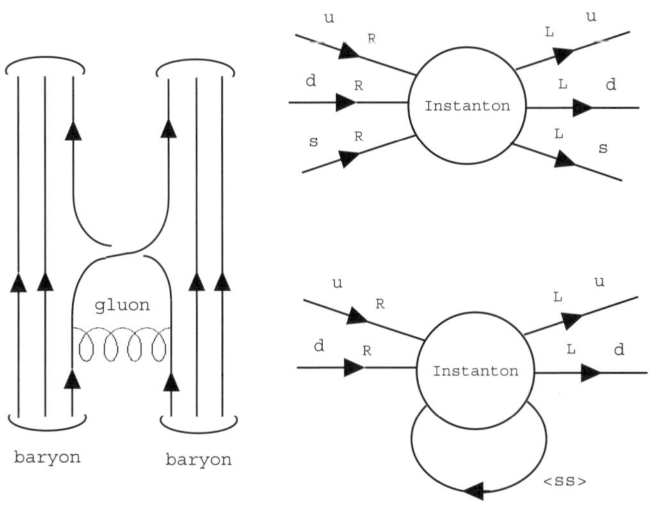

Fig. 3. Quark interaction diagrams for $6q$ dibaryon calculations [7, 28]. Left: one gluon exchange accompanied by quark exchange. Right: instanton contributions.

by Fujiwara and collaborators [30] do not use the instanton interactions, but add several meson exchanges to the gluon exchange interaction. None of these models produces a stable $6q$ dibaryon. This may be demonstrated for the H within several frameworks, including the QCM, as follows:

- In the QCM, owing to the repulsive instanton induced interaction, the H becomes barely bound or unbound, perhaps a $\Lambda\Lambda$ resonance below the $N\Xi$ threshold [27].

- In QCD sum rule calculations, where m_H is correlated with m_{nn}, the H might be bound, although with a large uncertainty, $B_H = 40 \pm 70$ MeV [31].
- Particle stability of ${}_{\Lambda\Lambda}^{\ 6}$He rules out $B_H > 7\,\mathrm{MeV}$, otherwise ${}_{\Lambda\Lambda}^{\ 6}$He would have disintegrated into $H + {}^4$He, at odds with its observed weak decay.

4. Pion Assisted Dibaryons

Here we sketch the recent calculation [20] of a $(I, J^P) = (\frac{3}{2}, 2^+)$ $\pi\Lambda N$ quasibound state, driven by the two-body πN resonance $\Delta(1232)$ and the $\pi\Lambda - \pi\Sigma$ resonance $\Sigma(1385)$, for a $\Lambda N - \Sigma N$ ${}^3S_1 - {}^3D_1$ coupled channels configuration. The YN interaction was taken from a chiral QM calculation, with parameters constrained by acceptable low-energy parameters [32]. The πB interactions were taken in a rank-one separable form

$$\langle p \,|\, V_{\pi B, \pi B'} \,|\, p' \rangle = (\gamma_B \gamma'_B)^{\frac{1}{2}} g_{\pi B}(p) g_{\pi B'}(p'). \tag{3}$$

The $\pi N\Delta(1232)$ p-wave form factor $g_{\pi N}$ was obtained from a very good fit of the P_{33} phase shift to a functional form with three well constrained parameters (in addition to $\gamma_{\pi N}$):

$$g_{\pi N}(p) = p[e^{-p^2/\beta^2} + Ap^2 e^{-p^2/\alpha^2}], \tag{4}$$

$$A = 0.2\,\mathrm{fm}^2, \quad \beta = 1.31\,\mathrm{fm}^{-1}, \quad \alpha = 3.21\,\mathrm{fm}^{-1}. \tag{5}$$

The r.m.s. momentum associated with $g_{\pi N}(p)$ is $\langle p^2_{\pi N} \rangle^{\frac{1}{2}} = 5.55\,\mathrm{fm}^{-1} = 1095\,\mathrm{MeV/c}$, where

$$\langle p^2_{\pi N} \rangle = \frac{\int_0^\infty g_{\pi N}(p)\, p^2\, d^3p}{\int_0^\infty g_{\pi N}(p)\, d^3p}. \tag{6}$$

This high-momentum value does not rule out a spatial size of order 1 fm for $\Delta(1232)$. Indeed, if $\tilde{g}_{\pi N}(r)$ is the Fourier transform of $g_{\pi N}(p)$, for $\ell = 1$, then $\langle r^2_{\pi N} \rangle^{\frac{1}{2}} = 0.875\,\mathrm{fm}$. The $\pi Y\Sigma(1385)$ coupled-channel p-wave form factor was fitted to the position, width and decay branching ratios of $\Sigma(1385)$, using the form

$$g_{\pi Y}(p) = p(1 + Ap^2)e^{-p^2/\alpha^2}, \tag{7}$$

leaving room for gridding over one of four fit parameters which was chosen to be A. The sensitivity of the quasibound calculation to values of A in the range 1.0–1.8 fm^2 is demonstrated by the last four columns in Table 3. It is by far the strongest sensitivity in this calculation, remarkably so for a

Table 3. $\pi\Lambda N$ binding energy (in MeV) calculated [20] for four πY interaction models specified by $\langle p_{\pi Y}^2\rangle^{\frac{1}{2}}$ and six chiral QM versions of the ${}^3S_1 - {}^3D_1$ YN interaction fitted to given ΛN scattering length a and effective range r_0. The momentum $p_{\rm lab}(\delta = 0)$ is the Λ laboratory momentum where the ${}^3S_1$ ΛN phase shift changes sign. The last row corresponds to switching off the YN interaction.

YN interaction			$\langle p_{\pi Y}^2\rangle^{\frac{1}{2}}$ (fm^{-1})			
a (fm)	r_0 (fm)	$p_{\rm lab}(\delta = 0)$ (MeV/c)	3.91	3.76 $B_{\pi\Lambda N}$	3.60 (MeV)	3.48
−1.35	3.39	987	99	65	30	6
−1.40	3.32	1011	99	66	30	6
−1.64	3.09	1146	102	68	32	8
−1.71	3.03	1198	102	68	33	9
−1.78	2.98	1272	103	69	33	9
−1.86	2.93	1446	104	69	34	10
—	—	—	120	84	47	21

fairly small variation over the range of r.m.s. momenta associated with the p-wave form factor $g_{\pi Y}(p)$.

Irrespective of which πY model is chosen, the YN interaction always produces repulsion, thus lowering the calculated binding energy, as demonstrated in the last line of Table 3 which corresponds to switching off the YN interaction. This repulsive YN effect owes its origin to the high-momentum components of the πB form factors which within the three-body calculation highlight the short-range repulsive region of the YN interaction. All in all, the calculations described above leave sufficient room for a quasibound $S = -1$ dibaryon, here denoted $\mathcal{D}$, decaying to a d-wave $I = \frac{3}{2}$ ΣN scattering state and perhaps also to $(\pi\Lambda N)_{I=\frac{3}{2}}$ if it corresponds to $\pi\Sigma N$ quasibound state above the $\pi\Lambda N$ threshold. To search for $\mathcal{D}$, the following reactions are possible:

$$K^- + d \to \mathcal{D}^- + \pi^+, \quad \pi^- + d \to \mathcal{D}^- + K^+, \tag{8}$$

$$p + p \to \mathcal{D}^+ + K^+. \tag{9}$$

5. Kaon Assisted Dibaryons

The low-energy $\bar{K}N$ s-wave interaction is particularly strong in the $I = 0$ channel, leading to a $\bar{K}N - \pi\Sigma$ QBS, the $\Lambda(1405)$, nominally 27 MeV below the K^-p threshold and with a decay width of about 50 MeV to

Table 4. Calculated K^-pp binding energies (B_{K^-pp}), mesonic (Γ_m) & nonmesonic (Γ_{nm}) widths (in MeV).

	Single $\bar{K}N - \pi\Sigma$ QBS pole					Two poles
	Variational		Faddeev			Variational
	[33, 34]	[35]	[36]	[37]		[38]
B_{K^-pp}	48	40–80	50–70	44–58	9–16	17–23
Γ_m	61	40–85	90–110	34–40	34–46	40–70
Γ_{nm}	12	∼20				4–12

the $\pi\Sigma$ channel. The lightest $\bar{K}NN$ dibaryon configuration maximizing the strongly attractive $I=0$ $\bar{K}N$ interaction is $[\bar{K}(NN)_{I=1}]_{I=1/2,J^\pi=0^-}$, loosely denoted as K^-pp. Several few-body calculations of K^-pp are summarized in Table 4.

The table supports the expectation that the K^-pp system is bound, although there are marked differences between the values calculated for the binding energy B_{K^-pp}. The input to all of the listed calculations is constrained by requiring that the $\Lambda(1405)$ $\pi\Sigma$ resonance position, and as much as possible also its shape, are reproduced. The models that achieve it with one $\bar{K}N - \pi\Sigma$ QBS pole, necessarily at or in the immediate vicinity of $\sqrt{s} = 1405$ MeV, produce K^-pp binding in the range 40–80 MeV, considerably higher than the 10–20 MeV range in models with two $\bar{K}N-\pi\Sigma$ QBS poles. Among the listed calculations, only Refs. [37, 38] used chiral $\bar{K}N - \pi\Sigma$ models that produce two such poles, but in the first listed Faddeev calculation of Ref. [37] the energy dependence of the $\bar{K}N - \pi\Sigma$ coupled channels input was suppressed in favor of fixed threshold values, thus making it effectively a single-pole calculation.[a] The shallow QBS of the last two columns are primarily related to the pole position of the chiral $\bar{K}N$ amplitude which resonates at 1420 MeV (close to the upper of the two $\bar{K}N - \pi\Sigma$ poles) in chiral models, whereas in single-pole models the $\bar{K}N$ amplitude resonates necessarily at 1405 MeV. This correlation with the resonance behavior of the $\bar{K}N$ amplitude has been verified in the variational calculations of Ref. [35] and in the coupled-channel Faddeev study of Ref. [39]. A notable feature of the K^-pp single-pole coupled-channel

[a]The second listed Faddeev calculation of Ref. [37], in addition to a relatively shallow K^-pp QBS listed in the column before last, also produced a deeply lying and very broad K^-pp QBS (unlisted in the table) in the range of $B_{K^-pp} = (67–89)$ MeV and $\Gamma_m = (244–320)$ MeV.

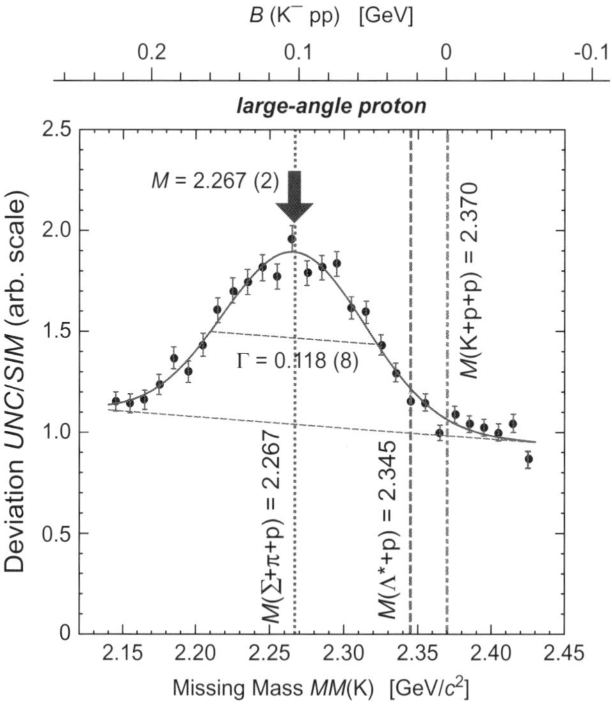

Fig. 4. DISTO reanalysis of K^+ missing-mass spectrum in $pp \to K^+\Lambda p$ at $T_p = 2.85\,\text{GeV}$ [41].

calculations [35, 36] in Table 4 is that the explicit use of the $\pi\Sigma N$ channel adds about $20 \pm 5\,\text{MeV}$ to the K^-pp binding energy with respect to that calculated using effective $\bar{K}N$ potential within a $\bar{K}NN$ single-channel calculation [40].

A reanalyzed DISTO spectrum of $pp \to K^+\Lambda p$ from Ref. [41] is shown in Fig. 4. The authors claim that the peak structure of the outgoing Λp gives evidence for a K^-pp quasibound state decaying via $K^-pp \to \Lambda p$. The location of the peak practically on top of the $\pi\Sigma N$ threshold, and its large width, are at odds with any of the few-body calculations listed in Table 4, posing a problem for a K^-pp QBS interpretation. At present, besides ongoing $p(p, K^+)$ measurements at GSI to improve on the DISTO data, the K^-pp system will be explored at J-PARC in the $^3\text{He}(K^-, n)$ and $d(\pi^+, K^+)$ reactions.

6. Summary and Outlook

In this Festschrift contribution I have reviewed the state of the art in dibaryons with strangeness, including K^-pp for which extensive experimental searches are underway. A new class of strange dibaryons which are termed meson assisted strange dibaryons was highlighted. K^-pp is just one example, where the strong s-wave K^-p interactions stabilize the initially unstable pp dibaryon. Pion assisted dibaryons $\pi BB'$ offer more possibilities by making use of the strong p-wave πB and $\pi B'$ resonances classified in the SU(3) $10_{\rm f}$ representation and its extensions into charm $\mathcal{C} \neq 0$. Possible candidates are as follows [19].

- $\mathcal{S} \neq 0$, $\mathcal{C} = 0$:

$$\mathcal{S} = -1\text{: } \pi\Lambda N, \quad \mathcal{S} = -2\text{: } \pi\Xi N, \quad \mathcal{S} = -3\text{: } \pi\Lambda\Xi, \tag{10}$$

- $\mathcal{C} = +1$:

$$\pi N\Lambda_c(2286), \quad \pi N\Xi_c(2470), \quad \pi N\Omega_c(2700), \tag{11}$$

$$\pi\Lambda\Lambda_c(2286), \quad \pi\Lambda\Xi_c(2470), \quad \pi\Lambda\Omega_c(2700), \tag{12}$$

$$\pi\Xi\Lambda_c(2286), \quad \pi\Xi\Xi_c(2470), \quad \pi\Xi\Omega_c(2700). \tag{13}$$

- $\mathcal{C} = +2$:

$$\pi\Lambda_c(2286)\Xi_c(2470), \quad \pi\Lambda_c(2286)\Omega_c(2700), \quad \pi\Xi_c(2470)\Omega_c(2700). \tag{14}$$

Note the appearance of the $\frac{1}{2}^+$ Ω_c baryon, of quark structure ssc. In the case of charmed baryons, the p-wave non-charmed SU(3)-$10_{\rm f}$ $\frac{3}{2}^+$ resonances are replaced by charmed SU(3)-$6_{\rm f}$ members of the SU(4)-$20_{\rm f}$ extension:

$$\Sigma(1385) \to \Sigma_c(2520), \quad \Xi(1530) \to \Xi_c(2645), \quad \Omega(1670) \to \Omega_c(2770). \tag{15}$$

Here we limited listing to $\mathcal{C} = +1$ baryons. For a future charmed bound-state study, note that the $\pi N\Lambda_c(2286)$ threshold lies *below* $N\Sigma_c(2455)$, where $\Sigma_c(2455)$ is the lowest lying known Σ_c, with assumed $J^P = \frac{1}{2}^+$. Therefore, if $\pi N\Lambda_c(2286)$ is bound, it will decay only by weak interactions.

Realistic calculations of pion assisted dibaryons have been reported so far only for the $\pi\Lambda N$ system with $(I = \frac{3}{2}, J^P = 2^+)$, where its coupling to the higher channel $\pi\Sigma N$ was considered while the coupling to the lower nonpionic channel ΣN was ignored [19,20]. This system is a good candidate for a quasibound dibaryon either below or above the $\pi\Lambda N$ threshold, but its precise location depends sensitively on the poorly known $\pi\Lambda\Sigma(1385) - \pi\Sigma\Sigma(1385)$ form factor. We note that the $\pi\Lambda N$ threshold is about 130 MeV below the $\Sigma(1385)N$ threshold and 230 MeV below the $\Sigma\Delta(1232)$ threshold, these latter thresholds being relevant in the discussion of the leading $6q$ $\mathcal{S} = -1$ dibaryon configuration of Table 2. In this respect a $\pi\Lambda N$ dibaryon, if established, is the next one in excitation energy to the lowest $\mathcal{S} = -1$ thresholds of ΛN and ΣN.

Acknowledgments

Gerry Brown has played a leading role in the development of nuclear physics and its derivatives. On this occasion of his 85th birthday Festschrift, I am pleased to acknowledge many inspiring suggestions he made to me throughout my own career. The work on pion assisted dibaryons is coauthored by Humberto Garcilazo who has contributed significantly to its development. I wish to also thank Wolfram Weise with whom I enjoyed several inspiring discussions on the K^-pp dibaryon and its implications to $\bar{K}$-nuclear physics. This research is partially supported by the EU Initiative FP7, HadronPhysics2, under Project 227431.

References

[1] M. Bashkanov *et al.* (CELSIUS/WASA Collaboration), *Phys. Rev. Lett.* **102**, 052301 (2009); see, however, T. Skorodko *et al.* (CELSIUS/WASA Collaboration), *Phys. Lett.* **B679**, 30 (2009) and *Phys. Lett.* **B695**, 115 (2011), and S. Dymov *et al.* (COSY-ANKE Collaboration), *Phys. Rev. Lett.* **102**, 192301 (2009).
[2] M. Oka and K. Yazaki, *Phys. Lett.* **B90**, 41 (1980).
[3] R.L. Jaffe, *Phys. Rev. Lett.* **38**, 195 (1977).
[4] B. Bassalleck, *Nucl. Phys.* **A639**, 401c (1998); B.A. Cole, *Nucl. Phys.* **A639**, 407c (1998); H.J. Crawford, *Nucl. Phys.* **A639**, 417c (1998).
[5] T. Goldman, K. Maltman, G.J. Stephenson, Jr., K.E. Schmidt and F. Wang, *Phys. Rev. Lett.* **59**, 627 (1987).
[6] M. Oka, K. Shimizu and K. Yazaki, *Phys. Lett.* **B130**, 365 (1983).
[7] M. Oka, *Phys. Rev. D* **38**, 298 (1988).
[8] T.H. Tan, *Phys. Rev. Lett.* **23**, 395 (1969).
[9] G. Toker, A. Gal and J.M. Eisenberg, *Phys. Lett.* **B88**, 235 (1979); *Nucl. Phys.* **A362**, 405 (1981).
[10] M. Torres, R.H. Dalitz and A. Deloff, *Phys. Lett.* **B174**, 213 (1986).
[11] P.J. Mulders, A.T. Aerts and J.J. de Swart, *Phys. Rev. D* **21**, 2653 (1980); A.T. Aerts and C.B. Dover, *Phys. Lett.* **B146**, 95 (1984).
[12] K. Johnston *et al.*, *Phys. Rev. C* **46**, R1573 (1992).

[13] R.E. Chrien, C.B. Dover and A. Gal, *Czech. J. Phys.* **42**, 1089 (1992).
[14] A. Budzanowski *et al.* (COSY-HIRES Collaboration), *Phys. Lett.* **B692**, 10 (2010).
[15] A. Budzanowski *et al.* (COSY-HIRES Collaboration), *Phys. Lett.* **B687**, 31 (2010).
[16] R. Siebert *et al.*, *Nucl. Phys.* **A567**, 819 (1994).
[17] P.J. Mulders, A.T. Aerts and J.J. de Swart, *Phys. Rev. Lett.* **40**, 1543 (1978).
[18] R. Bilger, H.A. Clement and M.G. Schepkin, *Phys. Rev. Lett.* **71**, 42 (1993) and **72**, 2972 (1994).
[19] A. Gal and H. Garcilazo, *Phys. Rev. D* **78**, 014013 (2008).
[20] H. Garcilazo and A. Gal, *Phys. Rev. C* **81**, 055205 (2010).
[21] W. Weise, *Nucl. Phys.* **A835**, 51 (2010).
[22] V.G.J. Stoks and Th.A. Rijken, *Phys. Rev. C* **59**, 3009 (1999).
[23] J. Haidenbauer and U.-G. Meißner, *Phys. Lett.* **B684**, 275 (2010).
[24] Th.A. Rijken, M.M. Nagels and Y. Yamamoto, *Nucl. Phys.* **A835**, 160 (2010); Th.A. Rijken and Y. Yamamoto, *Nucl. Phys.* **A804**, 51 (2008).
[25] H. Polinder, J. Haidenbauer and U.-G. Meißner, *Nucl. Phys.* **A779**, 244 (2006).
[26] H. Polinder, J. Haidenbauer and U.-G. Meißner, *Phys. Lett.* **B653**, 29 (2007).
[27] S. Takeuchi and M. Oka, *Phys. Rev. Lett.* **66**, 1271 (1991); M. Oka and S. Takeuchi, *Nucl. Phys.* **A524**, 649 (1991).
[28] M. Oka, *Hyperfine Interactions* **103**, 275 (1996) [arXiv:hep-ph/9510354].
[29] N.I. Kochelev, *Yad. Fiz.* **41**, 456 (1985) [*Sov. J. Nucl. Phys.* **41**, 291 (1985)].
[30] Y. Fujiwara, Y. Suzuki and C. Nakamoto, *Prog. Part. Nucl. Phys.* **58**, 439 (2007).
[31] N. Kodama, M. Oka and T. Hatsuda, *Nucl. Phys.* **A580**, 445 (1994).
[32] H. Garcilazo, T. Fernández-Carames and A. Valcarce, *Phys. Rev. C* **75**, 034002 (2007); H. Garcilazo, A. Valcarce and T. Fernández-Carames, *Phys. Rev. C* **76**, 034001 (2007).
[33] T. Yamazaki and Y. Akaishi, *Phys. Lett.* **B535**, 70 (2002).
[34] Y. Akaishi and T. Yamazaki, *Phys. Rev. C* **65**, 044005 (2002) [Eq. (29)].
[35] S. Wycech and A.M. Green, *Phys. Rev. C* **79**, 014001 (2009).
[36] N.V. Shevchenko, A. Gal and J. Mareš, *Phys. Rev. Lett.* **98**, 082301 (2007).
[37] Y. Ikeda, H. Kamano and T. Sato, *Prog. Theor. Phys.* **124**, 533 (2010).
[38] A. Doté, T. Hyodo and W. Weise, *Nucl. Phys.* **A804**, 197 (2008); *Phys. Rev. C* **79**, 014003 (2009).
[39] N.V. Shevchenko, A. Gal, J. Mareš and J. Révai, *Phys. Rev. C* **76**, 044004 (2007).
[40] Y. Ikeda and T. Sato, *Phys. Rev. C* **79**, 035201 (2009); see also Y. Ikeda and T. Sato, *Phys. Rev. C* **76**, 035203 (2007).
[41] T. Yamazaki *et al.*, *Phys. Rev. Lett.* **104**, 132502 (2010).

Chapter 8

Lattice Nuclear Force

T. Hatsuda*

We derive an energy-independent and non-local potential for the baryon-baryon interaction from the Nambu-Bethe-Salpeter amplitude on the lattice. The method is applied to the nucleon-nucleon interactions with the (2+1)-flavor lattice QCD simulations. The central and tensor potentials are obtained as leading order terms of the velocity expansion of the non-local potential. The central potential has a repulsive core surrounded by an attractive well, while the tensor potential has intermediate and long range attractions. Validity of the velocity expansion is tested by the nucleon-nucleon interaction with finite relative momentum. Interactions among octet-baryons in the flavor SU(3) limit are also studied in full QCD simulations to unravel the origin of the repulsive core.

1. Introduction

Understanding of the nuclear force from quantum chromodynamics (QCD) is one of the most challenging problems in nuclear and particle physics. Experimentally, a large number of proton-proton and neutron-proton scattering data as well as deuteron properties have been accumulated and summarized e.g. in the Nijmegen database [1]. Below the pion production threshold, the notion of the NN potential (either in the coordinate space or in the momentum space) is useful in the sense that it can be used not only to describe the two-body system but also to study the nuclear many-body problems through ab-initio calculations [2].

*Department of Physics, The University of Tokyo, Tokyo 113-0033, Japan and IPMU, The University of Tokyo, Chiba, 277-8583, Japan.

The phenomenological NN potentials in the coordinate space are known to reflect some characteristic features of the NN interaction [3]:

(i) The long range part of the nuclear force (the relative distance $r > 2\,\text{fm}$) is dominated by the one-pion exchange [4]. Because of the pion's Nambu-Goldstone character, it couples to the spin-isospin density of the nucleon and hence leads to the tensor force.
(ii) The medium range part ($1\,\text{fm} < r < 2\,\text{fm}$) receives significant contributions from the exchange of multi-pions. In particular, the spin-isospin independent attraction of about 50–100 MeV in this region plays an essential role to nuclear binding.
(iii) The short range part ($r < 1\,\text{fm}$) is best described by a strong repulsive core [5, 6]. Such a short range repulsion is relevant for the maximum mass of neutron stars.
(iv) There is also a strong attractive spin-orbit force in the isospin 1 channel at medium and short distances. This leads to the $^3P_2$ neutron pairing in neutron matter and hence the neutron superfluidity inside neutron stars [7].

Several high precision NN forces are now available to fit neutron-proton and proton-proton scattering data (about 4500 data points) with $\chi^2/\text{dof} \sim 1$. However, they have typically 20–40 fitting parameters: e.g. CD Bonn potential, AV18 potential and N^{3}LO chiral effective field theory have 38, 40, and 24 parameters, respectively [8]. If one tries to extend these to hyperon-nucleon and hyperon-hyperon interactions, the task becomes extremely tough since the number of parameters increase and the scattering data are scarce. In this situation, it is highly desirable to study the general baryon-baryon interactions from the first principle lattice QCD simulations, since all the hadronic interactions in QCD are controlled only by the QCD scale parameter (Λ_{QCD}) and the quark masses (m_u, m_d, m_s) whose values are pretty well determined [9].

A theoretical framework to study the hadron-hadron interaction using lattice QCD was first proposed by Lüscher [10]: For two hadrons in a finite box with a size $L \times L \times L$ in the periodic boundary condition, an exact relation between the energy spectra in the box and the elastic scattering phase shift at these energies was derived. If the range of the hadronic interaction R is sufficiently smaller than the size of the box $R < L/2$, the behavior of the equal-time Nambu-Bethe-Salpeter (NBS) amplitude $\psi(\boldsymbol{r})$ in the interval $R < |\boldsymbol{r}| < L/2$ under the periodic boundary condition has sufficient information to relate the phase shift and the two-particle spectrum. This Lüscher's method bypasses the difficulty to treat the real-time scattering process on the Euclidean lattice.

Recently, an alternative approach to the hadron interactions in lattice QCD was proposed [11–13]. The starting point is the same equal-time NBS amplitude $\psi(\boldsymbol{r})$: Instead of looking at the amplitude outside the range of the interaction, the internal region $|\boldsymbol{r}| < R$ is considered and an energy-independent non-local potential $U(\boldsymbol{r}, \boldsymbol{r}')$ is defined from $\psi(\boldsymbol{r})$. Since $U(\boldsymbol{r}, \boldsymbol{r}')$ in QCD is a localized function in space due to confinement of quarks and gluons, it receives finite volume effect only weakly. Therefore, once U, although it is not a direct physical observable, is determined on the lattice, one may simply use the Schrödinger equation in the infinite space to calculate observables such as the scattering phase shifts, bound state spectra etc. Moreover, the potential would be a smooth function of the quark masses: This is in sharp contrast to the scattering length which shows a singular behavior around the quark mass corresponding to the formation of the two-body bound states such as the deuteron. Similar situation is well-known in BEC-BCS crossover of cold fermionic atoms [14].

In this article, we will show some recent results of the nuclear force on the lattice (or the lattice nuclear force in short) after a brief introduction to our basic formulation.

2. Deriving the *NN* Potential on the Lattice

2.1. *NBS wave function on the lattice*

In field theory, the best analogue of the two-particle wave function is the equal-time Nambu-Bethe-Salpeter (NBS) amplitude or the "NBS wave function": Let us consider an exact six-quark state $|\mathcal{E}\rangle$ which has total energy $\mathcal{E}$, total three-momentum zero and total electric charge $+e$ in a finite box. Then we define the NBS wave function by

$$\psi(\boldsymbol{r}) = \langle 0 \,|\, n_\beta(\boldsymbol{x}+\boldsymbol{r}, t=0) p_\alpha(\boldsymbol{x}, t=0) \,|\, \mathcal{E}\rangle. \tag{1}$$

The local composite operators for the proton and the neutron are denoted by $p_\alpha(\boldsymbol{x}, t)$ and $n_\beta(\boldsymbol{y}, t)$ with spinor indices α and β. One should keep in mind that $|\mathcal{E}\rangle$ is *not* a simple superposition of a product state $|\mathrm{p}\rangle \otimes |\mathrm{n}\rangle$, since there are complicated exchanges of quarks and gluons between the two composite particles. The NBS wave function $\psi(\boldsymbol{r})$ can be regarded as a probability amplitude in $|\mathcal{E}\rangle$ to find "neutron-like" three-quarks located at point $\boldsymbol{x}+\boldsymbol{r}$ and "proton-like" three-quarks located at point $\boldsymbol{x}$.

The spatial extent of the NN interaction in QCD is short ranged and is exponentially suppressed beyond the distance $R \sim 2\,\mathrm{fm}$. Therefore, the spatial part of the NBS wave function in the "outer region" satisfies the Helmholtz equation,

$$(\nabla^2 + k^2)\psi(\boldsymbol{r}) = 0 \quad (|\boldsymbol{r}| > R), \tag{2}$$

up to an exponentially small correction in L. Here the "asymptotic momentum" k is determined by the asymptotic behavior of the wave function in the outer region.

An important property of the NBS wave function $\psi(\boldsymbol{r})$ is that its asymptotic behavior at large $|\boldsymbol{r}|$ in the infinite volume limit reproduces the correct phase shift obtained from the S-matrix of the elastic NN scattering. This can be shown explicitly by using the Nishijima-Zimmermann-Haag(NHZ)'s reduction formula [15] for the products of local composite operators. (See Appendix A of [13] for the proof.) To define the NBS wave function on the lattice, we start with the four-point function

$$\mathcal{G}(\boldsymbol{r}, t - t_0) = \langle 0\,|\, n_\beta(\boldsymbol{x}+\boldsymbol{r}, t) p_\alpha(\boldsymbol{x}, t) S(t_0)\,|\,0\rangle \to \psi(\boldsymbol{r})\, e^{-\mathcal{E}_0(t-t_0)} \;\; (t \gg t_0), \tag{3}$$

where $\mathcal{E}_0$ is the lowest energy state created by the source operator $S(t_0)$.

2.2. *Non-local potential and the velocity expansion*

To define the NN potential from the NBS wave function, let us introduce the following local function:

$$K_E(\boldsymbol{r}) = \frac{1}{2\mu}(\nabla^2 + k^2)\psi_{E}(\boldsymbol{r}) \equiv (E - H_0)\psi_{E}(\boldsymbol{r}). \tag{4}$$

In the second equality, we introduce an "effective center of mass energy", $E = k^2/(2\mu)$, and the free Hamiltonian $H_0 = -\nabla^2/(2\mu)$, with $\mu = m_N/2$ being the reduced mass of the two nucleons. They are introduced only to make a formal resemblance with the Schrödinger type equation and have nothing to do with non-relativistic approximation. Hereafter, we put the suffix E to the NBS wave function to emphasize its E-dependence. Since the "plane-wave" part of the NBS wave function in the outer region ($r > R$) is projected away by the operator $E - H_0$, the function $K_E(\boldsymbol{r})$ is non-vanishing only in the inner region ($r < R$). Note also that the Fourier transform of $K_E(\boldsymbol{r})$ is essentially the half off-shell T-matrix.

We can rewrite Eq. (4) in two equivalent ways:

$$(E - H_0)\psi_{_E}(\boldsymbol{r}) = U_E(\boldsymbol{r})\psi_{_E}(\boldsymbol{r}) = \int U(\boldsymbol{r}, \boldsymbol{r}')\psi_{_E}(\boldsymbol{r}')d\boldsymbol{r}'. \tag{5}$$

The first equality is just a definition of the energy-dependent local potential, $U_E(\boldsymbol{r}) = K_E(\boldsymbol{r})/\psi_{_E}(\boldsymbol{r})$. On the other hans, the energy-independent non-local potential, $U(\boldsymbol{r}, \boldsymbol{r}')$, is defined from $U_E(\boldsymbol{r})$ through a self-consistent equation,

$$U(\boldsymbol{r}, \boldsymbol{r}') = \langle \boldsymbol{r} \,|\, \hat{U} \,|\, \boldsymbol{r}' \rangle = \sum_E \int_{-\infty}^{+\infty} \frac{dt}{2\pi} U_E(\boldsymbol{r}) \langle \boldsymbol{r} \,|\, e^{i(\hat{H}_0 + \hat{U} - E)t} \,|\, \boldsymbol{r}' \rangle. \tag{6}$$

Carrying out the t integration formally, one may also write Eq. (6) as $\hat{U} = \sum_E \hat{U}_E \delta(E - \hat{H}_0 - \hat{U})$. In the above formulas, $\sum_E$ stands for the summation (integration) over the discrete (continuum) energies. In particular, E is always discrete on the lattice with a finite volume. Eliminating the E-dependence of the potential through Eq. (6) has been discussed in a transparent manner by Królikowski and Rzewuski [16] long time ago: their motivation was to prove the equivalence between the multiple-time Nambu-Bethe-Salpeter type equation with an E-dependent kernel and the equal-time Schrödinger type equation with an E-independent potential. Essentially the same method was rediscovered and discussed in [11–13] in the context of the NBS wave function on the lattice.

If we further focus on the low-energy scattering with E sufficiently smaller than the intrinsic scale of the system or the scale of the non-locality of the potential in Eq. (5), the velocity expansion of $U(\boldsymbol{r}, \boldsymbol{r}')$ in terms of its non-locality is useful [17]: For example, the potential with hermiticity, rotational invariance, parity symmetry, and time-reversal invariance may be expanded as [18]

$$U(\boldsymbol{r}, \boldsymbol{r}') = V(\boldsymbol{r}, \boldsymbol{v})\delta(\boldsymbol{r} - \boldsymbol{r}'), \tag{7}$$

$$V(\boldsymbol{r}, \boldsymbol{v}) = \underbrace{V_C(r) + V_T(r)S_{12}}_{\text{LO}} + \underbrace{V_{LS}(r)\boldsymbol{L}\cdot\boldsymbol{S}}_{\text{NLO}} + \underbrace{O(\boldsymbol{v}^2)}_{\text{N}^2\text{LO}} + \cdots, \tag{8}$$

where $\boldsymbol{v} = \boldsymbol{p}/\mu$ and $\boldsymbol{L} = \boldsymbol{r} \times \boldsymbol{p}$ with $\boldsymbol{p} = -i\nabla$, and $S_{12} = 3(\boldsymbol{\sigma}_1 \cdot \boldsymbol{r})(\boldsymbol{\sigma}_2 \cdot \boldsymbol{r})/r^2 - \boldsymbol{\sigma}_1 \cdot \boldsymbol{\sigma}_2$. Each coefficient of the expansion is a local potential and can be determined successively by measuring the NBS wave functions for several different energies. The central potential V_C and the tensor potential V_T are classified as the leading order (LO) potentials since they are of $O(\boldsymbol{v}^0)$. The next-to-leading (NLO) potential of $O(\boldsymbol{v})$ is the spin-orbit potential $V_{LS}(r)$. The LO and NLO potentials are phenomenologically known to be the dominant interactions governing the nucleon interactions at low energies.

An advantage of defining the potential from the NBS wave functions in the "inner region" is that the effect of the lattice boundary is exponentially suppressed for finite range interactions: Then one can first make appropriate extrapolation of $U(\boldsymbol{r}, \boldsymbol{r}')$ or $V(\boldsymbol{r}, \boldsymbol{v})$ to $L \to \infty$, and then solve the Schrödinger equation using the extrapolated potential to calculate the observables such as the phase shifts and binding energies in the infinite volume. This is in contrast to the Lüscher's approach [10] in which the wave functions in the "outer region" suffering from the boundary conditions is ingeniously utilized to probe the scattering observables. Apparently, the two approaches are the opposite sides of a same coin.

2.3. *Interpolating operator and the potential*

In Eq. (1), simplest interpolating operators for the neutron and the proton written in terms of the up-quark $u(x)$ and the down-quark $d(x)$ would be

$$n_\beta(x) = \varepsilon_{abc}(u_a(x)C\gamma_5 d_b(x))d_{c\beta}(x), \quad p_\alpha(x) = \varepsilon_{abc}(u_a(x)C\gamma_5 d_b(x))u_{c\alpha}(x), \tag{9}$$

where $x = (\boldsymbol{x}, t)$ and the color indices are denoted by a, b and c. The charge conjugation matrix in the spinor space is denoted by C. The local operators given above are most convenient for relating the NBS wave function to the four-point Green's function and the scattering observables at $L \to \infty$ through the NZH reduction formula.

In principle, one may choose any composite operators with the same quantum numbers as the nucleon to define the NBS wave function. Different interpolating operators lead to different NBS wave functions and different NN potentials. However, they lead to the same physical observables by construction. Analogous situation can be seen in quantum mechanics where the unitary transformations modify both the wave function and the potential in such a way that observables are unchanged. Even more direct analogy is in field theory for point-like particles: Field re-definition modifies the vertices and propagators in the Feynmann rule, while the on-shell S-matrix is not affected by such changes.

2.4. *Central and tensor forces*

In the LO of the velocity expansion in Eq. (8), we have the central potential $V_C(r)$ and the tensor potential $V_T(r)$, so that the Shrödinger equation reads

$$(E - H_0)\psi_E(\boldsymbol{r}) = (V_C(r) + V_T(r)S_{12})\psi_E(\boldsymbol{r}). \tag{10}$$

The central potential acts separately on the orbital S-state and the D-state, while the tensor potential provides a coupling between these two. Therefore, a coupled-channel Schrödinger equation is obtained from Eq. (10) by using the projection operators $\mathcal{P}$ and $\mathcal{Q}$ to the S-state and D-state, respectively. Eventually we calculate V_C and V_T from the following formula where the quantities in the right hand side are all known on the lattice:

$$\begin{pmatrix} V_C \\ V_T \end{pmatrix} = \begin{pmatrix} \mathcal{P}\psi_E & \mathcal{P}S_{12}\psi_E \\ \mathcal{Q}\psi_E & \mathcal{Q}S_{12}\psi_E \end{pmatrix}^{-1} \begin{pmatrix} E - H_0 & 0 \\ 0 & E - H_0 \end{pmatrix} \begin{pmatrix} \mathcal{P}\psi_E \\ \mathcal{Q}\psi_E \end{pmatrix}. \tag{11}$$

3. Numerical Results in Quenched and Full QCD Simulations

3.1. *LO potentials*

To demonstrate whether the formalism discussed in the previous section indeed works, we first carried out a quenched QCD simulations with the standard plaquette gauge action and the standard Wilson quark action on a $32^3 \times 48$ lattice [11–13]. The lattice spacing is $a = 0.137$ fm which corresponds to the spatial size $L = 4.4$ fm. The light quark masses are chosen so that we have $m_\pi = 731$, 529 and 380 MeV and $m_N = 1558$, 1334 and 1197 MeV, respectively. Periodic or anti-periodic boundary conditions are imposed on the quark field along the spatial direction.

As for full QCD with the vacuum polarization of quarks included, we use the PACS-CS gauge configurations in (2+1)-flavor QCD generated by the Iwasaki gauge action and the $O(a)$-improved Wilson quark (clover) action on a $32^3 \times 64$ lattice [20]. The lattice spacing is $a = 0.091$ fm which corresponds to the spatial size $L = 2.9$ fm. The light quark masses are chosen so that we have $m_\pi = 701$, 570 and 411 MeV and $m_N = 1583$, 1412 and 1215 MeV, respectively. Also, Periodic boundary condition is imposed on the quark field along the spatial direction.

Shown in Fig. 1(a) are the LO potentials (V_C for $^1S_0$ and $^3S_1$ channels and V_T determined from $^3S_1 - ^3D_1$ channel) in (2+1)-flavor QCD for $m_\pi =$ 701 MeV. Even with such a large quark mass, there is a clear evidence of

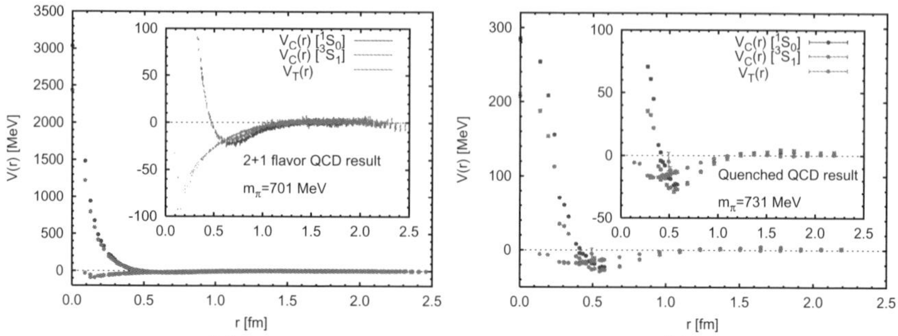

(a) (b)

Fig. 1. (a) LO potentials in (2+1)-flavor QCD for $m_\pi = 701\,\text{MeV}$ [21]. (b) LO potentials in quenched QCD for $m_\pi = 731\,\text{MeV}$ [13].

the repulsive core surrounded by attractive well for central potential and an evidence of a mild tensor force [21]. They have qualitative similarity with phenomenological potentials. We show in Fig. 1(b) the LO potentials in quenched QCD with $m_\pi = 731\,\text{MeV}$ for comparison [13]. Although the qualitative structure of the potentials are the same, the magnitude of the repulsive core and the tensor force are relatively weak in quenched QCD.

Shown in Figs. 2(a)–(c) are the quark mass dependence of the LO potentials in (2+1)-flavor QCD [21]. As the quark mass decreases, the repulsive core in (a,b) and the tensor force in (c) become stronger and the attractive well in (a,b) becomes wider. We have fitted these potentials and have calculated the NN scattering phase shift by solving the Schrödinger equation. We found that deuteron bound state does not appear for these quark masses, so that further reduction of the quark mass would be necessary to obtain the realistic lattice potentials.

3.2. *Convergence of the velocity expansion*

So far, the potentials are derived with the periodic boundary condition in the spatial direction for the quark fields. This leads to the the "effective center of mass energy" $E = k^2/(2\mu)$ almost zero. To study the convergence of the velocity expansion of the non-local potential in Eq. (8), we compare the local $^1S_0$ potential (in quenched QCD with $m_\pi = 529\,\text{MeV}$) obtained at $E \simeq 0\,\text{MeV}$ under the periodic boundary condition and that obtained at $E \simeq 45\,\text{MeV}$ under the anti-periodic boundary condition [22]. Good agreement between the two as shown in the left panel of Fig. 3 indicates that a $O(\boldsymbol{v}^2)$ term in the N^{2}LO level is rather small in this energy interval.

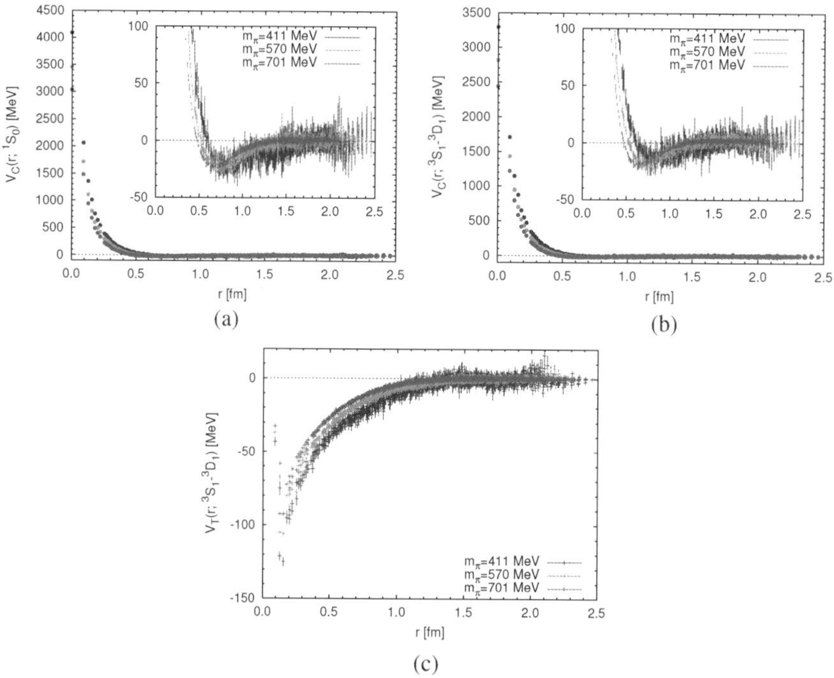

Fig. 2. Quark mass dependence of the LO potentials in (2+1)-flavor QCD. (a) The central potential in the spin-singlet channel, (b) the central potential in the spin-triplet channel, and (c) the tensor potential in the spin-triplet channel [21].

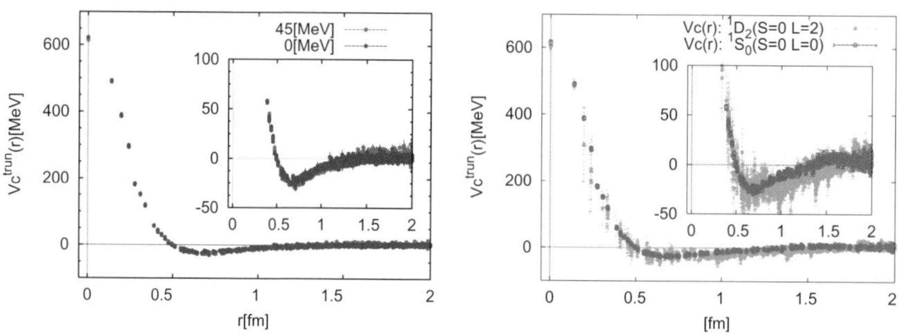

Fig. 3. (Left) A comparison of the ${}^1S_0$ central potentials obtained at different energies ($E \sim 0\,\mathrm{MeV}$ vs. $E \sim 45\,\mathrm{MeV}$). (Right) A comparison of the spin-singlet central potential at $E \sim 0\,\mathrm{MeV}$ with different orbital angular momenta ($L = 0$ vs. $L = 2$). Figures are taken from [22].

Shown in the right panel of Fig. 3 is a test for a different $O(\boldsymbol{v}^2)$ term in the N^{2}LO level [22]: In this case, local potentials determined at the same energy ($E \simeq 0\,\text{MeV}$) with different orbital angular momenta ($L = 0, 2$) in the spin-singlet channel are compared. Again, within statistical errors, the effect of the N^{2}LO term is likely to be small.

4. Hyperon Interactions

To study the origin of the repulsive core in the NN interaction, let us consider the S-wave interaction between octet baryons in the flavor SU(3) limit. In this case, two baryon states with a given angular momentum are labeled by the irreducible flavor multiplets as

$$\mathbf{8} \otimes \mathbf{8} = \underbrace{\mathbf{27} \oplus \mathbf{8}_s \oplus \mathbf{1}}_{\text{symmetric}} \oplus \underbrace{\mathbf{10}^* \oplus \mathbf{10} \oplus \mathbf{8}_a}_{\text{anti-symmetric}} . \tag{12}$$

Here "symmetric" and "anti-symmetric" stand for the symmetry under the flavor exchange of two baryons. For the system in the orbital S-wave, the Pauli principle between two baryons imposes $\mathbf{27}$, $\mathbf{8}_s$ and $\mathbf{1}$ to be spin singlet ($^1S_0$) while $\mathbf{10}^*$, $\mathbf{10}$ and $\mathbf{8}_a$ to be spin triplet ($^3S_1$). Since there are no mixings among different multiplets in the SU(3) limit, one can define the corresponding potentials as

$$^1S_0 : V^{(\mathbf{27})}(r), \quad V^{(\mathbf{8}_s)}(r), \quad V^{(\mathbf{1})}(r), \tag{13}$$

$$^3S_1 : V^{(\mathbf{10}^*)}(r), \quad V^{(\mathbf{10})}(r), \quad V^{(\mathbf{8}_a)}(r) . \tag{14}$$

Potentials among octet baryons, both the diagonal part ($B_1B_2 \to B_1B_2$) and the off-diagonal part ($B_1B_2 \to B_3B_4$), are obtained by suitable combinations of $V^{(\alpha)}(r)$ with $\alpha = \mathbf{27}, \mathbf{8}_s, \mathbf{1}, \mathbf{10}^*, \mathbf{10}, \mathbf{8}_a$.

In this SU(3) study, we employ the gauge configurations on a $16^3 \times 32$ lattice generated by CP-PACS and JLQCD Collaborations with the renormalization group improved Iwasaki gauge action and the non-perturbatively $O(a)$ improved Wilson quark action. The lattice spacing and the lattice volume are $a = 0.121(2)\,\text{fm}$ and $L = 1.93(3)\,\text{fm}$, respectively. These configurations are provided by Japan Lattice Data Grid (JLDG) and International Lattice Data Grid (ILDG) [23].

Figure 4 shows the NBS wave functions as a function of the relative distance between two baryons at $m_\pi = 835\,\text{MeV}$ [24]. To draw all data in a same scale, they are normalized to 1/2 for the singlet channel and to 1 for other channels at the maximum distance. The wave functions in Fig. 4 show

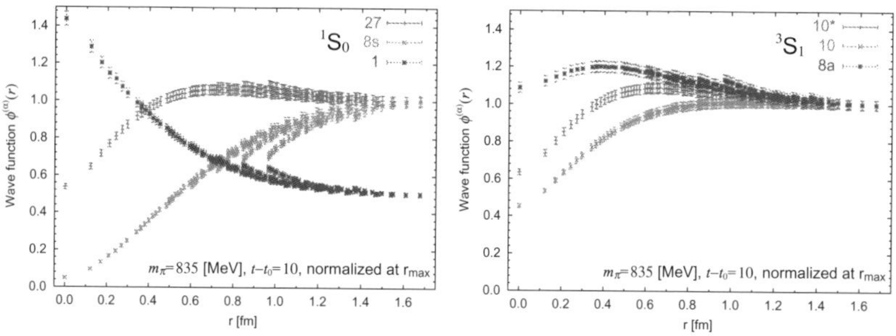

Fig. 4. NBS wave functions at $m_\pi = 835\,\text{MeV}$, normalized to 1/2 for the singlet channel and to 1 for other channels at the maximum distance [24].

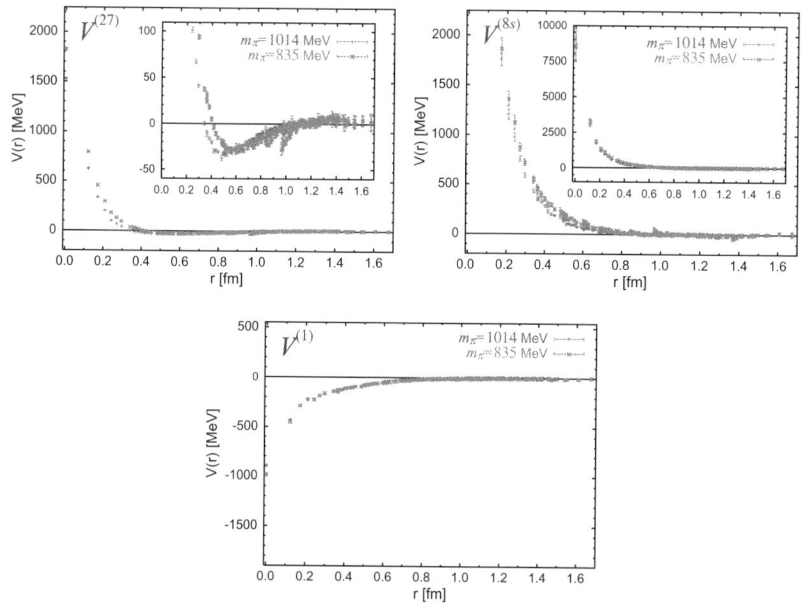

Fig. 5. The three independent BB potentials in the $^1S_0$ channel in the flavor SU(3) limit, extracted from the lattice QCD simulation at $m_\pi = 1014\,\text{MeV}$ (bars) and $m_\pi = 835\,\text{MeV}$ (crosses).

characteristic flavor dependence: In particular, a strong suppression at short distance appears in the $\mathbf{8}_s$ channel, while a strong enhancement appears in the **1** channel. Similar results are obtained for $m_\pi = 1014\,\text{MeV}$ too.

Figure 5 shows the resulting three independent BB potentials in the $^1S_0$ channel in the flavor basis obtained from the NBS wave functions. Bars (crosses) correspond to the pion mass 1014 MeV (835 MeV): Although there is a tendency that the magnitude (range) of the potentials becomes larger at short distances (longer at large distances) for lighter quark mass, the differences are not substantial for the present heavy quark masses. Left panels of Fig. 5 show $V^{(\mathbf{27})}$ which corresponds to NN $^1S_0$ potential. It has a repulsive core at short distance and an attractive pocket as we have shown already in quenched and (2+1)-flavor simulations. The middle panel of Fig. 5 corresponding to $V^{(\mathbf{8}_s)}$ has a very strong repulsive core among all channels. In contrast, the right panel of Fig. 5 corresponding to $V^{(\mathbf{1})}$ shows attraction for all distances.

These features are consistent with what has been observed in phenomenological quark models [25]. In particular, the potential in the $\mathbf{8}_s$ channel in quark models becomes strongly repulsive at short distance since the six quarks cannot occupy the same orbital state due to quark Pauli blocking. On the other hand, the potential in the **1** channel does not suffer from the quark Pauli blocking and can become attractive due to short range gluon exchange. Such an agreement between the lattice data and the phenomenological models suggests that the quark Pauli blocking plays an essential role for the repulsive core in BB systems as suggested long time ago in [26]. One can also confirm the idea of the Pauli blocking by considering the meson-baryon interaction such as charmonium-nucleon potential [27] and kaon-nucleon potential [28] within the present lattice approach. Generalization of the baryon-baryon interaction to the case with explicit SU(3) breaking is also under way [29, 30].

5. Summary and Concluding Remarks

In this paper, we have discussed the basic notion of the nucleon-nucleon potential and its field-theoretical derivation from the equal-time Nambu-Bethe-Salpeter wave function in QCD. By construction, the non-local potential defined through the projection of the wave function to the interaction region (the inner region) correctly reproduces the asymptotic form of the wave function in the region beyond the range of the nuclear force (the outer region). Thus the observables such as the phase shifts and the binding energies can be calculated after extrapolating the potential to the infinite volume limit. Non-locality of the potential can be taken into account successively by making its velocity expansion, which introduces the velocity-dependent local potentials. The leading-order terms of such velocity expansion for the nucleon-nucleon interaction are the central and the tensor potentials.

To show how this formulation works, some results in the quenched and (2+1)-flavor lattice QCD simulations are shown for relatively heavy pion masses, $m_\pi \sim 400, 500, 700\,\mathrm{MeV}$. We found that the NN potential calculated on the lattice at low energy shows all the characteristic features expected from the empirical NN potentials obtained from the experimental NN phase shifts, namely the attractive well at long and medium distances and the repulsive core at short distance for the central potential. As for the tensor potential obtained from the coupled channel treatment of the $^3S_1$-state and the $^3D_1$-state, we found appreciable attraction at long and medium distances.

As the quark mass decreases, the repulsive core and attractive well in the central potential, and the attractive well in the tensor potential tend to be enhanced. To make the deuteron bound state, however, it is necessary to go the lighter quark masses. We have also shown that the derivative expansion in terms of the local and energy-independent potentials works well at low energies for at least the quark masses studies above.

There are a number of directions to be investigated on the basis of our approach. Among others, the most important direction is to carry out (2+1)-flavor simulations with a large volume (e.g. $L = 6\,\mathrm{fm}$) at physical quark mass ($m_\pi = 135\,\mathrm{MeV}$) to extract the realistic NN potentials. This will be indeed started soon as a first priority simulation at 10 PFlops national supercomputer "KEI" which will have full operation in 2012 at Advanced Institute for Computational Science (AICS) in Kobe, Japan [31]. Simulations of the three nucleons on the lattice are also a challenging problem to be studied in relation to the attractive binding of finite nuclei and to the repulsive effect in high density matter relevant for neutron stars. Study along this line has been recently started [32].

If it turns out that the program described in this paper indeed works in lattice QCD with the physical quark mass, it would be a major step toward the understanding of atomic nuclei and neutron stars from the fundamental law of the strong interaction, the quantum chromodynamics.

Acknowledgments

I stayed in Stony Brook as a Gerry Brown's posdoc for two years starting from March 1988. Soon after I arrived at Stony Book, Gerry gave me a copy of his book "The Nucleon-Nucleon Interaction". It is my great pleasure to present some of our recent progresses on lattice nuclear force in this volume to celebrate his 85th birthday. This work was supported by Grant-in-Aid for Scientific Research on Innovative Areas (No. 2004: 20105001,20105003). I thank the members of HAL QCD collaboration (S. Aoki, T. Doi, Y. Ikeda, T. Inoue, N. Ishii, K. Murano, H. Nemura and K. Sasaki) for fruitful discussions. Also, I thank Y. Fujiwara, C. Nakamoto, M. Oka, T. Takatsuka, R. Tamagaki and H. Toki for stimulating discussions and encouragements.

References

[1] http://nn-online.org/.
[2] G.E. Brown, T.T.S. Kuo, J.W. Holt and S. Lee, *The Nucleon-Nucleon Interaction and the Nuclear Many-Body Problem* (World Scientific, Singapore, 2010).
[3] G.E. Brown and A.D. Jackson, *Nucleon-nucleon Interaction* (North-Holland, Amsterdam, 1976). R. Machleidt and I. Slaus, *J. of Phys. G* **27**, R69 (2001).
[4] H. Yukawa, *Proc. Math.-Phys. Soc. Jpn.* **17**, 48 (1935).
[5] R. Jastrow, *Phys. Rev.* **81**, 165 (1951).
[6] Y. Nambu, *Phys. Rev.* **106**, 1366 (1957).
[7] R. Tamagaki, *Soryushiron Kenkyu (Kyoto)* **40**, D26 (1969); *Prog. Theor. Phys.* **44**, 905 (1970); M. Hoffberg *et al.*, *Phys. Rev. Lett.* **24**, 775 (1970).
[8] R. Machleidt, arXiv:0704.0807 [nucl-th].
[9] G. Colangelo *et al.*, arXiv:1011.4408 [hep-lat].
[10] M. Lüscher, *Nucl. Phys. B* **354**, 531 (1991).
[11] N. Ishii, S. Aoki and T. Hatsuda, *Phys. Rev. Lett.* **99**, 022001 (2007) [nucl-th/0611096].
[12] S. Aoki, T. Hatsuda and N. Ishii, *Comput. Sci. Dis.* **1**, 015009 (2008) [arXiv:0805.2462 [hep-ph]].
[13] S. Aoki, T. Hatsuda and N. Ishii, *Prog. Theor. Phys.* **123**, 89 (2010) [arXiv:0909.5585 [hep-lat]].
[14] T. Koehler, K. Goral and P.S. Julienne, *Rev. Mod. Phys.* **78** 1311, (2006) [cond-mat/0601420].
[15] K. Nishijima, *Phys. Rev.* **111**, 995 (1958); W. Zimmermann, *Nuovo Cim.* **10**, 597 (1958); R. Haag, *Phys. Rev.* **112**, 669 (1958); See also, W. Zimmermann, MPI-PAE/PTh-61/87 (1987), unpublished.
[16] W. Królikowski and J. Rzewuski, *Nuovo Cimento* **4**, 1212 (1956).
[17] R. Tamagaki and W. Watari, *Prog. Theor. Phys.* Suppl. No. 39, 23 (1967).
[18] S. Okubo and R.E. Marshak, *Ann. of Phys.* **4**, 166 (1958).
[19] S. Aoki *et al.* (CP-PACS Collaboration), *Phys. Rev. D* **71**, 094504 (2005).
[20] S. Aoki *et al.* [PACS-CS Collaboration], *Phys. Rev. D* **79**, 034503 (2009) [arXiv:0807.1661 [hep-lat]].
[21] N. Ishii (for PACS-CS Collaboration and HAL-QCD Collaboration), PoS **LAT2009**, 019 (2009) [arXiv:1004.0405 [hep-lat]].

[22] K. Murano, N. Ishii, S. Aoki and T. Hatsuda, Lattice 2010 Proceedings.
[23] CP-PACS and JLQCD Collaboration, http://www.jldg.org/ildg-data/CPPACS+JLQCDconfig.html.
[24] T. Inoue *et al.* (HAL QCD collaboration), *Prog. Theor. Phys.* **124**, 591 (2010) [arXiv:1007.3559 [hep-lat]]; arXiv:1007.3559 [hep-lat].
[25] M. Oka, K. Shimizu and K. Yazaki, *Nucl. Phys. A* **464**, 700 (1987); *Prog. Theor. Phys. Suppl.* **137**, 1 (2000); Y. Fujiwara, Y. Suzuki and C. Nakamoto, *Prog. Part. Nucl. Phys.* **58**, 439 (2007) [arXiv:nucl-th/0607013]; Y. Fujiwara and Y. Suzuki, *Nucl. Phys. News* **18N2**, 17 (2008).
[26] S. Otsuki, M. Yasuno and R. Tamagaki, *Prog. Theor. Phys. Suppl.* Extra Number **1965**, 578 (1965); S. Machida and M. Namiki, *Prog. Theor. Phys.* **33**, 125 (1965); V.G. Neudachin, Yu. F. Smirnov and R. Tamagaki, *Prog. Theor. Phys.* **58**, 1072 (1977).
[27] T. Kawanai and S. Sasaki, *Phys. Rev. D* **82**, 091501 (2010), [arXiv:1009.3332 [hep-lat]]; arXiv:1011.1322 [hep-lat].
[28] Y. Ikeda (for HAL QCD Collaboration), arXiv:1002.2309 [hep-lat].
[29] H. Nemura, N. Ishii, S. Aoki and T. Hatsuda, *Phys. Lett. B* **673**, 136 (2009) [arXiv:0806.1094 [nucl-th]].
[30] H. Nemura (for HAL QCD Collaboration and PACS-CS Collaboration), PoS **LAT2009**, 152 (2009) [arXiv:1005.5352 [hep-lat]].
[31] http://www.aics.riken.jp/index.html.
[32] T. Doi (for HAL QCD Collaboration), arXiv:1011.0657 [hep-lat].

Chapter 9

Density-Dependent Nuclear Interactions and the Beta Decay of ^{14}C: Chiral Three-Nucleon Forces and Brown-Rho Scaling*

J.W. Holt†, N. Kaiser† and W. Weise†

We study the role of density-dependent low-momentum nucleon-nucleon interactions in describing the anomalously long beta decay lifetime of ^{14}C. We approach this problem both from the perspective of chiral effective field theory, in which genuine three-body forces generate an effective density-dependent two-body interaction, as well as from the perspective of Brown-Rho scaling, in which the masses and form factor cutoffs in one-boson-exchange interactions are modified in a dense nuclear medium due to the partial restoration of chiral symmetry. The beta decay transition of ^{14}C to the ground state of ^{14}N is calculated within the shell model using a model space consisting of two $0p$-shell holes within a closed ^{16}O core. The effective $0p$-shell interaction is calculated up to second order in perturbation theory with single-particle energies extracted from experiment. We find that both three-nucleon forces and Brown-Rho scaling medium modifications give qualitatively similar results not only for the ground state to ground state Gamow-Teller transition but also for Gamow-Teller transitions from excited states of ^{14}C to the ground state of ^{14}N. In this way, it is observed that at a low-momentum scale of $\Lambda_{\text{low}-k} = 2.1\,\text{fm}^{-1}$, medium-modifications of the nuclear force play an essential role in increasing the lifetime of ^{14}C from a few minutes to an archaeologically long one of 5730 years.

*Work supported in part by BMBF, GSI and by the DFG cluster of excellence: Origin and Structure of the Universe.
†Physik Department, Technische Universität München, D-85747 Garching, Germany.

1. Introduction

In recent years our understanding of nuclear structure physics, and the nuclear force in particular, has benefited greatly by incorporating aspects of the fundamental theory of strong interactions, QCD. Although QCD is highly nonperturbative at the energy scales characteristic for nucleons bound within nuclei, one can exploit the symmetry breaking pattern of QCD, namely the explicit and spontaneous breaking of chiral symmetry, to constrain the underlying effective theory for interacting nucleons. At low energies, this is accomplished through chiral effective field theory [1–10], in which the light pseudoscalar Nambu-Goldstone bosons (e.g. pions) generated by the spontaneous breaking of chiral symmetry comprise the low-energy mesonic degrees of freedom relevant for nuclear interactions. All other heavier mesons are integrated out and their effects are absorbed in a series of short-range contact interactions.

The above description utilizes the spontaneous breaking of chiral symmetry at low energies. However, as the temperature and/or density of a hadronic medium increases, the order parameter for chiral symmetry breaking (the scalar quark condensate $\langle \bar{q}q \rangle$) decreases in magnitude and is expected to vanish at an energy density around $1\,\mathrm{GeV/fm^3}$, signaling the restoration of chiral symmetry. The vacuum structure of QCD, encoded in various quark and gluon condensates, is expected to play an important role in determining the properties (e.g. masses) of light hadrons [11–13], with the scalar quark condensate being of primary importance. The work of Brown and Rho [12, 14] was one of the first attempts to connect medium-dependent meson masses to the partial restoration of chiral symmetry as the temperature and/or density changes. By imposing scale invariance — a symmetry of classical, massless QCD — on chiral effective Lagrangians, Brown and Rho obtained a scaling law in which the masses of hadrons, except the pion, decrease with the cube root of the scalar quark condensate.

A natural testing ground for Brown-Rho scaling is heavy ion collisions, which produce hot/dense hadronic matter in the vicinity of chiral restoration. For a recent review see Refs. [15, 16]. However, even for densities at the center of heavy nuclei, the scalar quark condensate is expected to decrease in magnitude by about 30% as found from model studies [17], low-density expansions [18], and calculations within in-medium chiral effective field theory [19]. The decreasing of meson masses at these densities would have important consequences for nuclear structure, since the intermediate-range attraction and strong short-distance repulsion in one-boson-exchange models of the nucleon-nucleon potential arise largely from the exchange of the σ and ω mesons, respectively. Moreover, phenomena that are sensitive to the tensor component of the nucleon-nucleon interaction would be affected by the decreasing of the ρ meson mass. It was therefore hoped that evidence for the partial restoration of chiral symmetry at finite density could already be found by studying the properties of normal nuclei.

Brown and Rho proposed employing the hidden local symmetry approach [20, 21] as a framework for studying mass scaling effects. In this formalism vector mesons, such as the ρ and ω, are treated as gauge bosons that acquire mass through the Higgs mechanism. Presently a consistent hidden local symmetry approach incorporating baryons has yet to be fully developed. This limits the applicability of the formalism to dense nuclear systems. A more practical approach has been to directly modify the masses of mesons in one-boson exchange interactions [22–26]. This leads to a density-dependent nucleon-nucleon interaction that can be used in nuclear structure studies. One aims to achieve a qualitative understanding of the effects of in-medium hadron masses on nuclear structure.

Not only can the mass of a hadron be affected by a background medium, but more generally the entire spectral function will be modified due to Pauli blocking effects, collisional broadening, and the opening of additional decay channels in a hadronic medium [27–31]. Indeed the study of spectral functions has become the dominant focus both experimentally and theoretically for probing the tendency towards chiral symmetry restoration below the transition point. The importance of resonance-hole excitations in modifying the spectral features of the vector mesons ρ and ω has been studied in detail.

Chiral effective field theory, representing the interface between QCD and nuclear physics at low energy and momentum scales, takes a different approach by integrating out mesons much heavier than pions. The short-distance dynamics is encoded in contact terms. Intermediate-range effects that would emerge from σ exchange and the ρ-exchange tensor force, are now generated by explicit two-pion exchange processes, subject to Pauli principle corrections in the nuclear medium. Resonance-hole effects are incorporated through genuine three-nucleon forces.

With the development of highly-accurate methods for solving the nuclear many-body problem for light nuclei, as well as infinitely-extended nuclear matter, it is now well established that free-space two-nucleon interactions alone are insufficient to describe the properties of dense nuclear systems. In particular, calculations using only nonrelativistic two-body forces have been unable to accurately reproduce the binding energies and spectra of light nuclei [32–34], the saturation binding energy and density of isospin-symmetric nuclear matter [35–39], and finally the nucleon-deuteron scattering differential cross sections at intermediate energies [40–43]. In all of these cases, the errors arising from the solution of the nuclear many-body problem are small compared to the uncertainties in the underlying nuclear force model, which suggests the need to include explicitly degrees of freedom beyond nucleons and mesons, or equivalently the need to include three-body forces.

In this work we study the unique problem of the beta decay lifetime of ^{14}C, which has been shown [25, 44] to be highly sensitive to density-dependent effects in the nuclear interaction. Describing the beta decay lifetime of ^{14}C has long been a challenge in nuclear structure physics, because the anomalously small Gamow-Teller (GT) transition matrix element $M_{GT} \simeq 0.002$ results from the cancellation between terms that are otherwise on the order of 1. The valence particles of ^{14}C inhabit a region with a large nuclear density, and therefore medium modifications to the nucleon-nucleon interaction could be sizable. In Ref. [25] it was suggested that a decreasing in-medium tensor force from Brown-Rho scaling plays a significant role in the suppression of the Gamow-Teller transition. This was inspired by early work [45, 46] that highlighted the sensitivity of the beta decay lifetime to the tensor component of the nuclear interaction. In contrast, a later study including three-nucleon forces [44] found qualitatively similar effects that resulted from additional short-distance repulsion coming from the contact three-body force. Within Brown-Rho-scaled nucleon-nucleon interaction models such an effect is generated by the decreasing of the ω meson mass. We wish to revisit the problem of the ^{14}C beta decay lifetime and suggest that, indeed, in both models of the low-momentum density-dependent nucleon-nucleon interaction, additional short-distance repulsion generates significant suppression of the Gamow-Teller transition.

This paper is organized as follows. In Sec. 2 we describe the framework of chiral effective field theory, and how a density-dependent two-nucleon force can be constructed from the chiral three-body interaction by summing over the filled Fermi sea of nucleons. In Sec. 3 we motivate Brown-Rho scaling and study the effect that such hadronic medium modifications have on the nuclear force. In Sec. 4 we apply both models to the problem of the extremely long beta decay lifetime of ^{14}C, which was suggested to be particularly sensitive to the density dependence of the nuclear force. We highlight the important role played by the short-distance repulsion (increasing with the nuclear density) in both approaches to the in-medium nuclear interaction. We end with a summary and conclusions.

2. Density-Dependent Nucleon-Nucleon Interaction from Chiral Effective Field Theory

Chiral effective field theory has proven to be a highly useful framework for studying the nuclear interaction problem (for current reviews, see Refs. [10, 47]). By implementing the spontaneous and explicit breaking of chiral symmetry in QCD, one constructs a low-energy effective field theory of interacting pions and nucleons order-by-order in powers of Q/Λ_χ, where Q represents a nucleon momentum or the pion mass. The long- and intermediate-range parts of the nuclear force are generated by 1π, 2π (and more) exchange processes, whereas the short-distance dynamics due to heavy mesons and baryon resonances are integrated out and their effects are encoded in nucleon-nucleon contact terms. Currently the nuclear interaction has been calculated up to 4th order (N^{3}LO) in the chiral power counting [6–9]. It is at this order (and beyond) that it is possible to achieve an agreement with empirical NN scattering phase shifts that is comparable to previous high-precision NN potentials [48–50]. By adjusting the 29 parameters (mostly low-energy constants) that occur at this order, the 1999 database for np and pp elastic scattering up to $E_{\rm lab} = 290\,$MeV can be fit with a χ^2/dof of 1.1 for np scattering and 1.5 for pp scattering. Furthermore, the experimental deuteron binding energy, charge radius, and electric quadrupole moment are very well reproduced by the chiral N^{3}LO potential [9].

One decisive practical advantage of the chiral effective field theory approach is that two- and many-body forces are generated consistently within the same framework. Chiral three-nucleon forces first appear at N^{2}LO. There are three different components, which we show in diagrams (a), (b), and (c) in Fig. 1. The long-range, two-pion exchange interaction is given by

$$V_{3N}^{(2\pi)} = \sum_{i\neq j\neq k} \frac{g_A^2}{8f_\pi^4} \frac{\vec{\sigma}_i \cdot \vec{q}_i \, \vec{\sigma}_j \cdot \vec{q}_j}{(\vec{q_i}^{\,2} + m_\pi^2)(\vec{q_j}^{\,2} + m_\pi^2)} F_{ijk}^{\alpha\beta} \tau_i^\alpha \tau_j^\beta , \tag{1}$$

where $g_A = 1.27$, $f_\pi = 92.4\,$MeV, $m_\pi = 138.04\,$MeV is the average pion mass, $\vec{q}_i = \vec{p_i}' - \vec{p}_i$ is the difference between the final and initial momentum

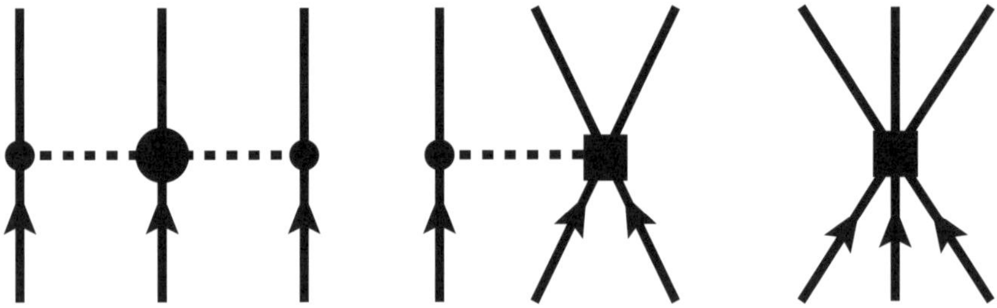

(a) (b) (c)

Fig. 1. The leading-order contributions to the chiral three-nucleon force: (a) the long-range 2π-exchange force $V_{3N}^{(2\pi)}$, (b) the medium-range 1π-exchange force $V_{3N}^{(1\pi)}$, and (c) the short-range contact interaction $V_{3N}^{(\mathrm{ct})}$.

of nucleon i, and

$$F_{ijk}^{\alpha\beta} = \delta^{\alpha\beta}(-4c_1 m_\pi^2 + 2c_3 \vec{q}_i \cdot \vec{q}_j) + c_4 \epsilon^{\alpha\beta\gamma} \tau_k^\gamma \vec{\sigma}_k \cdot (\vec{q}_i \times \vec{q}_j). \tag{2}$$

The three low-energy constants of the long-range 3NF, namely $c_1 = -0.76\,\mathrm{GeV}^{-1}$, $c_3 = -4.78\,\mathrm{GeV}^{-1}$, and $c_4 = 3.96\,\mathrm{GeV}^{-1}$ appear in the two-pion exchange component of the NN interaction and therefore can be determined from fits to low-energy NN phase shifts [51]. The medium-range, one-pion exchange three-nucleon interaction is proportional to the low-energy constant c_D/Λ_χ and has the analytic form

$$V_{3N}^{(1\pi)} = -\sum_{i\neq j\neq k} \frac{g_A c_D}{8 f_\pi^4 \Lambda_\chi} \frac{\vec{\sigma}_j \cdot \vec{q}_j}{\vec{q}_j^{\,2} + m_\pi^2} \vec{\sigma}_i \cdot \vec{q}_j \, \vec{\tau}_i \cdot \vec{\tau}_j, \tag{3}$$

with $\Lambda_\chi = 700\,\mathrm{MeV}$. The filled black square in diagram (b) of Fig. 1 symbolizes the corresponding two-nucleon one-pion contact interaction. Finally, the short-distance contact three-body force is proportional to the low-energy constant c_E/Λ_χ and is given by

$$V_{3N}^{(\mathrm{ct})} = \sum_{i\neq j\neq k} \frac{c_E}{2 f_\pi^4 \Lambda_\chi} \vec{\tau}_i \cdot \vec{\tau}_j. \tag{4}$$

In Eqs. (1)–(4) the nucleon labels i, j, k can take the values 1, 2, 3 which results in six possible permutations for each sum. The low-energy constants of the medium-range ($V_{3N}^{(1\pi)}$) and short-range ($V_{3N}^{(\mathrm{ct})}$) three-body forces are fit to properties of few-nucleon systems, such as the triton binding energy together with the *nd* doublet scattering length [52], the ^{4}He binding energy [53], or the binding energies and spectra of light nuclei [34]. The N^{3}LO three-body interaction contains numerous one-loop

Table 1. The values of the low-energy constants c_D and c_E of the chiral three-nucleon interaction fit to the binding energies of $A = 3, 4$ nuclei for different values of the momentum cutoff $\Lambda_{\text{low}-\text{k}}$.

$\Lambda_{\text{low}-k}$	c_D	c_E
$2.1\,\text{fm}^{-1}$	-2.062	-0.625
$2.3\,\text{fm}^{-1}$	-2.785	-0.822

diagrams but no additional low-energy constants and is currently under construction [47].

In this work we utilize renormalization group techniques to decimate the chiral N^3LO two-body interaction down to an energy scale in the vicinity of $\Lambda = 2.0\,\text{fm}^{-1}$, following the details provided in Refs. [54, 55]. Such a decimation procedure has the desirable feature that the resulting "low-momentum interaction", referred to as $V_{\text{low}-k}$, is nearly unique regardless of which underlying high-precision NN potential is employed, and therefore $V_{\text{low}-k}$ provides a model-independent description of the nucleon-nucleon interaction below $\Lambda = 2.0\,\text{fm}^{-1}$. The analogous evolution of realistic three-body forces down to this scale has yet to be achieved. However, since such a decimation would change primarily the short-distance part of the three-body interaction, a practical alternative has been to treat the low-energy constants c_D and c_E as functions of the variable $\Lambda_{\text{low}-k}$. By fitting $c_D(\Lambda_{\text{low}-k})$ and $c_E(\Lambda_{\text{low}-k})$ to the binding energies of ^{3}H and ^{4}He, one obtains [56] the values shown in Table 1 for two different choices of the low-momentum cutoff.

We now show how to derive from the leading-order chiral three-nucleon interaction, Eqs. (1)–(4), an effective density-dependent in-medium NN interaction. We keep our discussion brief and refer the interested reader to the original references [44, 57]. We consider the on-shell scattering of two nucleons in isospin-symmetric (spin-saturated) nuclear matter of density $n = 2k_f^3/3\pi^2$ in the center-of-mass frame, $N_1(\vec{p}) + N_2(-\vec{p}) \to N_1(\vec{p}+\vec{q}) + N_2(-\vec{p}-\vec{q})$, where k_f is the Fermi momentum. The magnitude of the in- and out-going nucleon momenta is $|\vec{p}| = p = |\vec{p}+\vec{q}|$, and $q = |\vec{q}|$ is the magnitude of the momentum transfer. The resulting interaction in momentum-space has the following (general) form

$$V(\vec{p},\vec{q}) = V_C + \vec{\tau}_1 \cdot \vec{\tau}_2\, W_C + [V_S + \vec{\tau}_1 \cdot \vec{\tau}_2\, W_S]\, \vec{\sigma}_1 \cdot \vec{\sigma}_2$$
$$+ [V_T + \vec{\tau}_1 \cdot \vec{\tau}_2\, W_T]\, \vec{\sigma}_1 \cdot \vec{q}\, \vec{\sigma}_2 \cdot \vec{q}$$

$$+ [V_{SO} + \vec{\tau}_1 \cdot \vec{\tau}_2 \, W_{SO}] \; i(\vec{\sigma}_1 + \vec{\sigma}_2) \cdot (\vec{q} \times \vec{p})$$
$$+ [V_Q + \vec{\tau}_1 \cdot \vec{\tau}_2 \, W_Q] \; \vec{\sigma}_1 \cdot (\vec{q} \times \vec{p}) \, \vec{\sigma}_2 \cdot (\vec{q} \times \vec{p}). \quad (5)$$

The subscripts refer to the central (C), spin-spin (S), tensor (T), spin-orbit (SO), and quadratic spin orbit (Q) components of the NN interaction, each of which occurs in an isoscalar (V) and an isovector (W) version.

In Figs. 2 and 3 we show the six topologically distinct contributions to the in-medium density-dependent nucleon-nucleon interaction to one-loop order. The long-range two-pion exchange three-nucleon force generates the diagrams shown in Fig. 2. Tadpole diagrams obtained by contracting the in- and outgoing lines at a πNN vertex vanish, which leaves only three topologically distinct nonvanishing contributions. The short double-line on a nucleon propagator symbolizes the filled Fermi sea of nucleons, which introduces the "medium insertion" $-2\pi\delta(k_0)\,\theta(k_f - |\vec{k}\,|)$ in the in-medium nucleon propagator. Diagram (1) in Fig. 2 represents one-pion exchange with a Pauli-blocked pion self-energy correction. The intermediate nucleon

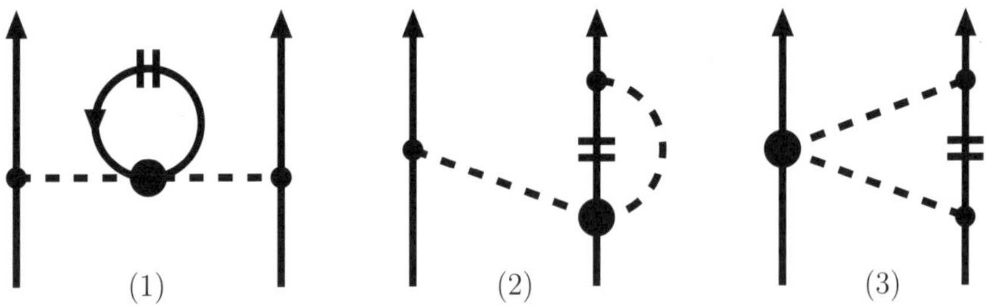

Fig. 2. In-medium NN interaction generated by the two-pion exchange component of the chiral three-nucleon interaction. The short double-line symbolizes the filled Fermi sea of nucleons. Reflected diagrams of (2) and (3) are not shown.

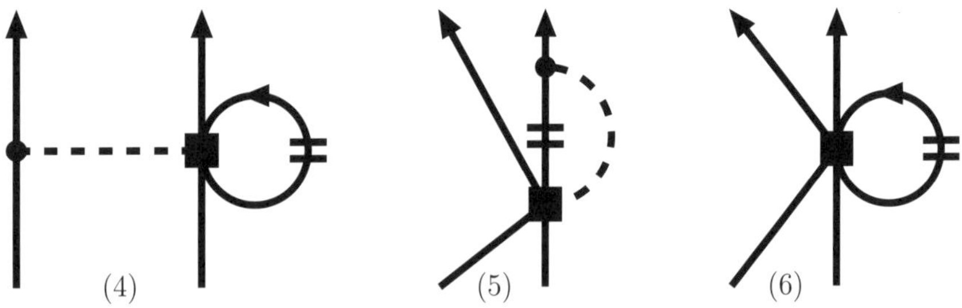

Fig. 3. In-medium NN interaction generated by the one-pion exchange component and the short-range component of the chiral three-nucleon interaction.

momentum is independent of the momentum transfer $\vec{q}$ carried by the pions and therefore the integration over the filled Fermi sea of nucleons produces a correction proportional to the density n:

$$V_{NN}^{\text{med},1} = \frac{g_A^2 M_N n}{8\pi f_\pi^4} \vec{\tau}_1 \cdot \vec{\tau}_2 \frac{\vec{\sigma}_1 \cdot \vec{q}\, \vec{\sigma}_2 \cdot \vec{q}}{(m_\pi^2 + q^2)^2} (2c_1 m_\pi^2 + c_3 q^2) \,. \tag{6}$$

Since $c_{1,3} < 0$, this term gives a (large) enhancement of the bare 1π-exchange:

$$V_{NN}^{(1\pi)} = -\frac{g_A^2 M_N}{16\pi f_\pi^2} \vec{\tau}_1 \cdot \vec{\tau}_2 \frac{\vec{\sigma}_1 \cdot \vec{q}\, \vec{\sigma}_2 \cdot \vec{q}}{m_\pi^2 + q^2}. \tag{7}$$

In part this enhancement can be interpreted in terms of the reduced in-medium pion decay constant, $f_{\pi,s}^{*2} = f_\pi^2 + 2c_3\rho$, which is associated with the spatial components of the axial current. Diagram (2) of Fig. 2 represents a vertex correction to one-pion exchange. Fermi sphere integrals, $\int_{|\vec{k}|\le k_F} d^3k$, over a single static pion propagator give rise to p-dependent functions $\Gamma_j(p)$, where the dependence on k_F is implicit. For the sake of brevity and continuity in the current discussion, we have left the explicit expressions for the $\Gamma_j(p)$ in the Appendix. Summing diagram (2) of Fig. 6 together with the three reflected diagrams yields the following contribution

$$\begin{aligned} V_{NN}^{\text{med},2} &= \frac{g_A^2 M_N}{32\pi^3 f_\pi^4} \vec{\tau}_1 \cdot \vec{\tau}_2 \frac{\vec{\sigma}_1 \cdot \vec{q}\, \vec{\sigma}_2 \cdot \vec{q}}{m_\pi^2 + q^2} \\ &\times \Bigg\{ -4c_1 m_\pi^2 [\Gamma_0(p) + \Gamma_1(p)] - (c_3 + c_4)[q^2(\Gamma_0(p) + 2\Gamma_1(p) \\ &+ \Gamma_3(p)) + 4\Gamma_2(p)] + 4c_4 \left[\frac{2k_f^3}{3} - m_\pi^2 \Gamma_0(p) \right] \Bigg\}. \end{aligned} \tag{8}$$

By analyzing the momentum- and density-dependent factors in Eq. (8) relative to $V_{NN}^{(1\pi)}$, one finds that this contribution corresponds to a (large) reduction of the 1π-exchange in the nuclear medium. Approximately, this feature can be interpreted in terms of a quenched nucleon axial-vector constant g_A^*.

Diagram (3) in Fig. 2 represents Pauli-blocked 2π-exchange. The Fermi sphere integrals over two pion propagators give rise to p- and q-dependent functions $G_j(p, q)$ which we tabulate in the Appendix. Diagram (3) taken together with the corresponding reflected diagram then produces the contribution

$$\begin{aligned}V_{NN}^{\text{med},3} &= \frac{g_A^2 M_N}{64\pi^3 f_\pi^4}\{-12c_1 m_\pi^2[2\Gamma_0(p) - (2m_\pi^2 + q^2)G_0(p,q)] \\ &- c_3[8k_f^3 - 12(2m_\pi^2 + q^2)\Gamma_0(p) - 6q^2\Gamma_1(p) \\ &+ 3(2m_\pi^2 + q^2)^2 G_0(p,q)] + 4c_4\, \vec{\tau}_1 \cdot \vec{\tau}_2\, (\vec{\sigma}_1 \cdot \vec{\sigma}_2\, q^2 - \vec{\sigma}_1 \cdot \vec{q}\, \vec{\sigma}_2 \cdot \vec{q}) \\ &\times G_2(p,q) - (3c_3 + c_4 \vec{\tau}_1 \cdot \vec{\tau}_2)\, i(\vec{\sigma}_1 + \vec{\sigma}_2) \cdot (\vec{q} \times \vec{p})[2\Gamma_0(p) \\ &+ 2\Gamma_1(p) - (2m_\pi^2 + q^2) \times (G_0(p,q) + 2G_1(p,q))] \\ &- 12c_1 m_\pi^2\, i(\vec{\sigma}_1 + \vec{\sigma}_2) \cdot (\vec{q} \times \vec{p})[G_0(p,q) + 2G_1(p,q)] \\ &+ 4c_4\, \vec{\tau}_1 \cdot \vec{\tau}_2\, \vec{\sigma}_1 \cdot (\vec{q} \times \vec{p})\, \vec{\sigma}_2 \cdot (\vec{q} \times \vec{p})[G_0(p,q) \\ &+ 4G_1(p,q) + 4G_3(p,q)]\}. \qquad (9)\end{aligned}$$

In comparison to the analogous 2π-exchange interaction in vacuum (see Sec. 4.2 in Ref. [58]) the Pauli blocking in the nuclear medium has generated additional spin-orbit terms, $i(\vec{\sigma}_1 + \vec{\sigma}_2) \cdot (\vec{q} \times \vec{p})$, and quadratic spin-orbit terms, $\vec{\sigma}_1 \cdot (\vec{q} \times \vec{p})\, \vec{\sigma}_2 \cdot (\vec{q} \times \vec{p})$, written in the last three lines of Eq. (9).

Next we consider the 1π-exchange component of the chiral three-nucleon interaction proportional to the parameter c_D/Λ_χ. Contracting two external lines results in two (nonvanishing) topologies. By closing a nucleon line at the contact vertex, one obtains a vertex correction to 1π-exchange:

$$V_{NN}^{\text{med},4} = -\frac{g_A M_N c_D n}{32\pi f_\pi^4 \Lambda_\chi}\, \vec{\tau}_1 \cdot \vec{\tau}_2\, \frac{\vec{\sigma}_1 \cdot \vec{q}\, \vec{\sigma}_2 \cdot \vec{q}}{m_\pi^2 + q^2}. \qquad (10)$$

We note that this contribution is proportional to the nuclear density n due to the q-independent Fermi sphere integral. Since c_D is negative, $V_{NN}^{\text{med},4}$ reduces the bare 1π-exchange, roughly by about 16% at normal nuclear matter density $\rho_0 = 0.16\,\text{fm}^{-3}$. Diagram (5) in Fig. 3 represents Pauli-blocked (pionic) vertex corrections to the short-range NN interaction. The corresponding contribution to the density dependent in-medium NN interaction reads:

$$\begin{aligned}V_{NN}^{\text{med},5} &= \frac{g_A M_N c_D}{64\pi^3 f_\pi^4 \Lambda_\chi}\left\{\vec{\tau}_1 \cdot \vec{\tau}_2 \left[2\vec{\sigma}_1 \cdot \vec{\sigma}_2\, \Gamma_2(p) + \left(\vec{\sigma}_1 \cdot \vec{\sigma}_2 \left(2p^2 - \frac{q^2}{2}\right)\right.\right.\right. \\ &\left. + \vec{\sigma}_1 \cdot \vec{q}\, \vec{\sigma}_2 \cdot \vec{q} \left(1 - \frac{2p^2}{q^2}\right) - \frac{2}{q^2}\, \vec{\sigma}_1 \cdot (\vec{q} \times \vec{p})\, \vec{\sigma}_2 \cdot (\vec{q} \times \vec{p})\right) \\ &\left.\left. \times (\Gamma_0(p) + 2\Gamma_1(p) + \Gamma_3(p))\right] + 4k_f^3 - 6m_\pi^2 \Gamma_0(p)\right\}. \qquad (11)\end{aligned}$$

Due to the special spin and isospin structure of this term, it gives nonvanishing contributions only to relative S-states as well as to the $^3S_1-{}^3D_1$ mixing matrix element.

Finally, we discuss the short-range component of the chiral 3N interaction, represented by a three-nucleon contact-vertex proportional to c_E/Λ_χ. By closing one nucleon line (see diagram (6) in Fig. 3) one obtains the following contribution to the in-medium NN-interaction:

$$V_{NN}^{\mathrm{med},6} = -\frac{3M_N c_E n}{8\pi f_\pi^4 \Lambda_\chi}, \tag{12}$$

which simply grows linearly in density n and is independent of spin, isospin and nucleon momenta.

In Fig. 4 we show the contribution from each of these components of the density-dependent NN interaction to the $^1S_0$ and $^3S_1$ diagonal partial wave matrix elements (see Ref. [57] for an analysis of higher partial waves). The black solid line gives the matrix elements of $V_{\mathrm{low}-k}$ obtained from the Idaho N^{3}LO chiral NN potential for a momentum cutoff of $\Lambda_{\mathrm{low}-k} = 2.1\,\mathrm{fm}^{-1}$. The various curves labeled $i = 1, 2, \ldots, 6$ show the matrix elements of $V_{\mathrm{low}-k} + V_{NN}^{\mathrm{med},i}$. We see that the effects from the long-range two-pion

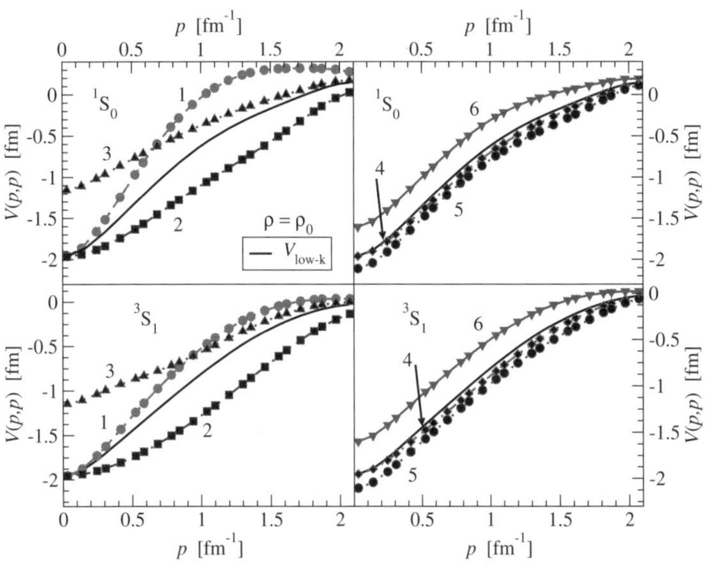

Fig. 4. Effect (at normal nuclear matter density, $n_0 = 0.16\,\mathrm{fm}^{-3}$) of the six different terms $V_{NN}^{\mathrm{med,i}}$ on the S-wave components of the free-space nucleon-nucleon potential $V_{\mathrm{low}-k}$ for a low-momentum cutoff of $\Lambda_{\mathrm{low-k}} = 2.1\,\mathrm{fm}^{-1}$.

exchange part of the chiral three-nucleon force are quite large. However, the pion self-energy correction $V_{NN}^{\text{med},1}$ largely cancels the effects of the vertex correction $V_{NN}^{\text{med},2}$. Since $V_{NN}^{\text{med},1}$ and $V_{NN}^{\text{med},2}$ are simply modifications to one-pion exchange, a similar cancellation occurs in all higher partial waves as well. The final component, $V_{NN}^{\text{med},3}$, that comes from the long-range three-body force has a much richer spin and isospin structure, and at large distances (small momenta) the overall effect is strongly repulsive.

The two terms $V_{NN}^{\text{med},4}$ and $V_{NN}^{\text{med},5}$ arise from the medium-range one-pion exchange three-body force proportional to the low-energy constant c_D. In Fig. 4 we see that the effects are much weaker than from the long-range chiral three-nucleon force. Since $V_{NN}^{\text{med},4}$ is again just a reduction of one-pion exchange, it has nonzero contributions in all partial waves. The second term arising from the mid-range three-nucleon force, $V_{NN}^{\text{med},5}$, is also relatively small and gives an additional (equal) attraction in the two relative S-states, as well as the coupled ${}^3S_1 - {}^3D_1$ mixing matrix element not shown in the figure. The final contribution $V_{NN}^{\text{med},6}$ is a pure contact interaction and therefore contributes only to the two relative S-states. The interaction is spin and isospin independent and therefore contributes to both of these partial waves equally. Indeed, the additional repulsion is approximately twice as large as the attraction coming from $V_{NN}^{\text{med},5}$, so taking these two terms together essentially results in a weaker short-range repulsive interaction.

In Fig. 5 we show the density dependence (for $n = 0, n_0/2$, and n_0) of the diagonal S-wave matrix elements of $V_{\text{low}-k} + V_{NN}^{\text{med}}$. Overall, we find that the effects are repulsive and quite large in comparison to the free-space

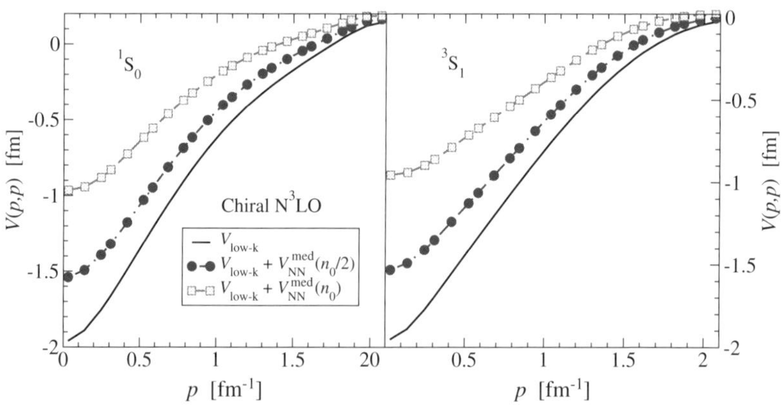

Fig. 5. Partial wave matrix elements for relative S-states from $V_{\text{low}-k} + V_{NN}^{\text{med}}$. The low-momentum decimation scale is taken to be $\Lambda_{\text{low}-k} = 2.1\,\text{fm}^{-1}$.

low-momentum NN interaction. The same holds in most higher partial waves as well, with the exception that the tensor component of the in-medium interaction becomes more attractive with the density [57]. The large S-wave repulsion in the three-body forces increases with the density and thereby provides the mechanism for saturation of low-momentum interactions [59]. In Sec. 4 we study the role that these density-dependent effects play in describing the anomalously long beta decay lifetime of ^{14}C and make a comparison with analogous interactions obtained from the Brown-Rho scaling hypothesis.

3. Brown-Rho Scaled Nucleon-Nucleon Interactions

In this section we summarize Brown-Rho scaling and describe a model approach for implementing these effects in nuclear interactions at densities close to that of saturated nuclear matter. The starting point is the scaling relation for (light) hadron masses derived by Brown and Rho [12]:

$$\sqrt{\frac{g_A}{g_A^*}}\frac{M_N^*}{M_N} \simeq \frac{m_\sigma^*}{m_\sigma} \simeq \frac{m_\rho^*}{m_\rho} \simeq \frac{m_\omega^*}{m_\omega} \simeq \frac{f_\pi^*}{f_\pi} = \Phi(n), \tag{13}$$

which was obtained by implementing scale invariance in chiral effective Lagrangians following the approach of Campbell *et al.* [60]. In Eq. (13) the asterisk refers to an in-medium value of the given quantity, $f_\pi^* = f_{\pi,t}^*$ is related to the time component of the axial vector current, and Φ (to be discussed in more detail shortly) is a function of the nuclear density n. Absent from the scaling hypothesis in Eq. (13) is the pion mass, which in the chiral limit of massless bare quarks would be zero due to the Goldstone boson nature of pions. The nonzero pion mass is due rather to the explicit breaking of chiral symmetry by the quark mass term in the QCD Lagrangian and is therefore not as sensitive to the background medium as are the masses of other hadrons. This is supported empirically from measurements of energy shifts in deeply bound pionic atoms [61,62].

From Eq. (13), hadronic masses are predicted to scale with the pion decay constant f_π, the hadronic observable connected to the spontaneous breaking of chiral symmetry. The connection between f_π and the fundamental order parameter $\langle\bar{q}q\rangle$ of QCD is given by the Gell-Mann–Oakes–Renner relation [17,63]

$$f_\pi^2 m_\pi^2 = -m_q\langle\bar{u}u + \bar{d}d\rangle, \tag{14}$$

where m_q is the average (current) mass for up u and down d quarks. Assuming that the pion mass is protected by chiral invariance, this relation would produce

$$\frac{f_\pi^*}{f_\pi} = \left(\frac{\langle \bar{q}q\rangle^*}{\langle \bar{q}q\rangle}\right)^{1/2}, \tag{15}$$

where $\langle \bar{q}q\rangle^* = \langle \Omega|\bar{q}q|\Omega\rangle$ and $|\Omega\rangle$ denotes the ground state of the dense medium. In the limit of low densities the scalar quark condensate is related to the nuclear density by [18, 64]

$$\frac{\langle \bar{q}q\rangle^*}{\langle \bar{q}q\rangle} = 1 - \frac{\sigma_{\pi N}}{f_\pi^2 m_\pi^2} n + \cdots , \tag{16}$$

where $\sigma_{\pi N}$ is the pion-nucleon sigma term that describes how the nucleon mass varies with the masses of bare quarks. From a dispersion relation analysis of low-energy πN scattering, one can deduce $\sigma_{\pi N} \simeq 45\,\mathrm{MeV}$ [65]. Thus, at nuclear matter saturation density one would obtain from Eqs. (13), (15), and (16) that $\Phi(n_0) \simeq 0.83$. In Brown-Rho scaling one therefore assumes a linear decrease of hadronic mass with the density:

$$\frac{m^*}{m} \simeq 1 - \frac{1}{2}\frac{\sigma_{\pi N}}{f_\pi^2 m_\pi^2} n \simeq 1 - 0.17\frac{n}{n_0}. \tag{17}$$

Such a relationship has also be deduced from in-medium QCD sum rule analyses, which can provide important constraints on the form the vector meson spectral functions [13, 28, 66].

Brown-Rho scaling was one of the first attempts to study — through changes in the symmetry breaking pattern of the ground state — the effects of a hot/dense medium on the properties of an embedded hadron. Subsequent work had the broader goal to obtain the full spectral function of in-medium hadrons, with special emphasis on the vector mesons: ρ and ω. Within hadronic many-body calculations, self-energy corrections $\Sigma_{\rm med}$ to the meson propagator $D(\omega, \vec{q})$ defined by

$$D(\omega, \vec{q}) = \frac{1}{\omega^2 - \vec{q}^{\,2} - m^2 - \Sigma_{\rm vac}(q) - \Sigma_{\rm med}(\omega, \vec{q})} \tag{18}$$

result in modifications to the spectral function $A(\omega, \vec{q})$, which is proportional to the imaginary part of the propagator:

$$A(\omega, \vec{q}) = \frac{1}{\pi}\frac{\mathrm{Im}\,\Sigma(\omega, \vec{q})}{(\omega^2 - \vec{q}^{\,2} - m^2 + \mathrm{Re}\,\Sigma(\omega, \vec{q}))^2 + \mathrm{Im}\,\Sigma^2(\omega, \vec{q})}. \tag{19}$$

The real part of the self-energy then contributes to the effective mass of the particle while the imaginary part leads to collisional broadening of the

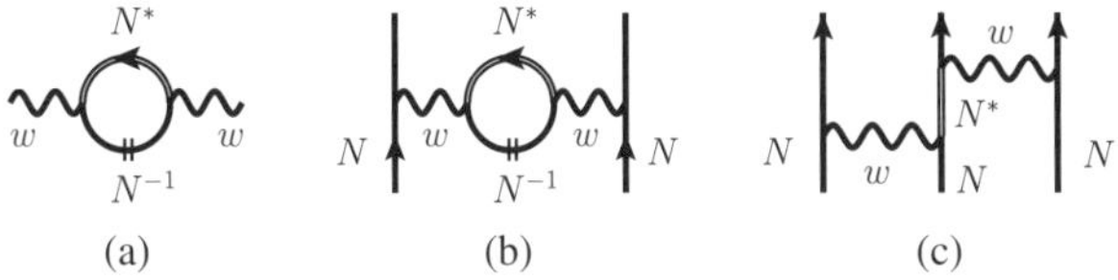

Fig. 6. The contribution of resonance-hole excitations to (a) the in-medium ω meson propagator, (b) the in-medium NN interaction resulting from ω meson exchange, and (c) the analogous contribution to the three-nucleon force.

width. In a background medium of cold nuclear matter, contributions to $\Sigma_{\rm med}$ arise from the coupling of mesons to nucleon-hole and resonance-hole excitations [28, 30, 67, 68] as shown in Fig. 6(a). Calculations are simplified by invoking the following general result that holds at low-densities (corresponding to small Fermi momenta) and which relates the in-medium self-energy to the total forward scattering amplitude $T(\vec{q}, \omega)$ of a vector meson on a nucleon

$$\Sigma_{\rm med}(\omega, \vec{q}) = \int_{p_N < k_f} \frac{d^3 p_N}{(2\pi)^3 2E_N} T(q, p_N) \simeq \frac{n}{8m_N} T(\vec{q}, \omega), \qquad (20)$$

where n is the nucleon density and one assumes that the scattering amplitude varies slowly with momentum. The result of such studies was that medium modifications to vector meson spectral functions in nuclear matter are dominated by subthreshold resonances, such as the $D_{13}(1520)$ resonance which produces an enhancement in the low-energy region of the ρ and ω meson spectral functions. Most calculations concluded that the spectral functions of vector mesons in cold nuclear matter are largely broadened with little shift of the peak mass. However, within an in-medium QCD sum rule approach it has been found [66] that the first moment of the ρ-meson spectral function can decrease (similarly to that in Eq. (13)) despite the fact that the mass peak changes very little.

Three-nucleon forces and hadronic medium modifications provide two complementary (and partly equivalent) frameworks for understanding and implementing density-dependence in the nucleon-nucleon interaction. The role of baryonic resonances are important for each: on the one hand they give rise to genuine three-nucleon forces, and on the other hand they provide the main mechanism for altering mesonic spectral functions in medium. The connection can be visualized in the three diagrams of Fig. 6, which show (a) a process contributing to the modification of the ω-meson spectral function in medium, (b) the effect that such a process would have on the one-boson-exchange NN interaction due to ω meson exchange, and

(c) the corresponding three-nucleon force obtained by opening the nucleon hole line in diagram (b). At present chiral three-nucleon forces are better constrained experimentally and provide a complete basis for including effects of the medium, while the properties of hadrons in medium are less constrained but have a more direct connection to questions regarding chiral symmetry restoration in dense nuclear systems. In the rest of this section we consider a specific model for implementing Brown-Rho scaling medium modifications in one-boson-exchange interactions in order to make a quantitative comparison to models of the density-dependent NN interaction from chiral three-nucleon forces.

The scaling relation in Eq. (13) should be regarded as only an approximate, mean-field relation that could be altered through higher-order corrections. Indeed, it is well-established [22,23] that a naive implementation of Eq. (13) would yield far too much attraction due to the decreasing of the effective scalar σ meson mass. A more accurate treatment of the intermediate-range attraction in the NN potential would be to replace the "ficticious" σ meson by a pair of correlated pions (in the $J^\pi = 0^+$ channel). This has been done in Ref. [69], where a meson exchange model for the S-wave $\pi\pi$ interaction was developed and found to be dominated by the exchange of the ρ meson as well as repulsive $\pi\pi$ contact interactions constrained by chiral symmetry. With this model for the $\pi\pi$ interaction, the $\pi\pi \to N\bar{N}$ amplitude was then constructed, which can be related through crossing relations and a dispersion integral to the $\pi N \to \pi N$ amplitude from which one can derive the effective 2π-exchange NN potential.

In this formalism finite density effects in the scalar-isoscalar channel arise through the density-depedence of the ρ meson mass (m_ρ) and form factor cutoff (Λ_ρ) as well as the $\pi\pi$ contact interactions. Moreover, the in-medium pion propagator is dressed with Δ-hole excitations. In practice, this complicated treatment of correlated 2π exchange was modeled with the exchange of two scalar-isoscalar bosons with a form factor cutoff of $\Lambda_\sigma = 3$ GeV and masses and coupling constants shown in Table 2. Together with the scaling of the vector meson masses and form factor cutoffs,[b] namely

$$\frac{m_\rho^*}{m_\rho} = \frac{m_\omega^*}{m_\omega} = \frac{\Lambda^*}{\Lambda} = 1 - 0.15\frac{n}{n_0}, \tag{21}$$

[b]To the extent that the phenomenological form factors associated nucleon-meson interaction vertices are governed by the nucleon radius, they too will decrease with density due to the proposed increase in the nucleon size [70].

Table 2. Masses and coupling constants for the two zero-width bosons parameterizing medium-modified correlated 2π exchange in the Bonn-B potential [23].

n/n_0	m_{σ_1} (MeV)	$g^2_{\sigma_1}/4\pi$	m_{σ_2} (MeV)	$g^2_{\sigma_2}/4\pi$
0.0	640	4.36	377	0.640
0.5	538	3.41	295	0.450
0.75	471	2.88	249	0.309
1.0	395	2.53	220	0.264
1.25	365	2.35	200	0.234

the modifications to correlated 2π exchange form the basis of the in-medium Bonn-B potential [23]. The potential includes also the exchange of the π, η, and a_0 mesons as described in Ref. [71]. Using this model of the medium-modified (MM) Bonn-B potential, the properties of symmetric nuclear matter were studied [23], and it was found that medium effects in scalar-isoscalar exchange diminished with increasing density, thereby allowing for a reasonable description of nuclear saturation in a Dirac-Brueckner-Hartree-Fock calculation.

In Fig. 7 we show the on-shell matrix elements of the low-momentum MM Bonn-B potential in relative S-waves for three different densities: $n = \{0, n_0/2 \text{ and } n_0\}$. In comparison to the S-wave matrix elements found for the density-dependent chiral interaction, we see that the main difference lies in the low-momentum components of the interaction. Whereas the Pauli-blocked two-pion exchange diagram in Fig. 3 gives rise to a large repulsion at

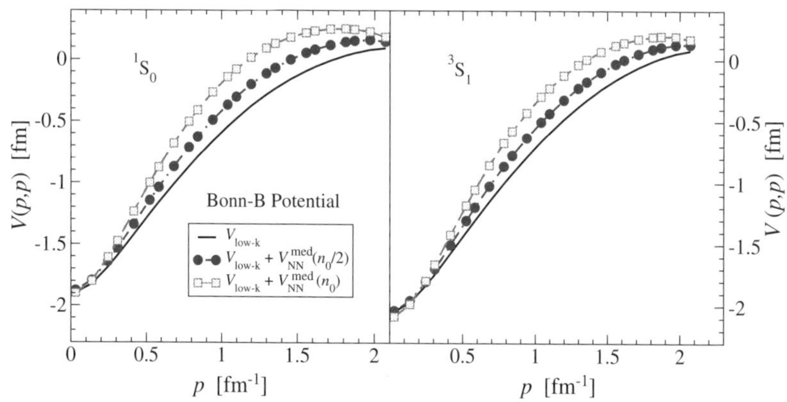

Fig. 7. Diagonal matrix elements for relative S waves obtained from the low-momentum medium-modified Bonn-B potential at three densities: $n = \{0, n_0/2, n_0\}$.

low momenta (see Fig. 5), the dropping of the effective σ_1 and σ_2 masses in the MM Bonn-B potential increases the medium- and long-range attraction. Indeed, at nuclear matter saturation density, the effective σ meson masses have dropped to 395 and 220 MeV (from 640 and 377 MeV respectively), which extends the range of the attraction significantly. This enhanced attraction is partially mitigated by a simultaneous drop in the effective coupling constants g_{σ_1} and g_{σ_2}. However, at higher momenta we see that the two theories lead to similar modifications in the S-wave interactions. In the density-dependent chiral interaction the short-range contact interaction proportional to c_E provides most of the additional repulsion, whereas in the MM Bonn-B potential an analogous effect is achieved through the dropping of the ω meson mass.

One of the original motivations for studying the beta decay of ^{14}C within the framework of Brown-Rho scaling was the sensitivity of the Gamow-Teller matrix element to the nuclear tensor force. In fact, employing a residual interaction consisting of only central and spin-orbit forces it is not possible to achieve a vanishing matrix element in a pure two p-hole configuration [45]. Jancovici and Talmi [46] showed that by including a strong tensor force one could construct an interaction which reproduces the lifetime of ^{14}C as well as the magnetic moment and electric quadrupole moment of ^{14}N, although agreement with the known spectroscopic data was unsatisfactory. The most important contributions to the tensor component of the nuclear potential arise from π and ρ meson exchange, which have opposite signs:

$$V_\rho^T(r) = -\frac{f_{N\rho}^2}{4\pi} m_\rho \vec{\tau}_1 \cdot \vec{\tau}_2\, S_{12} \left[\frac{1}{(m_\rho r)^3} + \frac{1}{(m_\rho r)^2} + \frac{1}{3m_\rho r}\right] e^{-m_\rho r},$$

$$V_\pi^T(r) = \frac{f_{N\pi}^2}{4\pi} m_\pi \vec{\tau}_1 \cdot \vec{\tau}_2\, S_{12} \left[\frac{1}{(m_\pi r)^3} + \frac{1}{(m_\pi r)^2} + \frac{1}{3m_\pi r}\right] e^{-m_\pi r}, \quad (22)$$

where $S_{12} = 3\vec{\sigma}_1 \cdot \hat{r}\, \vec{\sigma}_2 \cdot \hat{r} - \vec{\sigma}_1 \cdot \vec{\sigma}_2$ is the tensor operator and $f_{N\pi} = g_A m_\pi / 2f_\pi$. Since the ρ meson mass is expected to decrease substantially at nuclear matter density while the pion mass remains relatively constant, an unambiguous prediction of Brown-Rho scaling is the decreasing of the total tensor force at finite density. However, as we show in the next section, the additional short-range repulsive force resulting from the decreasing ω meson mass also plays an important role in suppressing the associated Gamow-Teller strength.

4. Beta Decay Lifetime of ^{14}C

The extremely long half-life of ^{14}C ($T_{1/2} \simeq 5730$ yr [72]) plays an essential role in archaeological dating methods, yet naively one would expect the lifetime to be nearly six orders of magnitude smaller. The 0^+ ground state of ^{14}C and the 1^+ ground state of ^{14}N to which it decays satisfy the selection for allowed Gamow-Teller transitions, and according to the Wigner spin-isospin SU(4) symmetry for light nuclei the associated transition matrix element would be unity. However, in order to reproduce the experimentally observed lifetime, the Gamow-Teller transition matrix element must be approximately $2 \cdot 10^{-3}$, which is expected to result from an accidental cancellation among otherwise large terms. This transition therefore provides a sensitive test for nuclear interaction models and many-body methods.

Early studies with phenomenological interactions highlighted the importance of the nuclear tensor force, without which it would not be possible to obtain a vanishing Gamow-Teller matrix element in a $0p^{-2}$ model space [45, 46]. Employing realistic NN interaction models did not improve the theoretical description significantly [73], but the use of larger model spaces in no-core shell model calculations appeared to generally diminish the Gamow-Teller matrix element [74]. Despite the long history of the problem, it was only recently suggested that the beta decay lifetime is particularly sensitive to the density-dependence of the nuclear force [25, 44]. The first of these studies employed the medium modified Bonn-B potential, described in the previous section, while the latter made use of the density-dependent chiral NN interaction constructed in Sec. 2. Indeed, since ^{14}C and ^{14}N lie just below the double shell-closure at ^{16}O, the density in the vicinity of the valence nucleons is fairly large. In Fig. 8 we plot twice the charge distribution of ^{14}N obtained from electron scattering experiments [75, 76] fit to the harmonic oscillator density distribution

$$n(r) \propto \left(1 + b\frac{r^2}{d^2}\right) e^{-r^2/d^2}, \tag{23}$$

where for ^{14}N, $b = 1.29$ and $d = 1.74\,\text{fm}$. Also plotted in the same figure is the square of the radial $0p$ wavefunction which peaks at a radius where the nucleon density is close to 90% that of saturated nuclear matter $n_0 = 0.16\,\text{fm}^{-3}$. The rest of this section is devoted to a systematic comparison between the two models of the in-medium nuclear interaction, which despite

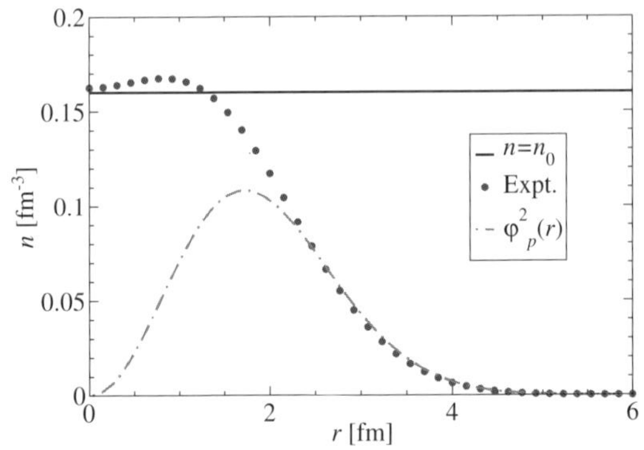

Fig. 8. Twice the charge distribution of ^{14}N taken from [75,76] together with the square of the radial $0p$-shell wavefunctions.

several qualitative differences in their density dependence, both predict that the beta decay transition of ^{14}C is strongly suppressed by medium effects in the nuclear force.

Nuclei with mass number $A > 13$ are currently beyond the reach of ab-initio methods, such as the no-core shell model [77] and quantum Monte Carlo techniques [78]. We approach this problem from shell model perturbation theory using a model space consisting of two $0p$ holes in a closed ^{16}O core. The many-body Schrödinger equation

$$H\Psi_n = E_n\Psi_n, \tag{24}$$

is replaced by the model space equation

$$H_{\text{eff}}\chi_m = E_m\chi_m, \tag{25}$$

where $H_{\text{eff}} = H_0 + V_{\text{eff}}$ and H_0 is the sum of the kinetic energy and the harmonic oscillator single particle potential with oscillator parameter $\hbar\omega =$ 14 MeV. We derive the shell model effective interaction V_{eff} following the formalism explained in Refs. [79,80]. The effective interaction is constructed from a folded diagram expansion in the so-called $\hat{Q}$-box, which consists of all irreducible valence-linked diagrams (usually truncated to a given order). In the present calculation we include hole-hole diagrams in the $\hat{Q}$-box up to second-order in perturbation theory, shown in Fig. 9. In this figure the wavy line represents the full interaction, either $V_{\text{low}-k} + V_{NN}^{\text{med}}$ or the low-momentum medium-modified Bonn-B potential, $V_{\text{Bonn-B}}^{\text{med}}$. Finally, we need the energy splitting between the two $0p$ orbitals, $\epsilon = e(p_{1/2}) - e(p_{3/2})$, which

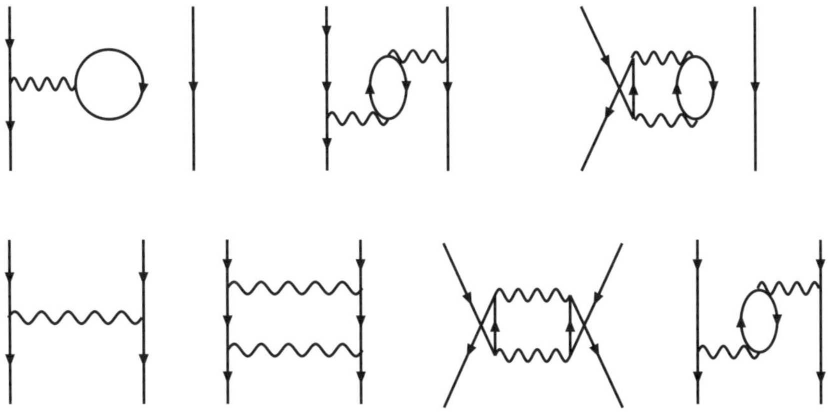

Fig. 9. Diagrams contributing to the effective interaction V_{eff} in our present calculation. Wavy lines represent the density-dependent in-medium nuclear interaction caculated in Secs. 2 and 3.

is most accurately obtained from the experimental excitation energy of the first $3/2^-$ state of ^{15}N located 6.3 MeV above the $1/2^-$ ground state.

We denote the relevant single-hole states by

$$|1\rangle = |0s_{1/2}^{-1}\rangle, |2\rangle = |0p_{3/2}^{-1}\rangle, |3\rangle = |0p_{1/2}^{-1}\rangle, \tag{26}$$

and the two-hole states coupled to good angular momentum J, isospin T, and parity π are denoted by $|\alpha\beta; J^\pi T\rangle$, where $\alpha, \beta = 2$ or 3. After constructing the effective interaction with the folded diagram method, we diagonalize the following matrix to obtain the eigenvalues and eigenstates with $J^\pi = 1^+, T = 0$ (relevant for ^{14}N):

$$\begin{bmatrix} & \vdots & \\ \cdots & \langle\alpha\beta; 1^+0|V_{\text{eff}}|\gamma\delta; 1^+0\rangle & \cdots \\ & \vdots & \end{bmatrix} + \begin{bmatrix} 0 & 0 & 0 \\ 0 & -\epsilon & 0 \\ 0 & 0 & -2\epsilon \end{bmatrix}, \tag{27}$$

For the $J^\pi = 0^+$, $T = 1$ states, we diagonalize a similar 2×2 matrix. The ground states for ^{14}C and ^{14}N are the lowest $J^\pi = 0^+$, $T = 1$ and $J^\pi = 1^+, T = 0$ states respectively, which we denote by

$$\begin{aligned} \psi_i &= a|22; 0^+1\rangle + b|33; 0^+1\rangle, \\ \psi_f &= x|22; 1^+0\rangle + y|23; 1^+0\rangle + z|33; 1^+0\rangle. \end{aligned} \tag{28}$$

From these wavefunctions one can derive the following expression for the reduced Gamow-Teller matrix element $M_{\rm GT}$:

$$M_{GT} = \sum_k \langle \psi_f || \sigma(k)\tau_+(k) || \psi_i \rangle = \frac{1}{\sqrt{6}}(-2\sqrt{5}ax + 2\sqrt{2}ay + 4by + 2bz), \quad (29)$$

where experimentally $M_{GT} \simeq 2 \cdot 10^{-3}$.

For orientation we first determine the low-lying states of ^{14}C and ^{14}N, as well as the Gamow-Teller transition matrix element connecting the two ground states, using the free-space low-momentum N^{3}LO potential at a resolution scale of $\Lambda_{{\rm low}-k} = 2.1\,{\rm fm}^{-1}$. We will see shortly that there is only a small difference when using the free-space Bonn-B potential. For the two 0^+ states of ^{14}C we find

$$\begin{aligned} \psi_1(0^+,1) &= 0.396|22;0^+,1\rangle + 0.918|33;0^+,1\rangle, \\ \psi_2(0^+,1) &= 0.918|22;0^+,1\rangle - 0.396|33;0^+,1\rangle, \end{aligned} \quad (30)$$

with energy splitting $\Delta E(0^+) = 12.9\,{\rm MeV}$. For the two lowest 1^+ states we find

$$\begin{aligned} \psi_1(1^+,0) &= 0.142|22;0^+,1\rangle - 0.681|23;0^+,1\rangle + 0.719|33;0^+,1\rangle, \\ \psi_2(1^+,0) &= 0.361|22;0^+,1\rangle - 0.663|23;0^+,1\rangle - 0.656|33;0^+,1\rangle, \end{aligned} \quad (31)$$

where the first excited state lies $\Delta E(1^+) = 2.76\,{\rm MeV}$ above the ground state. The second excited $J^\pi = 1^+, T = 0$ state lies nearly 16 MeV higher in energy than the ground state and will therefore be neglected throughout. With these wavefunctions, the ground state to ground state Gamow-Teller transition matrix element is found to be

$$M_{GT} = -0.90, \quad (32)$$

which is much too large to describe the known lifetime of ^{14}C. From the ground state wavefunctions in Eqs. (30) and (31) we see that the first three terms in Eq. (29) add coherently while the last gives a contribution of the opposite sign, though not enough to cancel the first three. In order for medium effects to suppress the transition, strength must therefore be shifted to the $|33\rangle$ component of both ground state wavefunctions.

We now implement density-dependence in the nuclear potential and study the effect on the model space interaction matrix elements. In Tables 3 and 4 we show the shell model matrix elements for the low-momentum interactions derived from the in-medium chiral N^{3}LO and

Table 3. Matrix elements (in units of MeV) between $0p^{-2}$ states coupled to $(J^\pi, T) = (0^+, 1)$ for the in-medium chiral interaction as well as the Brown-Rho-scaled Bonn-B potential at three different densities: $n = 0, 0.25n_0$, and n_0.

$J^\pi = 0^+, T = 1$			
n/n_0	$\langle 22\lvert V^{\rm med}_{\rm Chiral}\rvert 22\rangle$	$\langle 22\lvert V^{\rm med}_{\rm Chiral}\rvert 33\rangle$	$\langle 33\lvert V^{\rm med}_{\rm Chiral}\rvert 33\rangle$
0	−3.28	−3.77	−0.61
0.25	−2.64	−3.66	−0.05
1.0	−0.12	−2.88	1.91
n/n_0	$\langle 22\lvert V^{\rm med}_{\rm Bonn}\rvert 22\rangle$	$\langle 22\lvert V^{\rm med}_{\rm Bonn}\rvert 33\rangle$	$\langle 33\lvert V^{\rm med}_{\rm Bonn}\rvert 33\rangle$
0	−3.12	−3.65	−0.54
0.25	−2.72	−3.38	−0.33
1.0	−1.09	−2.21	0.48

Brown-Rho-scaled (BRS) Bonn-B potentials both in free space and at nuclear matter saturation density. Table 3 shows the matrix elements for the $J^\pi = 0^+, T = 1$ states, while Table 4 shows the matrix elements for the $J^\pi = 1^+, T = 0$ states. Overall, medium effects in both potentials lead to qualitatively similar changes in the shell model matrix elements, though the density-dependence of the in-medium chiral interaction is much stronger. This should not be surprising, given the stronger S-wave repulsion in the in-medium chiral interaction, as compared to the BRS Bonn-B potential, shown in Figs. 4 and 7.

At nuclear matter saturation density, medium effects on both potentials are quite large. Nevertheless, we can gain insight by studying the interactions at low densities and employing perturbation theory. One finds the following corrections to the ground state wavefunctions:

$$\psi_1^{(1)}(0^+,1) = \psi_1^{(0)}(0^+,1) - \frac{1}{12.9}[0.36\langle 22|V^{\rm med}|22\rangle + 0.69\langle 22|V^{\rm med}|33\rangle - 0.36\langle 33|V^{\rm med}|33\rangle]\psi_2^{(0)}(0^+,1) \tag{33}$$

and

$$\psi_1^{(1)}(1^+,0) = \psi_1^{(0)}(1^+,0) - \frac{1}{2.76}[0.05\langle 22|V^{\rm med}|22\rangle - 0.34\langle 23|V^{\rm med}|22\rangle + 0.45\langle 23|V^{\rm med}|23\rangle - 0.03\langle 23|V^{\rm med}|33\rangle + 0.17\langle 22|V^{\rm med}|33\rangle - 0.47\langle 33|V^{\rm med}|33\rangle]\psi_2^{(0)}(1^+,0). \tag{34}$$

Table 4. Matrix elements (in units of MeV) between $0p^{-2}$ states coupled to $(J^\pi, T) = (1^+, 0)$ for the in-medium chiral interaction as well as the Brown-Rho-scaled Bonn-B potential at three different densities: $n = 0, 0.25n_0$, and n_0.

$J^\pi = 1^+, T = 0$			
n/n_0	$\langle 22 \vert V^{\rm med}_{\rm Chiral} \vert 22\rangle$	$\langle 23 \vert V^{\rm med}_{\rm Chiral} \vert 22\rangle$	$\langle 23 \vert V^{\rm med}_{\rm Chiral} \vert 23\rangle$
0.0	−1.19	3.89	−5.20
0.25	−0.61	3.74	−4.81
1.0	1.29	2.81	−3.03
n/n_0	$\langle 23 \vert V^{\rm med}_{\rm Chiral} \vert 33\rangle$	$\langle 22 \vert V^{\rm med}_{\rm Chiral} \vert 33\rangle$	$\langle 33 \vert V^{\rm med}_{\rm Chiral} \vert 33\rangle$
0.0	1.36	1.66	−1.67
0.25	1.50	1.33	−1.00
1.0	1.92	0.44	0.72
n/n_0	$\langle 22 \vert V^{\rm med}_{\rm Bonn} \vert 22\rangle$	$\langle 23 \vert V^{\rm med}_{\rm Bonn} \vert 22\rangle$	$\langle 23 \vert V^{\rm med}_{\rm Bonn} \vert 23\rangle$
0.0	−1.28	3.50	−4.91
0.25	−1.09	3.27	−4.56
1.0	−0.31	2.31	−3.11
n/n_0	$\langle 23 \vert V^{\rm med}_{\rm Bonn} \vert 33\rangle$	$\langle 22 \vert V^{\rm med}_{\rm Bonn} \vert 33\rangle$	$\langle 33 \vert V^{\rm med}_{\rm Bonn} \vert 33\rangle$
0.0	1.23	1.46	−1.68
0.25	1.24	1.38	−1.57
1.0	1.26	1.04	−1.09

A straightforward calculation using the matrix elements shown in Tables 3 and 4 for $n = 0.25n_0$ gives for the un-normalized perturbed wavefunctions arising from $V^{\rm med}_{\rm Chiral}$

$$\psi_0^{(1)}(0^+,1) = \psi_0^{(0)}(0^+,1) - 0.01\psi_1^{(0)}(0^+,1) \quad \text{and}$$
$$\psi_0^{(1)}(1^+,0) = \psi_0^{(0)}(1^+,0) + 0.04\psi_1^{(0)}(1^+,0), \tag{35}$$

We see that a positive $\langle 22; 0^+1 | V^{\rm med} | 22; 0^+1\rangle$ and a positive $\langle 22; 0^+1 | V^{\rm med} | 33; 0^+1\rangle$ lead to a stronger $|33; 0^+, 1\rangle$ component in the perturbed ^{14}C ground state wavefunction. One therefore expects quite generally that additional density-dependent repulsion would increase the $|33; 0^+1\rangle$ component in the ^{14}C ground state and therefore help suppress the beta decay of ^{14}C. Both the in-medium chiral nuclear interaction and the BRS Bonn-B potential share this feature.

For the ground state of ^{14}N the situation is more complicated due to the larger number of terms in the perturbed wavefunction of Eq. (34)

as well as the more complicated dependence of the shell model matrix elements on the perturbing potential. The two dominant terms are $\langle 33; 1^+0|V^{\rm med}|33; 1^+0\rangle$ and $\langle 23; 1^+0|V^{\rm med}|23; 1^+0\rangle$, which have the same sign as the perturbing potential in S-waves but enter with opposite sign in the perturbed wavefunction. At one-quarter of nuclear matter density, the sum of all six contributions to $V^{\rm med}_{\rm Chiral}$ yields $\langle 33; 1^+0|V^{\rm med}_{\rm Chiral}|33; 1^+0\rangle =$ 0.67 MeV and $\langle 23; 1^+0|V^{\rm med}_{\rm Chiral}|23; 1^+0\rangle = 0.39$ MeV. The remaining four terms together give a minute contribution to the perturbed wavefunction, and therefore we expect at low densities for the $|33; 1^+, 0\rangle$ component in the ground state of ^{14}N to decrease for the in-medium chiral interaction and consequently increase the magnitude of the Gamow-Teller transition matrix element. Since different components of the in-medium chiral interaction have different density-dependences, this behavior can change at higher densities. Indeed by inspecting the shell model matrix elements at nuclear matter saturation density we see that the opposite effect occurs and the GT matrix element is suppressed rather than enhanced (the perturbing potential is quite large and therefore our argument based on perturbation theory should be treated cautiously). In contrast, the BRS Bonn-B potential has $\langle 33; 1^+0|V^{\rm med}_{\rm Bonn}|33; 1^+0\rangle = 0.11$ MeV and $\langle 23; 1^+0|V^{\rm med}_{\rm Bonn}|23; 1^+0\rangle =$ 0.35 MeV, which suggests that even at low densities the beta decay should be suppressed.

We now consider the results of the full calculation, including all density-dependent contributions to second-order in the shell model effective interaction shown in Fig. 9. In Table 5 we show the results of our calculations for the expansion coefficients of the ground state wavefunctions of ^{14}C and ^{14}N, as well as the reduced GT matrix element, as a function of the nuclear density up to $n = n_0$. The full results agree well with the conclusions we obtained from arguments based on perturbation theory. The Gamow-Teller transition matrix element decreases immediately at low densities in the BRS

Table 5. The coefficients of the jj-coupled wavefunctions defined in Eq. (28) and the associated reduced GT matrix element as a function of the nuclear density n for both the Brown-Rho-scaled Bonn-B potential and the in-medium chiral nuclear interaction.

	In-medium chiral N^{3}LO						In-medium Bonn-B					
n/n_0	a	b	x	y	z	$M_{\rm GT}$	a	b	x	y	z	$M_{\rm GT}$
0.0	0.40	0.92	0.14	−0.68	0.72	−0.90	0.40	0.93	0.11	−0.62	0.78	−0.70
0.2	0.38	0.92	0.15	−0.69	0.71	−0.91	0.35	0.94	0.08	−0.55	0.83	−0.49
0.4	0.36	0.93	0.14	−0.67	0.73	−0.82	0.32	0.95	0.05	−0.50	0.86	−0.32
0.6	0.33	0.94	0.13	−0.65	0.75	−0.74	0.29	0.96	0.03	−0.45	0.89	−0.17
0.8	0.30	0.96	0.12	−0.62	0.78	−0.64	0.25	0.97	0.02	−0.40	0.92	−0.04
1.0	0.26	0.97	0.11	−0.59	0.80	−0.53	0.21	0.98	0.01	−0.36	0.93	0.07

Bonn-B potential, while only after $n \simeq 0.25n_0$ do the density-dependent corrections of the chiral nuclear interaction lead to a suppression of the matrix element.

The Gamow-Teller transition strength, $B(GT)$, is related to the reduced GT transition matrix element by

$$B(GT) = (g_A^*)^2 \frac{1}{2J_i+1} |M_{\rm GT}|^2 \simeq |M_{\rm GT}|^2, \tag{36}$$

where $J_i = 0$ and we have used the approximation that the in-medium axial vector coupling constant $g_A^* \simeq 1$. In general, the effective Gamow-Teller operator has additional terms in a nuclear medium [81, 82]:

$$\vec{\mathcal{O}}_{\rm GT,eff} = g_{LA}\vec{L} + g_A^*\,\vec{\sigma} + g_{PA}[Y_2, \vec{\sigma}], \tag{37}$$

where $\vec{L}$ is the orbital angular momentum operator and Y_2 denotes the rank-2 spherical harmonic $Y_{2,m}$. However, from theoretical calculations [83] of the effective GT operator in-medium, as well as beta decay calculations [84] performed with phenomenological shell model effective interations, it is known that g_{LA} and g_{PA} are almost negligible and that in light nuclei g_A^* is smaller by 15–20% from its free space value of $g_A = 1.27$ (measured in free neutron beta decay). In general, one must calculate both the effective interaction $V_{\rm eff}$ and the effective Gamow-Teller operator $\vec{\mathcal{O}}_{\rm GT,eff}$. In our calculations we assume a 20% reduction of g_A in medium and therefore set $g_A^* = 1.0$.

In Fig. 10 we plot the $B(GT)$ strengths for transitions from the low-lying states of ^{14}C to the ground state of ^{14}N for both density-dependent NN interactions. These are compared to the experimental values [85] obtained from the charge-exchange reaction $^{14}\mathrm{N}(d, {}^2\mathrm{He})^{14}\mathrm{C}$. From Fig. 10 one sees that the most dramatic effect is a reduction in the ground state to ground state Gamow-Teller transition strength in both models. In the Brown-Rho-scaled Bonn-B potential also the first excited $(0^+, 1)$ state of ^{14}C is decreased toward the experimental value. Such an effect can be obtained from in-medium chiral interactions, but only if the strength of the contact three-body force is increased [44]. In the present version of the in-medium chiral nuclear interaction, there appears to be very little effect on the other Gamow-Teller transition strengths.

Besides comparing the Gamow-Teller strengths within the two models, it is useful to study the energy splitting between the ground state of ^{14}N and the first excited $(0^+, 1)$ state (the isobaric analogue of the ^{14}C ground state), which is experimentally observed to lie at 2.3 MeV. In Fig. 11 we plot this

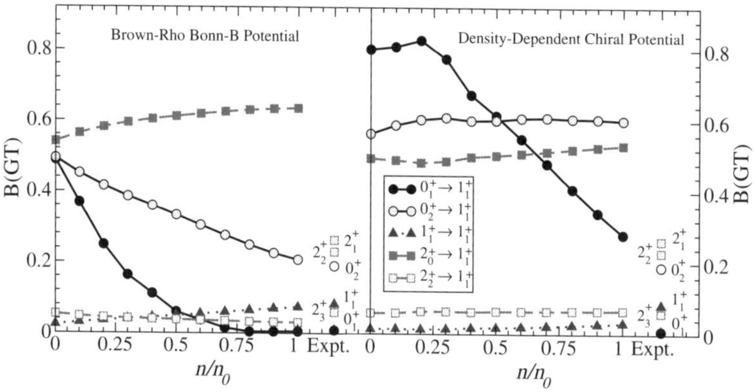

Fig. 10. The $B(GT)$ values for transitions from low-lying states of ^{14}C to the ground state of ^{14}N as a function of the nuclear density. The experimental values are from [85]. Note that there are three experimental low-lying 2^+ states compared to two theoretical 2^+ states in the $0p^{-2}$ shell model configuration.

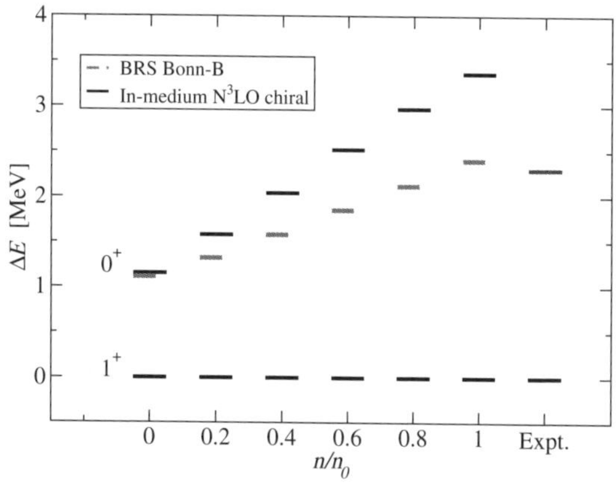

Fig. 11. The splitting between the 1^+ ground state of ^{14}N and the first excited 0^+ state as a function of the nuclear density for both the medium-modified Bonn-B potential and the in-medium chiral nuclear interaction. Also shown is the experimental energy splitting $\Delta E_{0^+-1^+} = 2.3\,\text{MeV}$.

energy splitting as a function of the nuclear density for the in-medium chiral nuclear interaction and the BRS Bonn-B potential. It is clear that free-space nucleon-nucleon interactions systematically underpredict this level splitting and that additional density-dependent repulsion leads to an enhancement of the splitting. The in-medium chiral nuclear interaction is much more repulsive than the BRS Bonn-B potential and therefore leads to a larger increase in the splitting which overpredicts the value at higher densities.

Nevertheless, we find that qualitatively the two models for the density-dependent NN interaction are consistent with one another and improve the agreement with experiment.

In order to better understand the relationship between the in-medium chiral nuclear interaction and the BRS Bonn-B potential, we first study in detail the six different one-loop contributions to $V_{NN}^{\rm med}$ arising from the leading-order chiral three-nucleon force. In Tables 6 and 7 we show the

Table 6. Matrix elements (in units of MeV) between $0p^{-2}$ states coupled to $(J^\pi, T) = (0^+, 1)$ for the six density-dependent contributions to the in-medium chiral nuclear interaction at nuclear matter saturation density n_0.

	$J^\pi = 0^+, T = 1 \quad (n = n_0)$		
	$\langle 22 \vert V^{\rm med}_{\rm Chiral} \vert 22\rangle$	$\langle 22 \vert V^{\rm med}_{\rm Chiral} \vert 33\rangle$	$\langle 33 \vert V^{\rm med}_{\rm Chiral} \vert 33\rangle$
1	3.88	0.80	3.32
2	−3.18	−0.71	−2.68
3	2.15	0.32	1.92
4	−0.56	−0.12	−0.47
5	−0.89	−0.64	−0.44
6	1.75	1.24	0.88

Table 7. Matrix elements (in units of MeV) between $0p^{-2}$ states coupled to $(J^\pi, T) = (1^+, 0)$ for the six density-dependent contributions to the in-medium chiral nuclear interaction at nuclear matter saturation density n_0.

	$J^\pi = 1^+, T = 0 \quad (n = n_0)$		
	$\langle 22 \vert V^{\rm med}_{\rm Chiral} \vert 22\rangle$	$\langle 23 \vert V^{\rm med}_{\rm Chiral} \vert 22\rangle$	$\langle 23 \vert V^{\rm med}_{\rm Chiral} \vert 23\rangle$
1	4.25	−0.65	1.67
2	−3.15	0.73	−1.47
3	1.36	−0.76	1.53
4	−0.54	0.13	−0.27
5	−0.47	0.56	−1.02
6	1.04	−1.09	1.73
	$\langle 23 \vert V^{\rm med}_{\rm Chiral} \vert 33\rangle$	$\langle 22 \vert V^{\rm med}_{\rm Chiral} \vert 33\rangle$	$\langle 33 \vert V^{\rm med}_{\rm Chiral} \vert 33\rangle$
1	2.14	−1.67	3.28
2	−1.44	1.47	−2.65
3	0.00	−0.86	1.70
4	−0.23	0.25	−0.47
5	0.100	0.14	−0.35
6	−0.01	−0.55	0.88

most important shell model matrix elements for each of these components at nuclear matter saturation density n_0. Again, we let perturbation theory arguments direct our attention to the relevant physics. As previously mentioned, in order to suppress the Gamow-Teller transition matrix element, it is necessary to increase the $|33; J^\pi T\rangle$ components of the ground state wavefunctions for both ^{14}C and ^{14}N. From Eq. (33) it follows that only $V_{\rm NN}^{\rm med,1}$, $V_{\rm NN}^{\rm med,3}$, and $V_{\rm NN}^{\rm med,6}$ contribute in the desired way. The remaining three contributions to $V_{\rm NN}^{\rm med}$ are attractive and decrease the strength of the $|33; 0^+, 1\rangle$ component of the ^{14}C ground state wavefunction. The structure of the ^{14}N perturbed ground state is most strongly affected by the $\langle 33; 1^+0|V^{\rm med}|33; 1^+0\rangle$ and $\langle 23; 1^+0|V^{\rm med}|23; 1^+0\rangle$ matrix elements. A secondary role is played by the $\langle 22; 1^+0|V^{\rm med}|23; 1^+0\rangle$ and $\langle 22; 1^+0|V^{\rm med}|33; 1^+0\rangle$ components. From Table 7 we see that out of the three contributions that shift strength to the $|33; 0^+1\rangle$ component of the ^{14}C ground state, only the three-nucleon contact interaction proportional to c_E also strongly shifts strength to the $|33; 1^+0\rangle$ component of ^{14}N:

$$\begin{aligned} V_{\rm NN}^{\rm med,1} &: \psi_0^{(1)}(1^+, 0) = \psi_0^{(0)}(0^+, 1) + 0.25\psi_1^{(0)}(0^+, 1), \\ V_{\rm NN}^{\rm med,3} &: \psi_0^{(1)}(1^+, 0) = \psi_0^{(0)}(1^+, 0) - 0.03\psi_1^{(0)}(1^+, 0), \\ V_{\rm NN}^{\rm med,6} &: \psi_0^{(1)}(1^+, 0) = \psi_0^{(0)}(1^+, 0) - 0.25\psi_1^{(0)}(1^+, 0). \end{aligned} \tag{38}$$

A full treatment shows that $V_{\rm NN}^{\rm med,3}$ alone has little effect on the ground state wavefunction of ^{14}N and in fact actually decreases the strength in the $|33; 1^+0\rangle$ component of the wavefunction. Therefore, the strong short-distance repulsion generated by the three-nucleon contact interaction is the dominant term contributing to the suppression the Gamow-Teller transition matrix element.

In the BRS Bonn-B potential one might expect analogously that the scaling of the ω-meson mass plays a similar role. We therefore have performed a similar calculation using a BRS Bonn-B potential with no ω-meson mass scaling. In Fig. 12 we have plotted the $B(GT)$ values of the ground state to ground state transition using medium-modified interactions with the short-distance repulsion removed (the three-body contact term in the case of the in-medium chiral interaction and the dropping ω meson mass in the case of the BRS Bonn-B potential). We find that in both cases there is in fact a strong enhancement of the Gamow-Teller transition. Therefore, despite the fact that these two density-dependent interactions are behaving very differently at long distances (as inferred from the low-momentum components of their interactions in Figs. 4 and 7), they both include a

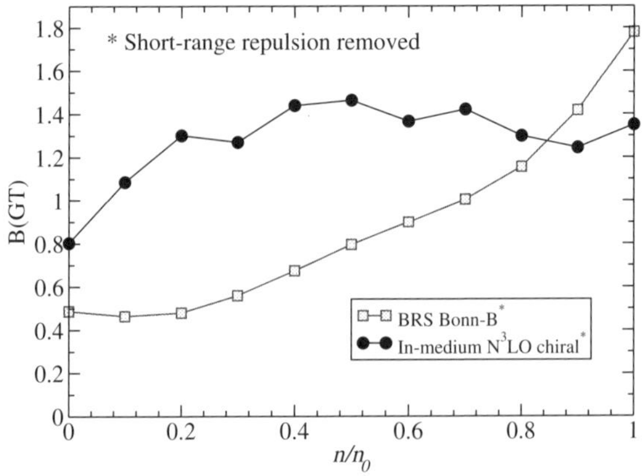

Fig. 12. The $B(GT)$ values for the ^{14}C ground state to ^{14}N ground state transition as a function of the density for the medium-modified Bonn-B potential and the in-medium chiral nucleon-nucleon potential. The asterisk denotes that the additional density-dependent short-range repulsion has been removed from both potentials: the ω meson scaling in the case of the MM Bonn-B potential and the short-range contact three-nucleon force proportional to c_E in the in-medium chiral interaction.

strong short-distance repulsion that dramatically suppresses the Gamow-Teller transition matrix element.

5. Summary and Conclusions

We have studied the effect of density-dependent low-momentum nucleon-nucleon interactions on the beta decay lifetime of ^{14}C. The in-medium chiral nuclear interaction was obtained from the leading-order three-nucleon force by contracting one pair of external lines and summing over the filled Fermi sea of nucleons, whereas the Brown-Rho-scaled Bonn-B potential was obtained by decreasing the masses and form factor cutoffs of the vector mesons and treating microscopically effective σ meson exchange as correlated two-pion exchange. Although both density-dependent interactions introduce additional strong repulsion at short distances, they differ largely at intermediate and long distances. The Brown-Rho scaled Bonn-B potential enhances the attraction at these distances (through a decreasing "σ" meson mass), but the Pauli-blocked two-pion exchange diagram obtained from the long-range chiral three-nucleon force is strongly repulsive in this region. Nevertheless, we find that in both models of the in-medium nuclear interaction the ground state to ground state Gamow-Teller transition strength is strongly suppressed at high densities. In contrast, the

GT strengths from the ground state of ^{14}N to the excited states of ^{14}C exhibit only a small density dependence. From perturbative considerations we have found that, in general, short-range repulsive interactions lead to a strong suppression of the ground state to ground state transition. Moreover, we have explicitly removed the source of short-range repulsion in both interactions and found that this leads rather to an enhancement of the Gamow-Teller strength. We therefore suggest that the common short-distance behavior of these two interactions is responsible for their qualitatively similar effects on the beta decay lifetime of ^{14}C.

Dedication

The authors of this work represent three generations of former students and postdocs who have been immensely inspired by Gerry Brown's physics intuition and ideas. We gratefully dedicate this paper to Gerry on the occasion of his 85th birthday.

Appendix: Explicit Expressions for the Fermi Sphere Integrals in V_{NN}^{med}

In this appendix we tabulate the different functions that contribute to the density-dependent expressions in Eqs. (6)–(12). The functions $\Gamma_j(p)$ result from Fermi sphere integrals over a single static pion propagator and have the explicit forms:

$$\Gamma_0(p) = k_f - m_\pi \left[\arctan\frac{k_f + p}{m_\pi} + \arctan\frac{k_f - p}{m_\pi}\right] + \frac{m_\pi^2 + k_f^2 - p^2}{4p} \ln\frac{m_\pi^2 + (k_f + p)^2}{m_\pi^2 + (k_f - p)^2}, \tag{A.1}$$

$$\Gamma_1(p) = \frac{k_f}{4p^2}(m_\pi^2 + k_f^2 + p^2) - \Gamma_0(p) - \frac{1}{16p^3}[m_\pi^2 + (k_f + p)^2] \times [m_\pi^2 + (k_f - p)^2] \ln\frac{m_\pi^2 + (k_f + p)^2}{m_\pi^2 + (k_f - p)^2}, \tag{A.2}$$

$$\Gamma_2(p) = \frac{k_f^3}{9} + \frac{1}{6}(k_f^2 - m_\pi^2 - p^2)\Gamma_0(p) + \frac{1}{6}(m_\pi^2 + k_f^2 - p^2)\Gamma_1(p), \tag{A.3}$$

$$\Gamma_3(p) = \frac{k_f^3}{3p^2} - \frac{m_\pi^2 + k_f^2 + p^2}{2p^2}\Gamma_0(p) - \frac{m_\pi^2 + k_f^2 + 3p^2}{2p^2}\Gamma_1(p). \tag{A.4}$$

The $G_j(p,q)$ functions result from Fermi sphere integrals over the product of two static pion propagators. Angular integrations are performed analytically, and the remaining relevant momentum integrals are given by

$$G_{0,*,**}(p,q) = \frac{2}{q}\int_0^{k_f} dk\, \frac{\{k,k^3,k^5\}}{\sqrt{A(p)+q^2k^2}} \ln\frac{q\,k+\sqrt{A(p)+q^2k^2}}{\sqrt{A(p)}}, \qquad \text{(A.5)}$$

where $A(p) = [m_\pi^2 + (k+p)^2][m_\pi^2 + (k-p)^2]$. The remaining functions $G_j(p,q)$ are related to those above through the equations

$$G_1(p,q) = \frac{\Gamma_0(p) - (m_\pi^2+p^2)G_0(p,q) - G_*(p,q)}{4p^2-q^2}, \qquad \text{(A.6)}$$

$$G_{1*}(p,q) = \frac{3\Gamma_2(p) + p^2\Gamma_3(p) - (m_\pi^2+p^2)G_*(p,q) - G_{**}(p,q)}{4p^2-q^2}, \qquad \text{(A.7)}$$

$$G_2(p,q) = (m_\pi^2+p^2)G_1(p,q) + G_*(p,q) + G_{1*}(p,q), \qquad \text{(A.8)}$$

$$G_3(p,q) = \frac{\frac{1}{2}\Gamma_1(p) - 2(m_\pi^2+p^2)G_1(p,q) - 2G_{1*}(p,q) - G_*(p,q)}{4p^2-q^2}. \qquad \text{(A.9)}$$

In this chain of equations the functions indexed with an asterisk play only an auxiliary role for the construction of $G_{1,2,3}(p,q)$. We note that all functions $G_j(p,q)$ are non-singular at $q = 2p$ (corresponding to scattering in backward direction). For notational simplicity, the k_f-dependence of $\Gamma_j(p)$ and $G_j(p,q)$ has been suppressed.

References

[1] S. Weinberg, *Phys. Lett. B* **251**, 288 (1990).
[2] S. Weinberg, *Nucl. Phys.* **B363**, 3 (1991).
[3] C. Ordonez and U. van Kolck, *Phys. Lett. B* **291**, 459 (1992).
[4] N. Kaiser, R. Brockmann and W. Weise, *Nucl. Phys.* **A625**, 758 (1997).
[5] E. Epelbaum, W. Glöckle and U.-G. Meißner, *Nucl. Phys.* **A637**, 107 (1998).
[6] N. Kaiser, *Phys. Rev. C* **64**, 057001 (2001).
[7] N. Kaiser, *Phys. Rev. C* **65**, 017001 (2001).
[8] D.R. Entem and R. Machleidt, *Phys. Rev. C* **66**, 014002 (2002).
[9] D.R. Entem and R. Machleidt, *Phys. Rev. C* **68**, 041001(R) (2003).
[10] E. Epelbaum, *Prog. Part. Nucl. Phys.* **57**, 654 (2006).
[11] C. Adami, T. Hatsuda and I. Zahed, *Phys. Rev. D* **43**, 921 (1991).
[12] G.E. Brown and M. Rho, *Phys. Rev. Lett.* **66**, 2720 (1991).
[13] T. Hatsuda and S.H. Lee, *Phys. Rev. C* **46**, R34 (1992).
[14] G.E. Brown and M. Rho, *Phys. Rept.* **269**, 333 (1996).

[15] R. Rapp and J. Wambach, *Adv. Nucl. Phys.* **25**, 1 (2000).
[16] R. Rapp, J. Wambach and H. van Hees, arXiv:0901.3289.
[17] M. Lutz, S. Klimt and W. Weise, *Nucl. Phys.* **A542**, 521 (1992).
[18] E.G. Drukarev and E.M. Levin, *Nucl. Phys.* **A511**, 679 (1990); **A516**, 715 (1990).
[19] N. Kaiser, P. de Homont and W. Weise, *Phys. Rev. C* **77**, 025204 (2008).
[20] M. Bando, T. Kugo and K. Yamawaki, *Phys. Rept.* **164**, 217 (1988).
[21] M. Harada and K. Yamawaki, *Phys. Rept.* **381**, 1 (2003).
[22] G.E. Brown, H. Müther and M. Prakash, *Nucl. Phys.* **A506**, 565 (1990).
[23] R. Rapp, R. Machleidt, J.W. Durso and G.E. Brown, *Phys. Rev. Lett.* **82**, 1827 (1999).
[24] J.W. Holt, G.E. Brown, J.D. Holt and T.T.S. Kuo, *Nucl. Phys.* **A785**, 322 (2007).
[25] J.W. Holt, G.E. Brown, T.T.S. Kuo, J.D. Holt and R. Machleidt, *Phys. Rev. Lett.* **100**, 062501 (2008).
[26] L.-W. Siu, J.W. Holt, T.T.S. Kuo and G.E. Brown, *Phys. Rev. C* **79**, 054004 (2009).
[27] S. Leupold, V. Metag and U. Mosel, *Int. J. Mod. Phys. E* **19**, 147 (2010).
[28] F. Klingl, N. Kaiser and W. Weise, *Nucl. Phys.* **A624**, 527 (1997).
[29] M. Post and U. Mosel, *Nucl. Phys.* **A688**, 808 (2001).
[30] W. Peters, M. Post, H. Lenske, S. Leupold and U. Mosel, *Nucl. Phys.* **A632**, 109 (1998).
[31] R. Rapp and J. Wambach, *Adv. Nucl. Phys.* **25**, 1 (2000).
[32] A. Nogga, H. Kamada and W. Glöckle, *Phys. Rev. Lett.* **85**, 944 (2000).
[33] S.C. Pieper, K. Varga and R.B. Wiringa, *Phys. Rev. C* **66**, 044310 (2002).
[34] P. Navrátil, V.G. Gueorguiev, J.P. Vary, W.E. Ormand and A. Nogga, *Phys. Rev. Lett.* **99**, 042501 (2007).
[35] B.D. Day, *Rev. Mod. Phys.* **39**, 719 (1967).
[36] H.Q. Song, M. Baldo, G. Giansiracusa and U. Lombardo, *Phys. Rev. Lett.* **81**, 1584 (1998).
[37] I.E. Lagaris and V.R. Pandharipande, *Nucl. Phys.* **A359**, 331 (1981); **A359**, 349 (1981).
[38] S.D. Yang, J. Heyer and T.T.S. Kuo, *Nucl. Phys.* **A448**, 420 (1986).
[39] R.B. Wiringa, V. Fiks and A. Fabrocini, *Phys. Rev. C* **38**, 1010 (1988).
[40] N. Sakamoto *et al.*, *Phys. Lett. B* **367**, 60 (1996).
[41] H. Sakai *et al.*, *Phys. Rev. Lett.* **84**, 5288 (2000).
[42] H. Witala, W. Glöckle, D. Hüber, J. Golak and H. Kamada, *Phys. Rev. Lett.* **81**, 1183 (1998).
[43] S. Nemoto, K. Chmielewski, S. Oryu and P.U. Sauer, *Phys. Rev. C* **58**, 2599 (1998).
[44] J.W. Holt, N. Kaiser and W. Weise, *Phys. Rev. C* **79**, 054331 (2009).
[45] D.R. Inglis, *Rev. Mod. Phys.* **25**, 390 (1953).
[46] B. Jancovici and I. Talmi, *Phys. Rev.* **95**, 289 (1954).
[47] E. Epelbaum, H.-W. Hammer and U.-G. Meißner, *Rev. Mod. Phys.* **81**, 1773 (2010).
[48] R. Machleidt, *Phys. Rev. C* **63**, 024001 (2001).

[49] V.G.J. Stoks, R.A.M. Klomp, C.P.F. Terheggen and J.J. de Swart, *Phys. Rev. C* **49**, 2950 (1994).
[50] R.B. Wiringa, V.G.J. Stoks and R. Schiavilla, *Phys. Rev. C* **51** 38 (1995).
[51] M.C.M. Rentmeester, R.G.E. Timmermans and J.J. de Swart, *Phys. Rev. C* **67**, 044001 (2003).
[52] E. Epelbaum, A. Nogga, W. Glöckle, H. Kamada, U.-G. Meißner and H. Witala, *Phys. Rev. C* **66**, 064001 (2002).
[53] A. Nogga, P. Navratil, B.R. Barrett and J.P. Vary, *Phys. Rev. C* **73**, 064002 (2006).
[54] S.K. Bogner, T.T.S. Kuo, L. Coraggio, A. Covello and N. Itaco, *Phys. Rev. C* **65**, 051301(R) (2002).
[55] S.K. Bogner, T.T.S. Kuo and A. Schwenk, *Phys. Rep.* **386**, 1 (2003).
[56] A. Nogga, S.K. Bogner and A. Schwenk, *Phys. Rev. C* **70**, 061002(R) (2004).
[57] J.W. Holt, N. Kaiser and W. Weise, *Phys. Rev. C* **81**, 024002 (2010).
[58] N. Kaiser, R. Brockmann and W. Weise, *Nucl. Phys.* **A625**, 758 (1997).
[59] S.K. Bogner, A. Schwenk, R.J. Furnstahl and A. Nogga, *Nucl. Phys.* **A763** (2005) 59.
[60] B.A. Campbell, J. Ellis and K.A. Olive, *Nucl. Phys.* **B345**, 57 (1990).
[61] K. Suzuki *et al.*, *Phys. Rev. Lett.* **92**, 072302 (2004).
[62] E.E. Kolomeitsev, N. Kaiser and W. Weise, *Phys. Rev. Lett.* **90**, 092501 (2003).
[63] M. Gell-Mann, R.J. Oakes and B. Renner, *Phys. Rev.* **175**, 2195 (1968).
[64] T.D. Cohen, R.J. Furnstahl and D.K. Griegel, *Phys. Rev. Lett.* **67**, 961 (1991).
[65] J. Gasser, H. Leutwyler and M.E. Sainio, *Phys. Lett.* **B253**, 252 (1991).
[66] Y. Kwon, M. Procura and W. Weise, *Phys. Rev. D* **78**, 055203 (2008).
[67] M. Post and U. Mosel, *Nucl. Phys.* **A699**, 169 (2002).
[68] B. Friman and H.J. Pirner, *Nucl. Phys.* **A617**, 496 (1997).
[69] H.-C. Kim, J.W. Durso and K. Holinde, *Phys. Rev. C* **49**, 2355 (1994).
[70] M. Rho, *Phys. Rev. Lett.* **54**, 767 (1985).
[71] R. Machleidt, *Adv. Nucl. Phys.* **19**, 189 (1989).
[72] F. Ajzenberg-Selove, J.H. Kelley and C.D. Nesaraja, *Nucl. Phys.* **A523**, 1 (1991).
[73] L. Zamick, *Phys. Lett.* **21**, 194 (1966).
[74] S. Aroua, P. Navrátil, L. Zamick, M.S. Fayache, B.R. Barrett, J.P. Vary and K. Heyde, *Nucl. Phys.* **A720**, 71–83 (2003).
[75] L.A. Schaller, L. Schellenberg, A. Ruetschi and H. Schneuwly, *Nucl. Phys.* **A343**, 333 (1980).
[76] W. Schütz, *Z. Phys.* **A273**, 69 (1975).
[77] P. Navrátil, J.P. Vary and B.R. Barrett, *Phys. Rev. C* **62**, 054311 (2000).
[78] S.C. Pieper and R.B. Wiringa, *Ann. Rev. Nucl. Part. Sci.* **51** 53 (2001).
[79] T.T.S. Kuo and E. Osnes, *Lecture Notes in Physics Vol. 364* (Springer-Verlag, New York, 1990).
[80] M. Hjorth-Jensen, T.T.S. Kuo and E. Osnes, *Phys. Rep.* **261**, 126 (1995), and references therein.
[81] A. Arima and L.J. Huang-Lin, *Phys. Lett.* **B41**, 429 (1972); **B41**, 435 (1972).

[82] B. Castel and I.S. Towner, *Modern Theories of Nuclear Moments* (Clarendon Press, Oxford, 1990).
[83] I.S. Towner, *Phys. Rept.* **155**, 263 (1987).
[84] B.A. Brown and B.H. Wildenthal, *Nucl. Phys.* **A474**, 290 (1987).
[85] A. Negret *et al.*, *Phys. Rev. Lett.* **97**, 062502 (2006).

Chapter 10

Effective Field Theory and High-Precision Calculations of Nuclear Electroweak Processes

Kuniharu Kubodera* and Mannque Rho†

High-precision calculations of electroweak processes in light nuclei are of great importance for multiple reasons. Gerry Brown together with Dan-Olof Riska published in 1972 a famous calculation on radiative capture of a thermal neutron on a proton, and their work was a precursor of subsequent great developments in high-precision calculations of electroweak processes in light nuclei. The application of effective field theory to nuclear systems is a prominent example of these developments. We present an overview of it with main emphasis placed on those topics in which we ourselves have been involved.

1. Introduction

In 1972 Dan-Olof Riska and Gerry Brown published their celebrated calculation [1] on radiative capture of a thermal neutron on a proton: $np \to \gamma d$. Until their work appeared, the $\sim$10% discrepancy between the experimental and theoretical values of the radiative np capture rate had been a persistent puzzle. Dan-Olof and Gerry demonstrated that this discrepancy could be removed by including the contributions of the exchange currents derived by Chemtob and Rho [2],[a] and this was the first convincing case that pointed to the existence of the exchange currents in nuclei. The 1972 Riska-Brown paper was a precursor of subsequent important developments in high-precision calculations of electroweak processes in light

*Department of Physics and Astronomy, University of South Carolina, Columbia, SC 29208, USA.
†Institut de Physique Théorique, CEA Saclay, 91191 GIf-sur-Yvette Cédex, France, & Department of Physics, Hanyang University, Seoul 133-791, Korea.
[a]See M. Rho, chapter 17 in this volume.

nuclei. On the occasion of Gerry's 85th birthday, we would like to give a succinct survey of these developments; we shall concentrate here mostly on those topics we ourselves have worked on, and so this note is not meant to be a comprehensive survey of the field.

High-precision calculations of electroweak processes in light nuclei are of great importance for multiple reasons. First of all, light nuclei, for which nuclear many-body-problem aspects are relatively well controlled, are expected to provide clean testing grounds for our understanding of the behavior of the hadronic electroweak currents in nuclei. In this connection, we should emphasize that the recent progress in formalisms and numerical techniques for solving the multi-nucleon Schrödinger equation has reached such a level [3] that the wave functions of low-lying levels for light nuclei can now be obtained practically with no approximation (once the validity of a given nuclear Hamiltonian is assumed). This situation frees us from the "usual" nuclear physics complications that arise as a result of truncating the nuclear Hilbert space down to certain model space (such as shell-model configurations, cluster-model trial functions, RPA, etc.).

We should also mention that nuclear electroweak processes in light nuclei are of particular importance in connection with astrophysics and particle physics. They appear in many astrophysical phenomena, prominent examples being stellar evolution (including solar nuclear fusion), big-bang nucleo-synthesis, supernova nucleo-synthesis, etc. Furthermore, in terrestrial experiments designed to detect astrophysical neutrinos, weak-interaction reactions on light nuclei play important roles. It is also to be noted that neutrino-nucleus reactions feature importantly in neutrino-oscillation experiments, one of the most hotly pursued frontiers in particle physics. Reliable information on the cross sections for these processes is obviously in great demand. However, for many of the processes in question, it is extremely difficult, or even impossible at present, to experimentally measure their cross sections. We must therefore totally rely on theoretical estimates, and this implies that we need a theoretical framework that allows us to arrive at estimates with reliable error assessments attached. Fortunately, the past two decades have witnessed significant progress toward this goal.

Recent developments in high-precision calculations of electroweak reactions in light nuclei encompass two lines of studies. One is the elaboration and expansion of the phenomenological potential model, which has come to be known as the standard nuclear physics approach (SNPA). Another line is the successful application of effective field theory (EFT) to nuclear physics. In the sense that the former is phenomenological in nature whereas the latter is based on a systematic expansion that is formally in conformity with the tenets of EFT, these two approaches are distinct and, perhaps because of this, it happens rather often that practitioners of one approach (SNPA or EFT) ignore the existence of and progress in the other approach. In our view, however, this does not seem an ideal or even productive situation. As discussed below, the forte of EFT is that it gives a clear-cut guiding principle on how to tame short-range physics. But, as far as long-range nuclear correlations are concerned, EFT does not necessarily lead to improvement over what has already been achieved in SNPA. And, as we will illustrate with concrete examples later, the high-precision calculation of an observable often requires that both short-range physics and long-range nuclear correlations are under control. So, at least at present, it appears to be a reasonable strategy to take advantage of the availability and the power of both SNPA and EFT while assuring, as best one can, consistency between the two. This viewpoint led us to propose the "more effective" effective field theory (MEEFT) approach, also called EFT*. It should be emphasized here that Tae-Sun Park has played a particularly important role in the developments of this approach. The nature of EFT limits us to consider only those processes which involve relatively low energies and momenta but, within this limitation, MEEFT has registered remarkable successes in giving reliable estimates for electroweak transition rates of current interest. We wish to describe here some of these results; primarily we shall be concerned with the following processes:

pp-fusion:	$p + p \to d + e^+ + \nu_e$
ν-*d* reactions:	$\nu_e + d \to e^- + p + p, \quad \nu + d \to \nu + n + p$
μ-*d* capture:	$\mu^- + d \to \nu_\mu + n + n$
Radiatice π-capture:	$\pi^- + d \to \gamma + n + n.$
Tritium β-decay:	$t \to {}^3\mathrm{He} + e^- + \bar{\nu}_e$
μ-^{3}He capture:	$\mu^- + {}^3\mathrm{He} \to \nu_\mu + t$
Hep reaction	${}^3\mathrm{He} + p \to {}^4\mathrm{He} + \nu_e + e^+$
Hen reaction	${}^3\mathrm{He} + n \to {}^4\mathrm{He} + \gamma$

This list may look like a haphazardly picked collection but, as will be explained below, these processes are interrelated in the context of demonstrating the important progress achieved in MEEFT. Specific motivations for considering these reactions will be described later for each case as the individual processes are discussed. Just to set the stage, however, we make some remarks here on the following three processes: *pp*-fusion, Hep and the νd reaction.

The pp fusion reaction, $pp \to de^+\nu_e$, is the primary solar nuclear reaction that essentially controls the burning rate of the sun, and hence the exact value of its cross section is a crucial input for any further developments of solar models. According to the latest review [4], this rate is required to be known with better than ~1% precision in order to give impact to the existing models. Meanwhile, the Hep reaction, $^3\mathrm{He} + p \to {}^4\mathrm{He} + e^+ + \nu_e$ is important in a different context. Although Hep plays only a very minor role in the pp-chain solar burning, it generates highest-energy solar neutrinos whose spectrum extends beyond the maximum energy of the ^{8}B neutrinos. So, even though the flux of the Hep neutrinos is small, it can distort the higher end of the ^{8}B neutrino spectrum [5], and this distortion can affect the interpretation of the results of a Super-Kamiokande experiment [6]. On the theoretical side, as explained in more detail below, Hep represents a unique case in which the leading-order contribution is drastically suppressed. This implies that Hep offers a good place for investigating higher order terms in EFT expansion. As for the ν-d reactions, their primary importance is connected to the Sudbury Neutrino Observatory (SNO) experiments, in which the solar neutrino flux was monitored using a heavy water Cerenkov counter. As is well known, the SNO experiments [7] have established that the total solar neutrino flux (summed over all neutrino flavors) agrees with the prediction of the standard solar model [8], whereas the electron neutrino flux from the sun is significantly smaller than the total solar neutrino flux. These results of the SNO experiments have given "smoking-gun" evidence for the transmutation of solar electron neutrinos into neutrinos of other flavors. Obviously, a precise knowledge of the ν-d reaction cross sections is of primary importance in interpreting the SNO data.

Before going into the description of achievements obtained in nuclear EFT, it seems useful to take a quick look at the latest lattice QCD calculation [9] of the nucleon weak form factors defined as

$$\begin{aligned} \langle n \,|\, V^\mu \,|\, p\rangle &= \bar{u}_n \left[F_V(q^2)\gamma^\mu + \frac{iF_M(q^2)}{2m_N}\sigma^{\mu\nu}q_\nu \right] u_p, \\ \langle n \,|\, A^\mu \,|\, p\rangle &= \bar{u}_n \left[F_A(q^2)\gamma^\mu\gamma^5 + \frac{F_P(q^2)}{m_\mu}q^\mu\gamma^5 \right] u_p, \end{aligned} \tag{1}$$

where $q = p_n - p_p$. The calculation [9] was carried out for the u/d quark mass that corresponded to $m_\pi = 0.33\,\mathrm{GeV} \sim 0.67\,\mathrm{GeV}$, and the form factors were calculated for $0.2\,\mathrm{GeV}^2 < q^2 < 0.75\,\mathrm{GeV}^2$. Here is a partial summary of the results: (1) $F_V(q^2)$ is described by the conventional dipole form, but the dipole mass M_1 is found to be 27% smaller than the empirical value; (2) $F_M(q^2)$ is described by the dipole form, but the corresponding dipole mass M_2 is found to be 26% smaller than the empirical value; (3) $g_A/g_V = 1.19$, which is 7% smaller than the experimental value $(g_A/g_V)_{\rm exp} = 1.2695 \pm 0.0029$ (PDG08); (4) $F_A(q^2)$ is found to have the dipole form but the axial dipole mass M_A is found to be 37% larger than $M_A^{\rm exp}$; (5) $\frac{(q^2-m_\pi^2)F_P(q^2)}{2m_N F_A(q^2)} \equiv \alpha(q^2)$ is found to be constant in conformity with pion-pole dominance, but $\alpha = 0.8 \sim 0.9$ (instead of $\alpha = 1$ expected from PCAC). These features lead us to expect that it will be a while before nuclear elecroweak processes can be calculated with required precision by lattice QCD.

2. Theoretical Frameworks

• Standard nuclear physics approach (SNPA)

The phenomenological potential picture has been highly successful in describing many different nuclear phenomena. In this picture an A-nucleon system is described by a Hamiltonian of the form

$$H = \sum_i^A t_i + \sum_{i<j}^A V_{ij} + \sum_{i<j<k}^A V_{ijk} + \cdots\,, \tag{2}$$

where t_i is the kinetic energy of the i-th nucleon, V_{ij} is a phenomenological two-body potential, V_{ijk} is a phenomenological three-body potential, and so on. Once H is specified, the A-body nuclear wave function $|\Psi\rangle$ is obtained by solving *exactly or essentially exactly* the Schrödinger equation

$$H|\Psi\rangle = E|\Psi\rangle. \tag{3}$$

There is much freedom in choosing possible forms of V_{ij} apart from a well-established requirement that, as the inter-nucleon distance r_{ij} becomes sufficiently large, V_{ij} should approach the one-pion exchange Yukawa potential. For the model-dependent short-range part of V_{ij}, one assumes certain functional forms and fix the parameters appearing therein by demanding that the solutions of Eq. (3) for the $A = 2$ case reproduce the nucleon-nucleon scattering data (typically up to the pion-production threshold energy) as well as the deuteron properties. There exist several so-called *high-precision phenomenological* N-N potentials that can reproduce all the existing two-nucleon data with normalized χ^2 values close to 1. These potentials differ significantly in the ways they parameterize short-range physics and, as a consequence, they exhibit substantial difference in their off-shell behavior. An important question therefore is to what extent the model-dependent short-range part affects the reliabilities of theoretical values of various observables in $A = 2$ and higher systems calculated with these phenomenological potentials.

Nuclear responses to external electroweak probes in normal circumstances are given, to good approximation, by one-body terms, which are also called the impulse approximation (IA) terms. To achieve higher accuracy, however, we must also consider exchange current (EXC) terms, which represent the contributions of nuclear responses involving two or more nucleons. If for some reason the IA contributions are suppressed, it becomes particularly important to consider the EXC contributions. These exchange currents are usually taken to be two-body operators, and they are derived from one-boson exchange diagrams; the vertices featuring in the relevant diagrams are determined to satisfy the requirements of low-energy theorems and current algebra [2, 10]. A formalism based on this picture is commonly called the *standard nuclear physics approach* (SNPA). (This is also called a potential model in the literature.) Schematically, an electroweak nuclear matrix element in SNPA is given by

$$\mathcal{M}_{fi}^{\rm SNPA} = \left\langle \Psi_f \left| \sum_{\ell}^{A} \mathcal{O}_\ell + \sum_{\ell<m}^{A} \mathcal{O}_{\ell m} + \sum_{\ell<m<n}^{A} \mathcal{O}_{\ell mn} \right| \Psi_i \right\rangle, \tag{4}$$

where the initial (final) nuclear wave function, Ψ_i (Ψ_f), is a solution of Eq. (3), and $\mathcal{O}_\ell$, $\mathcal{O}_{\ell m}$ and $\mathcal{O}_{\ell mn}$ are, respectively, one-body, two-body and three-body transition operators for a given electroweak process. SNPA has been used extensively for describing electroweak processes in light nuclei, and general good agreement found between theory and experiment [3] strongly suggests that SNPA captures much of the physics involved.

• **Effective field theory (EFT)**

Even though SNPA has been scoring great successes in describing a vast variety of data, it is still important from the fundamental point of view to raise the following issues. First, given the fact that hadrons and hadronic systems (such as nuclei) are governed by quantum chromodynamics (QCD), it is desirable to establish connections between SNPA with QCD. In this connection we note that, whereas chiral symmetry is known to be a fundamental symmetry of QCD, the formulation of SNPA is largely disjoint from this symmetry. Secondly, in introducing a phenomenological SNPA Lagrangian for describing the nuclear interaction and nuclear responses to the electroweak currents, it is not clear whether there is any identifiable expansion parameter that helps us to control the possible forms of terms in the Lagrangian and that gives a general measure of errors in our calculation.

An EFT-based approach [11] is a powerful tool to address these issues, and it has been studied with great intensity; for reviews, see [12–16]. The basic idea of EFT is simple. To describe phenomena characterized by a typical energy-momentum scale Q, we expect that we can exclude from our Lagrangian those degrees of freedom that belong to energy-momentum scales much higher than Q. So we introduce a cut-off scale Λ that is sufficiently larger than Q and we categorize our fields (to be generically represented by ϕ) into two groups: high-frequency fields $\phi_{\rm H}$ and low-frequency fields $\phi_{\rm L}$. By *integrating out* $\phi_{\rm H}$, we arrive at an *effective* Lagrangian that only contains $\phi_{\rm L}$ as explicit dynamical variables. The effective Lagrangian $\mathcal{L}_{\rm eff}$ is related to the original Lagrangian $\mathcal{L}$ as

$$\int [d\phi] e^{{\rm i}\int d^4x \mathcal{L}[\phi]} = \int [d\phi_{\rm H}][d\phi_{\rm L}] e^{{\rm i}\int d^4x \mathcal{L}[\phi_{\rm H},\phi_{\rm L}]} \equiv \int [d\phi_{\rm L}] e^{{\rm i}\int d^4x \mathcal{L}_{\rm eff}[\phi_{\rm L}]}. \quad (5)$$

$\mathcal{L}_{\rm eff}$ defined by Eq. (5) inherits the symmetries (and the pattern of symmetry breaking, if there is any) of the original Lagrangian $\mathcal{L}$. It can be shown that $\mathcal{L}_{\rm eff}$ is given by the sum of all possible monomials of $\phi_{\rm L}$ and their derivatives that are consistent with the symmetry requirements dictated by $\mathcal{L}$. Because a term involving n derivatives scales like $(Q/\Lambda)^n$, the terms in $\mathcal{L}_{\rm eff}$ can be organized into a perturbative series wherein Q/Λ serves as an expansion parameter. The coefficients of terms in this expansion are called low-energy constants (LECs). If all the LEC's up to a specified order n can be fixed either from theory or from fitting to the experimental values of relevant observables, $\mathcal{L}_{\rm eff}$ serves as a complete (and hence model-independent) Lagrangian to the given order of expansion. The results obtained are ensured to have accuracy of order $(Q/\Lambda)^{n+1}$. In applying EFT to nuclear systems, the underlying Lagrangian $\mathcal{L}$ is the QCD Lagrangian $\mathcal{L}_{\rm QCD}$, whereas, for the typical nuclear physics energy-momentum scale $Q \ll \Lambda_\chi \sim 1\,{\rm GeV}$, the effective degrees of freedom that should feature in $\mathcal{L}_{\rm eff}$ are hadrons rather than the quarks and gluons. It is not trivial to apply the formal definition in Eq. (5) to derive $\mathcal{L}_{\rm eff}$ written in terms of hadrons starting from $\mathcal{L}_{\rm QCD}$; this is because the hadrons cannot be simply identified with the low-frequency field, ϕ_L in Eq. (5). At present, the best one could do is to resort to symmetry considerations and the above-mentioned expansion scheme in the spirit of Weinberg's folk theorem [17]. Here chiral symmetry plays an important role. We know that chiral symmetry is spontaneously broken, generating the pions as Nambu-Goldstone bosons.[b] This feature can be taken into account by assigning suitable chiral transformation properties to the Goldstone bosons and writing down all possible chiral-invariant terms up to a specified chiral order [18]. The above consideration presupposes exact chiral symmetry in $\mathcal{L}_{\rm QCD}$. In reality, however, $\mathcal{L}_{\rm QCD}$ contains small but finite quark mass terms, which explicitly break chiral symmetry and lead to a non-vanishing value of the pion mass m_π. Again, there is a well-defined method to determine what terms are needed in the Goldstone boson sector to represent the effect of explicit chiral symmetry breaking [18]. These considerations lead to an EFT called chiral perturbation theory (χPT) [17, 19]. The successes of χPT in the meson sector are well documented; see, e.g. [12].

A serious problem we encounter in extending χPT to the nucleon sector is that, as the nucleon mass $m_{\rm N}$ is comparable to the cut-off scale Λ_χ, a simple application of expansion in Q/Λ does not work. This difficulty can be circumvented by employing heavy-baryon chiral perturbation theory (HBχPT), which essentially consists in shifting the reference point of the nucleon energy from 0 to $m_{\rm N}$ and in integrating out the small component of the nucleon field as well as the anti-nucleonic degrees of freedom. Thus an effective Lagrangian in HBχPT contains as explicit degrees of freedom the pions and the large components of the redefined nucleon field. HBχPT has as expansion parameters Q/Λ_χ, m_π/Λ_χ and $Q/m_{\rm N}$. As $m_{\rm N} \approx \Lambda_\chi$, it is convenient to combine chiral and heavy-baryon expansions and introduce the chiral index $\bar{\nu}$ defined by $\bar{\nu} = d + (n/2) - 2$. Here n is the number of fermion lines that participate in a given vertex, and d is the number of derivatives (with m_π counted as one derivative). A similar power counting scheme can be introduced for Feynman diagrams as well [11]. The contribution of a Feynman diagram that contains N_A nucleons, N_E external fields, L loops and N_C disjoint parts is shown to scale like $(Q/\Lambda)^\nu$, where the chiral index ν is defined as

$$\nu = 2L + 2(N_C - 1) + 2 - (N_A + N_E) + \sum_i \bar{\nu}_i, \tag{6}$$

[b]We limit ourselves here to SU(2)×SU(2) chiral symmetry.

with the summation running over all the vertices contained in the Feynman diagram. HBχPT has been applied with great success to the one-nucleon sector [12].

However, HBχPT cannot be applied in a straightforward manner to nuclei, which contain more than one nucleon. The reason is that nuclei involve very low-lying excited states, and the presence of this small energy scale spoils the original counting rule [11]. A method to avoid this difficulty was proposed by Weinberg [11]. Classify Feynman diagrams into two categories, irreducible and reducible diagrams. Irreducible diagrams are those in which every intermediate state has at least one meson in flight; all others are classified as reducible diagrams. One then applies the above chiral counting rules only to irreducible diagrams. The contributions of all the two-body irreducible diagrams (up to a specified chiral order) are treated as an effective potential (to be denoted by V_{ij}^{EFT}) acting on nuclear wave functions. Meanwhile, the contributions of reducible diagrams can be incorporated [11] by solving the Schrödinger equation

$$H^{\mathrm{EFT}}|\Psi\rangle^{\mathrm{EFT}} = E|\Psi\rangle^{\mathrm{EFT}}, \tag{7}$$

where

$$H^{\mathrm{EFT}} = \sum_{i}^{A} t_i + \sum_{i<j}^{A} V_{ij}^{\mathrm{EFT}} + \sum_{i<j<k}^{A} V_{ijk}^{\mathrm{EFT}} + \cdots . \tag{8}$$

We may refer to this two-step procedure as *nuclear* χPT or, more specifically, *nuclear* χPT in the Weinberg scheme.[c]

To apply nuclear χPT to a nuclear electroweak process, we derive a nuclear transition operator $\mathcal{T}$ from a complete set of all the irreducible diagrams (up to a given chiral order ν) involving the relevant external current(s) [20–22]. Thus

$$\mathcal{T}^{\mathrm{EFT}} = \sum_{i}^{A} \mathcal{O}_i^{\mathrm{EFT}} + \sum_{i<j}^{A} \mathcal{O}_{ij}^{\mathrm{EFT}} + \cdots . \tag{9}$$

To preserve consistency in chiral counting, the matrix element of $\mathcal{T}^{\mathrm{EFT}}$ must be calculated with the use of nuclear wave functions generated with the use of nuclear interactions that represent all the irreducible A-nucleon diagrams

[c]This is often called the Λ-counting scheme [14].

up to ν-th order. Thus a transition matrix in nuclear EFT is given by

$$\begin{aligned}\mathcal{M}_{fi}^{\mathrm{EFT}} &= \langle \Psi_f^{\mathrm{EFT}} \,|\, \mathcal{T}^{\mathrm{EFT}} \,|\, \Psi_i^{\mathrm{EFT}} \rangle \\ &= \left\langle \Psi_f^{\mathrm{EFT}} \left| \sum_i^A \mathcal{O}_i^{\mathrm{EFT}} + \sum_{i<j}^A \mathcal{O}_{ij}^{\mathrm{EFT}} + \cdots \right| \Psi_i^{\mathrm{EFT}} \right\rangle, \end{aligned} \tag{10}$$

where the superscript, "EFT", means that the relevant quantities are calculated according to EFT as described above. If this program is carried out rigorously, it would constitute an *ab initio* calculation. Let us call it "AB-EFT". It is to be noted that in EFT we know exactly at what chiral order three-body operators start to contribute to $\mathcal{T}$, and that, to chiral orders relevant to most applications described below, there is no need for three-body operators. With this understanding, we have kept only one- and two-body operators in Eq. (10). The unambiguous classification of transition operators according to their chiral orders is a great advantage of EFT, which is missing in Eq. (4).

We should point out that there exists an alternative form of nuclear EFT called the PDS scheme, proposed by Kaplan, Savage and Wise [23], which uses a counting scheme (often called Q-counting) different from the Weinberg scheme. Although many important results have been obtained in the PDS scheme (for a review, see e.g. [15]), we concentrate here primarily on the Weinberg scheme.

We also remark that, for very low energy nuclear phenomena whose typical energy-momentum scale is $Q \ll m_\pi$, even the pions can be treated as "heavy" particles and can be eliminated from $\mathcal{L}_{\mathrm{eff}}$, with the choice of $\Lambda \approx m_\pi$. This leads to pionless EFT [23–25]. Here there is no manifest chiral symmetry since there is no explicit pion. In this regime the NN interactions and electroweak currents are described by point-like contact terms.

• **Hybrid EFT**

In the above we emphasized the formal merits of nuclear EFT. In actual calculations, however, we need to consider the following two points. First, it is still a big challenge to generate, strictly within the EFT framework, nuclear wave functions the accuracy of which is comparable to that of SNPA wave functions. Secondly, whereas a chiral Lagrangian, $\mathcal{L}_{\mathrm{eff}}$, is definite only when the values of all the relevant LECs are fixed, this condition cannot be readily satisfied in many cases. A pragmatic solution to the first difficulty [26–31] is to use in Eq. (10) wave functions obtained in SNPA; this eclectic approach is called hybrid EFT. A nuclear transition matrix element in hybrid EFT is given by

$$\mathcal{M}_{fi}^{\text{hyb-EFT}} = \left\langle \Psi_f^{\text{SNPA}} \left| \sum_{\ell}^{A} \mathcal{O}_\ell^{\text{EFT}} + \sum_{\ell<m}^{A} \mathcal{O}_{\ell m}^{\text{EFT}} \right| \Psi_i^{\text{SNPA}} \right\rangle, \qquad (11)$$

Because, as mentioned, the NN interactions that generate SNPA wave functions reproduce accurately the two-nucleon data, the adoption of Eq. (11) is almost equivalent to using the empirical data themselves to control the initial and final nuclear wave functions. From a purely formal point of view, hybrid EFT may be deemed as a "regression". But, if our goal is to obtain a transition matrix element as accurately as possible with the maximum help of available empirical input, hybrid EFT does have a legitimate status. It offers more flexibility and predictive power than the AB-EFT, and it can be highly accurate so long as the off-shell problem and the contributions of three-body (and higher-body) terms are properly addressed. The issue of possible unknown LECs will be discussed in connection with MEEFT in the next subsection.

• **MEEFT**

A great merit of hybrid EFT is that it can be used for complex nuclei ($A = 3, 4, \ldots$) with essentially the same accuracy and ease as for the $A = 2$ system.[d] We reemphasize in this connection that the contributions of transition operators involving three or more nucleons, which can in principle contribute in A-nucleon systems ($A \geq 3$), are intrinsically suppressed according to chiral counting, and hence, up to a certain chiral order, a transition operator in an A-nucleon system consists of the same EFT-based 1-body and 2-body terms as used for the two-nucleon system. Then, since SNPA provides high-quality wave functions for the A-nucleon system, one can calculate $\mathcal{M}_{fi}^{\text{hyb-EFT}}$ with precision comparable to that for the corresponding two-nucleon case.

In most cases the one-body operator, $\mathcal{O}_\ell^{\text{EFT}}$, is free from unknown LECs. So we may concentrate on the two-body operator, $\mathcal{O}_{\ell m}^{\text{EFT}}$. Suppose that $\mathcal{O}_{\ell m}^{\text{EFT}}$ under consideration contains an LEC (call it κ) that cannot be determined with the use of $A = 2$ data alone. It can happen that an observable (call it Ω) in a A-body system ($A \geq 3$) is sensitive to κ and that the experimental value of Ω is known with sufficient accuracy. We then can determine κ by calculating $\mathcal{M}_{fi}^{\text{hyb-EFT}}$ responsible for Ω and adjusting

[d] Here we are ignoring "purely technical" complications that can grow in actual numerical calculations for higher-A systems.

κ to reproduce the experimental value of Ω. Once κ is determined this way, we can make *predictions* for any other observables for any other nuclear systems that are controlled by the same transition operators. When hybrid EFT is used in this manner, it is called *more effective* effective field theory (MEEFT) [30, 31].[e]

MEEFT is a powerful formalism for correlating various observables in different nuclei, using the transition operators controlled by EFT. A further practical advantage of MEEFT is that, since correlating the observables in neighboring nuclei is expected to serve as an additional renormalization, the possible effects of higher chiral order terms and/or off-shell ambiguities can be significantly suppressed with the use of MEEFT. We will come back to this later, when we discuss concrete examples.

3. Examples

• ***pp*-fusion, νd reactions and μd capture**

A common feature of these reactions is that a precise knowledge of the Gamow-Teller (GT) transition matrix elements is required in estimating their cross sections. We show here, following Ref. [31], that the idea of MEEFT can be used very neatly for this group of reactions. The 1-body IA operators for the GT transition can be fixed unambiguously from the available 1-body data. As for the 2-body operators, to next-to-next-to-next-to-leading order (N^3LO) in chiral counting, there appears one unknown LEC that at present cannot be determined from data for the $A = 2$ systems. This unknown LEC, denoted by $\widehat{d}_R$ in [27], parametrizes the strength of contact-type four-nucleon coupling to the axial current. Park *et al.* [31] noticed that the same LEC, $\widehat{d}_R$, also features as the only unknown parameter in the calculation of the tritium β-decay rate Γ^t_β, and they proposed to use MEEFT to place a constraint on $\widehat{d}_R$ from the experimental information on Γ^t_β. Since the empirical value of Γ^t_β is known with high precision, and since the accurate wave functions of ^3H and ^3He are available from SNPA [32], we can determine $\widehat{d}_R$ with sufficient accuracy for our purposes. Once the value of $\widehat{d}_R$ is determined this way, we can carry out parameter-free MEEFT calculations for *pp*-fusion [31], ν-d reactions [33], and μ-d

[e]It is also called EFT*. MEEFT should be distinguished from an earlier naive hybrid EFT model wherein the short-range terms were dropped altogether using an intuitive argument based on short-range NN repulsion.

capture. (An on-going attempt to deduce $\widehat{d}_R$ from μ-d capture will be discussed later.)

We should mention the important role of momentum cutoff in MEEFT. As emphasized before, the effective Lagrangian $\mathcal{L}_{\text{eff}}$ is, by construction, valid only below the specified cutoff scale Λ. Needless to say, this basic constraint should be respected in our nuclear EFT calculations, and for that we must make sure that nuclear intermediate states involved in the computation of Eq. (10) do not get out of this constrained world. Since, however, the use of V^{phenom} introduces high momentum components above Λ_{QCD}, we need to introduce momentum cutoff at Λ_{NN} to eliminate high-momentum components. It is reasonable to implement this constraint by requiring that the two-nucleon relative momentum should be smaller than Λ_{NN}. Park *et al.* [31] used a Gaussian cutoff function proportional to $\exp(-\vec{p}^2/\Lambda_{NN}^2)$ but its detailed form should not be too important. As a reasonable range of the value of Λ_{NN} we may choose: $500\,\text{MeV} \lesssim \Lambda_{NN} \lesssim 800\,\text{MeV}$, where the lower bound is dictated by the requirement that Λ_{NN} should be sufficiently larger than the pion mass (to fully accommodate pion physics), while the upper bound reflects the fact that our EFT is devoid of the ρ meson. Mismatch between Ψ^{EFT} and Ψ^{SNPA} only affects short-distance behavior, and we expect that Λ_{NN}-independence is a measure of model independence of an MEEFT calculation.

For a given value of Λ_{NN} within the above range, $\widehat{d}_R$ is tuned to reproduce Γ_β^t, and then the rates for *pp*-fusion, the νd reactions and μd capture are calculated. Before giving an account of the individual results, we should point out a notable common feature. Although the optimal value of $\widehat{d}_R$ varies significantly as a function of Λ_{NN}, the observables (in our case the above three reaction rates) exhibit remarkable stability against the variation of Λ_{NN} (within the above-discussed physically reasonable range). This stability may be taken as an indication that the use of MEEFT for inter-correlating the observables in neighboring nuclei *effectively* renormalizes various effects, such as the contributions of higher-chiral order terms, mismatch between the SNPA and EFT wave functions, etc. This stability is essential in order for MEEFT to maintain its predictive power.

3.1. *The pp-fusion S-factor*

The S-factor is related to the cross section σ as

$$\sigma(E) = \frac{S(E)}{E} \exp\{-2\pi\eta(E)\},$$

where E is the incident energy in the center-of-mass system, and $\eta(E) \equiv \frac{Z_1 Z_2 e^2}{\hbar v}$ is the Coulomb penetration factor. The S-factor for pp-fusion at threshold is related to the transition matrix element $\mathcal{M}$ as [4]

$$S_{pp}(0) = 6\pi^2 m_p \, \alpha \ln 2 \, \frac{1}{\gamma^3} (g_A/g_V)^2 \frac{f^R_{pp}}{(ft)_{0^+ \to 0^+}} |\mathcal{M}|^2 \tag{12}$$

where $\gamma = (2\mu B_d)^{1/2} = 0.2316\,\mathrm{fm}^{-1}$ is the deuteron binding wave number, and f^R_{pp} is the phase space factor including radiative corrections. As mentioned, the solar model and stellar evolution theory require 1% precision in S_{pp} [4].

It is useful to first recall what was achieved in SNPA. A benchmark calculation of S_{pp} in SNPA was carried out by Schiavilla *et al.* [34]. These authors took full advantage of the fact, first pointed out in Ref. [35], that the Gamow–Teller (GT) matrix elements for pp-fusion and tritium β-decay are closely correlated. For each of the two-body GT transition operators with different r-dependences (r = relative distance of the two nucleons), the transition density for pp-fusion is found to be accurately proportional to that for tritium β-decay. This means that a single multiplicative constant can match them all. Schiavilla *et al.* [34] fine-tuned the $N\Delta$-axial coupling constant, which featured in a dominant two-body axial current diagram (Δ-particle excitation diagram), and which was only loosely controlled from other sources of information. After this fine-tuning, S_{pp} was calculated with estimated uncertainties of ~0.3%; here the error estimate came from the range of variation in $S_{pp}(0)$ for the five different high-precision phenomenological potentials used in [34]. A further examination of this error estimation will be described below.

Park *et al.* [31] carried out an MEEFT calculation of $S_{pp}(0)$ and obtained

$$S_{pp}(0) = 3.94 \times (1 \pm 0.008) \times 10^{-25}\,\mathrm{MeV\,b}. \tag{13}$$

$S_{pp}(0)$ is found to change only by ~0.1% against changes in Λ_{NN}, ensuring the robustness of the MEEFT calculation in this respect. The convergence pattern in chiral expansion for the pp-fusion transition amplitude is reported to be [31]: $\mathcal{M} \sim (1 + 0 + 0.1\% + 0.9\% + 0.4\%)$. It is noted that the 2nd and 3rd order terms are accidentally small, but that the 4th and 5th order are roughly in conformity with expansion in $m_\pi/\Lambda_\chi \approx \frac{1}{7}$. The 3-body currents contribute at the sixth order. Uncertainties due to the higher order contributions estimated from the size of the 5th-order contribution are 0.4% (in amplitude), or 0.8% (in probability).[f] The MEEFT result, Eq. (13), is consistent with that obtained in SNPA [34]. Meanwhile, the fact that MEEFT allows us to make error estimation is a notable advantage over SNPA. It is worth emphasizing that the accuracy achieved in Eq. (13) represents an improvement by a factor of ∼5 over the previous results based on a naive hybrid EFT [27].

The calculation of S_{pp} in pionless EFT was performed up to NNLO by Kong and Ravndal [36], and up to fifth order by Butler and Chen [37]; see also Ando *et al.* [38]. Here again, up to the order considered, there is only one unknown LEC, denoted by $L_{1,A}$, which (analogously to $\hat{d}^R$) parameterizes the strength of four-nucleon contact axial-vector coupling. The results of Refs. [36, 37] are: (i) the central value of $S_{pp}(0)$ is consistent with those obtained in SNPA and MEEFT; (ii) the theoretical uncertainty in $S_{pp}(0)$ is ∼3%, which is significantly larger than the results obtained in SNPA and MEEFT. One thing to be noted here is that the PDS scheme [23] (used in pionless EFT) adopts an expansion scheme for transition amplitudes themselves, without employing the concepts of the potential or wave functions. In certain contexts this feature may be an advantage, but its disadvantage in the present context is that one cannot readily relate the transition matrix elements for an A-nucleon system with those for the neighboring nuclei; in PDS, each nuclear system requires a separate parameterization. Therefore, L_{1A} in pionless EFT needs to be determined from the empirical value of a 2-body observable, which however is at present available only with large experimental error. This situation is reflected in the rather large 3% uncertainty in $S_{pp}(0)$ calculated in pionless EFT.

In a very recent review [4], there has been an attempt to deduce the most updated estimate for $S_{pp}(0)$, and the recommended value is

$$S_{pp}(0) = 4.01 \times (1 \pm 0.009) \times 10^{-25}\,\text{MeV b}. \tag{14}$$

This estimate arises as follows. In Eq. (12), the matrix element $\mathcal{M}$ is assigned the value corresponding to the central value obtained in SNPA [34] and MEEFT [31]; these two approaches give exactly the same value. For the other inputs that enter into Eq. (12), the most updated values are used; some of them slightly differ from the values used in the earlier

[f] In [31], the relative error of 0.4% was assigned to $S_{pp}(0)$. However, it has been found subsequently that the 0.4% uncertainty should have been assigned to the transition amplitude, not to the transition probability; we therefore double the error estimate quoted in [31] and adopt 0.008 in Eq. (13).

publications. The relative uncertainty of 0.009 in Eq. (14) is dominated by the 0.8% uncertainty in $|\mathcal{M}|^2$ estimated from the MEEFT result [31] and the 0.5% uncertainty coming from the experimental error in the nn scattering length a_{nn}. These uncertainties and other smaller uncertainties are added in quadrature. Equation (14) indicates that $S_{pp}(0)$ now can be estimated within 1% precision.

3.2. ν-d reactions

As mentioned, ν-d reactions are important in connection with the SNO experiments on neutrino oscillations. Within SNPA a detailed calculation of the ν-d cross sections, $\sigma(\nu d)$, was carried out in Refs. [39–41]. Nakamura *et al.* [41], following Schiavilla *et al.*'s approach [34], used the two-body current whose strength was fine-tuned to reproduce the tritium β-decay rate. With the use of this fine-tuned two-body exchange current, Nakamura *et al.* calculated $\sigma(\nu d)$ for four types of reactions: charged-current and neutral-current reactions on d for both neutrino and anti-neutrino cases. Meanwhile, Butler, Chen and Kong (BCK) [42] carried out a pionless-EFT calculation of the ν-d cross sections. The results obtained agree with those of Nakamura *et al.* [41] in the following sense. A pionless-EFT calculation, as mentioned in the discussion of pp-fusion, involves one unknown LEC, $L_{1\mathrm{A}}$, which represents the strength of a four-nucleon axial-current coupling term. BCK determined $L_{1\mathrm{A}}$ by requiring that the νd cross sections of Nakamura *et al.* be reproduced by their pionless-EFT calculation. With the value of $L_{1\mathrm{A}}$ fine-tuned this way, the $\sigma(\nu d)$'s obtained by BCK show a perfect agreement with those of Nakamura *et al.* for all the four reactions mentioned above, and for the entire solar neutrino energy range, $E_\nu \lesssim 20\,\mathrm{MeV}$. Moreover, the optimal value, $L_{1\mathrm{A}} = 5.6\,\mathrm{fm}^3$, found by BCK [42] is consistent with the order of magnitude of $L_{1\mathrm{A}}$ expected from the naturalness argument (based on a dimensional analysis), $|L_{1\mathrm{A}}| \leq 6\,\mathrm{fm}^3$. The fact that an EFT calculation (with one parameter fine-tuned) reproduces the results of SNPA very well strongly suggests the robustness of the SNPA results for $\sigma(\nu d)$.

Even though it is reassuring that the ν-d cross sections calculated in SNPA and pionless-EFT agree with each other (in the sense explained above), it is desirable to carry out an EFT calculation that is free from any adjustable LEC. Fortunately, MEEFT allows us to carry out an EFT-controlled parameter-free calculation of the ν-d cross sections, because the only LEC appearing therein, $\widehat{d}_R$, has already been determined from Γ_β^t [31]. Ando *et al.* [33] took advantage of this situation and carried out an MEEFT calculation of $\sigma(\nu d)$. The results were found to agree within 1% with $\sigma(\nu d)$'s obtained in SNPA [41]. These results show that the ν-d cross sections used in interpreting the SNO experiments [7] are reliable at the 1% precision level.

There have been subsequent attempts to calibrate L_{1A} from reactor $\bar{\nu}$-deuteron breakup reactions [43], and also from a "self-calibrating" analysis of the SNO data [44]. The $\sigma(\nu d)$'s corresponding to the range of L_{1A} obtained in these analyses are consistent with the $\sigma(\nu d)$'s obtained in SNPA and MEEFT. However, the resulting constraints on L_{1A} are not stringent. We have mentioned that both L_{1A} and $\widehat{d}_R$ represent the strength of axial-current-four-nucleon contact coupling. It is to be noted, however, that L_{1A} belongs to pion-less EFT, while $\widehat{d}_R$ to pion-ful EFT. In the pion-ful EFT, because of the strong tensor force, the exchange current involving the deuteron d-state is important, and the s-wave exchange current arising from the $\widehat{d}_R$ term is separate from this tensor-force effect. By contrast, in the pion-less EFT, the explicit d-wave term is a higher-order correction, and hence the s-wave L_{1A} term must subsume the strong tensor-force contributions. It would be informative to investigate the relation between L_{1A} and $\widehat{d}_R$ from this perspective.

- **Off-shell effects**

In introducing hybrid EFT, we have replaced $|\Psi\rangle^{\rm EFT}$ for the initial and final nuclear states in Eq. (10) with the corresponding $|\Psi\rangle^{\rm SNPA}$; see Eq. (11). This replacement may bring in a certain degree of model dependence, called the off-shell effect, because the phenomenological NN interactions are constrained only by the on-shell two-nucleon observables.[g] This off-shell effect, however, is expected to be small for the reactions under consideration, since they involve low momentum transfers and hence are not extremely sensitive to the short-range behavior of the nuclear wave functions. One way to quantify this expectation is to compare a two-nucleon relative wave function generated by the phenomenological potential with that generated by an EFT-motivated potential. Phillips and Cohen [45] made such a comparison in their analysis of the 1-body operators responsible for electron-deuteron Compton scattering, and showed that a hybrid EFT should work well up to momentum transfer 700 MeV. A similar conclusion is expected to

[g] In a consistent theory, physical observables are independent of field transformations that lead to different off-shell behaviors, and therefore the so-called off-shell effect is not really a physical effect. In an approximate theory, observables may exhibit superficial dependence on off-shell behavior, and it is customary to refer to this dependence as an off-shell effect.

hold for a two-body operator, so long as its radial behavior is duly "smeared-out" reflecting a finite momentum cutoff. Thus, hybrid EFT as applied to low energy phenomena is expected to be practically free from the off-shell ambiguities. The off-shell effect should be even less significant in MEEFT, wherein an additional "effective" renormalization is likely to be at work (see subsection 3.1).

We now discuss briefly another interesting development, due to Tom Kuo and his colleagues [46], which can shed much light on the reliability of a hybrid EFT or MEEFT calculation. As mentioned, a "realistic phenomenological" nuclear interaction, V_{ij} in Eq. (2), is determined by solving the Schrödinger equation, Eq. (3), for the $A = 2$ system and fitting the results to the full set of two-nucleon data up to the pion production threshold energy. So, physically, V_{ij} should reside in a momentum regime below a certain cutoff, Λ_c. In the conventional treatment, however, the existence of this cutoff scale is ignored, and Eq. (3) is solved, allowing the entire momentum range to participate. Kuo *et al.* proposed to construct an *effective low-momentum* potential V_{low-k} by eliminating (or integrating out) from V_{ij} the momentum components higher than Λ_c, and calculated V_{low-k}'s corresponding to many well-established examples of V_{ij}'s. Remarkably, it was found that all these V_{low-k}'s give identical results for the half-off-shell T-matrices, even though the ways short-range physics is encoded in these V_{ij}'s are highly diverse. This implies that the V_{low-k}'s are free from the off-shell ambiguities, and therefore the use of V_{low-k}'s is essentially equivalent to employing $V_{ij}^{\rm EFT}$ (that appeared in Eq. (8)), which by construction should be model-independent. Now, as mentioned, our MEEFT calculation has a momentum-cutoff regulator, and this essentially ensures that the matrix element, $\mathcal{M}_{fi}^{\text{hyb-EFT}}$, in Eq. (11) is only sensitive to the half-off-shell T-matrices that are controlled by V_{low-k} instead of V_{ij}. Therefore, we can expect that the MEEFT results reported here are essentially free from the off-shell ambiguities.

Recently, Mosconi *et al.* [47] have compared νd cross sections calculated for various models that differ in the way the "nuclear PCAC requirement" is implemented, and concluded from this comparison that the $\sigma(\nu d)$'s obtained in SNPA, MEEFT and pionless EFT have uncertainties as large as 2–3%. Mosconi *et al.* have reached this conclusion by examining the scatter of unconstrained calculations of $\sigma(\nu d)$. However, all state-of-the-art calculations use Γ_β^t to reduce two-body current and other uncertainties. Once this constraint is imposed, the scatter in the calculated values of $\sigma(\nu d)$ should be significantly reduced.

3.3. μd capture and MuSun experiment

Although it is likely that the determination of $\widehat{d}_R$ from Γ^t_β is good enough for practical purposes, it is worthwhile to study a possibility to fix $\widehat{d}_R$ with the use of an observable belonging to the $A = 2$ systems. A promising candidate is the μd-capture process, $\mu^- + d \to \nu_\mu + n + n$. Even though rather large energy-momentum transfers caused by the disappearance of a μ^- seem to make the applicability of EFT here a delicate issue, we can show [48] that, as far as the hadron sector is concerned, μ-d capture is in fact a reasonably "gentle" process. This is because: (1) the ν_μ carries away most of the energy, and (2) there is a large enhancement of the transition amplitude in a kinematic region where the relative motion of the final two nucleons is low enough to justify the use of EFT. Thus μd capture can be used for controlling $\widehat{d}_R$, if the quality of experimental data improves sufficiently. The MuSun experiment at PSI (proposed in 2008) [49] aims to determine the μd capture rate $\Gamma_{\mu d}$ with ~1.5% accuracy; for a review, see [50]. This will enable us to determine $\widehat{d}^R$ within $A = 2$ system. In this connection, it is important to point out that $g_P \equiv F_A(q^2 = -0.88\, m_\mu^2)$ is totally under control by now. Previously, the main goal of μ-nucleus capture experiments was to obtain information on g_P given the fact that μ capture goes much faster on nuclei (with high Z) than on a proton. This goal, however, has been perennially elusive because of nuclear physics complications. It is therefore highly noteworthy that a HBχPT calculation [51] gives a robust prediction, $g_P = 8.26 \pm 0.23$, and that this value is in good agreement with the experimental value, $g_P = 7.7 \pm 1.1$, obtained in the MuCap Experiment at PSI [52]. The fact that the 1-body current is well known allows us to concentrate on the 2-body part.

Recent SNPA calculations of μ-d capture can be found in Refs. [53–55]; According to Tatara *et al.* [53], the hyperfine-doublet capture rate Γ_d (by far the dominant capture channel) is $\Gamma_d = 397\text{–}400\,\text{s}^{-1}$, and the results are stable against the use of different NN potentials available at the time of their work. The latest calculation by Ricci *et al.* [55] reports $\Gamma_d = 416\,\text{s}^{-1}$, and that there is non-negligible NN-potential dependence between the earlier and new-generation realistic phenomenological potentials. Thus the results obtained within SNPA exhibit some scatter, and although Ricci *et al.* offer rather detailed discussion of its possible origin, further investigation seems needed. An MEEFT calculation of Γ_d was carried out by Ando *et al.* [48], yielding the value $\Gamma_d = 386\,\text{s}^{-1}$ with a high degree of stability against variations in Λ_{NN}. A pionless EFT calculation was performed by Chen *et al.* [56] but this approach is reliable only for limited kinematical regions, because of the very low momentum cutoff inherent with pionless EFT.

Comparison with experiment is complicated by the existence of two barely overlapping experimental results: $\Gamma_d^{\rm exp} = 470 \pm 29\,{\rm s}^{-1}$ [57], and $\Gamma_d^{\rm exp} = 409 \pm 40\,{\rm s}^{-1}$ [58]. All the above-mentioned theoretical values are consistent with the value of $\Gamma_d^{\rm exp}$ reported in [58], and disagree with that given in [57].

Since the MuSun experiment envisages to measure Γ_d with 1.5% precision, it is important to match the accuracy of theoretical prediction. Most recently, a highly elaborate calculation of Γ_d has been performed in both SNPA and MEEFT by Marcucci *et al.* [59].[h] The nuclear wave functions are generated from the Argonne V_{18} potential (for SNPA) or from a chiral N3LO two-nucleon potential [60] (for MEEFT). As in the pp-fusion calculation, one parameter needs to be fixed (the N-to-Δ axial coupling constant in SNPA or $\widehat{d}_R$ in MEEFT). This is fixed by reproducing the tritium β-decay rate. The model dependence with respect to the adopted interactions and current (including cutoff dependence in the case of MEEFT) is found to be weak. Marcucci *et al.* give

$$\Gamma_d = 392 \pm 2.3\,{\rm s}^{-1}. \tag{15}$$

It is interesting to note that this result is close to that obtained by Ando *et al.* [48]. We also remark that what Marcucci *et al.* refer to as MEEFT is actually "less hybrid" in nature than the MEEFT used by Ando *et al.* in that the former employs the chiral N3LO two-nucleon potential instead of a phenomenological NN potential. When the MuSun result becomes available, one can use it in two ways — compare it with the theoretical prediction in Eq. (15) based on Γ_β^t, or use it to determine $\widehat{d}_R$ from the $A = 2$ system (without referring to the $A = 3$ system).

3.4. *Hep and Hen reactions*

• Hep reaction

As mentioned, the Hep reaction, $^3{\rm He} + p \to {}^4{\rm He} + \nu_e + e^+$, produces solar neutrinos of the highest maximum energy ($E_\nu^{\rm max} = 18.8\,{\rm MeV}$), although the flux is extremely low. An accurate estimation of this cross section is a particularly challenging task because: (1) the contribution of

[h]The authors of this work use the term EFT* instead of MEEFT.

the leading-order 1-body GT operator is highly suppressed due to the approximate wave function orthogonality [61], and (2) there is a strong cancellation between the 1-body and 2-body GT matrix elements [32, 35]. Reflecting this situation, estimates of $S_{\rm Hep}(0)$ changed by orders of magnitude from Salpeter's original value, $S_{\rm Hep}(0)_{\rm Salpeter} = 630 \times 10^{-20}$ keV-b to modern values; for a review, see [62]. Park *et al.* [31] carried out an MEEFT calculation of $S_{\rm Hep}(0)$ with the use of $\widehat{d}_R$ determined from Γ^t_β (the same procedure as used in calculating $S_{pp}(0)$), and obtained

$$S_{\rm Hep}(0) = (8.6 \pm 1.3) \times 10^{-20}\,\text{keV b}, \tag{16}$$

where the error spans the range of the Λ_{NN} dependence for $\Lambda_{NN} =$ 500–800 MeV. Again, the MEEFT result agrees with that obtained in SNPA by Marcucci *et al.* [32]: $S_{\rm Hep}(0)_{\rm SNPA} = 9.6 \times 10^{-20}$ keV-b. The above-mentioned large cancellation between the 1-body and 2-body contributions in this case amplifies the cutoff dependence of $S_{\rm Hep}(0)$, but the error quoted in Eq. (16) is still small enough for the purpose of analyzing the existing Super-Kamiokande data [6].

• Hen reaction

The Hen reaction, $^3\text{He} + n \to {}^4\text{He} + \gamma$, is very similar to Hep except, of course, the weak interaction is replaced by the EM interaction. Thus, just like the Gamow-Teller matrix element for Hep, the leading-order 1-body M1 matrix element for Hen is highly suppressed due to the pseudo-orthogonality of the initial and final nuclear wave functions. Moreover, there is strong cancellation between the hindered 1-body contribution and the 2-body exchange-current contribution, leading to a further suppression of the Hen reaction rate. Thus, like Hep, Hen is sensitive to higher order terms in chiral expansion. Meanwhile, the experimental value of $\sigma_{\rm Hen}$ is known rather accurately: $\sigma^{\rm exp}_{\rm Hen} = (54 \pm 6)\,\mu b$ [63], or $\sigma^{\rm exp}_{\rm Hen} = (55 \pm 3)\,\mu b$ [64]. Therefore, Hen provides a good testing ground for the validity of MEEFT as used for Hep. A very elaborate MEEFT calculation of Hen was recently carried out by Lazauskas, Song and Park [65]. Up to N^{3}LO considered by these authors, the M1 operators contain two LECs, which can be determined from the magnetic moments, $\mu(^3\text{H})$ and $\mu(^3\text{He})$. The initial and final nuclear wave functions are obtained from the rigorous Faddev-Yakubowski equations [66] for five sets of high-precision phenomenological potentials. Lazauskas *et al.*'s calculation gives

$$\sigma_{\rm Hen} = (38 \sim 58)\,\mu b, \tag{17}$$

with high stability against changes in Λ_{NN} in the range $500 \sim 900\,\mathrm{MeV}$. The authors emphasized the importance of long-range nuclear correlation effects in obtaining an accurate estimate of σ_{Hen}. To explain this point, it is useful to first discuss what Song *et al.* [67] pointed out in their MEEFT study of the M1 properties of the $A = 3$ nuclei. They noticed that the M1 matrix elements (MEs) in the $A = 3$ nuclei calculated in MEEFT for various realistic potentials have strong correlation with the triton binding energies B_3 calculated with the corresponding realistic potentials. In fact, these quantities fall on a well-defined single curve. Meanwhile, since B_3 governs the long distance contribution to the MEs, the model dependence (that is, variations in MEs calculated for nuclear potentials that give different values for B_3) cannot be lifted by renormalizing the local (or short-range) operators. However, the use of the empirically found correlation curve between the MEs and B_3 allows us to drastically reduce scatters in the calculated MEs. This is achieved by introducing the constraint that one should accept only those values of the MEs which, along the correlation curve, have values of B_3 consistent with its experimental value. This constraint is found to essentially eliminate the model dependence in the MEs [67]. Now, Lazauskas *et al.* [65] noticed that a very similar feature exists also for Hen. Defining the quantity ζ by $\zeta \equiv [q(a_{n\mathrm{He}^3}/r^2_{\mathrm{He}^4}]^{-2.75}$ ($q =$ photon momentum), Lazauskas *et al.* found that σ_{Hen} and ζ are so well correlated that the σ_{Hen}-ζ plot forms a sharply defined straight line. Then, again, by accepting only those values of σ_{Hen} which are consistent with the empirical value of ζ, one can deduce the model dependence of σ_{Hen} significantly. The estimate of σ_{Hen} given in Eq. (17) was obtained through this method.[i] The fact that this theoretical value is in good agreement with the above quoted experimental value renders strong support for the basic soundness of the MEEFT approach. We consider it noteworthy that MEEFT works well *also* for a highly suppressed process wherein higher-chiral order terms become important; for another remarkable example, see Ref. [68]. The examples of Hen and the MEs in the $A = 3$ systems also demonstrate that it is profitable to avail ourselves of the merit of EFT (to control short-range physics) and

[i]The main point here is *not* a linear relation between σ_{Hen} and ζ; after all, the latter is a rather "esoteric" quantity, introduced just to "dramatize" the correlation between σ_{Hen} and binding properties of the relevant nuclei. The message here is that, to get a reliable estimate of σ_{Hen}, one must employ, even among so-called high-precision nuclear potentials, only those that can reproduce the binding properties of the $A = 3$ and 4 nuclei with needed accuracy. This criterion does not require the introduction of ζ.

the strength of SNPA (to control long-range nuclear correlations), as we do in MEEFT.

3.5. μ-^{3}He *capture*

Ackerbauer *et al.* [69] carried out a high-precision measurement of the capture rate for $\mu^- \mathrm{He}^3 \to \nu_\mu t$. Achieving astonishing accuracy of 0.3%, they reported the experimental value $\Gamma(\mu^3\mathrm{He})^{\mathrm{exp}} = 1494(4)\,\mathrm{s}^{-1}$. A detailed MEEFT calculation of this capture rate was performed by Gazit [70]. Here again, the only unknown LEC, $\widehat{d}_R$, was fixed by fitting to Γ^t_β for each given value of the cutoff parameter Λ_{NN} (ranging 500 $\sim$ 800 MeV), and then $\Gamma(\mu^3\mathrm{He})$ was calculated. The results are summarized as $\Gamma(\mu^3\mathrm{He}) = 1499(2)_\Lambda(3)_{NM}(5)_t(6)_{RC}$, where the first error comes from the Λ_{NN} dependence, the second error from uncertainties in extrapolating the nucleon weak form factors to finite momentum transfer, and in the choice of specific nuclear models, the third error from experimental accuracy for Γ^t_β, and the fourth error from uncertainty in radiative corrections. The theoretical value (with total error $\sim$1%) agrees well with $\Gamma(\mu^3\mathrm{He})^{\mathrm{exp}}$, providing a yet another case to support the validity of MEEFT; see also Gazit *et al.* [71], where a full EFT calculation for tritium β-decay and the tritium binding energy is presented.

In Ref. [59], quoted earlier in connection to S_{pp} calculations, Marcucci *et al.* also carried out an elaborate calculation of $\Gamma(\mu^3\mathrm{He})$ in both SNPA and MEEFT, and they found excellent agreement between the two approaches. Combining the results coming from these two methods, Marcucci *et al.* arrived at a highly accurate estimate $\Gamma(\mu^3\mathrm{He}) = 1484 \pm 13\,\mathrm{s}^{-1}$. Again, the theoretical value agrees very well (within quoted errors) with the above experimental value.

3.6. *Neutron-neutron scattering length* a_{nn}

In the construction of a high-precision phenomenological NN potential, V_{NN}^{phenom}, the experimental errors in the nn scattering length, a_{nn}, are embedded in the overall χ^2-value, and it is not obvious to what extent the existing samples of V_{NN}^{phenom} directly reflect the uncertainties in a_{nn}. Meanwhile, low-energy electroweak transitions in lightest nuclei are in general rather sensitive to the precise value of a_{nn}. For instance, in [4], uncertainty in the pp-fusion S-factor,$S_{pp}(0)$, due to the existing uncertainty in a_{nn} is estimated to be 0.5%, which is not devastatingly large but rather significant.

Apart from this point, it is of general importance to determine, as accurately as possible, the empirical value of a_{nn}, one of the fundamental parameters characterizing the NN interaction. Since extraction of a_{nn} from the analysis of $nd \to nnp$ reaction at present involves some inconsistencies [72–74], the standard value $a_{nn} = -18.63 \pm 0.27(\text{exp}) \pm 0.30(\text{theor})$ fm [75] is based on the $\pi^- d \to \gamma nn$ reaction. The theoretical framework responsible for this result uses phenomenological NN interactions, and intrinsic short-distance physics uncertainties in this framework limit the theoretical precision. Gardestig and Phillips [76, 77] investigated the possibility of using EFT to reduce theoretical uncertainties in the extraction of a_{nn}. They have shown that short-range physics involved in the reaction can be parameterized using the same LEC, $\widehat{d}_R$, as introduced in [31], and that, if for the sake of demonstration the currently accepted value of $\widehat{d}_R$ is used, the theoretical uncertainty in a_{nn} can be reduced down to $\Delta a_{nn}(\text{theor}) = 0.05$ fm. This is quite an impressive reduction (by a factor of 6) of the existing theoretical uncertainty.

3.7. *Radiative corrections*

At the level of a few percent precision, radiative corrections (RCs) become an important issue. For instance, in μp capture, the electroweak RC in the semileptonic reaction and the QED corrections to the muonic atom wave function are found to increase the capture rate by $2.8 \pm 0.4\%$ [78]. The best available estimates of RCs for single-nucleon processes have been given by Marciano and Sirlin [79] for neutron β decay, and by Czarnecki *et al.* [78] for μp capture. In these works RCs of order α is decomposed into so-called outer and inner corrections [80]. The outer correction is a universal function of the lepton energy and is independent of the details of hadron physics, whereas the inner correction is affected by short-distance physics and the hadron structure. The inner corrections coming from γ-(weak boson) loop diagrams are divided into high-momentum and low-momentum parts. The former is calculated in the current quark picture, while the latter is evaluated with the use of the phenomenological weak form factors of the nucleon. As regards RCs in multinucleon systems, all existing works (except those related to the $0^+ \to 0^+$ Fermi transitions), concentrate on the single-nucleon RC diagrams because the genuinely two-body-type RCs are expected to be small. Kurylov *et al.* [81] carried out extensive studies of RCs in the $A = 2$ systems based on the Marciano-Sirlin approach [79], and their results have been widely used in analyzing the $A = 2$ weak-interaction observables (including those discussed earlier in this article). According to [81], with the use of G_F from μ-decay, and the "effective" g_A obtained from neutron β-decay, radiative corrections specific to the νd reaction are estimated to range from 3 to 4% as a function of the neutrino energy (with 0.1% uncertainty).

Although the existing estimates of RCs based on the Marciano–Sirlin approach [79] are believed to be reliable at the level of quoted accuracy, it is worth noting that HBχPT offers a model-independent framework to calculate RCs, insofar as all the relevant LECs are known. The first HBχPT evaluation of RCs for neutron β decay was carried out to order α by Ando *et al.* [82]. In this approach the short-distance physics contributions can be condensed into two LECs, one related to the Fermi constant G_F and the other to the axial coupling constant g_A. If one can determine these LECs from empirical data, one would have a totally model-independent way to evaluate RCs to neutron β decay. A similar situation is expected to exist for any low-energy weak-interaction processes on a single nucleon. Even if one cannot determine these LECs at present, it seems of great interest to correlate RCs for various weak processes on a single nucleon taking advantage of the fact that we know exactly what LECs are involved.

4. Summary

Despite the limited scope of topics covered, we hope we have succeeded in conveying the message that MEEFT is a highly reliable predictive framework for computing transition amplitudes for a large class of electroweak processes in light nuclei. We remark that, in each of the cases for which both SNPA and MEEFT calculations have been performed, it has been found that the results of the two methods agree very well.

Although MEEFT has been registering and will continue to score significant successes, we should mention that great progress is being made in attempts to carry out full EFT (or "AB-EFT") calculations in which both the transition operators and nuclear wave functions are derived from HBχPT; see. e.g. [83–88]. We believe that, as increasingly higher order terms are included in AB-EFT calculations, their results will come increasingly closer to those of MEEFT, which are anchored on accurate experimental inputs. Although the developments in AB-EFT calculations are enormous and highly impressive, we will not discuss them here; we trust they will be covered in Ulf Meissner's contribution to this volume.

As mentioned, Riska and Brown's ground-breaking work on the $np \to \gamma d$ reaction almost three decades ago was a precursor of explosive developments in *high-precision* calculations of electroweak processes in light nuclei. We hope Gerry is happy to see the enormous growth of the field in which he has played such an important role.

Acknowledgments

It is our great pleasure to contribute this article to the Festschrift in honor of Gerry Brown's 85th birthday. A detailed account of MR's life-long scientific collaboration with Gerry is given in another article contributed to this volume. KK started to work with Gerry Brown in 1986, when Gerry kindly invited him to spend a sabbatical year at Stony Brook. KK wishes to express his deepest gratitude to Gerry for having been an inexhaustible source of inspiration and encouragement for him ever since. This article is largely based on the work done in collaboration with Tae-Sun Park, to whom we owe our sincerest thanks. The work of KK is partially supported by the US National Science Foundation under grant number PHY-0758114 and that of MR by the WCU Project of the Korean Ministry of Education, Science and Technology (R33-2008-000-10087-0).

References

[1] D.-O. Riska and G.E. Brown, *Phys. Lett. B* **38**, 193 (1972).
[2] M. Chemtob and M. Rho, *Nucl. Phys. A* **163**, 1 (1971).
[3] See, e.g. J. Carlson and R. Schiavilla, *Rev. Mod. Phys.* **70**, 743 (1998); S.C. Pieper and R.B. Wiringa, *Ann. Rev. Nucl. Part. Sci.* **51**, 53 (2001); A. Kievsky *et al.*, *J. Phys. G* **35**, 063101 (2008).
[4] E. Adelberger *et al.*, 'Solar fusion cross sections II: pp-chain and CNO cycle', arXiv:1004.2314 (to appear in Rev. Mod. Phys.).
[5] V.A. Kuzmin, *Phys. Lett.* **17**, 27 (1965).
[6] S. Fukuda *et al.*, *Phys. Rev. Lett.* **86**, 5651 (2001).
[7] Q. Ahmad *et al.*, *Phys. Rev. Lett.* **87**, 071301 (2001); *Phys. Rev. Lett.* **89**, 011301 (2002); *Phys. Rev. Lett.* **89**, 011302 (2002); B. Aharmim *et al.*, *Phys. Rev. Lett.* **101**, 111301 (2008).
[8] J. Bahcall, M. Pinsonneault and S. Basu, *Astrophys. J.* **555**, 990 (2001).
[9] T. Yamazaki *et al.* (RBC-UKQCD Collaborations), *Phys. Rev. D* **79**, 114505 (2009).
[10] E. Ivanov and E. Truhlik, *Nucl. Phys. A* **316**, 451,437 (1979).
[11] S. Weinberg, *Phys. Lett. B* **251**, 288 (1990); *Phys. Lett. B* **295**, 114 (1992).
[12] V. Bernard, N. Kaiser and U.-G. Meissner, *Int. J. Mod. Phys. E* **4**, 193 (1995).
[13] U. van Kolck, *Prog. Part. Nucl. Phys.* **43**, 337 (1999).

[14] G.P. Lepage, in: Proceedings of the Second Workshop on Effective Field Theories in Nuclear Physics, eds. P. Bedaque *et al.,* (World Scientific, 1999), p. 353.
[15] S. Beane *et al.,* in "*At the Frontier of Particle Physics — Handbook of QCD*", ed. M. Shifman, vol. 1 (World Scientific, Singapore, 2001).
[16] G.E. Brown and M. Rho, *Phys. Rep.* **363**, 85 (2002).
[17] S. Weinberg, *Physica* **6A**, 327 (1979).
[18] See, e.g. H. Georgi, *Weak Interactions and Modern Particle Physics* (Benjamin, 1984).
[19] J. Gasser and H. Leutwyler, *Ann. Phys.* **158**, 142 (1984); J. Gasser, M. Sainio and S. Švarc, *Nucl. Phys. B* **307**, 779 (1988).
[20] M. Rho, *Phys. Rev. Lett.* **66**, 1275 (1991).
[21] T.-S. Park, D.-P. Min and M. Rho, *Phys. Rep.* **233**, 341 (1993).
[22] T.-S. Park, D.-P. Min and M. Rho, *Phys. Rev. Lett.* **74**, 4153 (1995).
[23] D. Kaplan, M. Savage and M. Wise, *Nucl. Phys. B* **478**, 629 (1996); *Nucl. Phys. B* **535**, 329 (1998).
[24] P.F. Bedaque, H.-W. Hammer and U. van Kolck, *Phys. Rev. Lett.* **82**, 463 (1999).
[25] J.-W. Chen, G. Rupak and M.J. Savage, *Nucl. Phys. A* **653**, 386 (1999).
[26] T.-S. Park, K. Kubodera, D.-P. Min and M. Rho, *Phys. Rev. C* **58**, 637 (1998).
[27] T.-S. Park, K. Kubodera, D.-P. Min and M. Rho, *Astrophys. J.* **507**, 443 (1998).
[28] T.-S. Park, K. Kubodera, D.-P. Min and M. Rho, *Nucl. Phys. A* **646**, 83 (1999).
[29] T.-S. Park, K. Kubodera, D.-P. Min and M. Rho, nucl-th/9904053; M. Rho, in: Proceedings of the Second Workshop on Effective Field Theories in Nuclear Physics, ed. P. Bedaque, M. Savage, R. Seki, U. van Kolck (World Scientific, Singapore, 1999), p. 225.
[30] T.-S. Park, K. Kubodera, D.-P. Min and M. Rho, *Nucl. Phys. A* **684**, 101c (2001).
[31] T.-S. Park, L. Marcucci, R. Schiavilla, M. Viviani, A. Kievsky, S. Rosati, K. Kubodera, D.-P. Min and M. Rho, *Phys. Rev. C* **67**, 055206 (2003).
[32] L. Marcucci *et al., Phys. Rev. Lett.* **84**, 5959 (2000); L. Marcucci, R. Sciavilla, M. Viviani, A. Kievsky, S. Rosati and J. Beacom, *Phys. Rev. C* **63**, 015801 (2001).
[33] S. Ando, Y.H. Song, T.-S. Park, H. Fearing and K. Kubodera, *Phys. Lett. B* **555**, 49 (2003).
[34] R. Schiavilla *et al., Phys. Rev. C* **58**, 1263 (1998).
[35] J. Carlson, D.-O. Riska, R. Schiavilla and R.B. Wiringa, *Phys. Rev. C* **44**, 619 (1991).
[36] X. Kong and F. Ravndal, *Phys. Rev. C* **64**, 044002 (2001).
[37] M. Butler and J.-W. Chen, *Phys. Lett. B* **520**, 87 (2001).
[38] S. Ando *et al., Phys. Lett. B* **668**, 187 (2008).
[39] K. Kubodera and S. Nozawa, *Int. J. Mod. Phys. E* **3**, 101 (1994).

[40] S. Nakamura, T. Sato, V. Gudkov and K. Kubodera, *Phys. Rev. C* **63**, 034617 (2001).
[41] S. Nakamura, T. Sato, S. Ando, T.-S. Park, V. Gudkov, F. Myhrer and K. Kubodera, *Nucl. Phys. A* **707**, 561 (2002).
[42] M. Butler and J.-W. Chen, *Nucl. Phys. A* **675**, 575 (2000); M. Butler, J.-W. Chen and X. Kong, *Phys. Rev. C* **63**, 035501 (2001).
[43] M. Butler, J.-W. Chen and P. Vogel, *Phys. Lett. B* **549**, 26 (2002).
[44] J.-W. Chen, K.M. Heeger and R.G.H. Robertson, *Phys. Rev. C* **67**, 025801 (2003).
[45] D.R. Phillips and T.D. Cohen, *Nucl. Phys. A* **668**, 45 (2000).
[46] S. Bogner, T.T.S. Kuo and L. Corragio, *Nucl. Phys. A* **684**, 432c (2001); S. Bogner, T.T.S. Kuo, L. Corragio, A Covello and N. Itaco, *Phys. Rev. C* **65**, 051301 (2002); A. Schwenk, G.E. Brown and B. Friman, *Nucl. Phys. A* **703**, 745 (2002).
[47] B. Mosconi, P. Ricci, E. Truhlik and P. Vogel, *Phys. Rev. C* **75**, 044610 (2007).
[48] S. Ando, T.-S. Park, K. Kubodera and F. Myhrer, *Phys. Lett. B* **533**, 25 (2002).
[49] V.A. Andreev *et al.* (MuSun Collaboration), arXiv:1004.1754.
[50] P. Kammel and K. Kubodera, *Annu. Rev. Nucl. Part. Sci.* **60**, 327 (2010).
[51] V. Bernard, N. Kaiser and U.-G. Meissner, *Phys. Rev. D* **50**, 6899 (1994).
[52] V.A. Andreev *et al.*, *Phys. Rev. Lett.* **99**, 032002 (2007).
[53] N. Tatara, Y. Kohyama and K. Kubodera, *Phys. Rev. C* **42**, 1694 (1990).
[54] J. Adam *et al.*, *Nucl. Phys. A* **507**, 675 (1990).
[55] P. Ricci, E. Truhlik, B. Mosconi and J. Smejkal, *Nucl. Phys. A* **837**, 110 (2010).
[56] J.-W. Chen, T. Inoue, X.-D. Ji and Y.-C. Li, *Phys. Rev. C* **72**, 061001 (2005).
[57] G. Bardin *et al.*, *Nucl. Phys. A* **453**, 591 (1986).
[58] M. Cargnelli *et al.*, Proc. *Yamada Conf. Nucl. Weak Process Nucl. Struc., 23rd*, p.115 (1989).
[59] L.E. Marcucci *et al.*, *Phys. Rev. C* **83**, 014002 (2011).
[60] D.R. Entem and R. Machleidt, *Phys. Rev. C* **68**, 041001 (2003).
[61] C. Werntz and J.G. Brennan, *Phys. Rev.* **157**, 759 (1967).
[62] K. Kubodera and T.-S. Park, *Annu. Rev. Nucl. Part. Sci.* **54**, 19 (2004).
[63] F.L.H. Wolfs *et al.*, *Phys. Rev. Lett.* **63**, 2721 (1989).
[64] R. Wervelman *et al.*, *Nucl. Phys. A* **526**, 265 (1991).
[65] R. Lazauskas, Y.-H. Song and T.-S. Park, *Phys. Rev. C* **83**, 034006 (2011); arXiv: 0905. 3119 [nucl-th].
[66] O.A. Yakubowsky, *Sov. J. Nucl. Phys.* **5**, 937 (1967).
[67] Y.-H. Song, R. Lazauskas and T.-S. Park, *Phys. Rev. C* **79**, 064002 (2009).
[68] T.-S. Park, K. Kubodera, D.-P. Min and M. Rho, *Phys. Lett. B* **472**, 232 (2000).
[69] P. Ackerbauer *et al.*, *Phys. Lett. B* **417**, 224 (1998).
[70] D. Gazit, *Phys. Lett. B* **666**, 472 (2008),
[71] D. Gazit, S.Quaglioni and P. Navrátil, *Phys. Rev. Lett.* **103**, 102502 (2009).
[72] V. Huhn *et al.*, *Phys. Rev. C* **63**, 014003 (2000).

[73] W. van Witch, V. Ruan and H. Witala, *Phys. Rev. C* **74**, 014001 (2006).
[74] D.E. Gonzales Trotter *et al.*, *Phys. Rev. C* **73**, 034001 (2006).
[75] Q. Chen *et al.*, *Phys. Rev. C* **77**, 054002 (2008).
[76] A. Gardestig and D.R. Phillips, *Phys. Rev. C* **73**, 014002 (2006); A. Gardestig and D.R. Phillips, *Phys. Rev. C* **74**, 017001 (2006).
[77] A. Gardestig, *J. Phys. G* **36**, 053001 (2009).
[78] A. Czarnecki, W.J. Marciano and A. Sirlin, *Phys. Rev. Lett.* **99**, 032003 (2007).
[79] W.J. Marciano and A. Sirlin, *Phys. Rev. Lett.* **56**, 22 (1986).
[80] A. Sirlin, *Phys. Rev.* **164**, 1767 (1967); A. Sirlin, *Rev. Mod. Phys.* **50**, 573 (1978).
[81] A. Kurylov, M.J. Ramsey-Musolf and P. Vogel, *Phys. Rev. C* **67**, 035502 (2003).
[82] S. Ando *et al.*, *Phys. Lett. B* **64**, 250 (2004).
[83] E. Epelbaum, H.-W. Hammer and U.-G. Meissner, *Rev. Mod. Phys.* **81**, 1773 (2009).
[84] S. Kölling *et al.*, *Phys. Rev. C* **80**, 045502 (2009).
[85] D. Rozpedzik, J. Golak, S. Kölling and E. Epelbaum, "Two-pion exchange currents in photodisintegration of the deuteron," arXiv:1010.6269 [nucl-th].
[86] D. Rozpedzik *et al.*, arXiv: 1103.4062 [nucl-th].
[87] L. Girlanda, S. Pastore, R. Schiavilla and M. Viviani, arXiv: 0912.2753 [nucl-th].
[88] L. Girlanda, A. Kievsky, L.E. Marcucci, S. Pastore, R. Schiavilla, and M. Viviani, *Phys. Rev. Lett.* **105**, 232502 (2010); arXiv: 1008.0356 [nucl-th].

Chapter 11

The V_{low-k} Low-Momentum Interaction and its Application to Finite-Nuclei and Neutron Stars

T.T.S. Kuo*

I have been working with Gerry for a long time, over 40 years, on a number of topics concerning the nucleon-nucleon interaction and the nuclear many-body problem. Time really goes by fast. In celeberating Gerry's birthday, I would really like to write here an extensive article recollecting the various projects which we have had fun to work on. But for practical reasons, let me here just review one recent topic which is also of interest to Gerry, namely the low-momentum interaction V_{low-k} and its application to finite nuclei and neutron stars. The V_{low-k} interaction works well for finite nuclei. But to satisfactorily describe the properties of large systems such as neutron stars, one may need to include the medium corrections from the Brown-Rho scaling and/or the three-nucleon force.

1. Introduction

The nucleon-nucleon interaction and the nuclear many-body problem are both difficult problems. Recalling his Princeton years, Gerry recently wrote [1]:

"One of the authors, Gerry Brown, arrived at Princeton in early September, 1964. The next morning, as he came to the Palmer Physics Laboratory, Eugene Wigner, who just preceded him, opened the door for him (It was a real contest to get ahead of Eugene and open the door for him which very few succeeded in doing.) Eugene asked Gerry, as we went into the building, what he planned to work on. 'I plan to work out the nucleon-nucleon interaction in nuclei.' Eugene said that it would take someone cleverer than him, to which Gerry replied that they probably

*Department of Physics and Astronomy, Stony Brook University, New York 11794, USA.

disagreed what it meant to 'work out'. Gerry wanted to achieve a working knowledge, sufficiently good to be able to work out problems in nuclear physics. What follows in this book is the story of what Gerry and Tom Kuo, with considerable help from their students, collaborators and friends, put together."

As indicated by the above Wigner-Gerry conversation, it would indeed be very hard to "work out" the nucleon-nucleon (NN) interaction in nuclei in a full-fledged way. But it may be much simpler to deal this problem in an "effective" way, namely using an "effective" theory. We are mostly interested only in certain low-energy properties of the many-body system. For them, a low-energy model space approach with a small number of active degrees of freedom may be adequate. Consider for example the nucleus ^{42}Ca which has 42 nucleons. Suppose we can describe it by a Schroedinger equation

$$H\Psi_n(A) = E_n P\Psi_n(A); \quad H = H_0 + V_{NN} \tag{1}$$

where H_0 is the one-body Hamiltonian and V_{NN} is the NN interaction. The above is a many-body problem of A particles with $A = 42$ (^{42}Ca). Many-body problem with $A = 3$ is already very difficult. To solve the $A = 42$ problem exactly is practically impossible. In addition, V_{NN} is a strong interaction which is still not fully understood. These difficulties suggest that one should probably use a different approach, such as an effective model-space theory, to deal the problem.

We consider an effective Hamiltonian $PH_{\rm eff}P = P(H_0 + V_{\rm eff})P$. Here P a model-space projection operator. For example for the Ca isotopes one may choose $P = |f_{7/2}\rangle\langle f_{7/2}|$, namely a model space consisted of a closed ^{40}Ca core with all the valence nucleons confined in the $f_{7/2}$ orbit. H_0 is given by a harmonic oscillator potential, and $V_{\rm eff}$ is treated as an empirical interaction chosen to fit the experiments.

The above $V_{\rm eff}$ approach has been very successful as illustrated by the pioneering work of Talmi [2] on the calcium isotopes ^{41}Ca–^{48}Ca. In this work these isotopes were treated as composed of n valence neutrons in the $f_{7/2}$ shell outside an inert ^{40}Ca core. For this model space, $V_{\rm eff}$ can be expressed in terms of 3 parameters. In fact the ground state energies of them are given by a very simple formula

$$\langle (f_{7/2})^n | H_{\rm eff} | (f_{7/2})^n \rangle = nC + \frac{n(n-1)}{2}\alpha + \left[\frac{1}{2}n\right]\beta \tag{2}$$

where $[\frac{1}{2}n]$ is a step function equal to $\frac{n}{2}$ if $n =$ even, and $\frac{n-1}{2}$ if $n =$ odd. The parameters C, α and β are adjusted, so as to best fit the experimental

Table 1. Binding energies of Ca isotopes in MeV.

Nucleus	Experimental	Calculated [2]
^{41}Ca	8.36	8.38
^{42}Ca	19.83	19.86
^{43}Ca	27.75	27.78
^{44}Ca	38.89	38.80
^{45}Ca	46.31	46.26
^{46}Ca	56.72	56.82
^{47}Ca		63.81
^{48}Ca	73.95	73.93

energies. The optimum values for them, in MeV, are determined as

$$C = 8.38 \pm 0.05, \quad \alpha = -0.21 \pm 0.01, \quad \beta = 3.33 \pm 0.12.$$

The binding energies given by these parameters are compared with experiments in Table 1. As seen the agreement is astonishingly excellent. The term C is just the neutron single particle (s.p.) energy which can extracted from the experimental binding energies of ^{41}Ca and ^{40}Ca. We note that the optimum value of C is practically the same as the experimental s.p. energy (8.38 vs. 8.36). C is actually not a parameter; it is the experimantal s.p. energy. Thus in this model space approach, there are only 2 parameters, α and β, which characterize the effective interaction V_{eff}.

The above example strongly supports the success of the model-space approach: by confining the nucleons in a small (manageable) space and treating V_{eff} as composed of adjustable parameters, experimental results can indeed be highly satisfactorily "worked out". A challenging task is then can we derive the empirical V_{eff} from an underlying nucleon-nucleon interaction V_{NN}? In Sec. 2 we shall describe a folded-diagram many-body method which provides a framework for deriving microscopically shell-model effective interactrions from the free NN potentials. In Sec. 3 we shall review some details of the low-momentum interaction V_{low-k}. Some applications of V_{low-k} to finite nuclei and neutron stars will be reported in Sec. 4. A summary and outlook will be presented in Sec. 5.

2. Folded-Diagram Effective Interaction Theory

In this section we describe a linked-diagram expansion for the model-space effective interaction. This expansion provides a microscopic framework for deriving the shell-model V_{eff} from V_{NN}. The full-space Schröedinger equation (1) can be reduced to model-space one

$$PH_{\text{eff}}P\Psi_m = E_m P\Psi_m; \quad m = 1, d \tag{3}$$

with $H_{\text{eff}} = H_0 + V_{\text{eff}}$ where V_{eff} is the effective interaction and H_0 denotes the s.p. energies. d is the dimension of the model space. $\{E_m\}$ is a subset of $\{E_N\}$ of Eq. (1). Consider the nucleus ^{18}O as an example. Suppose that H_0 is a harmonic oscillator s.p. Hamiltonian. Then the full-space wave functions Ψ_n of this nucleus consists of, in principle, shell-model wave functions $|2p0h\rangle$, $|3p1h\rangle$, ... and $|18p16h\rangle$. The dimension of this full shell model space is practically "infinitely" large. To simplify the problem, some truncation or renormalization of the full-space problem appears to be indispensible. To describe its low-energy states, this nucleus maybe treated as composed of an inert ^{16}O core with two valence neutrons in the $0d1s$ shell, corresponding to using a model space of

$$P = \Sigma_{2p'} |2p'0h\rangle\langle 2p'0h|, \tag{4}$$

where $2p'$ indicates the restriction that the two neutrons are confined in the $0d1s$ shell. The secular equation used in this shell model is of the form

$$PH_{\text{eff}}P\Psi_m = E_m P\Psi_m; \quad m = 1, d \tag{5}$$

with $H_{\text{eff}} = H_0 + V_{\text{eff}}$ where V_{eff} is the effective interaction and H_0 denotes the s.p. energies. d is the dimension of the model space.

There are two theoretical frameworks for deriving the above V_{eff}: the Feshbach method [4] and the folded-diagram method [5–10]. In the Feshbach formulation, the full-spce Schröedinger equation ($H\Psi_n = E_n\Psi_n$, $H = H_0 + V$) can be rewritten as

$$\begin{pmatrix} PHP & PHQ \\ QHP & QHQ \end{pmatrix} \begin{pmatrix} P\Psi \\ Q\Psi \end{pmatrix} = E \begin{pmatrix} P\Psi \\ Q\Psi \end{pmatrix} \tag{6}$$

where Q is the complement of P, namely $Q = 1 - P$. We can readily eliminate $Q\Psi$, obtaining a P-space equation

$$H_{\text{eff}}(E_n)P\Psi_n = E_n P\Psi_n \tag{7}$$

with

$$H_{\text{eff}}(E_n) = PHP + PHQ\frac{1}{E_n - QHQ}QHP. \tag{8}$$

In this way, the effective interaction is given by

$$PV_{\rm eff}(E_n)P = PVP + PHQ\frac{1}{E_n - QHQ}QHP. \quad (9)$$

The above is an interesting result. We see that now we only need to deal with an equation which is entirely contained in the P-space. However, there is a significant drawback in that the effective Hamiltonian is dependent on the eigenvalue E_n. In other words, we need to use different effective interactions for different eigenstates. This feature is clearly not convenient and desirable for applications.

Among the various folded-diagram methods mentioned earlier, the Kuo-Lee-Ratcliff (KLR) [8, 10] method is relatively more convenient for calculations, and it has been applied to shell-model effective interactions for a wide range of nuclei [11, 12]. It has been developed with the purpose of providing an energy-independent model-space effective interaction. It provides a formal framework for reducing the full-space many-body problem to a model-space one of the form

$$PH_{\rm eff}P\Psi_m = (E_m - E_0^{\rm core})P\Psi_m; \quad PH_{\rm eff}P = P[H_0^{\rm expt} + V_{\rm eff}]P \quad (10)$$

where $m = 1, d$, d being the dimension of the model space. Note that this equation calculates the energy difference among neighboring nuclei. For example, E_m is the energy of ^{19}O and $E_0^{\rm core}$ is the ground-state energy of ^{16}O. In addition, $H_0^{\rm expt}$ is the s.p. energy part which is extracted from experimental energy differences between neighboring nuclei such as ^{17}O and ^{16}O. The above $H_{\rm eff}$ is of the same form as the one commonly used in shell model calculations (such as the calculation of the Ca isotopes mentioned earlier), except for the difference that here $V_{\rm eff}$ is derived from the nucleon-nucleon interaction V_{NN} where in empirical shell model calculations it is determined mainly by fitting selected experimental data. This method has been rather extensively applied to nuclear structure calculations using realistic NN interactions. [11, 12, 48].

The model-space effective interaction given by the KLR method [8,10] is

$$V_{\rm eff} = \hat{Q} - \hat{Q}'\int\hat{Q} + \hat{Q}'\int\hat{Q}\int\hat{Q} - \hat{Q}'\int\hat{Q}\int\hat{Q}\int\hat{Q}\cdots. \quad (11)$$

The above is a folded-diagram expansion for the model-space effective interaction, which is energy-independent and is valence-linked. For the case of a one-dimensional model space, the above expansion reduces to the well-known Goldstone linked diagram expansion [10, 14]. The folded-diagram expansion is basically an extension of the Goldstone expansion to multi-dimensional model space.

Let us now explain the various terms of the above expansion. Each integral sign in the above denotes a "fold" [8]. For instance the last term in the equation is a three-fold term; it has three integral signs. $\hat{Q}$ represents a so-called $\hat{Q}$-box, which may be schematically written as

$$\hat{Q}(\omega) = \left[PVP + PVQ\frac{1}{\omega - QHQ}QVP\right]_{\text{linked}}. \tag{12}$$

(For convenience, we shall use V in place of V_{NN}.) In fact $\hat{Q}$-box is a irreducible vertex function where the intermediate states between any two vertices must belong to the Q-space. (Note that we use Q, without hat, to denote the Q-space projection operator.) It contains valence-linked diagrams only, as indicated by the subscript linked. The $\hat{Q}'$-box of Eq. (19) is defined as ($\hat{Q}$ -PVP).

For a neucleus of n valence nucleons, such as $n = 3$ for ^{19}O, the above V_{eff} contains 1-body (1b), 2-body (2b),... and up to n-body (nb) terms [10], namely

$$V_{\text{eff}} = V_{\text{eff}}(1b) + V_{\text{eff}}(2b) + V_{\text{eff}}(3b) + \cdots + V_{\text{eff}}(nb). \tag{13}$$

An important feature of the above V_{eff} is its "nucleus independence". As an example, the $V_{\text{eff}}(2b)$ for nuclei ^{A}O, $A = 1, 2, \ldots$ are all the same according to the above folded diagram method. This is a desirable feature, as it allows us to use the same 2b effective interaction derived for $A = 18$ for all the other sd-shell nuclei. However the many-body forces for different nuclei are different. For example ^{19}O and ^{20}O have the same $V_{\text{eff}}(3b)$, but ^{20}O has $V_{\text{eff}}(4b)$ while ^{19}O does not.

The above V_{eff} allows us to use the experimental s.p. energies. To illustrate, let us consider the 0d1s-shell nuclei. The energies of ^{17}O is given by $(H_0 + V_{\text{eff}}(1b))$ which corresponds to the experimental s.p. energies of ^{17}O. Since $V_{\text{eff}}(1b)$ is the same for all the 0d1s-shell nuclei, we can use these sp energies for all of them. Thus the model-space secular equation is of the form shown in Eq. (10). Note, however, when experimental s.p. energies are employed, the effective interaction is at least 2-body, namely $H_{\text{eff}} = H_0^{\text{expt}} + V_{\text{eff}}(nb); n \geq 2$. As indicated in Eq. (10), the energies given by H_{eff} is $E_m - E_0^{\text{core}}$. This subtraction is because the diagrams contained in the folded-diagram expansion are all valenced-linked, unlinked core-excitation diagrams having been removed [8, 10].

We now come to the calculation of the folded-diagram expansion for the effective interaction. It can summed up using either the Krenciglowa-Kuo (KLR) [16] or the Lee-Suzuki (LS) [17, 18] iteration methods. The KLR method employes a partial summation framework which has a clear diagrammatic struture, explicitly showing how the folded diagrams are generated and summed in each additional iteration. Another iteration method is the Lee-Suzuki method [17,18], which has been widely employed in shell-model calculations [11, 12]. Let us outline the Lee-Suzuki method: We define the effective interaction given by the nth iteration as R_n, and the initial condition is chosen as

$$R_1 = \hat{Q}(\omega = W_0) \tag{14}$$

where $\hat{Q}$ is the irreducible vertex function of Eq. (20). We consider a degenerate model space with $PH_0P = W_0$. The effective interaction for the second iteration is

$$R_2 = \frac{1}{1 - \frac{d\hat{Q}}{d\omega}\mid_{\omega=W_0}} \times [H_0 - W_0 + \hat{Q}(W_0)], \tag{15}$$

and for the nth iteration we have

$$R_n = \frac{1}{1 - \hat{Q}_1 - \sum_{m=2}^{n-1} \hat{Q}_m \prod_{k=n-m+1}^{n-1} R_k} \left[H_0 - W_0 + \hat{Q}(W_0)\right] \tag{16}$$

where $\hat{Q}_m$ is the energy derivative of the $\hat{Q}$-box as given in Eq. (24). The Lee-Suzuki method is convenient for calculations, and it usually converges after 3 or 4 iterations.

A difference between the above two iteration methods may be pointed out. Recall that for a model space of dimension d, $PH_{\text{eff}}P$ reproduces only d eigenstates of the full-space H. The question is then which d states of H are reproduced by H_{eff}? It is of interest that the Lee–Suzuki method reproduces the d lowest (in energy) states of H [17, 18] while the states reproduced by the Krenciglowa–Kuo method are those of the largest P-space overlaps [16]. The above methods were originally both fomulated for the case of a degenerate P-space, name PH_0P is degenerate. Iteration methods for summing the folded diagrams with non-degenerate PH_0P have also been developed [19, 20].

3. Low-Momentum Interaction V_{low-k}

Starting from a realistic NN potential V_{NN}, we may now proceed to calculate the effective interaction V_{eff} using the folded-diagram framework discussed earlier. But there are two difficulties we need to deal with beforehand. How to "tame" the NN potentials, and which of the different NN potentials one should use. Realistic NN potentials usually have "hard cores" (strong short range repulsions), and they need to be "tamed" before used in many-body calculations. There are a number of successful models for V_{NN}, such as the Paris [22], Bonn [23], CD-Bonn [24], Argonne [25], Nijmegen [26], and Idaho [27] potentials. A common feature of these potentials is that they all can reproduce the empirical deuteron properties and low-energy phase shifts very accurately. But they are, however, quite different from each other, as we shall discuss later. Certainly we would like to have a "unique" NN potential. (The gravitational potential $V(r) = -Gm_1m_2/r_{12}$ is unique as is the Coulomb potential $V(r) = kq_1q_2/r_{12}$.) We are confronted with the difficult question in deciding which, if any, of the existing potential models is the correct one forour many-body calculations.

A partial answer to the above question may be the low-momentum nucleon-nucleon potential V_{low-k} which was initially developed, about 10 years ago, at Stony Brook (USA) and Napoli (Italy) [28–33]. The V_{low-k} project was started in about 1997 when Kuo, Coraggio and Bogner were visiting Arturo Polls and Angela Ramos in Barcelona. An initial purpose for developing the V_{low-k} interaction was to have an energy-independent effective interaction which is more convenient for nuclear many-body calculations than the well-known Brueckner G-matrix. Because of its strong short-range repulsions, the bare interaction V_{NN} is not suitable for being directly used in such calculation; it first needs to be "tamed" into a smooth potential. The G-matrix is such a tamed interaction, and has been widely used for many many years. But G is not convenient to use, mainly because of its off-shell energy dependence. For example, to evaluate the matrix element $\langle k_1k_2|G(\omega)|k_3k_4\rangle$ for a certain vertex in a diagram we need to know the energy variable ω. But knowing the external indices (k_1, k_2, k_3, k_4) alone are in general not enough to determine ω; we need to have also information about what other particles in the diagram are doing. This makes the calculation using the G-matrix rather complicated, particularly for high-order diagrams. The V_{low-k} interaction does not have this off-shell energy dependence of the G-matrix. The Brueckner G-matrix had been the pillar in nuclear many body problems for 50 years or so. Maybe it is time to try something different!

The V_{low-k} interaction tames the "hard core" of V_{NN} in a different way, by integrating out the high-momentum ($k > \Lambda$) components of V_{NN} using a renormalization group (RG) method. Here Λ is the decimation scale. Certain low-energy physics, such as phase shifts, should be preserved by the integrating-out process. Thus, we have used the following T-matrix equivalence formalism. We start from the half-on-shell T-matrix

$$T(k', k, k^2) = V_{NN}(k', k) + P \int_0^{\infty} q^2 dq \frac{V_{NN}(k', q) T(q, k, k^2)}{k^2 - q^2}, \qquad (17)$$

where P denotes the principal value integration. We then define an effective low-momentum T-matrix by

$$T_{low-k}(p', p, k^2) = V_{low-k}(k', k) + P \int_0^{\Lambda} q^2 dq \frac{V_{low-k}(p', q) T_{low-k}(q, p, p^2)}{p^2 - q^2}, \qquad (18)$$

noting that the integration limit here is Λ. To preserve phase shifts, we require the following half-on-shell T-matrix equvalence:

$$T(p', p, p^2) = T_{low-k}(p', p, p^2); \; p', p \leq \Lambda, \qquad (19)$$

which ensures that the low energy ($E_{\rm lab} \leq 2\Lambda^2\hbar^2/m^2$) phase shifts of V_{NN} are preserved by V_{low-k}. The above equations define the effective low-momentum interaction V_{low-k}.

How to solve the above coupled equations was in fact a rather difficult task. It is a special inverse scattering problem, determining the V_{low-k} interaction backwards from the half-on-shell T-matrix. Also V_{low-k} has a special feature that it is nonvanishing only within the momentum model space $k < \Lambda$. The KLR folded diagram method [8, 10] was originally formulated for bound state problems. We have found that it can also be applied to the effective interaction for scattering problems.

In time dependent formulation, the half-on-shell T-matrix can be written as $\langle p' \,|\, VU(0, -\infty) \,|\, p \rangle$, U being the time evolution operator. In this way we can readily perform a diagrammatic analysis of the T-matrix. A general term of this T-matrix may be written as $\langle p' \,|\, (V + V\frac{1}{e(p)}V + V\frac{1}{e(p)}V\frac{1}{e(p)}V + \cdots) \,|\, p \rangle$ where $e(p) \equiv (\varepsilon_p - H_o)$. Note that the intermediate states (represented by 1 in the numerator) cover the entire space, and $1 = P + Q$ where P denotes the model space (momentum $\leq \Lambda$) and Q its complement. Expanding it out in terms of P and Q, a typical term of T is of the form $V\frac{Q}{e}V\frac{Q}{e}V\frac{P}{e}V\frac{Q}{e}V\frac{P}{e}V$. Note it has three segments partitioned by two $\frac{P}{e}$ propagators. Let us define a $\hat{Q}$-box as $\hat{Q} = V + V\frac{Q}{e}V + V\frac{Q}{e}V\frac{Q}{e}V + \cdots$,

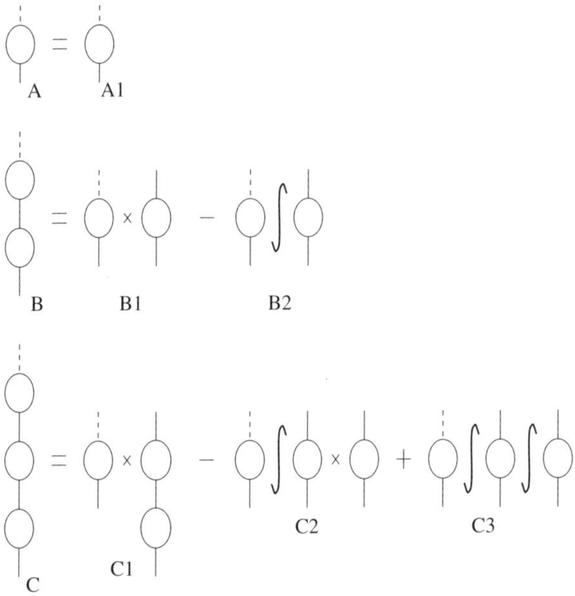

Fig. 1. Folded-diagram factorization of the half-on-shell T-matrix.

where all intermediate states belong to Q. One readily sees that the previous term is just a part of the three-$\hat{Q}$-box term, and in general we have $T = \hat{Q} + \hat{Q}\frac{P}{e}\hat{Q} + \hat{Q}\frac{P}{e}\hat{Q}\frac{P}{e}\hat{Q} + \cdots$.

This regrouping is depicted in Fig. 1, where each $\hat{Q}$-box is denoted by a circle and the solid line represents the propagator $\frac{P}{e}$. The diagrams A, B and C are respectively the one- and two- and three-$\hat{Q}$-box terms of T, and clearly $T = \mathrm{A} + \mathrm{B} + \mathrm{C} + \cdots$. Note the dashed vertical line is not a propagator; it is just a "ghost" line to indicate the external indices. We now perform a folded-diagram factorization for the T-matrix, following closely the KLR folded-diagram method [8, 10]. Diagram B of Fig. 1 is factorized into the product of two parts (see B1) where the time integrations of the two parts are independent from each other, each integrating from $-\infty$ to 0. In this way we have introduced a time-incorrect contribution which must be corrected. In other words B is not equal to B1, rather it is equal to B1 plus the folded-diagram correction B2. Note that the integral sign represents a generalized folding [8, 10].

Similarly we factorize the three-$\hat{Q}$-box term C as shown in the third line of Fig. 1. Higher-order $\hat{Q}$-box terms are also factorized following the same folded-diagram procedure. Let us now collecting terms in the figure in a "slanted" way. Let us denote the sum of terms A1, B2, C3... is as the low-momentum effective interaction

$$V_{low-k} = \hat{Q} - \hat{Q}' \int \hat{Q} + \hat{Q}' \int \hat{Q} \int \hat{Q} - \hat{Q}' \int \hat{Q} \int \hat{Q} \int \hat{Q} + \cdots, \tag{20}$$

which is in fact just the same as the KLR folded-diagram effective interaction of Eq. (11). The sum B1, C2, D3... is $V_{\text{eff}} \frac{P}{e} \hat{Q}$. Similarly the sum $C1 + D2 + E3 + \cdots$ is just $V_{low-k} \frac{P}{e} \hat{Q} \frac{P}{e} \hat{Q}$. (Note diagrams D1, D2, ..., E1, E2, ... are not shown in the figure.) Continuing this way, it is easy to see that the effective interaction required by Eqs. (17) to (19) is just the above folded-diagram effective interaction. The Lee-Suzuki [17, 18] or the Andreozzi [21] iteration methods are commonly used to calculate V_{low-k}.

We have just shown that the above V_{low-k} interaction preserves the half-on-shell T-matrix, which implies the preservation of not only the low-energy phase shifts but also the low-momentum components of the scattering wave functions. This effective interaction also preserve certain bound state properties of V_{NN}, since the expansion of Eq. (20) is the same as the KLR effective interaction, which was originally designed for bound-state problems [8, 10]. The V_{low-k} as given above is not Hermitian. There are a number of ways to transform it into a Hermitian interaction, as discussed by Holt *et al.* [34]

In Fig. 2 we compare some V_{NN} and V_{low-k} matrix elements calculated from the Paris [22] and Bonn-A [23] potentials. It is of interest that V_{NN} of these two potentials are largely different from either other, while the V_{low-k} derived from them are nearly the same. A main difference betwen the Paris and BonnA potentials is that the former has a much stronger "hard core", making their k-space matrix elements differ mainly by a large constant shift. In deriving V_{low-k}, these hard cores (high-k components) are integrated out. This makes the resulting V_{low-k}s nearly the same. In Fig. 3, we present V_{low-k}s derived from six NN potentials; they are nearly the same although the corresponding V_{NN}s are significantly different.

We have mentioned earlier that in using the V_{NN} interactions in nuclear many-body problems, there are two uncertainties: (i) How to tame the "hard cores" of V_{NN}, (ii) which V_{NN} one should choose among the various different V_{NN} models. The V_{low-k} approach provides an answer to both. It is a smooth potential (its high-k components have been integrated out), and it has the remarkable uniqueness feature, in the sense that for decimation scale $\Lambda \leq\sim 2fm^{-1}$ the V_{low-k}s derived from various realistic NN potentials are nearly the same [33]. The NN potentials are constrained by scattering data

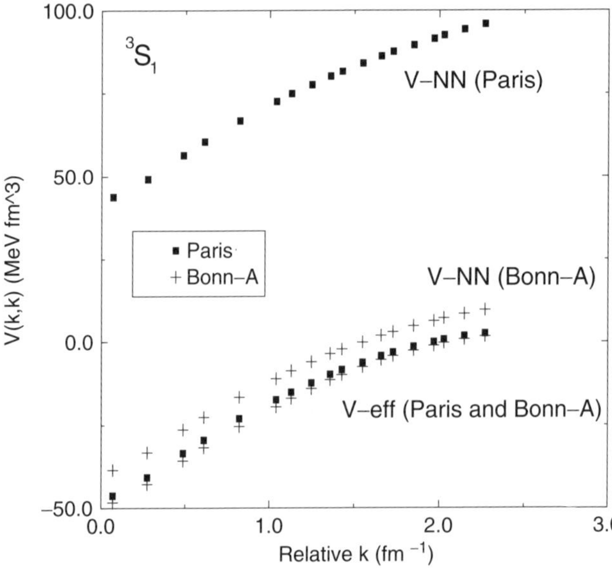

Fig. 2. Comparison of $V_{low-k}(k,k)$ (V-eff) and $V_{NN}(k,k)$.

up to $E_{\text{lab}} \sim 350\,\text{MeV}$, corrresponding to the above Λ value. Thus, we only know the low-momentum NN potentials up to this Λ. The NN potential beyond this Λ is largely unknown and is model dependent.

4. Application of V_{low-k} to Finite Nuclei and Neutron Stars

4.1. *Finite nuclei*

To illustrate the use of V_{low-k} in finite nuclei, we present in Figs. 5 and 6 shell model calculations of ^{18}O and ^{134}Te using V_{low-k} directly as the internucleon interaction, without first calculating the familiar G-matrix. [12, 30] As seen, the spectra of ^{18}O and ^{134} obtained with V_{low-k} are both in remarkably good agreement with experiments. Besides, both are in slightly better agreement with experiments than the corresponding G-matrix calculation. We have employed two values for the decimation scale Λ, namely 2.0 and $2.2\,\text{fm}^{-1}$. It is of interest that the results are rather insensitive to the decimation scale. It may be noted that ^{134}Te and ^{18}O are in rather different nuclear regions. Nevertheless, the same V_{low-k} appears to work equally well for both. In contrast, the G-matrix for these two

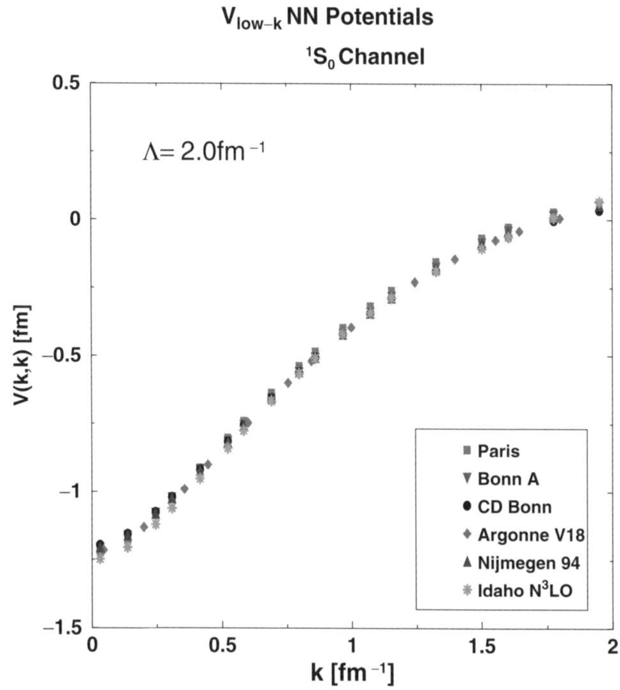

Fig. 3. Comparison of the $V_{low-k}(k,k)$s from six V_{NN} potentials.

nuclei would be very different due to the drastically different Pauli exclusion operators for them.

4.2. *Neutron stars*

The V_{low-k} interaction has given satisfactory results for a wide range of finite nuclei, from the light nuclei in the sd-shell to the heavy ones in the lead region [12]. Can this interaction also describe the properties of even larger nuclear systems such the neutron star? To investigate this question, Dong, Kuo and Machleidt [35] have carried out neutron-star calculations using the V_{low-k} interaction. We first calculate the nuclear matter equation (EOS) using the V_{low-k} inteaction with decimation scale $\Lambda = 3.5\,\text{fm}^{-1}$. This Λ is chosen since the central density of neutron stars is high, $\sim 8n_0$. The EOS is calculated using a ring-diagram formalism [35, 45] where the *pphh* ring digrams as shown in Fig. 6 are included to all orders. The CDbonn [24], BonnA [23], Nijmegen [26] and Argonne [25] NN potentials have been employed.

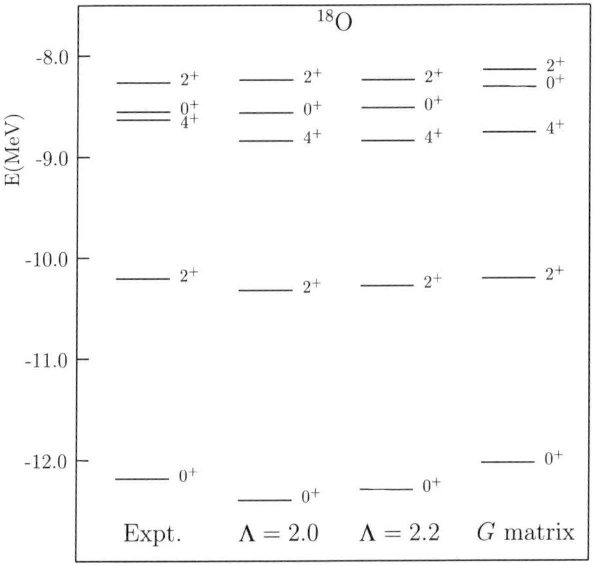

Fig. 4. Low lying states of ^{18}O.

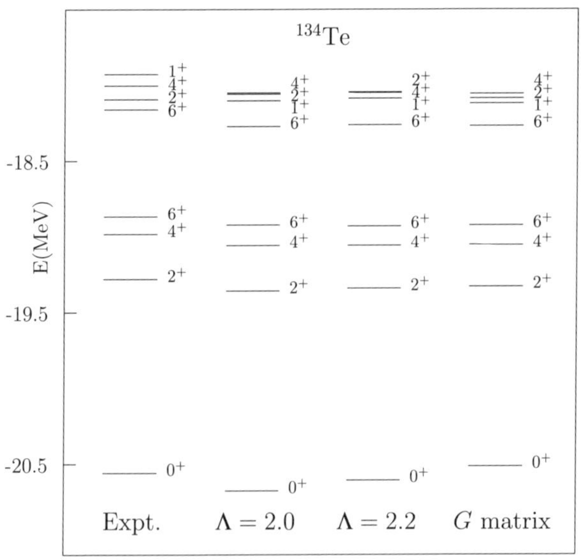

Fig. 5. Low lying states of ^{134}Te.

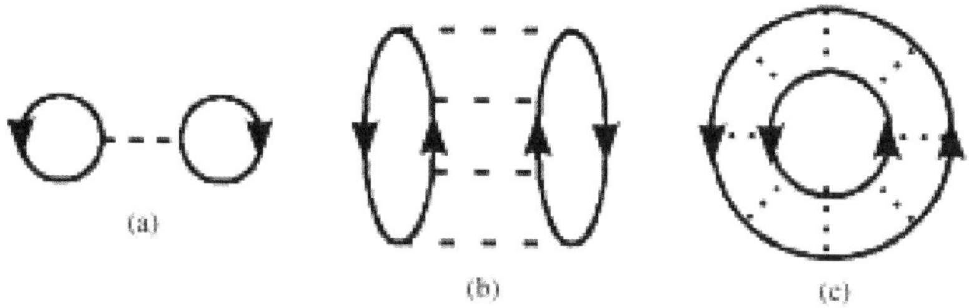

Fig. 6. Sample diagrams included in the all-order *pphh* ring-diagram equation of state for nuclear matter. Each dashed line represents a V_{low-k} vertex.

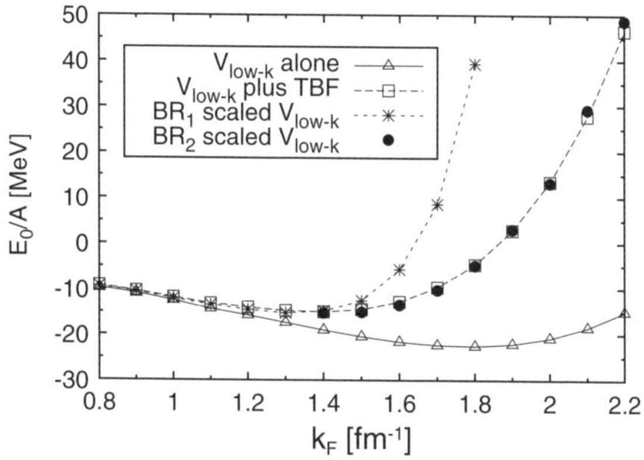

Fig. 7. Ring-diagram EOSs for symmetric nuclear matter given by V_{low-k} alone, V_{low-k} with linear (BR_1) and nonlinear (BR_2) scalings, and V_{low-k} plus the three-body force (TBF) V_{3b} of Eq. (13). $\Lambda = 3.5\,\text{fm}^{-1}$ used for all cases. See text for other explanations.

In Fig. 7, we present several calculations for the ring-diagram EOS of symmetric nuclear matter [35], using the V_{low-k} interaction obtained from the BonnA [23] potential. The resulting EOS, denoted as "V_{low-k} alone" in Fig. 6, has saturation energy E_0/A and density n_0 being both too large compared with the empirical values of, respectively, ($\sim -16\,\text{MeV}$ and $\sim 0.16\,\text{fm}^{-3}$). We have also calculated the neutron-matter EOS using the V_{low-k} interaction. Using this EOS, we then calculate the neutron-star's mass M and radius R by way of the Tolman–Oppenheimer–Volkov equations [35]. The neutron-star mass-radius trajectory so obtained is displayed in Fig. 8, labelled as "no TBF". It gives $M \simeq 1.2\,M_\odot$ and $R \simeq 7\,\text{km}$, both being too small compared with the empirical empirical values of M between $\sim$1.4 to $\sim$2.5 $M_\odot$, and R between $\sim$7 to $\sim$15 km [47].

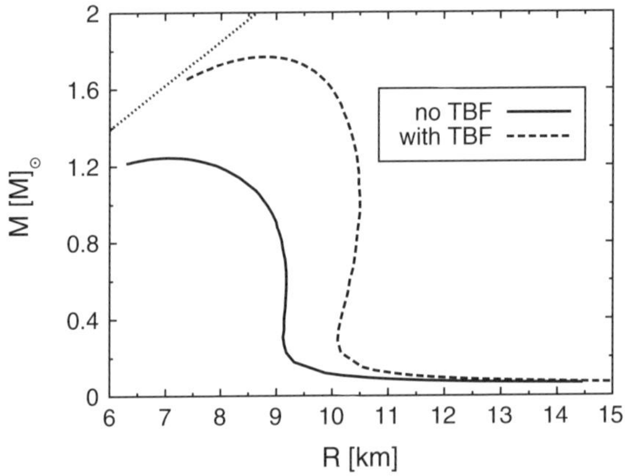

Fig. 8. Mass-radius trajectories of pure neutron stars from ring-diagram EOSs given by the CD-Bonn V_{low-k} interaction with and without the three-body force (TBF) V_{3b}. Causality limit is indicated by the straight line in the upper left corner. $M_\odot$ represents the solar mass.

The results given by the V_{low-k} interactions from the BonnA, CDBonn, Nijmegen and Argonne potentials are nearly the same.

The above discrepencies indicate a possible inadquacy of the NN interactions which we have employed. The V_{NN} interactions commonly employed are constrained by the free-space two-nulceon experimental data. In other words, these V_{NN}s are basically meant to be used for two nucleons in free space. The interaction for two nucleons in medium is likely to be different from that in free space, as suggested by the Brown-Rho (BR) scaling [36–40] where the masses of mesons in nuclear medium are scaled or 'dropped' compared to those in free space. Nucleon-nucleon interactions are mediated by meson exchange, and clearly the in-medium modifications of meson masses can significantly alter the NN interaction, leading to medium-dependent NN interactions. These modifications could arise from the partial restoration of chiral symmetry at finite density/temperature or from traditional many-body effects. Particularly important are the vector mesons, for which there is now evidence from both theory [37, 41, 42] and experiment [43, 44] that the masses may decrease by approximately 10–15% at normal nuclear matter density and zero temperature.

For densities below that of nuclear matter, a linear approximation for the in-medium mass scaling has been suggested [37], namely

$$\frac{m_V^*}{m_V} = 1 - C\frac{n}{n_0}, \tag{21}$$

where m_V^* is the vector meson mass in-medium, n is the local nuclear matter density and n_0 the nuclear matter saturation density. C is a constant of value ~0.10–0.15. The above scaling will be referred to as the linear BR scaling (BR_1).

We have applied the BR_1 scaling to the BonnA one-boson exchange potential with its (ω, ρ, σ) mesons scaled according to Eq. (21). The scaling parameters of $C_\omega = 0.128$, $C_\rho = 0.113$, and $C_\sigma = 0.102$ were chosen so that the resulting EOS gives satisfactory saturation density and energy. As shown in Fig. 7, this EOS (BR_1 scaled V_{low-k}) gives satisfactory saturation properties. But it is rather stiff, rising sharply after n_0. At the center of neutron stars, the density is $\sim 8n_0$. At such high density, the linear BR_1 scaling is clearly not suitable. How to scale the mesons at high densities is still an open question. Dong *et al.* [35] tried a non-linear scaling (to be denoted as the BR_2 scaling) of the form

$$\frac{m_V^*}{m_V} = 1 - C\left(\frac{n}{n_0}\right)^{1/3}. \tag{22}$$

Using the same scaling parameters used for BR_1, the resulting EOS (BR_2 scaled V_{low-k}) is also displayed in Fig. 7. For densities $\leq \sim n_0$ it is practically the same as the BR_1 EOS, but it is much softer than the BR_1 EOS for higher densities.

It has turned out that the above BR_2 EOS can be well reproduced by adding a well-known Skyrme density-dependent force [46] to the unscaled V_{low-k} interaction. The empirical Skyrme force is

$$V_{\text{skyrme}} = \sum_{i<j} V(i,j) + \sum_{i<j<k} V_{3b}(i,j,k). \tag{23}$$

Here $V(i,j)$ is a momentum dependent zero-range two-body force, and $V_{3b}(i,j,k)$ is a zero-range three-body force (TBF)

$$V_{3b}(i,j,k) = t_3 \delta(\vec{r}_i - \vec{r}_j)\delta(\vec{r}_j - \vec{r}_k). \tag{24}$$

For nucleons in a medium of density n, the above is expressed as a density-dependent two-body interaction

$$V_n(1,2) = \frac{1}{6} t_3 \delta(\vec{r}_1 - \vec{r}_2) n(\vec{r}_{av}), \tag{25}$$

where $\vec{r}_{av} = \frac{1}{2}(\vec{r}_1 + \vec{r}_2)$. The parameters t_3 is determined by fitting certain experimental data. The EOS calculated with the interaction given by the sum of V_{low-k} (unscaled) and the Skyrme $V_n(1,2)$ with $t_3 = 2000\,\text{MeV}$ is also shown in Fig. 7, labelled as "V_{low-k} plus TBF". It is of interest that this EOS is nearly identical to the one given by the BR_2-scaled V_{low-k}. It appears that the density dependent effects form the above non-linear BR scaling can be well reproduced by the empirical density-dependent Skyrme interaction.

In Fig. 8, our calculated mass-radius trajectories for pure-neutron stars are presented; one is obtained with V_{low-k} alone ("no TBF") and the other labelled "with TBF" is obtained with BR_2-scaled V_{low-k} or equivalently with the combination of the unscaled V_{low-k} and the Skyrme TBF. As seen, the Skyrme TBF (or the BR_2 scaling) has largely enhanced the neutron-star's maximum mass M and the associated radius R, increasing the mass from ~1.2 to ~1.8 $M_\odot$ and the radius from ~7 to ~9 km. The moments of inertia for neutron stars have also been calculated. For $M \geq 1.0\,M_\odot$, Lattimer and Schutz [47] have determined an empirical formula relating the moment of inertia I of neutron stars to their mass M and radius R, namely

$$I \approx (0.237 \pm 0.008) M R^2 \times \left(1 + 4.2 \frac{M}{M_\odot}\frac{km}{R} + 90\left(\frac{M}{M_\odot}\frac{km}{R}\right)^4\right). \tag{26}$$

We have computed I using our calculated M and R values (with TBF as in the top curve of Fig. 8) as inputs to Eq. (17). Results of this computation are shown in Fig. 9. As shown, the inclusion of the three-body force is very important in having good agreement with the empirical formula. As displayed in Table 2, the inclusion of TBF has increased (M, R, I) by about (50, 30, 200)% respectively. The density dependent effects from the BR scaling is clearly important for neutron-star properties. It have also

Table 2. Maximum mass and the corresponding radius and moment of inertia of β-stable neutron stars [35]. The three-body force TBF is included for the results in the first four rows, but is not in the last.

Potentials	$M[M_\odot]$	R[km]	$I[M_\odot \text{km}^2]$
CDBonn	1.80	8.94	60.51
Nijmegen	1.76	8.92	57.84
BonnA	1.81	8.86	61.09
Argonne V18	1.82	9.10	62.10
CDBonn($V_{3b} = 0$)	1.24	7.26	24.30

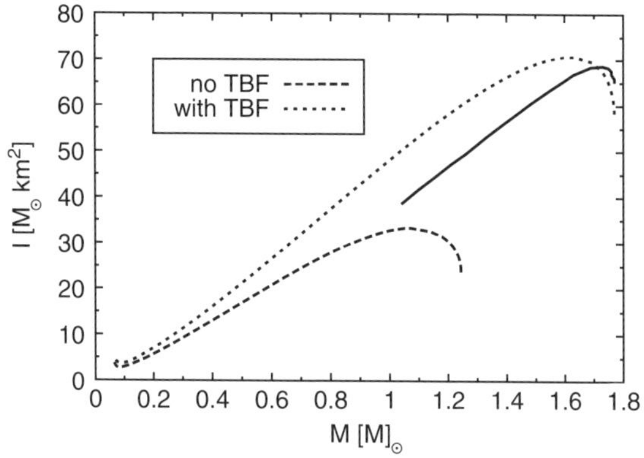

Fig. 9. Pure neutron stars' moments of inertia calculated from CDBonn potential with and without three-body force (TBF). Results from the empirical formula of Eq. (26) for $M \geq 1.0\,M_\odot$ with TBF are denoted by the solid line.

played an important role in explaining the long life time of the $^{14}N \to {}^{14}O$ β-decay [48].

5. Summary and Outlook

Let me quote a paragraph Gerry recently wrote [1] about the nucleon-nucleon interaction:

"Well, one can say to the young people that this is "much ado about nothing". In fact, long ago, when GEB was Professor at Princeton, Murph Goldberger (turning on its head Winston Churchill's famous quote about the R.A.F. during the Battle of Britain) claimed in reference to the nuclear interaction that 'never have so many contributed so little to so few'. Admittedly, at the time it was hard going."

Gerry was at Princeton from 1964 to 1968. The above refers to a conversation he had with Goldberger. The nucleon-nucleon interaction is clearly a difficult and not easy-going problem. Recently much progress has been made. A long standing problem about the NN interaction is that there are a number of V_{NN} models which are all different. Which one should use in nuclear many-body problems? The V_{low-k} interaction has turned out to be an essentialy unique interaction. It is obtained using a renormalization group approach where the high-momentum components of the various V_{NN} are integrated out. The low-energy properties of nuclear systems should not be sensitive to these high-k components, and this is a main reason for enacting this integrating-out procedure. The V_{low-k} interaction has worked well for the shell-model effective interactions for nucleons in dilute nuclear medium. But for nucleons embedded in dense nuclear medium such as neutron stars, this V_{low-k} interaction is inadequate. For them, one may need to include the three-nucleon force or the medium corrections to V_{NN} generated by the Brown-Rho scaling for the in-medium mesons. Again let me quote some paragraphs written by Gerry [1]:

"If we had a unique set of interactions, one for each angular momentum, spin, and isospin channel, it could be argued that it would be justified to stop there. However, since Brueckner came on the scene, Bethe reorganized the theory, Kuo and Brown wrote their paper that prepared the effective theory by using the Scott-Moszkowski separation method, and chiral invariance hit the scene. Chiral invariance doesn't do anything for Yukawa's pion exchange, because the pion gets most of its mass from somewhere outside of the low-energy system, maybe by coupling to the Higgs boson. But the masses of the other mesons drop with increasing density, like

$$m_\rho^* \cong m_\rho(1 - 0.2n/n_0) \quad \text{"Brown-Rho scaling"}$$

where n is the density and n_0 is nuclear matter saturation density. The change in masses of the scalar-σ and vector-ω mesons pretty much cancel each other in effects, the scalar exchange giving attraction and the vector repulsion. However, in the tensor force, the ρ-exchange "beats" against the pion exchange, the former cancelling more and more of the latter as the density increases. This decrease with density of the tensor force interaction has important effects:

1. It is responsible for saturation in the nuclear many-body system.
2. It converts an $\sim$ hour-long carbon-14 lifetime from a superallowed transition in the Wigner $SU(4)$ for p-shell nuclei into an archeologically long 5,700-year transition, and has many other effects.

"Brown-Rho scaling" is also important for neutron stars and may play an important role in turning them into black holes and for "cosmological natural selection". It must be admitted that the same effects could be given by three-body-forces, but Brown-Rho scaling has a deep connection with chiral restoration. We shall review these facets in detail.

Undoubtedly, much more is to come, but we believe that now is a good time to summarize the interesting history of the nucleon-nucleon interaction."

At the center region of neutron stars, the density becomes very high, $\sim 8n_0$. At such high densities, the above linear BR scaling is clearly not suitable. We have found that a nonlinear BR_2 scaling with the above n/n_0 factor changed to $(n/n_0)^{1/3}$ has provided satisfactory results for neutron stars. In fact the results given by the V_{low-k} with this BR_2 scaling and those by adding an empirical Skyrme density dependent force to the unscaled V_{low-k} are nearly the same. It is likely that the BR scaling may provide a microscopic foundation for the density-dependent empirical Skyrme force. Recently Holt, Kaiser and Weise [49] have derived a density dependent two-nucleon interaction from the chiral three-nucleon force by integrating out one of its participating nucleons over the Fermi sea. Further studies about the connection of the three-nucleon force and the density-dependent two-nucleon forces are of much interest.

Acknowledgment

I am especially grateful to Gerry for his guidance, encouragement and friendship throughout the long period, over 40 years, of our collaboration. It has been indeed my great pleasure to have worked with and learned a lot from him. Happy Birthday Gerry. I thank H. Dong for discussions and helping me preparing the present manuscript. This work is supported in part by the US Department of Energy under contract DE-FG02-88ER40388.

References

[1] G.E. Brown, T.T.S. Kuo, J.W. Holt and Sabine Lee (editors), 'The Nucleon-Nucleon Interaction and the Nuclear Many-Body Problems-Selected Papers of Gerald Brown and T.T.S. Kuo', World Scientific Pub. Co. (2010).
[2] A. de-Shalit and I. Talmi, Nuclear Shell Theory (Academic Press, New York and London, 1963).
[3] G.E. Brown and T.T.S. Kuo, *Nucl. Phys.* **A92**, 481 (1967).
[4] H. Feshbach, Ann. *Rev. Nucl. Sci.* **8**, 49 (1958).
[5] M. Morita, *Prog. Theor. Phys.* **29**, 351 (1963).
[6] B.H. Brandow, *Rev. Mod. Phys.* **39**, 771 (1967).
[7] M.B. Johnson and M. Baranger, *Ann. Phys. (N.Y.)* **62**, 172 (1971).
[8] T.T.S. Kuo, S.Y. Lee and K.F. Ratcliff, *Nucl. Phys.* **A176** (1971) 172.
[9] I. Lindgren, *J. Phys.* **B7**, 2441 (1974).
[10] T.T.S. Kuo and E. Osnes, Lecture Notes in Physics Vol. 364 (Springer-Verlag 1990) p. 1.
[11] M. Jorth-Jensen, T.T.S, Kuo and Eivind Osnes, **261**, 205 (1995).
[12] L. Coraggio, A. Covello, A. Gargano, N. Itako and T.T.S. Kuo, *Prog. Part. Nucl. Phys.* **62**, 135 (2009).
[13] J.W. Holt, G.E. Brown, T.T.S. Kuo, J.D. Holt and R. Machleidt, *Phys. Rev. Lett.* **100**, 062501 (2008).

[14] J. Goldstone, *Proc. Roy. Soc. (London)* **A293**, 267 (1957).
[15] J. Shurpin, T.T.S. Kuo and D. Strottman, *Nucl. Phys.* **A408**, 310 (1983).
[16] E.M. Krenciglowa and T.T.S. Kuo, *Nucl. Phys.* **A235**, 171 (1974).
[17] S.Y. Lee and K. Suzuki, *Phys. lett.* **91B**, 173 (1980).
[18] K. Suzuki and S.Y. Lee, *Prog. Theor. Phys.* **64**, 2091 (1980).
[19] K. Suzuki, R. Okamoto, P.J. Ellis and T.T.S. Kuo, *Nucl. Phys.* **A567**, 576 (1994).
[20] T.T.S. Kuo, F. Krmpotic, K. Suzuki and R. Okamoto, *Nucl. Phys.* **A582**, 126 (1995).
[21] F. Andreozzi, *Phys. Rev.* **C54**, 684 (1996).
[22] M. Lacombe *et al.*, *Phys. Rev.* **C21**, 861 (1980).
[23] R. Machleidt, *Adv. Nucl. Phys.* **19**, 189 (1989).
[24] R. Machleidt, *Phys. Rev. C* **63** 024001 (2001).
[25] R.B. Wiringa, V.G.J. Stoks and R. Schiavilla, *Phys. Rev. C* **51**, 38 (1995).
[26] V.G.J. Stoks, R.A.M. Klomp, C.P.F. Terheggen and J.J. de Swart, *Phys. Rev. C* **49**, 2950 (1994).
[27] D.R. Entem and R. Machleidt, *Phys. Rev. C* **68**, 041001 (2003).
[28] S.K. Bogner, T.T.S. Kuo and L. Coraggio, *Nucl. Phys.* **A684**, 432c (2001).
[29] T.T.S. Kuo, S.K. Bogner and L. Coraggio, *Nucl. Phys.* **A704**, 107c (2002).
[30] S.K. Bogner, T.T.S. Kuo, L. Coraggio, A. Covello and N. Itaco, *Phys. Rev. C* **65**, 051301(R). (2002).
[31] T.T.S. Kuo, S.K. Bogner, L. Coraggio, A. Covello, and N. Itaco, in *Challenges of Nuclear Structure* (Proceedings of the 7th International Seminar on Nuclear Structure; Maiori, Italy, May 27–31, 2001), ed. by A. Covello, p. 129, World Scientific Pub. Co. (2002).
[32] A. Schwenk, G.E. Brown, and B. Friman, *Nucl. Phys.* **A703**, 191 (2002).
[33] S.K. Bogner, T.T.S. Kuo and A. Schwenk, *Phys. Rep.* **386**, 1 (2003).
[34] J.D. Holt, T.T.S. Kuo and G.E. Brown, *Phys. Rev. C* **69**, 034329 (2004).
[35] H. Dong, T.T.S. Kuo and R. Machleidt, *Phys. Rev.* **C80**, 065803 (2009).
[36] G.E. Brown, M. Rho, *Phys. Rev. Lett.* **66**, 2720 (1991).
[37] T. Hatsuda and S.H. Lee, *Phys. Rev. C* **46**, R34 (1992).
[38] G.E. Brown and M. Rho, *Phys. Rep.* **269**, 333 (1996).
[39] G.E. Brown and M. Rho, *Phys. Rep.* **363**, 85 (2002).
[40] G.E. Brown and M. Rho, *Phys. Rept.* **396**, 1 (2004).
[41] M. Harada and K. Yamawaki, *Phys. Rept.* **381**, 1 (2003).
[42] F. Klingl, N. Kaiser and W. Weise, *Nucl. Phys.* **A624**, 527 (1997).
[43] D. Trnka *et al.*, *Phys. Rev. Lett.* **94**, 192303 (2005).
[44] M. Naruki *et al.*, *Phys. Rev. Lett.* **96**, 092301 (2006).
[45] L.W. Siu, J.W. Holt, T.T.S. Kuo and G.E. Brown, *Phys. Rev.* **C79**, 0540004 (2009).
[46] P. Ring and P. Schuck, *The Nuclear Many-Body Problem* (Springer-Verlag, New York, 1980), and references quoted therein.
[47] J.M. Lattimer and B.F. Schutz, *Astrophys. J.* **629**, 979 (2005).
[48] J.W. Holt, G.E. Brown, T.T.S. Kuo, J.D. Holt and R. Machleidt, *Phys. Rev. Lett.* **100**, 062501 (2008).
[49] J.W. Holt, N. Kaiser and W. Weise, *Phys. Rev.* **C79**, 054331 (2009); **C81**, 024002 (2010).

Chapter 12

What a Two Solar Mass Neutron Star Really Means

James M. Lattimer* and Madappa Prakash†

The determinations of neutron star masses are reviewed in light of a new measurement of $1.97\,M_\odot$ for PSR J1614-2230 and an estimate of $2.4\,M_\odot$ for the black widow pulsar. Using a simple analytic model related to the so-called maximally compact equation of state, model-independent upper limits to thermodynamic properties in neutron stars, such as energy density, pressure, baryon number density and chemical potential, are established which depend upon the neutron star maximum mass. Using the largest well-measured neutron star mass, $1.97\,M_\odot$, it is possible to show that the energy density can never exceed about 2 GeV, the pressure about 1.3 GeV, and the baryon chemical potential about 2.1 GeV. Further, if quark matter comprises a significant component of neutron star cores, these limits are reduced to 1.3 GeV, 0.9 GeV, and 1.5 GeV, respectively. We also find the maximum binding energy of any neutron star is about 25% of the rest mass. Neutron matter properties and astrophysical constraints additionally imply an upper limit to the neutron star maximum mass of about $2.4\,M_\odot$. A measured mass of $2.4\,M_\odot$ would be incompatible with hybrid star models containing *significant* proportions of exotica in the form of hyperons, Bose condensates or quark matter.

1. Introduction

One of the most fascinating stories in astrophysics concerns the accumulation of precisely measured neutron star masses. Gerry has had a long-time interest in these measurements, as for many years he has maintained that the neutron star maximum mass is no greater than $(1.5\text{–}1.6)\,M_\odot$. From a theoretical perspective, a maximum mass of $1.5\,M_\odot$ fits his fondness

*Department of Physics and Astronomy, Stae University of New York at Stony Brook, Stony Brook, NY 11794-3800, USA. E-mail: lattimer@mail.astro.sunysb.edu
†Department of Physics and Astronomy, Ohio University, Athens, Ohio 45701, USA. E-mail: prakash@harsha.phy.ohiou.edu

for the effects of kaon condensation proposed by Kaplan and Nelson [1]: the effective kaon mass falls with increasing density and the eventual onset of kaon condensation at a few times the nuclear saturation density softens the equation of state (hereafter, EOS) and leads to a rather small maximum mass. He takes as observational support of this thesis the "missing" neutron star in the remnant of SN 1987A, which had a baryon mass not in excess of $1.7\,\mathrm{M}_\odot$ based on estimates of the initial mass of the star and its ejected mass [2]. He reasons that the maximum stable mass of the proto-neutron star as it deleptonized must have eventually fallen below its actual mass, prompting a collapse to a black hole [3]. Of course, this had to have happened more than 12 seconds after the core collapse since neutrinos were observed during that time period and their emission is believed to abruptly cease [4] once the black hole's event horizon forms.

Measurements of Neutron Star Masses

The most accurate measurements of neutron star masses are for pulsars in bound binary systems. In these systems, five Keplerian parameters can be precisely measured by pulse-timing techniques [5], including the binary period P, the projection of the pulsar's semimajor axis on the line of sight $a_p \sin i$ (where i is the binary inclination angle), the eccentricity e, and the time and longitude of periastron T_0 and ω. Combining two of the observational parameters, one can form the mass function:

$$f_p = \left(\frac{2\pi}{P}\right)^2 \left(\frac{a_p \sin i}{c}\right)^3 \frac{\mathrm{M}_\odot}{\mathrm{T}_\odot} = \frac{(M_c \sin i)^3}{M^2}\,\mathrm{M}_\odot, \tag{1}$$

where $M = M_p + M_c$ is the total mass, M_p is the pulsar mass, and M_c is the companion mass (all measured in $\mathrm{M}_\odot$ units). The constant $\mathrm{T}_\odot = G\mathrm{M}_\odot/c^3$ is $4.9255\,\mu$s. The mass function f_p is also equal to the minimum possible mass M_c for the companion.

The inclination angle i is often the most difficult aspect to infer, but even if it was known *a priori* the above equation only specifies a relation between M_p and M_c unless the mass function f_c of the companion is also measurable. This occurs in the rare case when the companion is itself a pulsar or a star with an observable spectrum.

Fortunately, binary pulsars are compact systems and general relativistic effects can often be observed. These include the advance of the periastron of the orbit

$$\dot{\omega} = 3\left(\frac{2\pi}{P}\right)^{5/3} (M\mathrm{T}_\odot)^{2/3}(1-e^2)^{-1}, \tag{2}$$

the combined effect of variations in the tranverse Doppler shift and gravitational redshift around an elliptical orbit

$$\gamma = e\left(\frac{P}{2\pi}\right)^{1/3} \frac{M_c(M+M_c)}{M^{4/3}}\mathrm{T}_\odot^{2/3}, \tag{3}$$

and the orbital period decay due to the emission of gravitational radiation

$$\dot{P} = -\frac{192\pi}{5}\left(\frac{2\pi\mathrm{T}_\odot}{P}\right)^{5/3}\left(1+\frac{73}{24}e^2+\frac{37}{96}e^4\right)(1-e^2)^{-7/2}\frac{M_pM_c}{M^{1/3}}. \tag{4}$$

The inclination angle can be constrained by measurements of two or more of these effects. However, only in extremely compact systems is this precisely possible. Otherwise, additional effects, such as an eclipse or limits obtained from the lack of an eclipse, or Shapiro time delay, which is caused by the propagation of the pulsar signal through the gravitational field of its companion, must be observed. The Shapiro [6] delay in general relativity produces a delay in pulse arrival time [7, 8]

$$\delta_S(\phi) = 2M_c\mathrm{T}_\odot \ln\left[\frac{1+e\cos\phi}{1-\sin(\omega+\phi)\sin i}\right], \tag{5}$$

where ϕ is the true anomaly, the angular parameter defining the position of the pulsar in its orbit relative to the periastron. δ_S is a periodic function of ϕ with an approximate amplitude,

$$\Delta_S \simeq 2M_cT_\odot \left|\ln\left[\left(\frac{1+e\sin\omega}{1-e\sin\omega}\right)\left(\frac{1+\sin i}{1-\sin i}\right)\right]\right|, \tag{6}$$

which is large only for edge-on binaries, $\sin i \sim 1$ or those which have both large eccentricities and large magnitudes of $\sin\omega$. Only a fraction of pulsars in binaries have two or more sufficiently well-measured relativistic effects to enable precise measurements of the pulsar mass M_p.

In some cases, the companion star can be optically detected. In most cases, the companion is a white dwarf. It is possible to estimate white dwarf masses from observations of the optical flux, effective temperature and distance, with the latter being especially problematic. Another possibility is to measure the surface gravity from spectral measurements, but currently this is subject to large systematic effects.

Less accurate measurements of neutron star masses are possible in X-ray binaries, in which X-rays are emitted by matter accreting onto a neutron star from a companion star. Both X-ray and optical observations can yield both mass functions f_p and f_c, but f_p is subject to large uncertainties due to the faintness of optical radiation. In the case of the black widow pulsar, PSR B1957+20, optical observations of the companion yield both a mass function and an estimate of the inclination i from the shape of the light curve [9]. The inferred semimajor axis of the companion's orbit has to be corrected for the finite size of the companion, however (see below), introducing a source of systematic error in the pulsar mass determination of this system. Nevertheless, interesting lower limits to the pulsar mass are obtained [49].

Recent Measurements

The most recent information, as of November 2010, is summarized in Fig. 1 and Table 1 and collects information from Refs. 10 to 50. This compilation represents a significant update to the figure of observed masses and references in Ref. [52].

There is now ample observational support from pulsars for neutron stars with masses significantly greater than $1.5\,M_\odot$. These include PSR J1903+0327 which has a main-sequence companion; the pulsars I and J in the globular cluster Ter 5, the globular cluster binaries PSR J1748-2021B and PSR B1516+02B, and PSR J1614-2230, all with white-dwarf companions; and the black widow pulsar B1957+20. The inclination angles of the two binaries containing pulsars I and J in the globular cluster Ter 5 are unconstrained by observation, but if their inclinations are individually randomly directed, there is a 95% chance that at least one of these pulsars is greater than $1.68\,M_\odot$ [36]. Assumptions related to random orientation also have roles in the mass determinations for PSR J1748-2021B and PSR B1516+02B. One has to be careful about the assumption of random orientation, however, as the selection of potential candidates for follow-up observations necessary for accurate mass determinations are biased. The case of PSR J1909-3744 illustrates the point: initially estimated to have a mass $2.2 \pm 0.2\,M_\odot$, the mass was recently established to be $1.47\,M_\odot$ upon measurement of another relativistic parameter. Nevertheless, about 2 years ago, a measurement of the inclination of PSR J1903+0327 [46] using Shapiro time delay lead to an accurately determined mass of $1.67 \pm 0.01\,M_\odot$,

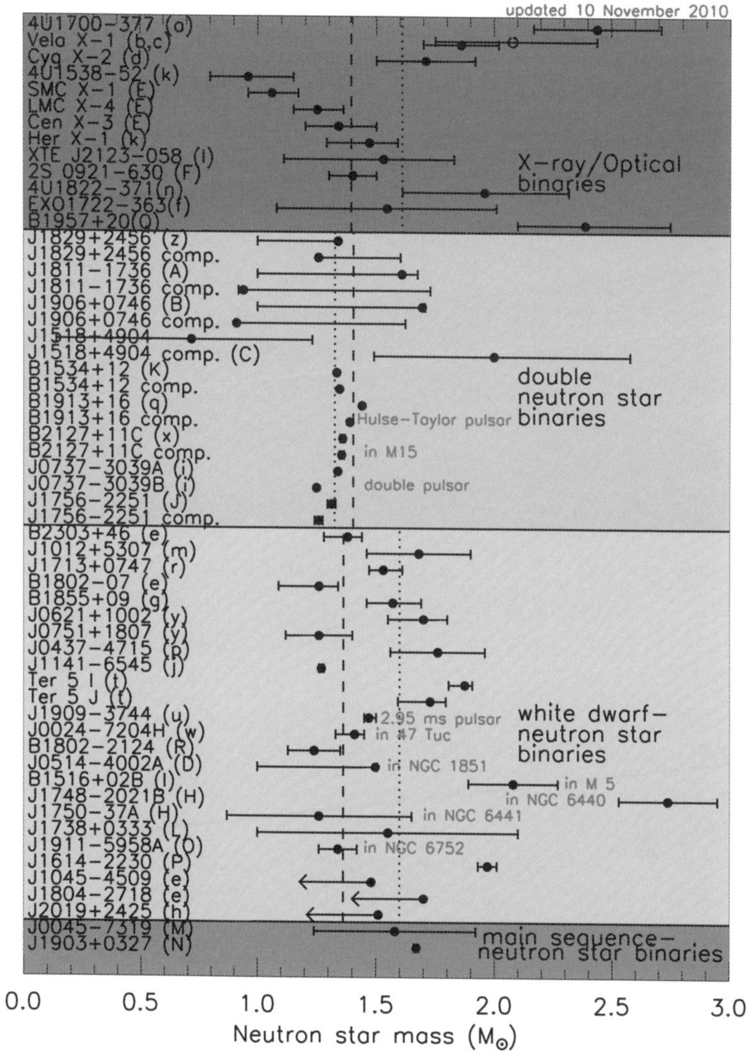

Fig. 1. Measured neutron star masses. References in parenthesis following source numbers are identified in Table 1.

replacing PSR B1913+16 as establishing the value for the minimum maximum mass.

But even more outstanding is the recent determination of $1.97 \pm 0.04\,M_{\odot}$ for the mass of PSR J1614-2230 [48], also using Shapiro time

Table 1. Neutron star mass measurements with 1σ uncertainties. Reference letters correspond to Fig. 1. For each group, mean and weighted masses are indicated.

Object	Mass ($M_\odot$)	Reference	Object	Mass ($M_\odot$)	Reference
X-Ray/Optical Binaries (*mean* = 1.609 $M_\odot$, *weighted mean* = 1.393 $M_\odot$)					
4U1700-377	$2.44^{+0.27}_{-0.27}$	a (10)	Vela X-1	$1.86^{+0.16}_{-0.16}$	b, c (11, 12)
Cyg X-2	$1.71^{+0.21}_{-0.21}$	d (13)	4U1538-52	$0.96^{+0.19}_{-0.16}$	k (14)
SMC X-1	$1.06^{+0.11}_{-0.10}$	E (15)	LMC X-4	$1.25^{+0.11}_{-0.10}$	E (15)
Cen X-3	$1.34^{+0.16}_{-0.14}$	E (15)	Her X-1	$1.47^{+0.12}_{-0.18}$	k (14)
XTE J2123-058	$1.53^{+0.30}_{-0.42}$	l (16, 17)	2S 0921-630	$1.4^{+0.1}_{-0.1}$	F (18)
4U 1822-371	$1.96^{+0.36}_{-0.35}$	n (19)	EXO 1722-363	$1.545^{+0.465}_{-0.465}$	f (47)
B1957+20	$2.39^{+0.36}_{-0.29}$	Q (49)			
Neutron Star – Neutron Star Binaries (*mean* = 1.325 $M_\odot$, *weighted mean* = 1.403 $M_\odot$)					
J1829+2456	$1.338^{+0.002}_{-0.338}$	z (20)	J1829+2456 (c)	$1.256^{+0.346}_{-0.003}$	z (20)
J1811-1736	$1.608^{+0.066}_{-0.608}$	A (21)	J1811-1736 (c)	$0.941^{+0.787}_{-0.021}$	A (21)
J1906+07	$1.694^{+0.012}_{-0.694}$	B (22)	J1906+07 (c)	$0.912^{+0.710}_{-0.004}$	B (22)
J1518+4904	$0.72^{+0.51}_{-0.58}$	C (23)	J1518+4904 (c)	$2.00^{+0.58}_{-0.51}$	C (23)
1534+12	$1.3332^{+0.0010}_{-0.0010}$	K (24)	1534+12 (c)	$1.3452^{+0.0010}_{-0.0010}$	K (24)
1913+16	$1.4398^{+0.0002}_{-0.0002}$	q (25)	1913+16 (c)	$1.3886^{+0.0002}_{-0.0002}$	q (25)
2127+11C	$1.358^{+0.010}_{-0.010}$	x (26)	2127+11C (c)	$1.354^{+0.010}_{-0.010}$	x (26)
J0737-3039A	$1.3381^{+0.0007}_{-0.0007}$	i (27)	J0737-3039B	$1.2489^{+0.0007}_{-0.0007}$	i (27)
J1756-2251	$1.312^{+0.017}_{-0.017}$	J (28)	J1756-2251 (c)	$1.258^{+0.017}_{-0.017}$	J (28)
Neutron Star – White Dwarf Binaries (*mean* = 1.599 $M_\odot$, *weighted mean* = 1.362 $M_\odot$)					
B2303+46	$1.38^{+0.06}_{-0.10}$	e (29)	J1012+5307	$1.68^{+0.22}_{-0.22}$	m (30)
J1713+0747	$1.53^{+0.08}_{-0.04}$	r (31, 51)	B1802-07	$1.26^{+0.08}_{-0.17}$	e (29)
B1855+09	$1.57^{+0.12}_{-0.11}$	g (32, 51)	J0621+1002	$1.70^{+0.10}_{-0.17}$	y (33)
J0751+1807	$1.26^{+0.14}_{-0.14}$	y (33)	J0437-4715	$1.76^{+0.20}_{-0.20}$	p (34)
J1141-6545	$1.27^{+0.01}_{-0.01}$	j (35)	Ter 5 I	$1.874^{+0.32}_{-0.068}$	t (36)
Ter 5 J	$1.728^{+0.066}_{-0.136}$	t (36)	J1909-3744	$1.47^{+0.03}_{-0.02}$	u (37)
J0024-7204H	$1.41^{+0.04}_{-0.08}$	w (38)	B1802-2124	$1.24^{+0.11}_{-0.11}$	R (50)
J0514-4002A	$1.497^{+0.008}_{-0.497}$	D (39)	B1516+02B	$2.08^{+0.19}_{-0.19}$	I (40)
J1748-2021B	$2.74^{+0.21}_{-0.21}$	H (41)	J1750-37A	$1.26^{+0.39}_{-0.39}$	H (41)
J1738+0333	$1.55^{+0.55}_{-0.55}$	L (42)	J1911-5958A	$1.34^{+0.08}_{-0.08}$	O (43)
J1614-2230	$1.97^{+0.04}_{-0.04}$	P (48)	J1045-4509	< 1.48	e (29)
J1804-2718	< 1.70	e (29)	J2019+2425	< 1.51	h (44)
Neutron Star – Main Sequence Binaries					
J0045-7319	$1.58^{+0.34}_{-0.34}$	M (45)	J1903+0327	$1.67^{+0.01}_{-0.01}$	N (46)

delay to measure the inclination. This 3.15 ms pulsar is in an 8.7 d nearly circular orbit about an $0.5\,\mathrm{M}_\odot$ companion, with $a_p \sin i = 11.3$ light-second and $\sin i = 0.99989$, i.e. it is almost edge on. The Shapiro time delay amplitude, from Eq. (6), becomes

$$\Delta_s \simeq 4M_cT_\odot \ln(2/\cos i) \simeq 48.8\,\mu\mathrm{s}. \tag{7}$$

By virtue of its accuracy, this mass measurement has become the new standard for the minimum value of the neutron star maximum mass.

In addition, a number of X-ray binaries seem to contain high-mass neutron stars: about $1.9\,\mathrm{M}_\odot$ in the case of Vela X-1 and $2.4\,\mathrm{M}_\odot$ in the case of 4U 1700-377. Nonetheless, the large systematic errors inherent in X-ray binary mass measurements warrant caution.

The black widow pulsar represents an intriguing case, with a best estimate of about $2.4\,\mathrm{M}_\odot$. This system has both pulsar timing and optical light curve information. It consists of a 1.6 ms pulsar in a nearly circular 9.17 h orbit around an extremely low mass companion, $M_c \simeq 0.03\,\mathrm{M}_\odot$. The pulsar is eclipsed for about 50–60 minutes of each orbit, but considering that $a_p \sin i = 0.089$ light-second $= 0.038\,\mathrm{R}_\odot$, and $a_c \sin i \sim 3\,\mathrm{R}_\odot$ is $M_p/M_c \simeq 80$ times larger, the eclipsing object has to be much larger ($\sim 3\,\mathrm{R}_\odot/10$) than the size of the companion star. It is believed that irradiation of the companion by the pulsar strongly heats its nearside to the point of ablation, leading to a comet-like tail and a large cloud of plasma which is responsible for the eclipsing. The pulsar is literally consuming its companion, hence the name "black widow", and has reduced its mass to a small fraction of its original mass. The irradiation also produces an enormous (factor of 100) variation in the brightness of the companion depending on how much of its nearside is visible during its orbit. The companion is bloated and nearly fills its Roche lobe. The companion's optical light curve allows the mass ratio M_p/M_c and the inclination angle i to be estimated. However, the large size of the companion means that the "center of light" of the system is not equivalent to its "center of mass": the optical light curve depends on the projected semi-major axis of the irradiated nearside of the companion rather than the projected semi-major axis of the center of mass of the companion. The extreme cases are either that the companion has zero radius or that it completely fills its Roche lobe, but estimates based on modeling considerably reduce the allowed range. The extremes lead to a range $1.7 < M_p/\mathrm{M}_\odot < 3.2$, but the likely value is $2.4 \pm 0.4\,\mathrm{M}_\odot$. It will clearly be valuable to extend observations and modeling of this system since a $2.4\,\mathrm{M}_\odot$ neutron star would have profound implications.

What Gerry Would Say

We would love to be able to argue with Gerry about these observations and their implications, and so we felt this was an ideal topic to explore in this contribution. We can, however, guess what Gerry would have said: "We can't tell nature how to behave; we have to explain it" and, as he often quoted Bismarck, "Was schert mich mein Geschwaetz von gestern?" ("Why do I care about my twaddle from yesterday?")

What We Say

One of the most surprising consequences of a large pulsar mass measurement is that it severely restricts values of the maximum central density and pressure in that star, and, by extension, that of any other neutron star as well. This realization was elaborated in Ref. [52], but it can be extended to restrict the possibilities for the presence of exotic matter, such as hyperons, Bose condensates and/or quark matter, in neutron star interiors. Moreover, we show here that upper limits to the chemical potentials of quark or baryonic matter can be established using the available mass data, and that these are smaller than often employed in calculations of the properties of quark matter. The consequences of well-determined large neutron star masses is the topic to which we now turn our attention.

2. Neutron Star Structure

The global properties of neutron stars are determined, assuming general relativity, by the Tolman-Oppenheimer-Volkov (TOV) equations [53]

$$\frac{dp}{dr} = -\frac{G(\varepsilon + p)(m + 4\pi r^3 p/c^2)}{r(rc^2 - 2Gm)}, \tag{8}$$

$$\frac{dm}{dr} = 4\pi r^2 \varepsilon/c^2, \tag{9}$$

$$\frac{d\mathcal{N}}{dr} = 4\pi r^2 n \left(1 - \frac{2Gm}{rc^2}\right)^{-1/2}, \tag{10}$$

where m is the enclosed gravitational mass, $\mathcal{N}$ is the enclosed baryon number, p is the pressure, ε is the total energy density, n is the baryon number density, and r is the radial variable. Choosing a value of the pressure or baryon or energy density at the star's center, and integrating to the surface where the pressure vanishes, results in a specific mass and radius,

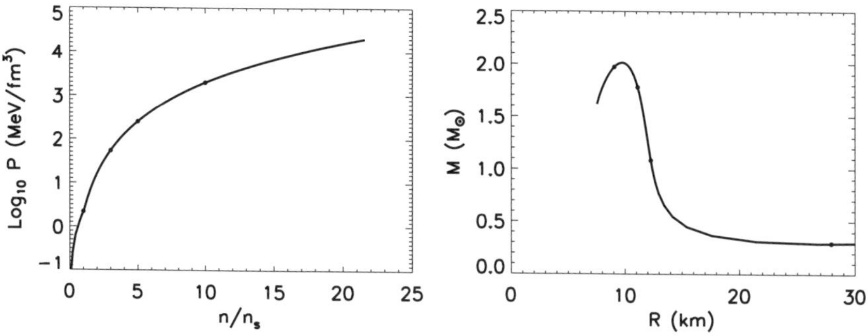

Fig. 2. Solution of the TOV equations. Integrations beginning at the star's center with pressures and baryon number densities (in units of the nuclear saturation density n_s) shown by the dots in the left panel result in specific masses and radii, shown by the corresponding dots in the right panel.

as shown in Fig. 2. A consequence of general relativity is the existence of a maximum mass: configurations with larger central pressures are dynamically unstable and collapse into black holes.

It is convenient to recast the TOV equations in terms of the dimensionless variable h defined by $dh = dp/(\varepsilon + p) = d\mu/\mu$ where $\mu = (\varepsilon + p)/n$ is the baryon chemical potential, which in beta equilibrium is the same as the neutron chemical potential. Therefore,

$$\mu = \mu_o e^h, \tag{11}$$

where $\mu_o = \varepsilon_o/n_o$, ε_o and n_o are the chemical potential, energy density and baryon number density, respectively, where $p = h = 0$. For a star with a normal crust, $\mu_o \simeq 930\,\text{MeV}$, the approximate mass-energy of iron nuclei per baryon. Note that the quantity h is related to the metric function ν, where $g_{tt} = e^\nu$ and $dh = -d\nu/2$. One finds

$$\frac{dr^2}{dh} = -\frac{2r^2c^2}{G}\left[\frac{rc^2 - 2Gm}{mc^2 + 4\pi r^3 p}\right], \tag{12}$$

$$\frac{dm}{dh} = -\frac{4\pi r^3 \varepsilon}{G}\left[\frac{rc^2 - 2Gm}{mc^2 + 4\pi r^3 p}\right], \tag{13}$$

$$\frac{d\mathcal{N}}{dh} = -\frac{4\pi r^{7/2} n c^3}{G}\left[\frac{\sqrt{rc^2 - 2Gm}}{mc^2 + 4\pi r^3 p}\right]. \tag{14}$$

Writing the TOV equations this way ensures that near both the center and the surface the right-hand sides are finite. The equations are integrated from a central value $h = h_c$ to $h = 0$. In order to obtain a specific total gravitational mass $M = m(h = 0)$, one iterates to find the appropriate value of h_c. For a given EOS, that is, the functions $\varepsilon(h)$ and $p(h)$, a specific value or h_c, $h_{\max}$, results in the maximum mass configuration where $M = M_{\max}$, $N = N_{\max}$ and $R = R_{\max}$ where $R = r(h = 0)$ is the stellar radius and $N = \mathcal{N}(h = 0)$ is the total baryon number. The baryon number is useful in the definition of the binding energy

$$BE = (Nm_b - M)c^2, \tag{15}$$

where m_b is the baryon mass.

It is an inescapable consequence of general relativity that all EOSs result in a maximum mass configuration, in contrast to Newtonian gravity for which no such limit exists. It is intuitive that the largest maximum masses will be a consequence of the *stiffest* EOSs, i.e. those with the largest pressures for a given density. It is commonly assumed that the maximum realistic limit to the pressure is obtained by assuming causality, in other words, that the sound speed is limited everywhere by light speed. The assumption of causality led Rhoades and Ruffini [54] to their estimate for the maximum neutron star mass. It is logical, therefore, that the maximum mass configuration so obtained must also have the largest possible values for the central pressure, energy density and chemical potential. The main motivation of this paper is to explore the consequences of this in light of new observations of large neutron star masses.

It was argued by Haensel and Zdunik [55] and by Koranda, Stergioulas and Friedman [56] that the most compact configurations result from the combination of a soft EOS at small densities and a stiff EOS at large densities. The extreme limit of this situation is shown in Fig. 3, where the pressure vanishes below the density ε_o and is assumed causal at *all* higher densities (the so-called maximally compact EOS):

$$p = 0 \quad \varepsilon \leq \varepsilon_o, \quad p = \varepsilon - \varepsilon_o \quad \varepsilon \geq \varepsilon_o. \tag{16}$$

Since ε_o is the only EOS parameter, the TOV equations scale with it according to [57]

$$\varepsilon \propto \varepsilon_o, \quad p \propto \varepsilon_o, \quad n \propto \varepsilon_o, \quad m \propto \varepsilon_o^{-1/2}, \quad r \propto \varepsilon_o^{-1/2}. \tag{17}$$

Following Lattimer, Prakash, Masak and Yahil [58] (hereafter LPMY), we wish to study a somewhat more general EOS, which encompasses the

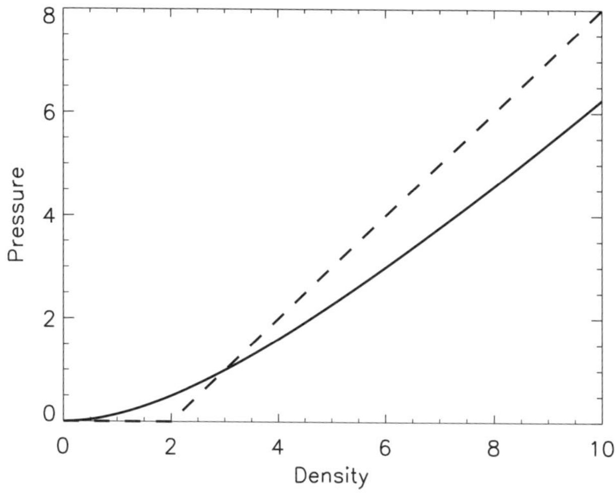

Fig. 3. The solid line is a typical neutron star EOS. The dashed line shows the maximally compact EOS formed by minimizing the pressure p at low energy density ε (i.e. $p = 0$) and maximizing p, subject to causality, at high density (i.e. $p = \varepsilon - \varepsilon_o$). p and ε are in the same arbitrary units, with ε_o in this example set to 2.

maximally compact EOS as a special case, namely,

$$p = s(\varepsilon - \varepsilon_o), \tag{18}$$

where s is a constant representing the square of the adiabatic sound speed. We can then express ε, p and the baryon number density n in terms of h as

$$\varepsilon = t^{-1}\varepsilon_o(s^{-1}e^{ht} + 1), \quad p = t^{-1}\varepsilon_o(e^{ht} - 1), \quad n = \varepsilon_o e^{h/s}/\mu_o \tag{19}$$

where $t = 1 + 1/s$. Using the dimensionless variables

$$x = r^2\epsilon_o G/c^4, \quad y = m\epsilon_o^{1/2}G^{3/2}/c^4, \quad z = \mathcal{N}\epsilon_o^{1/2}\mu_o G^{3/2}/c^6 \tag{20}$$

the scale-free TOV equations become

$$\frac{dx}{dh} = -\frac{2x(x^{1/2} - 2y)}{y + 4\pi x^{3/2}p/\varepsilon_o}, \tag{21}$$

$$\frac{dy}{dh} = 2\pi x^{1/2}\frac{\varepsilon}{\varepsilon_o}\frac{dx}{dh}, \tag{22}$$

$$\frac{dz}{dh} = 2\pi\frac{n\mu_o}{\varepsilon_o}\frac{x}{(x^{1/2} - 2y)^{1/2}}\frac{dx}{dh}. \tag{23}$$

Table 2. Dimensionless TOV eigenvalues for the maximum mass configurations for different s values. The case marked $1/3^*$ has $a_4 = 1/2$ and $a_2^{1/2}/B_{\rm eff}^{1/4} = 2.143$. The notation ** denotes self-bound stars with hadronic shells as described in the text. $h_{\rm max}, \varepsilon_{\rm max}$ and $p_{\rm max}$ are central values; $x_{\rm max}, y_{\rm max}$ and $z_{\rm max}$ are surface values.

s	$h_{\rm max}$	$\varepsilon_{\rm max}/\varepsilon_o$	$p_{\rm max}/\varepsilon_o$	$x_{\rm max}$	$y_{\rm max}$	$z_{\rm max}$	$\varepsilon_{\rm max}y_{\rm max}^2/\varepsilon_o$	$x_{\rm max}^{1/2}/y_{\rm max}$
1	0.8116	3.034	2.034	0.05779	0.08513	0.1127	0.02199	2.824
1**	0.855	3.00	2.00	0.0635	0.0854	0.109	0.0219	2.95
4/5	0.727	3.29	1.83	0.0535	0.0787	0.103	0.0204	2.94
2/3	0.663	3.55	1.70	0.0500	0.0732	0.0946	0.0190	3.05
1/2	0.569	4.01	1.51	0.0443	0.0644	0.0810	0.0166	3.27
1/3	0.4520	4.826	1.275	0.03649	0.05169	0.06212	0.01289	3.696
1/3*	0.4350	6.585	1.639	0.02696	0.04307	0.05134	0.01221	3.813
1/3**	0.491	4.68	1.23	0.0439	0.0522	0.0607	0.0128	4.01

For a given value of s, the value of $h_{\rm max}$ is the value of h_c that maximizes the total gravitational mass $y(h = 0)$. By virtue of Eq. (11), $h_{\rm max}$ then determines the maximum possible chemical potential for that EOS. Table 2 summarizes the properties of the maximum mass solutions for the cases $s = 1, 4/5, 2/3, 1/2$ and $s = 1/3$. The case $s = 1$ represents the maximally compact star; as discussed in §4, the case $s = 1/3$ is potentially applicable to quark stars or stars with quark cores. For each s are listed the central values of h, ε and p, and the surface values of x, y and z, together with other dimensionless quantities of interest.

We will demonstrate that the maximally compact and the $s = 1/3$ EOSs result in useful approximate limits for the maximum values of thermodynamic quantities in the interior of neutron and quark stars as well as their maximum masses.

3. Maximally Compact EOS

We first examine the maximally compact EOS, i.e. the case $s = 1$. Expressing the results for $s = 1$ shown in Table 2 in astrophysical units (note that $1\,{\rm MeV\,fm^{-3}} = 1.323 \times 10^{-6}\,{\rm km^{-2}}$, and $1\,{\rm M_\odot} = 1.477\,{\rm km}$), we find

$$\begin{aligned} M_{\rm max} &= y_{\rm max}c^4\epsilon_o^{-1/2}G^{-3/2} = 4.09\;(\varepsilon_s/\varepsilon_o)^{1/2}\,{\rm M_\odot}, \\ N_{\rm max}m_b &= z_{\rm max}m_bc^6\epsilon_o^{-1/2}\mu_o^{-1}G^{-3/2} = 5.41\,(m_bc^2/\mu_o)(\varepsilon_s/\varepsilon_o)^{1/2}\,{\rm M_\odot}, \\ R_{\rm max} &= x_{\rm max}c^2\epsilon_o^{-1/2}G^{-1/2} = 17.07\,(\varepsilon_s/\varepsilon_o)^{1/2}\,{\rm km}, \end{aligned}$$

$$\begin{aligned}\varepsilon_{\max} &= (e^{2h_{\max}}+1)\varepsilon_o/2 = 3.034\,\varepsilon_o,\\ p_{\max} &= (e^{2h_{\max}}-1)\varepsilon_o = 2.034\,\varepsilon_o,\\ \mu_{\max} &= \mu_o e^{h_{\max}} = 2.251\,\mu_o,\\ n_{\max} &= (\varepsilon_{\max}+p_{\max})e^{-h_{\max}}/\mu_o = 2.251\,(\varepsilon_o/\mu_o).\end{aligned} \tag{24}$$

where $\varepsilon_s \simeq m_b n_s \simeq 150\,\text{MeV fm}^{-3}$ is the mass-energy density at normal nuclear saturation baryon density $n_s \simeq 0.16\,\text{fm}^{-3}$. The result for $M_{\max}$ is very close to, but slightly smaller than, that discussed by Rhoades & Ruffini [54], the difference being their substitution of a low-density neutron star crust EOS for densities less than ε_f, where $\varepsilon_f > \varepsilon_o$ and $p_f > 0$. However, the difference is very small, showing the validity of the maximally compact EOS for an approximate limit.

The results for $\varepsilon_{\max}/\varepsilon_o$ and $p_{\max}/\varepsilon_o$ for the case $s = 1$ were previously obtained by Ref. [56]. What they did not apparently realize was the significance of the quantity $\varepsilon_{\max} y_{\max}^2/\varepsilon_o$, which, when expressed in astrophysical units for the maximally compact case $s = 1$, is

$$\varepsilon_{\max} M_{\max}^2 = 1.358 \times 10^{16}\,\text{g cm}^{-3}\,\text{M}_\odot^2. \tag{25}$$

This equation, of course, states that the central energy density of the maximum mass star is inversely proportional to the square of the maximum mass. Since the maximally compact EOS very closely predicts the largest value for the maximum mass, it seems to follow that this equation should also describe the maximum possible energy density as well. In other words, substituting the mass of the largest observed neutron star mass for $M_{\max}$ in the above expression not only determines the largest energy density in that star, but it might yield an approximate upper limit to the energy density to be found in *any* neutron star. Should a more massive star ever be measured, the upper energy density limit would have to decrease.

It is interesting that the analytic solution to the TOV equations for the energy density distribution

$$\varepsilon = \varepsilon_c[1-(r/R)^2], \tag{26}$$

i.e. the Tolman VII solution [63], coupled with the causality condition Eq. (27), predicts a value for $\varepsilon_{\max} M_{\max}^2$ to within a few percent of Eq. (25), even though the EOS resulting from Eq. (26) becomes acausal at sufficiently large densities. Reference [52] showed that the limit established from the Tolman VII solution is satisfied phenomenologically by all realistic EOSs,

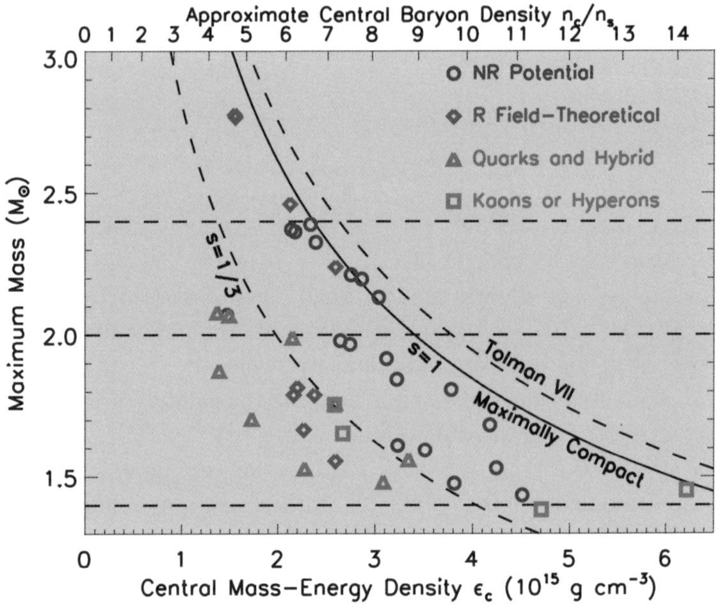

Fig. 4. The central density versus mass relations for the maximally compact EOS ($s = 1$, Eq. (25)) in comparison to Tolman VII + causality and the $s = 1/3$ (Eq. (39)) self-bound cases. Various points are results from different kinds of EOSs, as labelled. Horizontal dashed lines indicate masses of 1.4, 2.0 and 2.4 $M_\odot$, respectively.

as illustrated in Fig. 4. By comparison, the limit defined by Eq. (25) is slightly violated by two of the EOS models but nevertheless represents a realistic limit. As was the case for the estimate of the maximum neutron star mass, the slight violations result from the assumption of zero pressure at low densities.

Reference [56] argues that the maximally compact EOS generally predicts the smallest radius for a configuration with arbitrary mass. In this event, the minimum compactness R/M for any neutron star is determined by

$$R_{\rm max}/M_{\rm max} = 2.824\, G/c^2 \tag{27}$$

a result consistent with discussions by Refs. [55,58–60]. Modifying the EOS at low densities, i.e. for $\varepsilon < (2-3)\,\varepsilon_s$, does not significantly alter the values for $M_{\rm max}$, but the values for $R_{\rm max}$ are moderately increased [58].

Reference [56] also demonstrated that the axially symmetric structure equations in general relativity also scale exactly for the maximally compact

EOS in terms of ϵ_o. The maximum rotation rate, limited by mass-shedding at the equator, scales as $\sqrt{M_{\max}/R_{\max}^3} \propto \sqrt{\varepsilon_o}$ where the mass and radius refer to the values for maximum mass of the non-rotating configuration. As a result, they calculated the minimum spin period for a causally-limited EOS:

$$P_{\min} = 0.74 \left(\frac{\mathrm{M}_\odot}{M_{\max}}\right)^{1/2} \left(\frac{R_{\max}}{10\,\mathrm{km}}\right)^{3/2} \mathrm{ms} = 0.20 \frac{M_{\max}}{\mathrm{M}_\odot}\,\mathrm{ms} \tag{28}$$

Lattimer and Prakash [61] showed that for realistic equations of state the coefficient 0.74 ms is effectively replaced by 1.00 ms, a value almost exactly equal to the general relativistic result in the case that the gravitational potential of the rotating star is due to a point mass at the origin [62]. Furthermore, the revised formula accurately predicts the maximum rotation rate of any star, not just the maximum mass configuration, if $M_{\max}$ and $R_{\max}$ are replaced by their non-rotating values.

The maximally compact configuration is also expected to maximize the binding energy,

$$\mathrm{BE}_{\max} = \left(\frac{m_b c^2}{\mu_o} z_{\max} - y_{\max}\right) \frac{c^6}{\varepsilon_o^{1/2} G^{3/2}} = 0.337\, M_{\max} c^2. \tag{29}$$

The maximum binding energy is thus 25.2% of the rest mass, a fraction that is independent of ε_o and $M_{\max}$.

It is intersting that each of the cases shown in Table 2 yields limiting values for $h_{\max}$ and therefore $\mu_{\max}$ that are also independent of ε_o and therefore also $M_{\max}$. In the case of the maximally compact EOS, one finds the upper limit to the baryon chemical potential

$$\mu_{\max} = 2.251 \times 930\,\mathrm{MeV} = 2093\,\mathrm{MeV}. \tag{30}$$

These limits to the chemical potential, baryon number density, and binding energy have not been previously calculated. We next explore the implications of the large neutron star masses recently measured, and whether the limits obtained from the ideal $s = 1$ and $s = 1/3$ cases pose realistic upper bounds for thermodynamic quantities within neutron stars.

Implications of the newly-measured mass

The measured mass of 1.97 $\mathrm{M}_\odot$ can now be used to set specific limits because the value of ϵ_o is limited according to the first of Eqs. (24): $\varepsilon_o < 4.33\,\varepsilon_s$.

This implies that

$$\begin{aligned}\varepsilon_{\max} &< 3.034 \times 4.33\,\varepsilon_s = 13.1\,\varepsilon_s = 1.97\,\mathrm{GeV\,fm}^{-3},\\ p_{\max} &< 2.034 \times 4.33\,\varepsilon_s = 8.81\,\varepsilon_s = 1.32\,\mathrm{GeV\,fm}^{-3},\\ n_{\max} &< 2.251 \times 4.33\,n_s = 9.75\,n_s = 1.56\,\mathrm{fm}^{-3}.\end{aligned} \tag{31}$$

Recall that the upper limit to the chemical potential, $\mu_{\max} \simeq 2.23\varepsilon_s/n_s = 2.1\,\mathrm{GeV}$, is independent of the measured mass.

Finally, let us note the implications of a hypothetical measurement of $2.4\,\mathrm{M}_\odot$, for example, were the current estimate of the black widow pulsar to be confirmed. In this case, by virtue of the first of Eqs. (24), we find $\varepsilon_o < 2.92\,\varepsilon_s$. This in turn implies that

$$\begin{aligned}\varepsilon_{\max} &< 3.034 \times 2.92\,\varepsilon_s = 8.86\,\varepsilon_s = 1.33\,\mathrm{GeV\,fm}^{-3},\\ p_{\max} &< 2.034 \times 2.92\,\varepsilon_s = 5.94\,\varepsilon_s = 0.89\,\mathrm{GeV\,fm}^{-3},\\ n_{\max} &< 2.251 \times 2.92\,n_s = 6.57\,n_s = 1.05\,\mathrm{fm}^{-3}.\end{aligned} \tag{32}$$

4. Quark Matter EOSs

As at sufficiently high density a quark-hadron transition is likely to occur, it is reasonable to consider a quark matter EOS for the highest densities that might exist in neutron stars. In the case that quark matter has a lower energy per baryon at zero pressure than iron nuclei, a pure quark star is possible in which case the pressure vanishes at non-zero baryon density [57]. This is an example of a self-bound star. The radii of these stars decrease with decreasing mass in contrast to the case of normal neutron stars. The existence of a thin hadronic crust on such a star is possible, but this would not affect considerations of the maximum mass or central density, pressure or chemical potential of the star, although the mass-radius relation for small masses is dramatically altered.

The term hybrid star refers to the case in which the star's interior contains exotica such as hyperons, Bose condensates and /or quark matter whereas the star's exterior is made up of hadronic matter as in a normal neutron star. The presence of matter extending to vanishing density causes the radii of hybrid stars to increase with decreasing mass as for normal neutron stars. A quark matter EOS is obviously softer than the maximally compact EOS, and since the maximum mass depends primarily on the EOS at the highest densities, the maximal constraints for pure quark stars or for neutron stars with quark cores will be more severe than for the maximally compact case. A relevant question is whether or not the transition to quark matter is at low enough density to occur in a neutron star.

MIT Bag Model

Quark matter is likely to be highly interacting for densities near the transition density. Nevertheless, the leading order interaction terms are generally attractive so that a reasonable upper limit to the pressure of quark matter is that for non-interacting massless quarks,

$$p = \frac{1}{3}(\epsilon - 4B), \tag{33}$$

where B is the MIT bag constant (see, e.g. Ref. [64]). This is just the EOS of Eq. (18) with $s = 1/3$ and $\epsilon_o = 4B$. Even if QCD corrections are incorporated, the generic form in Eq. (18) is not greatly altered [65]. Additional corrections for a non-zero strange quark mass and for pairing introduces attractive contributions [65] and must lower the maximum mass so we ignore these for now. If the EOS of quark matter approaches the ideal of Eq. (33) at high density, and there is significant quark matter in a neutron star interior, we conjecture that limits observed in the case $s = 1/3$ could more realistically be expected to approximately constrain the maximum mass and maximum possible energy density, pressure and chemical potential than what is determined from the maximally compact EOS. Irrespective of the value for s, the relation $h = \ln(\mu/\mu_o)$ is still valid. The limits to mass and compactness were investigated for the case $s = 1/3$ by LPMY [58]. The results for the dimensionless integrations are contained in Table 2. All results for structural variables scale with ε_o just as in the maximally compact case.

From Table 2 it is clear for the case $s = 1/3$ that the maximum value of h, $h_{\max} = 0.4520$, is considerably smaller than its value for the maximally compact EOS. The maximum mass stars are also not as compact, the limiting R/M being 3.696. The various dimensionless limits for the case $s = 1/3$ include

$$M_{\max} = 2.48\,(\varepsilon_s/\varepsilon_o)^{1/2}\,\mathrm{M}_\odot,$$

$$N_{\max} m_b = 2.98\,(m_b c^2/\mu_o)(\varepsilon_s/\varepsilon_o)^{1/2}\,\mathrm{M}_\odot,$$

$$R_{\max} = 13.56\,(\varepsilon_s/\varepsilon_o)^{1/2}\,\mathrm{km},$$

$$\mu_{\max} = \mu_o e^{h_{\max}} = 1.571\,\mu_o,$$
$$\varepsilon_{\max} = (e^{2h_{\max}} + 1)\varepsilon_o/2 = 4.826\,\varepsilon_o,$$
$$p_{\max} = (e^{2h_{\max}} - 1)\varepsilon_o = 1.275\,\varepsilon_o,$$
$$n_{\max} = (\varepsilon_{\max} + p_{\max})/\mu_{\max} = 3.884\,(\varepsilon_o/\mu_o). \tag{34}$$

The results for $M_{\max}, R_{\max}$ and $\varepsilon_{\max}$ were previously obtained by Witten [57]. A result of great significance is that, under the reasonable assumption that $\varepsilon_o > \varepsilon_s$ (otherwise deconfined quark matter would be visible in atomic nuclei), the maximum mass of a pure quark star or a hybrid star with a substantial quark core can be no greater than $2.48\,M_\odot$. We will see that, in the case of pure quark stars, more realistic quark models generally have smaller pressures for a given energy density than the MIT bag model and therefore smaller maximum masses. In the case of hybrid stars, if quark matter exists in the cores, the limiting EOS is essentially that of pure quark matter. The magnitude of the transition density n_t depends sensitively on whether the hadronic EOS is soft or stiff at the relevant densities. For stiff hadronic EOSs n_t is often larger than the central density of the maximum mass star and quarks will not appear. In the case that n_t is less than several times n_s, we will show that the hadronic EOS at lower densities does not substantially alter the maximum mass. Thus, the observation of a $2.4\,M_\odot$ neutron star, with the certainty that the true maximum mass is larger, will greatly challenge the notion of the existence of quark matter in neutron star interiors.

It is important to note that the chemical potential $\mu = \mu_o \ln h$ refers to the baryon chemical potential and n refers to the number density of baryons. For neutral quark matter with three flavors, the quark chemical potential is 1/3 the baryon (neutron) chemical potential and the quark density is 3 times the baryon density. Nevertheless, the value of $n\mu$ is the same for both types of matter. The value of μ_o will depend on assumptions regarding the surface of quark matter stars. It is widely believed that a thin baryonic crust will exist, and in this case, the value of μ_o will be the same as previously considered: Iron at zero pressure with $\mu_o \simeq 930\,\text{MeV}$. If $\mu_o = 930\,\text{MeV}$, then we immediately find for the case $s = 1/3$ that $\mu_{\max} = 1.46\,\text{GeV}$. As we show below, this limit for a self-bound MIT bag quark star remains approximately valid for self-bound interacting quark stars and for hybrid quark stars with a hadronic exterior.

Importantly, the TOV equations do not depend on the value for μ_o as can be seen from Eq. (13), but absolute values for the number density and chemical potential do depend on this quantity. In order to discuss separate limits for the chemical potential or number density, one has to have a specific model for the dependence of the energy density on the chemical potential. For massless quarks in the MIT bag model with bag constant B, the relation is

$$\varepsilon = \frac{9}{4\pi^2(\hbar c)^3}\mu_q^4 + B, \tag{35}$$

where $\mu_q = \mu/3$ is the quark chemical potential. Each of the three flavors of quarks has the same chemical potential and density in the massless, neutral, case. We can therefore directly use the results from Table 2 using $\varepsilon_o = 4B$ to obtain

$$\begin{aligned} \mu_{qo} &= \left(\frac{\epsilon_o\pi^2}{3}\right)^{1/4}(\hbar c)^{3/4} = 268.6\left(\frac{B^{1/4}}{141\,\mathrm{MeV}}\right)\,\mathrm{MeV}, \\ n_{qo} &= \left(\frac{\varepsilon + p}{\mu}\right)_o = \left(\frac{3}{\pi^2}\right)^{1/4}\left(\frac{\epsilon_o}{\hbar c}\right)^{3/4} = 0.766\left(\frac{B^{1/4}}{141\,\mathrm{MeV}}\right)^3\,\mathrm{fm}^{-3}. \end{aligned} \tag{36}$$

Note that the limiting value for $\mu_{qo} < 310\,\mathrm{MeV}$ from the condition of stability of quark matter at zero pressure implies $B^{1/4} < 163\,\mathrm{MeV}$ for the MIT bag model [57].

Constraints from the 2-solar-mass star

We now see that the observed mass of $1.97\,\mathrm{M}_\odot$ suggests that

$$\begin{aligned} &\varepsilon_o < (2.48/1.97)^2\varepsilon_s = 1.59\,\varepsilon_s, \quad B^{1/4} < 146.3\,\mathrm{MeV} \\ &\mu_{qo} < \left(\frac{1.59\varepsilon_s\pi^2}{3}\right)^{1/4}(\hbar c)^{3/4} = 278.7\,\mathrm{MeV}, \\ &n_{qo} < \left(\frac{3}{\pi^2}\right)^{1/4}\left(\frac{1.59\varepsilon_s}{\hbar c}\right)^{3/4} = 0.856\,\mathrm{fm}^{-3}. \end{aligned} \tag{37}$$

Therefore, the MIT bag model (without strong interaction corrections) cannot explain the $1.97\,\mathrm{M}_\odot$ star unless strange quark matter is stable. The condition that $\mu_{qo} > 310\,\mathrm{MeV}$, necessary for iron to be the ground state of matter at zero pressure, implies that the maximum mass is $1.58\,\mathrm{M}_\odot$ if the MIT bag model is correct.

One also finds

$$\begin{aligned}\varepsilon_{\max} &= 1.59 \times 4.826\,\varepsilon_s = 7.669\,\varepsilon_s,\\ p_{\max} &= 1.59 \times 1.275\,\varepsilon_s = 2.026\,\varepsilon_s,\\ \mu_{q,\max} &= 1.571\mu_{qo} = 437.8\,\mathrm{MeV},\\ n_{b,\max} &= \frac{p_{\max}+\varepsilon_{\max}}{3\mu_{q,\max}} = 6.926\,n_s,\end{aligned} \tag{38}$$

and

$$\varepsilon_{\max} M_{\max}^2 = 7.91 \times 10^{15}\,\mathrm{g\,cm^{-3}}\,\mathrm{M}_\odot^2. \tag{39}$$

This relation is shown in Fig. 4. The $1.97\,\mathrm{M}_\odot$ star thus limits the energy density to $2.0 \times 10^{15}\,\mathrm{g\,cm^{-3}} = 1.13\,\mathrm{GeV\,fm^{-3}}$, just about half the value for the maximally compact EOS. Likewise, the maximum neutron chemical potential is $\mu_{\max} = 3\mu_{q,\max}$ or 1.29 GeV which is considerably less than the absolute upper limit of 2.09 GeV from causality alone.

Note that the reasonable constraint $\varepsilon_o > \varepsilon_s$ implies $B^{1/4} > 130.3\,\mathrm{MeV}$ in this model. The relatively narrow restrictions on the value of B imply that $248.2\,\mathrm{MeV} < \mu_{qo} < 278.7\,\mathrm{MeV}$. If one further assumes that the baryon density at the surface of the star is at q times the nuclear saturation density, or $n_{qo} = 3qn_s$, one has $1.26 < q < 1.78$. If a neutron star of $2.4\,\mathrm{M}_\odot$ were to be found, this range would shrink to an extreme degree.

Modifications to the Bag Model

In the model developed by Alford *et al.* [65], QCD corrections are included with a parameter a_4, and the effects of a finite strange quark mass and pairing are included with a term proportional to the parameter a_2:

$$\varepsilon = \frac{9a_4}{4\pi^2(\hbar c)^3}\mu_q^4 - \frac{3a_2}{4\pi^2(\hbar c)^3}\mu_q^2 + B_{\rm eff}, \tag{40}$$

where $B_{\rm eff}$ is an effective bag constant.

We first consider the case when $a_2 = 0$. Obviously, the introduction of the parameter a_4 and the replacement of B with $B_{\rm eff}$ will not affect the structure of the star and the dimensionless results from the case $s = 1/3$ still apply as for the MIT bag model. The introduction of the parameter a_4 affects the chemical potential and number densities, however. The quark chemical potential and quark number density at the surface of the star where $p = 0$ can now be determined for $a_2 = 0$ (see also Ref. [66]):

$$\mu_{qo} = \left(\frac{\epsilon_o \pi^2}{3a_4}\right)^{1/4} (\hbar c)^{3/4} = 268.6 \left(\frac{B_{\rm eff}^{1/4}}{141\,{\rm MeV}}\right) a_4^{-1/4}\,{\rm MeV},$$

$$n_{qo} = \left(\frac{\varepsilon + p}{\mu}\right)_o = \left(\frac{3a_4}{\pi^2}\right)^{1/4} \left(\frac{\epsilon_o}{\hbar c}\right)^{3/4} = 0.766 \left(\frac{B_{\rm eff}^{1/4}}{141\,{\rm MeV}}\right)^3 a_4^{1/4}\,{\rm fm}^{-3}. \tag{41}$$

In this case the observed mass of $1.97\,{\rm M}_\odot$ suggests that

$$\mu_{qo} < \left(\frac{1.59\varepsilon_s \pi^2}{3a_4}\right)^{1/4} (\hbar c)^{3/4} = 278.7\,a_4^{-1/4}\,{\rm MeV},$$

$$n_{qo} < \left(\frac{3a_4}{\pi^2}\right)^{1/4} \left(\frac{1.59\varepsilon_s}{\hbar c}\right)^{3/4} = 0.856\,a_4^{1/4}\,{\rm fm}^{-3}. \tag{42}$$

It is now evident that $a_4 < 0.656$ is required if both $\mu_{qo} = 310\,{\rm MeV}$ and $M_{\rm max} > 1.97\,{\rm M}_\odot$ in this model. Further, one has

$$\mu_{q,\rm max} = 1.571\mu_{qo} = 437.8\,a_4^{-1/4}\,{\rm MeV},$$

$$n_{b,\rm max} = \frac{p_{\rm max} + \varepsilon_{\rm max}}{3\mu_{q,\rm max}} = 6.926\,a_4^{1/4} n_s. \tag{43}$$

The introduction of the a_4 term does not alter the limiting energy density inferred from the observed $1.97\,{\rm M}_\odot$ star. However, the maximum baryon chemical potential $\mu_{\rm max} = 1.29/a_4^{1/4}\,{\rm GeV}$ increases moderately from the noninteracting quark case. Nevertheless, the condition that $\mu_{qo} = 310\,{\rm MeV}$ does not change the limiting value of $\mu_{\rm max} = 1.46\,{\rm GeV}$ from that case. We note that one expects from theories of interacting quark matter that $a_4 \sim 0.7$ [65].

The constraint $\varepsilon_o > \varepsilon_s$ implies $B_{\rm eff}^{1/4} > 130.3\,{\rm MeV}$ in this model, unchanged by the a_4 parameter. The restriction on the value of $B_{\rm eff}$ from the observed $1.97\,{\rm M}_\odot$ star now implies that $248.2 < \mu_{qo} a^{1/4} < 278.7\,{\rm MeV}$. Finally, assuming the baryon density at the surface of the star to be qn_s, one finds $1.26 < qa_4^{-1/4} < 1.78$.

We next have to determine how modifying the quark EOS by introducing the a_2 term changes these results. The a_2 term is generally a small correction to the quark EOS. In solving the TOV equation, we may once again scale energy and pressure with $\varepsilon_o = 4B_{\rm eff}$, but there is an additional parameter a_2. Furthermore, ε_o is no longer the value of ε when $p = 0$. So the introduction of a_2 is a moderate complication.

Although there are now 3 parameters affecting the quark EOS, there is a degeneracy in the parameters as far as structural considerations are concerned. We write Eq. (40) in the dimensionless form

$$\frac{\epsilon}{B_{\rm eff}} = \frac{9a_4}{4\pi^2}\left(\frac{\mu_q}{B_{\rm eff}^{1/4}}\right)^4 - \frac{3}{4\pi^2}\left(\frac{a_2^{1/2}}{B_{\rm eff}^{1/4}}\right)^2\left(\frac{\mu_q}{B_{\rm eff}^{1/4}}\right)^2 + 1,$$

$$\frac{p}{B_{\rm eff}} = \frac{3a_4}{4\pi^2}\left(\frac{\mu_q}{B_{\rm eff}^{1/4}}\right)^4 - \frac{3}{4\pi^2}\left(\frac{a_2^{1/2}}{B_{\rm eff}^{1/4}}\right)^2\left(\frac{\mu_q}{B_{\rm eff}^{1/4}}\right)^2 - 1. \qquad (44)$$

Therefore it is only necessary to consider as parameters the quantities a_4 and $a_2^{1/2}/B_{\rm eff}^{1/4}$ in order to establish scaled results for the maximum mass and central pressure and energy density. The value of $\mu_{qo}/B_{\rm eff}^{1/4}$ is a quadratic solution of the equation for $p = 0$ from Eq. (44). To compute numerical values for the mass, energy density, pressure and chemical potential, it is then additionally necessary to fix $B_{\rm eff}$. However, this is not necessary to establish the quantity $\varepsilon_{\rm max} y_{\rm max}^2/\varepsilon_o$ which is independent of $\varepsilon_o = 4B_{\rm eff}$.

We find that varying a_4 for fixed nonzero $a_2^{1/2}/B_{\rm eff}^{1/4}$ has small effects on eigenvalues $h_{\rm max}, x_{\rm max}, y_{\rm max}$ and $\varepsilon_{\rm max} y_{\rm max}^2/\varepsilon_o$. Likewise, there are similar small variations of these eigenvalues with varying $a_2^{1/2}/B_{\rm eff}^{1/4}$ for fixed values of a_4. In any event, the values of $h_{\rm max}, x_{\rm max}$ and $\varepsilon_{\rm max} y_{\rm max}^2/\varepsilon_o$ are largest for the case $a_4 = 1$ and $a_2 = 0$, as seen in Table 2. Furthermore, the increase in the upper limit to the chemical potential resulting from $a_4 < 1$ is approximately canceled by the a_2 contribution. Therefore, the upper limits to the central density and pressure established by the masssless quark bag model hold without regard to QCD corrections, pairing, or finite strange quark mass. This result is confirmed phenomenologically, as illustrated in Fig. 4 which indicates that the limiting density curve for $s = 1/3$, to within a few percent, bounds pure quark stars and hybrid stars containing both quark and hadronic matter. As we proceed to show, the addition of a hadronic mantle and/or crust may permit densities slightly larger than that predicted by Eq. (39), just as they increased the densities for some hadronic neutron stars in some cases beyond the $s = 1$ curve in Fig. 4.

Simple Hybrid Stars

The thermodynamic limits we have so far considered for quark matter apply to self-bound configurations with non-zero baryon density at their surface or those with a very thin hadronic crust. Hybrid stars consist of a quark core or a quark-hadron mixed phase together with a massive hadronic exterior. The presence of a mixed phase is possible only if the surface tension of quark matter is sufficiently large. We consider here the case where there is no mixed phase and the transition between quark and hadronic matter is a first-order transition at the point where μ and p in the two phases are equal, with values μ_t and p_t, respectively.

We will employ a straightforward example. A reasonable approximation for hadronic pressures in the vicinity of nuclear matter densities for neutron star matter is $p = K\varepsilon^2$, where $K \simeq 1.1 \cdot 10^{-4}\,\mathrm{fm}^3\,\mathrm{MeV}^{-1}$. For $h > h_t$ we will use the general form for self-bound matter, Eq. (18). Requiring μ and p to match at h_t, we find

$$\varepsilon = (e^{h/2} - 1)/K \quad h < h_t \tag{45}$$

$$\varepsilon = \varepsilon_0/t + e^{(h-h_t)t}[\varepsilon_o/t + (e^{h_t/2} - 1)^2/K]/s \quad h > h_t. \tag{46}$$

In general, ε and n will be discontinuous at h_t, but we can require them to be continuous if we set

$$\varepsilon_o = \frac{e^{h_t/2} - 1}{K}\left[1 - \frac{e^{h_t/2} - 1}{s}\right]. \tag{47}$$

This assumption does not have a significant effect on our results, and has the advantage of reducing the number of free parameters in this model to a single one: h_t, or, equivalently, p_t. In the case that h_t or p_t tends to zero, we recover the self-bound cases. $h_{\max}$ increases slowly with p_t, approximately as $\ln(1+p_t/p_s)$. For the case $p_t = 10p_s$, integration of the TOV equation with this EOS gives values for the quantities in Table 2 in the rows denoted **.

It is observed in both $s = 1$ and $s = 1/3$ cases with added shells and $p_t = 10p_s$ that $h_{\max}$ modestly increases by about 0.04, i.e. $\mu_{\max}$ is increased by about 4%. Further increasing p_t results in marginally larger values of $h_{\max}$. Because the self-bound cores in both the $s = 1$ and $s = 1/3$ cases are stiffer than is realistic for hadronic and quark matter, respectively, which tends to unrealistically enhance the values of $h_{\max}$, we conclude that the effects of adding a hadronic shell does not result in a significant increase in the central chemical potential, and the values established for the self-bound cases still hold approximately. However, values of $\varepsilon_{\max} y^2_{\max}/\varepsilon_o$ decrease with the addition of a hadronic shell, so our model is not sufficiently general to explain the small violations of the $s = 1$ and $s = 1/3$ curves seen in Fig. 4. We now compare these results, in the case for quark stars, with those from published models.

Published Models of Hybrid Stars

From many calculations in published literature it can be concluded that the density at which exotica appear is uncertain and depends sensitively on the nature of nucleonic interactions up to $(2\text{--}3)\,n_s$. In most cases, however, the EOS with exotica is softer than that without. Consequently, the maximum mass that a hybrid star can support is often smaller (with few exceptions) than that of a pure nucleonic star. In view of the observation of a $1.97\,\mathrm{M}_\odot$ star, the natural question to ask is whether or not exotica are even permitted within such stars.

At the current stage, it is difficult to rule out exotica as many model calculations indicate that a $2\,\mathrm{M}_\odot$ can indeed be supported by hybrid stars of different varieties. Indeed, Fig. 4 contains examples of hybrid quark stars exceeding this threshhold. As another example, Fig. 5 of Ref. [3] shows kaon-condensed stars of $2\,\mathrm{M}_\odot$ that would be compatible with the current observation. Ref. [67], which contrasts the MIT bag model with the Nambu Jona-Lasinio model for the treatment of quarks, presents results of hybrid stars containing nucleons and quarks, and nucleons, hyperons and quarks (see Figs. 4 of Ref. [67]) with maximum masses of $\sim 2\,\mathrm{M}_\odot$. The central densities of maximum mass configurations obtained there were in the range $(0.7\text{--}0.75)\,\mathrm{fm}^{-3}$ (see Fig. 5 of Ref. [67]), which is compatible with the limiting curve $s = 1/3$ in Fig. 4. It would be interesting to see whether or not Dyson-Schwinger -equation approaches [68] to the quark matter EOS yield similar results.

Recently, perturbative calculations of quark matter (without effects of pairing) including strong interaction effects to order α_s^2 and the effects of the strange quark mass have been performed by Kurkela *et al.* in Ref. [69]. As these authors point out, their results become reliable at the 10% level only for quark chemical potentials exceeding 1 GeV, a value far larger than can exist in pure quark or quark hybrid stars as we have shown (the maximum is about 0.5 GeV). Their calculations nevertheless result in quark hybrid stars with masses approaching $2\,\mathrm{M}_\odot$.

These examples show that it is at present not possible to rule out any form of exotica with the $1.97\,\mathrm{M}_\odot$ observation. It must be stressed, however, that many of these approaches suffer from a great deal of model dependence (even in cases in which a substantially lower than a $2\,\mathrm{M}_\odot$ maximum mass star is predicted). However, the situation would change dramatically in the event of a confirmed observation of a $2.4\,\mathrm{M}_\odot$ star. Such an observation would seem to be incompatible with quark or quark hybrid stars including interactions, and would severely restrict models containing other forms of exotica.

Gerry's Views on Hybrid Stars

Prior to his fascination with kaon condensates, Gerry, together with Hans Bethe and Jerry Coopertsein [70], had maintained that perturbative QCD cannot be trusted at densities near n_s where matter must be confined, and that any strangeness would appear in the form of Lambdas. To infer whether strangeness-rich quark matter can occur at high densities, they calculated the energy density of massless up, down and strange quarks to order α_s taking its density dependence as $\alpha_s = 2.2(n/n_s)^{1/3}$. However, as shown in Ref. [66], their choice turned out to be independent of α_s once thermodynamic consistency was imposed. The result of their calculations was that the transition to quark matter occurred at densities larger than those found in neutron stars.

A lesson lurks in Gerry's point of view. Most models of hyperon-filled stars have used hyperon-nucleon couplings based on simplistic quark counting in the non-perturbative regime. Analyses of laboratory experiments, however, indicate that at nuclear densities the Λ-nucleon potential is attractive, but the Σ^--nucleon potential is repulsive [71, 72]. Knorren, Prakash and Ellis [73] investigated the consequences of varying hyperon-nucleon couplings on the structural properties of neutron stars and concluded that, if only Lambdas were to exist in neutron stars, their effects on the maximum mass would be minimal. If such is indeed the case, then two birds can be killed with one stone: first, a star with a 2-solar mass could be easily attained, and second, an avenue for a fast cooling could arise even if the direct Urca process with nucleons could not take place. Even miniscule amounts of Lambdas, through their beta decays, can cool the stars rapidly [74]. In a final example, based on a field-theoretical model for the hadrons and an NJL model for the quarks, Reference [75] fits observed Λ and Ξ^- hyper-nuclei and is able to construct hybrid stars of $2\,M_\odot$. Further investigations to fortify these ideas appear promising.

5. Neutron Matter and Astrophysical Constraints

It is also possible to place constraints on the maximum mass by combining the condition of causality with laboratory data concerning the symmetry energy, neutron skins on neutron-rich nuclei, and the properties of pure neutron matter as well as with astrophysical measurements of neutron star radii and masses. Recently, estimates of neutron matter properties have been obtained by first principle Green's Function Monte Carlo calculations [76] and from chiral effective interactions [77]. These approaches yield similar results. Hebeler *et al.* [78] showed that such calculations restrict neutron star radii to be in the range of (11–12) km when the star's central density is at a density of $1.1\, n_s$, roughly the uppermost density for which the neutron matter calculations are reliable. With this constraint, it was possible to show that causality limits the maximum mass to about $2.9\, M_\odot$ (see Fig. 3 of Ref. [78]).

Steiner, Lattimer and Brown [79] demonstrated that estimates of masses and radii of neutron stars in photospheric radius expansion X-ray bursters and also quiescent, cooling neutron stars limit the radii of $1.5\, M_\odot$ stars to lie in the radius range (10.9–12.5) km. This radius range resulted from a Bayesian analysis of the observational data from three sources of both types. Encouragingly, this radius range is also consistent with the results of neutron matter simulations [76, 78]. With this mass-radius constraint, causality limits the maximum mass to about $2.4\, M_\odot$, about $0.5\, M_\odot$ smaller than that inferred from the neutron matter results themselves. This upper limit to the maximum mass is due to the fact that the radius does not rapidly change with mass for a significant part of the mass-radius relation, as can be observed in Fig. 2. Thus, a restriction of the radius at a specific mass translates into an upper mass limit when the causality condition (i.e. $M < Rc^2/(2.9G)$) is imposed.

6. Conclusions

Based on the observed largest well-measured mass, we have determined model-independent upper limits to the central energy density, central pressure and central chemical potential of neutron stars whether or not they are composed of hadrons, quarks, or both. If matter undergoes a quark-hadron transition within the density range permitted for neutron star interiors, these limits on central density and chemical potential are substantially lowered. Indeed, if it can be shown that the maximum mass of neutron stars exceeds $2.5\, M_\odot$, the possibility of a quark-hadron transition in neutron stars is actually excluded, and the transition in cold matter would have to occur at an energy density in excess of $8\varepsilon_s$ or about 1.2 GeV.

We have also demonstrated the existence of an upper limit to the baryon chemical potential, independent of mass observations, of 2.1 GeV. If quark matter exists in the cores of neutron stars, the upper limit is likely not greater than 1.5 GeV. We also find the maximum binding energy of any neutron star is about 25% of the rest mass.

The astronomical evidence currently favors the neutron star maximum mass to be between $2\,M_\odot$ and $2.5\,M_\odot$. A Bayesian analysis of the observed neutron star mass distribution suggests that this distribution shows no indication of being influenced by an absolute upper limit imposed by general relativity and causality [80]. In other words, this distribution shows no evidence of the existence of an abrupt upper mass cutoff; rather, the largest observed mass may simply be limited by the amount of mass it is reasonably possible for a neutron star to accrete in a binary. Supernova models generally imply that newly formed neutron stars are not more massive than about $1.5\,M_\odot$. In this case it is likely that the true maximum mass determined by the EOS is at least a few tenths of a solar mass larger than the largest observed mass. If so, the chemical potential, density and pressure limits we derived are somewhat conservative. In addition, our survey of EOSs with exotic matter indicates that stars containing *significant* proportions of this matter in the form of hyperons, Bose condensates or quark matter would be almost impossible.

Acknowlegements

This research was supported by the Department of Energy under the grants DE-AC02-87ER40317 (for JML) and DE-FG02-93ER40756 (for MP). We also value our many discussions with Gerry on topics related to the neutron star maximum mass.

References

[1] D. Kaplan and A. Nelson, *Phys. Lett.* **B175**, 57 (1986); **B179**, 409 (1986).
[2] G.E. Brown and H.A. Bethe, *Astrophys. J.* **423**, 659 (1994).
[3] V. Thorsson, M. Prakash and J.M. Lattimer, *Nucl. Phys. A* **572**, 693 (1994).
[4] A. Burrows, *Astrophys. J.* **334**, 891 (1988).
[5] R.N. Manchester and J.H. Taylor, Pulsars (W.H. Freeman, San Francisco, 1977).
[6] I.I. Shapiro, *Phys. Rev. Lett.* **26**, 789 (1964).
[7] T. Damour and N. Deruelle, *Ann. Inst. Henri Poincaré Phys. Theor.* **44**, 263 (1986).
[8] P.C.C. Freire and N. Wex, *MNRAS* **409**, 199 (2010).
[9] M.T. Reynolds, P.J. Callanan, A.S. Fruchter, M.A.P. Torres, M.E. Beer and R.A. Gibbons, *MNRAS* **379**, 1117 (2007).
[10] J.S. Clark, S.P. Goodwin, P.A. Crowther, L. Kaper, M. Fairbairn, N. Langer, and C. Brocksopp, *Astron. & Astrophys.* **392**, 909 (2002).
[11] O. Barziv, L. Karper, M.H. van Kerkwijk, J.H. Telging, and J. van Paradijs, *Astron. & Astrophys.* **377**, 925 (2001).

[12] H. Quaintrell, A.J. Norton, T.D.C. Ash, P. Roche, B. Willems, T.R. Bedding, I.K. Baldry and R.P. Fender, *Astron. & Astrophys.* **401**, 303 (2003).
[13] J. Casares, J.I. González Hernández, G. Israelian and R. Rebolo, *MNRAS* **401**, 2517 (2010).
[14] M.H. van Kerkwijk, J. van Paradijs, and E.J. Zuiderwijk, *Astron. & Astrophys.* **303**, 497 (1995).
[15] A.van der Meer, L. Kaper, M.H. van Kerkwijk, M.H.M. Heermskerk and E.P.J. van den Huevel, *Astron. & Astrophys.* **473**, 523 (2007).
[16] D.M. Gelino, J.A. Tomsick and W.A. Heindl, *Bull. Am. Astron. Soc.* **34**, 1199 (2003).
[17] J. Tomsick, private communication (2004).
[18] D. Steeghs and P.G. Jonker, *ApJLetters* **669**, 85 (2007).
[19] P.G. Jonker, M. van der Klis, P.J. Groot, *MNRAS* **339**, 663 (2003).
[20] D.J. Champion, D.R. Lorimer, M.A. McLaughlin, K.M. Xilouris, Z. Arzoumanian, P.C.C. Freire, A.N. Lommen, J.M. Cordes and F. Camilo, *MNRAS* **363**, 929 (2005).
[21] A. Corongiu , M. Kramer, A.G. Lyne, O. Löhmer, N. D'Amico, & A. Possenti, *Mem. S.A. It Suppl.* **5**, 188 (2004).
[22] D.R. Lorimer *et al. ApJ* **640**, 428 (2006).
[23] G.H. Janssen, B.W. Stappers, M. Kramer, D.J. Nice, A. Jessner, I. Cognard and M.B. Purver, *Astron. & Astrophys.* **490**, 753 (2008).
[24] I.H. Stairs, S.E. Thorsett, J.H. Taylor and A. Wolszczan, *ApJ* **581** 501 (2002).
[25] J.M. Weisberg, D.J. Nice & J.H. Taylor, *ApJ* **722**, 1030 (2010).
[26] B.A. Jacoby, P.B. Cameron, F.A. Jenet, S.B. Anderson, R.N. Murty and S.R. Kulkarni, *ApJLetters* **644**, L113 (2006).
[27] M. Kramer, I.H. Stairs, R.N. Manchester, M.A. McLaughlin, A.G. Lyne, R.D. Ferdman, M. Burgay, D.R. Lorimer, A. Possenti, N. D'Amico, J.M. Sarkisian, G.B. Hobbs, J.E. Reynolds, P.C.C. Freire and F. Camilo, *Science* **314**, 97 (2006).
[28] R.D. Ferdman, Ph.D. thesis, Univ. of British Columbia (2008).
[29] S.E. Thorsett and D. Chakrabarty, *Astrophys. J.* **512**, 288 (1999).
[30] Ch. Lange, F. Camilo, N. Wex, M. Kramer, D.C. Backer, A.G. Lyne and O. Doroshenko, *MNRAS* **326**, 274 (2001).
[31] E.M. Splaver, D.J. Nice, I.H. Stairs, A.N. Lommen and D.C. Backer, *ApJ* **620**, 405 (2005).
[32] D.J. Nice, E.M. Splaver and I.H. Stairs, in *Radio Pulsars*, ed. M. Bailes, D.J. Nice, and S.E. Thorsett (Ast. Soc. Pac. 302, San Francisco, 2003).
[33] D.J. Nice, I.H. Stairs and L.E. Kasian, in *40 YEARS OF PULSARS: Millisecond Pulsars, Magnetars and More*, AIP Conference Proceedings **983**, 453 (2008).
[34] J.P.W. Verbiest, M. Bailes, W. van Straten, G.B. Hobbs, R.T. Edwards, R.N. Manchester, N.D.R. Bhat, J.M. Sarkissian, B.A. Jacoby and S.R. Kulkarni, *ApJ* **679**, 675 (2008).
[35] N.D.R. Bhat, M. Bailes and J.P.W. Verbiest, *Phys. Rev. D* **77**, 124017 (2008).
[36] S.M. Ransom, J.W.T. Hessels, I.H. Stairs, P.C. Freire, F. Camilo, V.M. Kaspi and K.L. Kaplan, *Science* **307**, 892 (2005).

[37] A.W. Hotan, M. Bailes and S.M. Ord, *MNRAS* **369**, 1502 (2006).
[38] P.C.C. Freire, F. Camilo, M. Kramer, D.R. Lorimer, A.G. Lyne, R.N. Manchester & N.D'Amico, *MNRAS* **340**, 1359 (2003).
[39] P.C.C. Freire, S.M. Ransom and Y. Gupta, *ApJ* **662**, 1177 (2007).
[40] P.C.C. Freire, A. Wolszcan, M. van den Berg and J.W.T. Hessels, *ApJ* **679**, 1433 (2008).
[41] P.C.C. Freire, S.M. Ransom, S. Bégin, I.H. Stairs, J.W.T. Hessels, L.H. Frey and F. Camilo, *ApJ* **675**, 670 (2008).
[42] P.C.C. Freire, B.A. Jacoby and M. Bailes, in *40 YEARS OF PULSARS: Millisecond Pulsars, Magnetars and More*, AIP Conference Proceedings **983**, 488; arXiv:0711.1880 (2008).
[43] C.G. Bassa, M.H. van Kerkwijk, D. Koester and F. Verbunt, *Astron. & Astrophys.* **456**, 295 (2006).
[44] D.J. Nice, E.M. Splaver and I.H. Stairs, *ApJ* **549**, 516 (2001).
[45] D.J. Nice, *IAU Proceedings*, Sydney, Australia (2003).
[46] P.C.C. Freire, C.G. Bassa, N. Wex, I.H. Stairs, D.J. Champion, S.M. Ransom, P. Lazarus, V.M. Kaspi, J.W.T. Hessels, M. Kramer, J.M. Cordes, J.P.W. Verbiest, P. Podsiadlowski, D.J. Nice, J.S. Deneva, D.R. Lorimer, B.W. Stappers, M.A. McLaughlin and F. Camilo, *MNRAS*, in press; arXiv:1011.5809 (2010).
[47] A.B. Mason, A.J. Norton, J.S. Clark, I. Negueruela & P. Roche, *Astron. & Astrophys.* **509**, 79 (2010).
[48] P.B. Demorest, T. Pennucci, S.M. Ransom, M.S.E. Roberts and J.W.T. Hessels, *Nature*, **467**, 1081 (2010).
[49] M.H. van Kerkwijk, R. Breton & S.R. Kulkarni, submitted to *ApJ* (2010); arXiv:1009.5427.
[50] R.D. Ferdman, I.H. Stairs, M. Kramer, M.A. McLaughlin, D.R. Lorimer, D.J. Nice, R.N. Manchester, G. Hobbs, A.G. Lyne, F. Camilo, A. Possenti, P.B. Demorest, I. Cognard, G. Desvignes, G. Theureau, A. Faulkner, D.C. Backer, *ApJ* **711**, 764 (2010).
[51] T.M. Tauris and G.J. Savonije, *Astron. & Astrophys.* **350**, 928 (1999).
[52] J.M. Lattimer and M. Prakash, *Phys. Rev. Lett.* **94**, 111101 (2005).
[53] R.C. Tolman, *Proc. Nat. Acad. Sci. U.S.A.,* **20**, 3 (1934); J.R. Oppenheimer and G.M. Volkoff, *Phys. Rev.* **55**, 374 (1939).
[54] C.E. Rhoades Jr. and R. Ruffini, *Phys. Rev. Lett.* **32**, 324 (1974).
[55] P. Haensel and J.L. Zdunik, *Nature* **340**, 617 (1989).
[56] S. Koranda, N. Stergioulas and J.L. Friedman, *ApJ* **488**, 799 (1997).
[57] E Witten, *Phys. Rev. D* **30**, 272 (1984).
[58] J.M. Lattimer, M. Prakash. D. Masak and A. Yahil, *ApJ* **355**, 241 (1990).
[59] L. Lindblom, *ApJ* **278**, 364 (1984).
[60] N.K. Glendenning, *Phys. Rev. D* **46**, 4161 (1992).
[61] J.M. Lattimer and M. Prakash, *Science* **304**, 536 (2004).
[62] S.L. Shapiro, S.A. Teukolsky and I. Wasserman, *ApJ* **272**, 702 (1983).
[63] R.C. Tolman, Phys. Rev. **55**, 364 (1939).
[64] E. Fahri and R.L. Jaffe, Phys. Rev. **D30**, 2379 (1984).
[65] M.G. Alford, M. Braby, M. Paris and S. Reddy, *ApJ* **629**, 969 (2005).

[66] Manju Prakash, E. Baron and M. Prakash, *Phys. Lett. B* **243**, 175 (1990).
[67] A.W. Steiner, M. Prakash and J.M. Lattimer, *Phys. Lett.* **486**, 239 (2000).
[68] T. Klähn, C.D. Roberts, L. Chang, H. Chen and Yu-Xin Liu, *Phys. Rev. C* **82**, 035801 (2010).
[69] A. Kurkela, P. Romatschke and A. Vuorinen, *Phys. Rev. D* **81**, 105021 (2010).
[70] H.A. Bethe, G.E. Brown and G.E. Brown, *Nucl. Phys. A* **467**, 791 (1987).
[71] E. Friedman, A. Gal, and C.J. Batty, Ñucl. Phys. A **579**, 578 (1994).
[72] J. Mares, W. Friedman, A. Gal and B.K. Jennings, *Nucl. Phys. A* **594**, 311 (1995).
[73] R. Knorren, M. Prakash and P.J. Ellis, *Phys. Rev. C* **52**, 3470 (1995).
[74] M. Prakash, Manju Prakash, J.M. Lattimer and C.J. Pethick, *Astrophys. J. Lett.* **390**, L77 (1992).
[75] J.R. Stone, P.A.M. Guichon, H.M. Matevosyan and A.W. Thomas, *Nucl. Phys.*, A 792, 341 (2007).
[76] A. Gezerlis and J. Carlson, *Phys. Rev. C* **81**, 025803 (2010).
[77] K. Hebeler and A. Schwenk, *Phys. Rev. C* **82**, 014304 (2010).
[78] K. Hebeler, J.M. Lattimer, C.J. Pethick and A. Schwenk, *Phys. Rev. Lett.* **105**, 161102 (2010).
[79] A.W. Steiner, J.M. Lattimer and E.F. Brown, *ApJ* **722**, 33 (2010).
[80] B. Kiziltan, A. Kottas and S.E. Thorsett, submitted to ApJ; arXiv:1011.4291 (2010).

Chapter 13

Formation and Evolution of Black Hole Binaries in the Galaxy

Chang-Hwan Lee*

Gerry Brown has been known more as nuclear theorists. However, last ten years we have been working on the formation and evolution of black hole (BH) binaries. Two different types of BH binaries were being considered in our works; (1) BH-neutron star (NS) binaries, and (2) soft X-ray transient sources that are BH binaries with companions from main sequence to giant. In this review, I will summarize our recent works on black hole binaries. For the black hole-neutron star binaries, due to the supercritical accretion, we found a possibility that there might be more BH-NS binaries than NS-NS binaries. This possibility can be tested by the future gravitational wave observations. These binaries are also the possible sources of short-hard gamma-ray bursts. For the soft X-ray black hole binaries, we found the mass-period relation. As a byproduct, we predicted the spin of black holes in these systems, and a few of them were confirmed by the recent black hole spin observation. We also found that these black holes are the possible sources of long gamma-ray bursts that are caused by the rapidly spinning black holes.

1. First Contact to Black Hole Binaries

My first collaboration with Gerry Brown was on the kaon condensation in the core of neutron stars [1]. His first e-mail message to me was very short, asking the results of my calculation on kaon condensation. In fact, at that time, I just started Ph.D. course in Korea, and had not started the real calculation yet. Since then, our collaboration has been continued until very recently, and the research topic has been extended from kaon condensation

*Department of Physics, Pusan National University, Busan 609-735, Korea. E-mail: clee@pusan.ac.kr

in the core of neutron stars to black hole binaries [2–5]. One of the main issues on kaon condensation is the maximum mass of neutron stars, beyond which neutron stars cannot exist as stable stars and collapse into black holes.

At the extreme densities such as in the core of neutron stars, particles with strange quarks can appear because of the fermionic properties of nucleons that are composed of up and down quarks. The lightest particle with strange quark is kaon. In addition, kaon is meson so that they can condense in the ground state leading to the Bose-Einstein condensation [1]. The main question is whether kaon condensation can happen before other exotic states such as pion condensation and hyperons appear in the system. Since most of the theoretical approaches are based on the effective theory whose parameters are fixed by the nucleus properties up to normal nuclear matter density ($\rho_0 = 0.16\,\mathrm{fm}^{-3}$), there are large uncertainties in extending the theory up to very high densities beyond the normal nuclear matter density. We introduced a new approach that is based on the vector manifestation [6,7], in which the QCD symmetries are restored at the critical density. However, there still exist large uncertainties in the estimation of maximum neutron star mass.

The only theoretically confirmed possibility of compact stars with mass higher than the maximum mass of neutron star is black hole. So, it was quite natural to extend our work to black holes, some of which are observed in our Galaxy. Since those black hole systems that are observed are so complicated, there was a room for us to contribute. When I was a postdoctoral researcher in Stony Brook, Gerry Brown gave me an option to choose between nuclear physics and astrophysics. I took the risk to jump on to the black hole astrophysics that was completely new to me. I will summarize the outcome in this review.

Two different types of black hole binaries were being considered in our works; (1) black hole-neutron star binaries [3, 8], and (2) soft X-ray transient sources, black hole binaries with companions from main sequence to more evolved giants [2, 9, 10]. The first class of black holes are one of the main sources of gravitational waves that can be detected by the ground based gravitational wave observatories, such as LIGO and VIRGO. These are the possible sources of short-hard gamma-ray bursts [3]. The second class are the possible sources of long-duration gamma-ray bursts [11, 12]. Below, I will summarize our main results in these two black hole binaries.

2. Formation and Evolution of Double Compact Star Binaries

Neutron star masses strongly depend on the Fe-core mass of giants before the collapse. In order to form neutron star-neutron star binaries which are observed now, the progenitors of two neutron stars have be close in distance, and one has to know the Fe core masses in close binaries. Numerical calculations show that the Fe core masses in close binaries are less than $1.5M_\odot$ independently of the giant masses [3, 8]. Therefore, the new-born neutron star masses might be less than $1.5M_\odot$ independently of neutron star equation of states and the maximum mass of neutron stars. In order to understand the formation of double neutron star binaries, one has to consider the evolution process of neutron star progenitors in binaries [3, 8].

The lifetime τ of giants are inversely proportional to the mass by

$$\tau \propto \frac{1}{M^{2.5}}. \tag{1}$$

If the mass difference between the two progenitors of compact stars are bigger than 4%, the difference in lifetime is bigger than 10%. Stars with mass $8M_\odot$–$25M_\odot$ (progenitors of neutron stars) spend roughly 90% of their lifetime as main sequence. One can consider following two different cases.

a) Case 1: $\Delta M > 4\%$ and $\Delta T > 10\%$
Since the life is different by more than 10%, when the bigger giants evolves into neutron stars, the companion is still in main sequence. As in Fig. 1, when the second companion evolves into giants, the first-born neutron stars will go into the envelope of the expanding giants. Due to the interaction between the first-born neutron star and the giant envelope, some of the envelope will be accreted into the first-born neutron stars. Lee *et al.* [3] estimated the amount of mass accretion to be $\sim 0.7M_\odot$. This accretion is called super-critical mass accretion because the mass accretion rate is about 10^4 Eddington limit. In this case, the mass of the first-born neutron star $M_{\text{compact star, primary}}$ has to be $>1.8M_\odot$. As in Fig. 2 (upper dashed area), Lee *et al.* [3] estimated the possibility of having the mass difference larger than 4% to be 84%.

b) Case 2: $\Delta M < 4\%$ and $\Delta T < 10\%$
In this case, when the progenitor of first neutron star is in giants stage, the companion is also evolving into giants. Then, the expanding envelope of both progenitors overlap, and the system will go through common

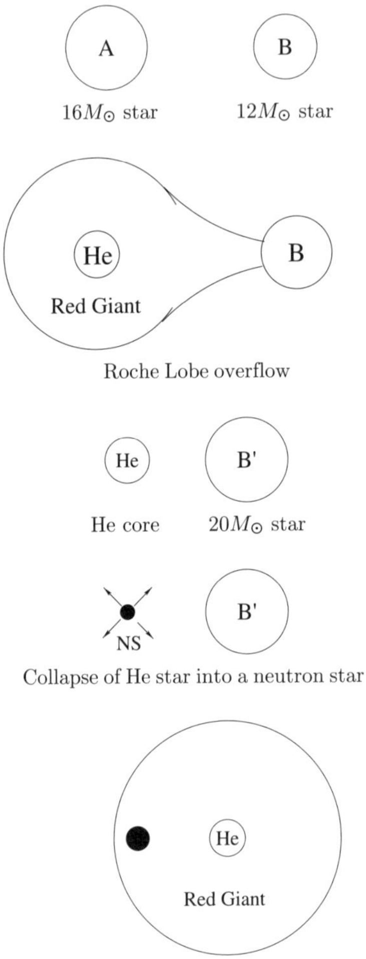

Fig. 1. Conventional model for double neutron star binary evolution [4]. The neutron star goes into common envelope evolution, dropping towards the star, the increased gravitational binding energy being used to expel the hydrogen envelope. The He star explodes into a neutron star and if the two neutron stars remain bound, this is a binary pulsar.

envelope phase. Since the system lose the hydrogen envelope in the common envelope phase, when the first neutron star is formed, there is no or little material to be accreted. In this case, masses of both neutron stars have to be less than $1.5M_\odot$. As in Fig. 2 (lower narrow area), the possibility of this case is $\sim$16%.

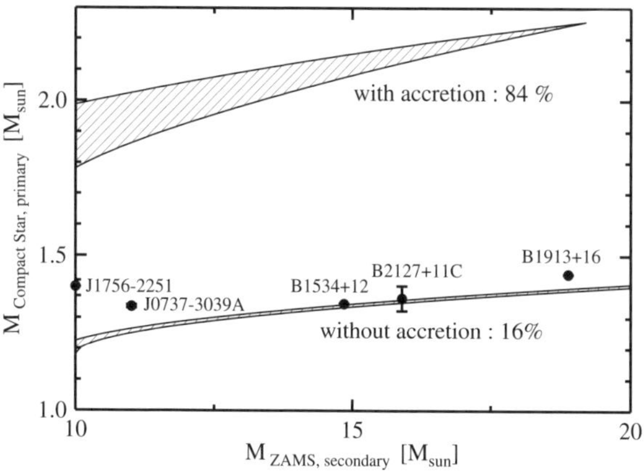

Fig. 2. Masses of primary (first-born) compact stars with and without accretion during H red giant stage of secondary star [3]. The filled circles show the primary (more massive) pulsar masses in double neutron star binaries. The corresponding secondary ZAMS masses were obtained from the secondary (less massive) pulsar masses using Eq. (8) of Lee *et al.* [3]. Note that the 84% corresponds to $M_{\text{Compact Star, primary}} > 1.8 M_\odot$. There is uncertainty in the final primary compact star masses due to the extra mass accretion, $\sim 0.1 M_\odot$–$0.2 M_\odot$, during He giant stage. This may increase the primary compact star masses for both "with-" and "without-accretion".

As summarized in Fig. 2, the possibility of having $M_{\text{compact star, primary}} > 1.8 M_\odot$ is 84%. But, the observed neutron star masses in double neutron star binaries lie only in the lower dashed area in which masses of both neutron stars are $<1.5 M_\odot$. Lee *et al.* [3] suggest that the first-born neutron star went into black hole in the process of supercritical accretion. This suggestion indicates that the neutron star masses in NS-NS binaries may not be in the indication of maximum neutron star mass.

There exists many theoretical neutron star equation of states [13] and the estimated maximum mass of neutron stars are in a wide range $1.5 M_\odot \sim 2.5 M_\odot$. However, due the uncertainty in the estimation of equation of states at high densities, one cannot make any firm conclusion on the maximum mass of neutron stars. Instead of reviewing different theoretical equations of states, I will briefly summarize the current estimation of neutron star masses.

a) *Neutron Star-White Dwarf Binaries*:
Many neutron stars with mass $\geq 1.5 M_\odot$ in these binaries have been reported as in Table 1. However, the neutron star masses in these binaries have to be determined by estimating the white dwarf mass via

Table 1. Neutron star masses in neutron-star, white dwarf binaries in which the central value of the neutron star mass is $\geq 1.5 M_\odot$ [14–17]. For J0751+1807, the first mass estimation [14] was $2.10 \pm 0.20 M_\odot$, but the value was reduced to $1.26^{+0.14}_{-0.12}$ two years later [15].

Name	Pulsar Mass ($M_\odot$)
J1713 + 0747	$1.54^{+0.007}_{-0.10}$
B1855 + 09	$1.57^{+0.12}_{-0.11}$
J0751 + 1807†	$2.10^{+0.20}_{-0.20} \Rightarrow 1.26^{+0.14}_{-0.12}$
J1804 − 2718	<1.70
J1012 + 5307	$1.68^{+0.22}_{-0.22}$
J0621 + 1002	$1.70^{+0.12}_{-0.29}$
J0437 − 4715	$1.58^{+0.18}_{-0.18}$
J2019 + 2425	<1.51
J1614 − 2230	$1.97^{+0.04}_{-0.04}$

Bayesian analysis. As in Table 1, due to the uncertainty in estimating the white dwarf mass, the first mass estimation [14] for J0751 + 1807 was $2.10 \pm 0.20 M_\odot$, but the value was reduced to $1.26^{+0.14}_{-0.12} M_\odot$ two years later [15]. This uncertainties indicate that the errors involved in neutron star-white dwarf binaries can be much larger than they quote. Note that the neutron star mass in J1614-2230 has been measured using Shapiro delay [16].

b) *X-ray Binaries*:
An X-ray pulsar consist of a neutron star and a helium star. Due to the much bigger radius of helium star (order of solar radius), the uncertainty in the mass estimation is even larger than that in neutron star-white dwarf binaries. I will give a few examples of the uncertainties in the mass estimation. In these binaries, the actual center of mass of the helium star can be different from the center of optical observation that was estimated by the red & blue shift in the spectral lines. Since the mass estimation of the neutron star is based on the binary motion of the system, the uncertainties in the binary separation (based on the center of mass) can lead large uncertainties in the mass estimation. Steiner *et al.* [18] showed that the different assumptions on the radius of photosphere can leads very different neutron star masses. Recently, many efforts have been put to improve the mass estimation in these binaries.

Table 2. Compilation of NS-NS binaries [13, 20–22]. For J0737−3039, both neutron stars are observed as pulsars [21].

Object	Mass ($M_\odot$)	Companion Mass ($M_\odot$)
J1518+49	$1.56^{+0.13}_{-0.44}$	$1.05^{+0.45}_{-0.11}$
B1534+12	$1.3332^{+0.0010}_{-0.0010}$	$1.3452^{+0.0010}_{-0.0010}$
B1913+16	$1.4408^{+0.0003}_{-0.0003}$	$1.3873^{+0.0003}_{-0.0003}$
B2127+11C	$1.349^{+0.040}_{-0.040}$	$1.363^{+0.040}_{-0.040}$
J0737−3039A	$1.337^{+0.005}_{-0.005}$	$1.250^{+0.005}_{-0.005}$ (J0737−3039B)
J1756−2251	$1.40^{+0.02}_{-0.03}$	$1.18^{+0.03}_{-0.02}$

c) *Double Neutron Star Binaries*:
In Table 2, masses of neutron stars in double neutron star binaries are summarized. The mass estimation in these binaries are most accurate because both stars are compact stars. All the well-measured masses are below $1.5M_\odot$. However, as discussed in the above, these masses may not be indications of the maximum mass of neutron stars because their masses can be the result of Fe core masses just before the collapse.

d) *Neutron Star-Main Sequence Binaries*:
Recently, neutron star mass in a neutron star-main sequence binary has been measured to be $1.74(4)M_\odot$ [19]. If this mass estimation is confirmed, maximum mass of neutron stars has to be $\geq 1.74M_\odot$. This is consistent with our discussion in the above that the maximum mass can lie above $1.5M_\odot$.

Double neutron star binaries are believed to be one of the major sources of gravitation wave radiation that can be detected in the ground-based gravitational wave observatory. In fact, the ground-based gravitational wave observatory is built based on the double neutron star binaries. According to the suggestion by Lee *et al.* [3], since black hole-neutron star binaries are $\sim$5 times more dominant than double neutron star binaries, the detection rate of gravitation wave will be increased by a factor of 10 (higher total mass will increase the detectability). If one can distinguish the black hole-neutron star binaries from double neutron star binaries in the gravitational wave observation, by obtaining the ratio between two systems, one can test the validity of the suggestion by Lee *et al.* [3].

In the evolution of first-born neutron stars in the double compact binaries, we assumed that the supercritical accretion is valid during the evolution. This argument is based on the fact the temperature of accreting neutron star (with 10^4 Eddington accretion rate) can reach 1 MeV with which the production of thermal neutrinos wins the production of thermal photons [23]. Since the generated neutrinos can leave the system without interaction with the accreting material, the pressure of the star can be reduced, allowing the further compression of the neutron star. However, one has to note that there are uncertainties on the validity of supercritical accretion with the mass accretion rate of 10^4 Eddington rate.

3. Spin of Black Holes in Soft X-Ray Transient Sources

Soft X-ray transient sources are black hole binaries with companions from main sequence to giants. Dozens of them have been found in our Galaxy [2]. Even though we are not experts on these binaries, when we jumped into this field, we are the first who tried to find a common evolution model that can explain all these binaries instead of focusing on individual systems. Gerry Brown's experiences in supernova and stellar evolution were essential in our development. In this section, I will briefly review our developments.

Lee *et al.* [2] discovered the mass-period relation in soft X-ray transient sources. Their argument is based on the tidal interaction between the black hole progenitors and their companions. They assumed that the spin periods of black hole progenitors are synchronized with the binary orbital periods due to the tidal locking. In addition, in order for the black hole to be formed, the initial separation between the black hole progenitor and the companion star has to be within the range that allows the Roche Lobe overflow only during the supergiant stage of the black hole progenitor [2]. If the binaries are too close, the binaries will merge without forming black hole binaries, and if the binaries are too wide, the close binaries that are observed today cannot be formed.

With above assumptions, Lee *et al.* [2] estimated the possible Kerr parameters of the black holes as a function of the preexplosion orbital period by assuming the angular momentum conservation as in Fig. 3. Kerr parameters $a_\star$ for two systems, GRO J1655−40 and 4U 1543−47, have been measured to be $a_\star = 0.65$–0.75 and $a_\star = 0.75$–0.85, respectively [24]. If we reconstruct their preexplosion orbital periods from current data, both system had the preexplosion period of $\sim$ half day [2, 25]. From Fig. 3, this preexplosion orbital period indicates that the corresponding Kerr parameter is $a_\star \simeq 0.8$. This is consistent with the current observation within the theoretical uncertainty [24].

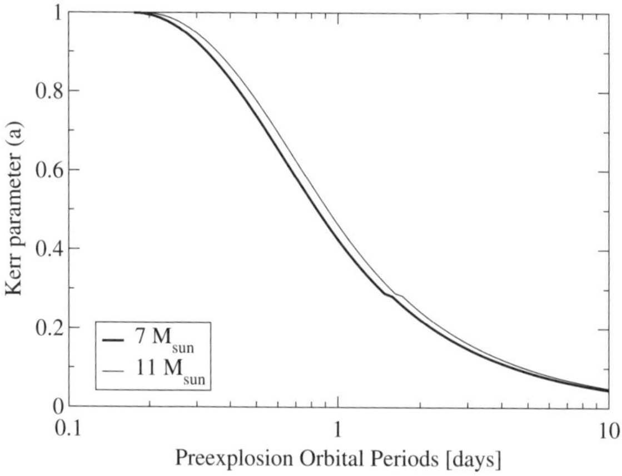

Fig. 3. The Kerr parameter of the black hole (here represented by a, but in the text referred as $a_\star$) resulting from the collapse of a helium star synchronous with the orbit, as a function of orbital period [2]. Thick (thin) solid line corresponds to initial $7M_\odot$ ($11M_\odot$) He star. Note that the result is almost insensitive to the mass of the helium star.

In Kerr geometry, there exist limits both on the value of the spin angular momentum and on the specific angular momentum that the incoming material can carry. In the process of black hole formation, those materials with specific angular momentum bigger than the critical angular momentum cannot directly fall into black hole [11]. Hence, the accretion disk will be formed around black holes. Since the falling material in He star with smaller orbital period can carry higher angular momentum, lots of them will not be accreted due to the limit on the specific angular momentum that can be carried into black hole. So, the smaller the preexplosion orbital period, the smaller the black hole mass. This is the essence of the mass-period relation we found [2]. In Fig. 4, the sudden drop in the black hole masses (dashed and dot-dashed lines) for the system with preexplosion orbital period less than a day is caused by the limit on the specific angular momentum. Since the orbital periods of those black hole that are observed today are not the same value at the time of black hole formation, one has to consider the evolution of the binary after black hole formation in order to understand the current orbital period. In Fig. 4, we summarized evolution tracks of a few systems after black hole formation [2]. Note that the consistency between the observation and our estimation on the black hole spin for GRO J1655−40 and 4U 1543−47 support our evolutionary scenario and the mass-period relation.

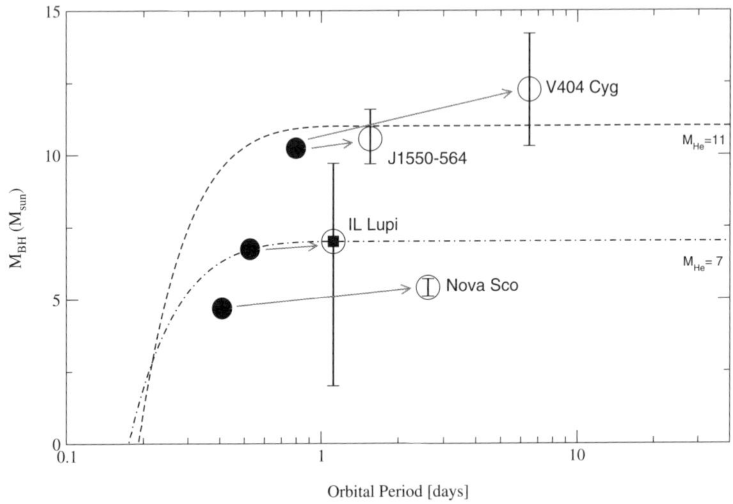

Fig. 4. Reconstructed pre-explosion orbital period vs. black hole masses of soft X-ray transients with evolved companions. The reconstructed pre-explosion orbital periods and black hole masses are marked by filled circles, and the current locations of binaries with evolved companions are marked by open circles. The dashed and dot-dashed lines are the resulting black hole masses with ideal polytropic He stars. Note that both IL Lupi and Nova Scorpii might have been evolved from $11M_\odot$ He stars. This figure is obtained from Fig. 11 of Lee *et al.* [2].

Table 3. Parameters at the time of black hole formation [25]. $E_{\rm BZ}$ is the rotational energy which can be extracted via Blandford-Znajek mechanism [26]. † Kerr parameter is the present one.

Name	M_{BH} $[M_\odot]$	M_d $[M_\odot]$	$a_\star$	$E_{\rm BZ}$ $[10^{51}$ ergs$]$
GRO J1655−40	∼5	1−2	0.8	∼430
4U 1543−47	∼5	1−2	0.8	∼430
XTE J1550−564	∼10	1−2	0.5	∼300
GS 2023+338	∼10	1−2	0.5	∼300
XTE J1819−254	6−7	∼10	0.2	10 ∼ 12
GRS 1915+105	6−7	∼10	0.2 (>0.98†)	10 ∼ 12
Cyg X−1	6−7	$\gtrsim$30	0.15	5 ∼ 6

In the above, we argued that the accretion disk has to be formed for the black hole binaries with high black hole spin. The magnetic field frozen in the disk can keep the strong magnetic field near the black hole rotation axis. In Table 3, we summarize the available rotational energy which can be extracted via Blandford-Znajek mechanism [25–27]. Those black holes with higher Kerr parameters might be the sources of long-duration gamma-ray bursts [11].

4. Future Prospects

In this review, I summarized our works on black hole binaries. For the double compact star binaries, we found that the black hole-neutron star binaries can be more dominant than double neutron star binaries if we allow the possibility of supercritical accretion [3]. This will increase the LIGO detectability by a factor of 10. In this scenario, if the original mass difference is bigger than 4%, the first-born neutron star will accrete more than $0.7 M_\odot$ via supercritical accretion. Based on this observation, we found that the maximum neutron star mass can lie between $1.5-1.8 M_\odot$ or above depending on the efficiency of supercritical accretion. The double neutron stars can be formed only if the initial mass difference between neutron star progenitors is less than 4% so that both progenitors evolve nearly at the same time and there is no time for the accretion. The neutron star masses in these binaries resulted from the Fe core masses before collapse. This is consistent with the fact that all well-measured neutron star masses in double neutron star binaries are $<1.5 M_\odot$.

For the soft X-ray transient sources, black hole binaries with companions from main sequence to giants, by assuming the tidal locking between the black hole progenitor and the companion, we found the black hole mass-period relation [2]. As a byproduct we found the spin of black holes as a function of the preexplosion orbital period. This prediction has been partially confirmed by Shafee *et al.* [24]. We discussed the possibility that those black hole binaries with higher spin, i.e. soft X-ray transient sources with orbital period less than half a day, are remnants of gamma-ray bursts [25].

We admit that there are large uncertainties in our scenario. But, at least, our predictions can be tested by future observations on neutron stars and the detections of gravitational-wave radiation and gamma-ray bursts which are caused by colliding compact star binaries.

Acknowledgments

My works on black hole binaries were initiated by Gerry Brown's suggestion and strong support during my postdoc years in Stony Brook. I really enjoyed staying in Stony Brook with him. He often invited me to his house for a dinner and treated me as a family member. Lots of our works have been developed in his house with hot meals he prepared. I would appreciate the warm hospitality he showed to me.

References

[1] C.-H. Lee, *Physics Reports* **275**, 255 (1996).
[2] C.-H. Lee, G.E. Brown and R.A.M.J. Wijers, *Astrophys. J.* **575**, 996 (2002).
[3] C.-H. Lee, H.-J. Park and G.E. Brown, *Astrophys. J.* **670**, 741 (2007).
[4] G.E. Brown, C.-H. Lee and M. Rho, *Phys. Rev. Letts.* **462**, 1 (2008).
[5] G.E. Brown, C.-H. Lee and M. Rho, *Phys. Rev. Letts.* **101**, 091101 (2008).
[6] M. Harada and K. Yamawaki, *Phys. Rev. Letts.* **86**, 757 (2001).
[7] G.E. Brown, C.-H. Lee and M. Rho, *Phys. Rev. Letts.* **96**, 062303 (2006).
[8] H.A. Bethe, G.E. Brown and C.-H. Lee, *Physics Reports* **442**, 5 (2007).
[9] G.E. Brown, C.-H. Lee and H.A. Bethe, *New Astronomy* **4**, 313 (1999).
[10] G.E. Brown, C.-H. Lee and T. Tauris, *New Astronomy* **6**, 331 (2001).
[11] G.E. Brown, C.-H. Lee, R.A.M.J. Wijers, H.K. Lee, G. Israelian and H.A. Bethe, *New Astronomy* **5**, 191 (2000).
[12] G.E. Brown, C.-H. Lee, R.A.M.J. Wijers and H.A. Bethe, (2000a), *Physics Reports* **333–334**, 471 (2000).
[13] J. Lattimer and M. Prakash, *Physics Reports* **442**, 166 (2007).
[14] D.J. Nice, E.M. Splaver, I.H. Stairs, O. Loehmer, A. Jessner, M. Kramer and J.M. Cordes, *Astrophys. J.* **634**, 1242 (2005).
[15] D.J. Nice, presentation at *40 Years of Pulsar*, McGill, Aug. 12–17 (2007).
[16] P.B. Demorest, T. Pennucci, S.M. Ransom, M.S.E. Roberts and J.W.T. Hessels, *Nature* **467**, 1081 (2010).
[17] J. Lattimer and M. Prakash, *Science* **304**, 536 (2004).
[18] A.W. Steiner, J.M. Lattimer, E.F. Brown, arXiv:1005.0811v2 [astro-ph.HE].
[19] D.J. Champion *et al.*, *Science* **320**, 1309 (2008).
[20] S.E. Thorsett and D. Chakrabarty, *Astrophys. J.* **512**, 288 (1999).
[21] A.G. Lyne *et al.*, *Science* **303**, 1153 (2004).
[22] A.J. Faulkner *et al.*, *Astrophys. J.* **618**, L119 (2004).
[23] G.E. Brown, C.-H. Lee and H.A. Bethe, *Astrophys. J.* **541**, 918 (2000).
[24] R. Shafee, J.E. McClintock, R. Narayan, S.W. Davis, L. Li and R.A. Remillard, *Astrophys. J.* **636**, L113 (2006).
[25] G.E. Brown, C.-H. Lee and E. Moreno Méndez, *Astrophys. J.* **671**, L41 (2007).
[26] H.K. Lee, R.A.M.J. Wijers and G.E. Brown, *Physics Reports* **325**, 83 (2000).
[27] G.E. Brown, C.-H. Lee and E. Moreno Méndez, *Astrophys. J.* **685**, 1063 (2008).

Chapter 14

Chiral Symmetry and the Nucleon-Nucleon Interaction

R. Machleidt* and D.R. Entem†

We summarize the current status of our understanding of nuclear forces based upon chiral symmetry — an idea that was advocated by Gerry Brown already several decades ago.

1. Introduction

Forty years ago, Gerry Brown published the Comment "Isn't it Time to Calculate the Nucleon-Nucleon Force" [1], and more than 30 years ago, he wrote a book chapter entitled "Chiral Symmetry and the Nucleon-Nucleon Interaction" [2]. In fact, as early as 1968, he had published, together with two co-workers, a paper [3] on three-nucleon forces, where the consequences of chiral symmetry were fully exploited. Nevertheless, it should take a few more *decades* until the problem of the nucleon-nucleon (NN) interaction was really solved taking chiral symmetry consistently into account — and this fact proves the very advanced nature of Gerry's ideas concerning nuclear forces. It is the purpose of this contribution to summarize where we are today in our understanding of nuclear forces, based upon concepts that Gerry advocated already many decades ago.

But let's first recall more of the interesting history of the idea of chiral symmetry. The modern understanding is that this symmetry arises because the up and down quarks happen to have relatively small masses. However, chiral symmetry and its significance for low-energy hadron (pion)

*Department of Physics, University of Idaho, Moscow, ID 83843, U.S.A. E-mail: machleid@uidaho.edu

†Grupo de Fisica Nuclear and IUFFyM, University of Salamanca, E-37008 Salamanca, Spain. E-mail: entem@usal.es

physics was discovered long before QCD. In 1960, based upon concepts proposed by Schwinger [4], Gell-Mann and Levy [5] developed the sigma model, which is a linear realization of chiral symmetry.[a] One major problem researchers had been struggling with in the 1950's was that the pion-nucleon scattering length came out two orders of magnitude too large when the (renormalizable) pseudo-scalar (γ_5) πN interaction was used. This unrealistic prediction was due to very large contributions from virtual anti-nucleon states (the so-called "pair terms" or "Z-graphs"). Similar problems occurred in the 2π-exchange contribution to the NN interaction. In the sigma model, the large pair terms are cancelled by processes involving the (fictitious) σ boson. In this way, the linear sigma model demonstrates how imposing chiral invariance fixes the problem with low-energy π-N scattering. However, the fictitious character of the σ particle as well as the reliance on the perfect cancelation of huge terms are uncomfortable features. In 1967, motivated by the current algebra approach to soft pion physics, Weinberg [7] worked out what has become known as the non-linear sigma model, which does not include a σ anymore and has pions and nucleons interact via pseudo-vector (derivative, $\gamma_5\gamma^\mu\partial_\mu$) coupling besides a new (non-linear) $\pi\pi NN$ term also involving a derivative ("Weinberg–Tomozawa term" [8, 9]). The derivative (equivalent to momentum) guarantees that the interaction vanishes when the momentum goes to zero providing a natural explanation for the weakness of the interaction by soft pions which does not rely on the cancellation of large terms. Following suggestions by Schwinger, Weinberg [10] developed, soon after, a general theory of non-linear realizations of chiral symmetry, which was further generalized in an elegant way by Callan, Coleman, Wess, and Zuimino [11].

However, ideas were needed for how to implement chiral symmetry consistently in the theory of pionic and nuclear interactions and how to deal with renormalization, since the derivative coupling is not renormalizable in the conventional sense. In his contribution to the "Festschrift" in honor of Schwinger of 1979 [12, 13], Weinberg proposed to consider the most general possible Langrangian including all higher-derivative terms that are consistent with chiral symmetry (besides the other commonly assumed symmetry principles). For this theory to be manageable, one needs to assume some sort of perturbative expansion such that only a finite number of terms contribute at a given order. This expansion is provided by powers of small external

[a]For a pedagogical introduction into chiral symmetry and the sigma model, see [6].

momenta over the chiral symmetry breaking scale, $\Lambda_\chi \sim 1\,\text{GeV}$. The higher-derivative terms supply the counterterms that make possible an order-by-order renormalization, which is the appropriate renormalization procedure for an effective field theory. Weinberg's suggestions were soon picked up by Gasser, Leutwyler, and associates who worked out, to one loop, the cases of $\pi\pi$ [14] and πN scattering [15] with great success.

But there was still the problem of the nuclear force which is more difficult, since nuclear interactions do not vanish in the chiral limit ($q \to 0$; $m_{u/d}, m_\pi \to 0$) and require a non-perturbative treatment because of the existence of nuclear bound states. In a series of papers published around 1990 [16–18], Weinberg picked up the nuclear force issue and suggested to calculate the NN potential perturbatively in the chiral expansion and then iterate it to all orders in a Schroedinger or Lippmann–Schwinger equation to obtain the nuclear amplitude. Here, the introduction of four-nucleon contact terms is crucial for renormalization.

Following the Weinberg proposal, pioneering work was performed by Ordóñez, Ray, and van Kolck [19, 20] who applied time-ordered perturbation theory to construct a NN potential up to next-to-next-to-leading order (NNLO). The results were encouraging and nuclear EFT quickly developed into one of the most popular branches of modern nuclear physics. The Munich group used covariant perturbation theory and dimensional regularization to calculated the perturbative NN amplitude without [21] and with $\Delta(1232)$-isobar degrees of freedom [22] at NNLO. Besides this, the Munich group worked out important loop contributions of higher order [23–27]. A relativistic approach was also taken by the Brazil group [28, 29]. The Bochum-Jülich group devised a method of unitarity transformations to eliminate the energy-dependence of time-ordered perturbation theory amplitudes and calculated the NN potentials up to NNLO [30, 31]. The Idaho group managed to construct a chiral NN potential at next-to-next-to-next-to-leading order (N^3LO) and showed that only at this order can one achieve the precision necessary for reliable few-nucleon and nuclear structure calculations [32–35]. Progress extended beyond the NN interaction, as nuclear many-body forces based upon chiral perturbation theory were also developed [18, 36–39].

During the past decade or so, chiral two-nucleon forces have been used in many microscopic calculations of nuclear reactions and structure [40–55] and the combination of chiral two- and three-nucleon forces has been applied in few-nucleon reactions [37, 56–64], structure of light- and medium-mass nuclei [65–68], and nuclear and neutron matter [69, 70] — with a great deal of success. The majority of nuclear structure calculations is nowadays based upon chiral forces.

This article is organized as follows. In Sec. 2, we sketch the essential ideas of an EFT for low-energy QCD. Section 3 provides an overview on nuclear forces derived from chiral EFT. The two-nucleon force is then discussed in detail in Sec. 4. Many-body forces are the subject of Sec. 5. Finally, Sec. 6 contains our conclusions.

2. Low-Energy QCD and Effective Field Theory

Quantum chromodynamics (QCD) is the theory of strong interactions. It deals with quarks, gluons and their interactions and is part of the Standard Model of Particle Physics. QCD is a non-Abelian gauge field theory with color $SU(3)$ the underlying gauge group. The non-Abelian nature of the theory has dramatic consequences. While the interaction between colored objects is weak at short distances or high momentum transfer ("asymptotic freedom"); it is strong at long distances ($\gtrsim 1\,\text{fm}$) or low energies, leading to the confinement of quarks into colorless objects, the hadrons. Consequently, QCD allows for a perturbative analysis at large energies, whereas it is highly non-perturbative in the low-energy regime. Nuclear physics resides at low energies and the force between nucleons is a residual color interaction similar to the van der Waals force between neutral molecules. Therefore, in terms of quarks and gluons, the nuclear force is a very complicated problem that, nevertheless, can be attacked with brute computing power on a discretized, Euclidean space-time lattice (known as lattice QCD). Advanced lattice QCD calculations [71, 72] are under way and will yield improving results in the near future. However, since these calculations are very time-consuming and expensive, they can only be used to check a few representative key-issues. For everyday nuclear structure physics, a more efficient approach is needed.

The efficient approach is an effective field theory. For the development of an EFT, it is crucial to identify a separation of scales. In the hadron spectrum, a large gap between the masses of the pions and the masses of the vector mesons, like $\rho(770)$ and $\omega(782)$, can clearly be identified. Thus, it is natural to assume that the pion mass sets the soft scale, $Q \sim m_\pi$, and the rho mass the hard scale, $\Lambda_\chi \sim m_\rho$, also known as the chiral-symmetry breaking scale. This is suggestive of considering an expansion in terms of the soft scale over the hard scale, Q/Λ_χ. Concerning the relevant degrees of freedom, we noticed already that, for the ground state and the low-energy excitation spectrum of an atomic nucleus as well as for conventional nuclear reactions, quarks and gluons are ineffective degrees of freedom, while nucleons and pions are the appropriate ones. To make sure that this EFT is not just another phenomenology, it must have a firm link with QCD. The link is established by having the EFT observe all relevant symmetries of the underlying theory. This requirement is based upon a 'folk theorem' by Weinberg [12]:

> If one writes down the most general possible Lagrangian, including *all* terms consistent with assumed symmetry principles, and then calculates matrix elements with this Lagrangian to any given order of perturbation theory, the result will simply be the most general possible S-matrix consistent with analyticity, perturbative unitarity, cluster decomposition, and the assumed symmetry principles.

Since the up and down quark masses are very small, one may assume (approximate) chiral symmetry which, however, is broken in two ways. It is explicitly broken, because the the up and down quark masses are not exactly zero. Moreover, we are also faced with so-called spontaneous symmetry breaking.

A (continuous) symmetry is said to be *spontaneously broken* if a symmetry of the Lagrangian is not realized in the ground state of the system. There is evidence that the (approximate) chiral symmetry of the QCD Lagrangian is spontaneously broken — for dynamical reasons of nonperturbative origin which are not fully understood at this time. The most plausible evidence comes from the hadron spectrum. From chiral symmetry, one would naively expect the existence of degenerate hadron multiplets of opposite parity, i.e. for any hadron of positive parity one would expect a degenerate hadron state of negative parity and vice versa. However, these "parity doublets" are not observed in nature. For example, take the ρ-meson which is a vector meson of negative parity ($J^P = 1^-$) and mass 776 MeV. There does exist a 1^+ meson, the a_1, but it has a mass of 1230 MeV and, thus, cannot be perceived as degenerate with the ρ. On the other hand, the ρ meson comes in three charge states (equivalent to three isospin states), the $\rho^\pm$ and the ρ^0, with masses that differ by at most a few MeV. In summary, in the hadron spectrum, $SU(2)_V$ (isospin symmetry) is well observed, while axial symmetry is broken: $SU(2)_L \times SU(2)_R$ is broken down to $SU(2)_V$.

A spontaneously broken global symmetry implies the existence of (massless) Goldstone bosons [73, 74]. The Goldstone bosons are identified with the isospin triplet of the (pseudoscalar) pions, which explains why pions are so light. The pion masses are not exactly zero because the up and down quark masses are not exactly zero either. Thus, pions are a truly remarkable species: they reflect spontaneous as well as explicit symmetry breaking. Goldstone bosons interact weakly at low energy. They are degenerate with the vacuum and, therefore, interactions between them must vanish at zero momentum and in the chiral limit ($m_\pi \to 0$).

The next step is to build the most general Lagrangian consistent with the (broken) symmetries discussed above. An elegant formalism for the construction of such Lagrangians was developed by Callan, Coleman, Wess, and Zumino (CCWZ) [11] who worked out the group-theoretical foundations of non-linear realizations of chiral symmetry. It is characteristic for these non-linear realizations that, whenever functions of the Goldstone bosons appear in the Lagrangian, they are always accompanied with at least one space-time derivative.

As discussed, the relevant degrees of freedom are pions (Goldstone bosons) and nucleons. Since the interactions of Goldstone bosons must vanish at zero momentum transfer and in the chiral limit ($m_\pi \to 0$), the low-energy expansion of the Lagrangian is arranged in powers of derivatives and pion masses. The hard scale is the chiral-symmetry breaking scale, $\Lambda_\chi \approx 1\,\text{GeV}$. Thus, the expansion is in terms of powers of Q/Λ_χ where Q is a (small) momentum or pion mass. This is chiral perturbation theory (ChPT).

The effective Lagrangian can formally be written as,

$$\mathcal{L}_{\text{eff}} = \mathcal{L}_{\pi\pi} + \mathcal{L}_{\pi N} + \mathcal{L}_{NN} + \cdots, \tag{1}$$

where $\mathcal{L}_{\pi\pi}$ deals with the dynamics among pions, $\mathcal{L}_{\pi N}$ describes the interaction between pions and a nucleon, and $\mathcal{L}_{NN}$ provides contact interactions between two nucleons. The individual Lagrangians are organized as follows:

$$\mathcal{L}_{\pi\pi} = \mathcal{L}_{\pi\pi}^{(2)} + \mathcal{L}_{\pi\pi}^{(4)} + \cdots, \tag{2}$$

$$\mathcal{L}_{\pi N} = \mathcal{L}_{\pi N}^{(1)} + \mathcal{L}_{\pi N}^{(2)} + \mathcal{L}_{\pi N}^{(3)} + \cdots, \tag{3}$$

and

$$\mathcal{L}_{NN} = \mathcal{L}_{NN}^{(0)} + \mathcal{L}_{NN}^{(2)} + \mathcal{L}_{NN}^{(4)} + \cdots, \tag{4}$$

where the superscript refers to the number of derivatives or pion mass insertions (chiral dimension) and the ellipsis stands for terms of higher dimensions. See Ref. [33] for a concise summary of the explicit expressions for these Lagrangians.

3. Nuclear Forces from Chiral EFT: Overview

3.1. *Chiral perturbation theory and power counting*

As discussed, effective field theories (EFTs) are defined in terms of effective Langrangians which are given by an infinite series of terms with increasing number of derivatives and/or nucleon fields, with the dependence of each term on the pion field prescribed by the rules of broken chiral symmetry. Applying this Lagrangian to a particular process, an unlimited number of Feynman graphs can be generated. Therefore, we need a scheme that makes the theory manageable and calculable. This scheme which tells us how to distinguish between large (important) and small (unimportant) contributions is chiral perturbation theory (ChPT), and determining the power ν of the expansion has become known as power counting.

Nuclear potentials are defined as sets of irreducible graphs up to a given order. The power ν of a few-nucleon diagram involving A nucleons is given in terms of naive dimensional analysis by:

$$\nu = -2 + 2A - 2C + 2L + \sum_i \Delta_i, \tag{5}$$

with

$$\Delta_i \equiv d_i + \frac{n_i}{2} - 2, \tag{6}$$

where C denotes the number of separately connected pieces and L the number of loops in the diagram; d_i is the number of derivatives or pion-mass insertions and n_i the number of nucleon fields (nucleon legs) involved in vertex i; the sum runs over all vertices contained in the diagram under consideration. Note that $\Delta_i \geq 0$ for all interactions allowed by chiral symmetry. For an irreducible NN diagram ("two-nucleon force", $A = 2$, $C = 1$), Eq. (5) collapses to

$$\nu = 2L + \sum_i \Delta_i. \tag{7}$$

In summary, the chief point of the ChPT expansion is that, at a given order ν, there exists only a finite number of graphs. This is what makes the theory calculable. The expression $(Q/\Lambda_\chi)^{\nu+1}$ provides a rough estimate of the relative size of the contributions left out and, thus, of the accuracy at order ν. In this sense, the theory can be calculated to any desired accuracy and has predictive power.

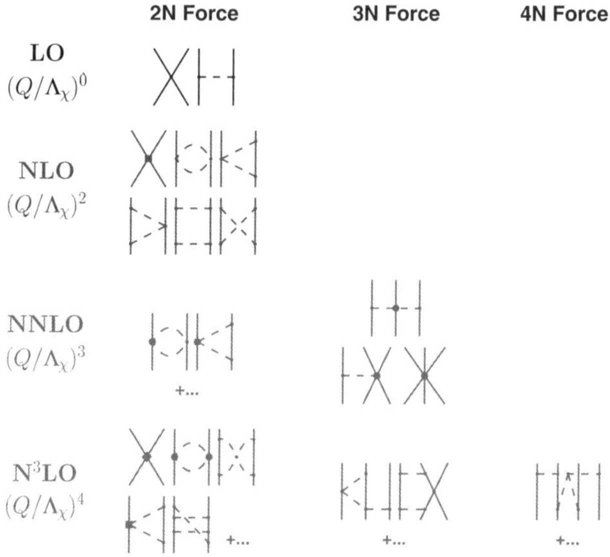

Fig. 1. Hierarchy of nuclear forces in ChPT. Solid lines represent nucleons and dashed lines pions. Small dots, large solid dots, solid squares, and solid diamonds denote vertices of index $\Delta = 0$, 1, 2, and 4, respectively. Further explanations are given in the text.

3.2. *The hierarchy of nuclear forces*

Chiral perturbation theory and power counting imply that nuclear forces emerge as a hierarchy controlled by the power ν, Fig. 1.

In lowest order, better known as leading order (LO, $\nu = 0$), the NN amplitude is made up by two momentum-independent contact terms ($\sim Q^0$), represented by the four-nucleon-leg graph with a small-dot vertex shown in the first row of Fig. 1, and static one-pion exchange (1PE), second diagram in the first row of the figure. This is, of course, a rather crude approximation to the two-nucleon force (2NF), but accounts already for some important features. The 1PE provides the tensor force, necessary to describe the deuteron, and it explains NN scattering in peripheral partial waves of very high orbital angular momentum. At this order, the two contacts which contribute only in S-waves provide the short- and intermediate-range interaction which is somewhat crude.

In the next order, $\nu = 1$, all contributions vanish due to parity and time-reversal invariance.

Therefore, the next-to-leading order (NLO) is $\nu = 2$. Two-pion exchange (2PE) occurs for the first time ("leading 2PE") and, thus, the creation of a more sophisticated description of the intermediate-range interaction is starting here. Since the loop involved in each pion-diagram implies already $\nu = 2$ [cf. Eq. (7)], the vertices must have $\Delta_i = 0$. Therefore, at this order, only the lowest order πNN and $\pi\pi NN$ vertices are allowed which is why the leading 2PE is rather weak. Furthermore, there are seven contact terms of $\mathcal{O}(Q^2)$, shown by the four-nucleon-leg graph with a solid square, which contribute in S and P waves. The operator structure of these contacts include a spin-orbit term besides central, spin-spin, and tensor terms. Thus, essentially all spin-isospin structures necessary to describe the two-nucleon force phenomenologically have been generated at this order. The main deficiency at this stage of development is an insufficient intermediate-range attraction.

This problem is finally fixed at order three ($\nu = 3$), next-to-next-to-leading order (NNLO). The 2PE involves now the two-derivative $\pi\pi NN$ seagull vertices (proportional to the c_i LECs) denoted by a large solid dot in Fig. 1. These vertices represent correlated 2PE as well as intermediate $\Delta(1232)$-isobar contributions. It is well-known from the meson phenomenology of nuclear forces [75, 76] that these two contributions are crucial for a realistic and quantitative 2PE model. Consequently, the 2PE now assumes a realistic size and describes the intermediate-range attraction of the nuclear force about right. Moreover, first relativistic corrections come into play at this order. There are no new contacts.

The reason why we talk of a hierarchy of nuclear forces is that two- and many-nucleon forces are created on an equal footing and emerge in increasing number as we go to higher and higher orders. At NNLO, the first set of nonvanishing three-nucleon forces (3NF) occur [36, 37], cf. column "3N Force" of Fig. 1. In fact, at the previous order, NLO, irreducible 3N graphs appear already, however, it has been shown by Weinberg [18] and others [36, 77, 78] that these diagrams all cancel. Since nonvanishing 3NF contributions happen first at order $(Q/\Lambda_\chi)^3$, they are very weak as compared to 2NF which start at $(Q/\Lambda_\chi)^0$.

More 2PE is produced at $\nu = 4$, next-to-next-to-next-to-leading order (N^3LO), of which we show only a few symbolic diagrams in Fig. 1. Two-loop 2PE graphs show up for the first time and so does three-pion exchange (3PE) which necessarily involves two loops. 3PE was found to be negligible at this order [23, 24]. Most importantly, 15 new contact terms $\sim Q^4$ arise and are represented by the four-nucleon-leg graph with a solid diamond. They include a quadratic spin-orbit term and contribute up to D-waves. Mainly due to the increased number of contact terms, a quantitative description of the two-nucleon interaction up to about 300 MeV lab. energy is possible, at N^{3}LO (for details, see below). Besides further 3NF, four-nucleon forces (4NF) start at this order. Since the leading order 4NF come into existence one order higher than the leading 3NF, 4NF are weaker than 3NF. Thus, ChPT provides a straightforward explanation for the empirically known fact that 2NF $\gg$ 3NF $\gg$ 4NF ....

4. The Nucleon-Nucleon Interaction

4.1. *Definition of the chiral NN potential*

The previous section has provided us with an overview. In this section, we will now discuss the NN interaction in more detail. In terms of naive dimensional analysis or "Weinberg counting", the various orders of the irreducible graphs which define the chiral NN potential are given by:

$$V_{\rm LO} = V_{\rm ct}^{(0)} + V_{1\pi}^{(0)} \tag{8}$$

$$V_{\rm NLO} = V_{\rm LO} + V_{\rm ct}^{(2)} + V_{1\pi}^{(2)} + V_{2\pi}^{(2)} \tag{9}$$

$$V_{\rm NNLO} = V_{\rm NLO} + V_{1\pi}^{(3)} + V_{2\pi}^{(3)} \tag{10}$$

$$V_{\rm N^3LO} = V_{\rm NNLO} + V_{\rm ct}^{(4)} + V_{1\pi}^{(4)} + V_{2\pi}^{(4)} + V_{3\pi}^{(4)} \tag{11}$$

where the superscript denotes the order ν of the low-momentum expansion. LO stands for leading order, NLO for next-to-leading order, etc. Contact potentials carry the subscript "ct" and pion-exchange potentials can be identified by an obvious subscript.

The one-pion exchange (1PE) potential reads

$$V_{1\pi}(\vec{p}\,',\vec{p}) = -\frac{g_A^2}{4f_\pi^2}\,\boldsymbol{\tau}_1\cdot\boldsymbol{\tau}_2\,\frac{\vec{\sigma}_1\cdot\vec{q}\,\vec{\sigma}_2\cdot\vec{q}}{q^2+m_\pi^2}\,, \tag{12}$$

where $\vec{p}\,'$ and $\vec{p}$ designate the final and initial nucleon momenta in the center-of-mass system and $\vec{q}\equiv\vec{p}\,'-\vec{p}$ is the momentum transfer; $\vec{\sigma}_{1,2}$ and $\boldsymbol{\tau}_{1,2}$ are the spin and isospin operators of nucleon 1 and 2; g_A, f_π, and m_π denote axial-vector coupling constant, the pion decay constant, and the pion mass, respectively. Since higher order corrections contribute only to mass and coupling constant renormalizations and since, on shell, there are no relativistic corrections, the on-shell 1PE has the form Eq. (12) up to all orders.

Multi-pion exchange, which starts at NLO and continues through all higher orders, involves divergent loop integrals that need to be regularized.

An elegant way to do this is dimensional regularization which (besides the main nonpolynomial result) typically generates polynomial terms with coefficients that are, in part, infinite or scale dependent [21]. One purpose of the contacts is to absorb all infinities and scale dependencies and make sure that the final result is finite and scale independent. This is the renormalization of the perturbatively calculated NN amplitude (which, by definition, is the "NN potential"). It is very similar to what is done in the ChPT calculations of $\pi\pi$ and πN scattering, namely, a renormalization order by order, which is the method of choice for any EFT. Thus, up to this point, the calculation fully meets the standards of an EFT and there are no problems. The perturbative NN amplitude can be used to make model independent predictions for peripheral partial waves [21, 22, 33]. A concise summary of the explicit expressions for the 2π exchange contributions up to order $\mathrm{N^3LO}$ can be found in Ref. [33].

4.2. *Regularization and renormalization*

For calculations of the structure of nuclear few and many-body systems, the lower partial waves are the most important ones. The fact that in S waves we have large scattering lengths and shallow (quasi) bound states indicates that these waves need to be treated nonperturbatively. Following Weinberg's prescription [16], this is accomplished by inserting the potential V into the Lippmann–Schwinger (LS) equation:

$$T(\vec{p}\,',\vec{p}) = V(\vec{p}\,',\vec{p}) + \int d^3p''\, V(\vec{p}\,',\vec{p}\,'')\, \frac{M_N}{p^2 - p''^2 + i\epsilon}\, T(\vec{p}\,'',\vec{p})\,, \quad (13)$$

where M_N denotes the nucleon mass.

In general, the integral in the LS equation is divergent and needs to be regularized. One way to do this is by multiplying V with a regulator function

$$V(\vec{p}\,',\vec{p}) \longmapsto V(\vec{p}\,',\vec{p})\; e^{-(p'/\Lambda)^{2n}}\; e^{-(p/\Lambda)^{2n}}\,. \quad (14)$$

Typical choices for the cutoff parameter Λ that appears in the regulator are $\Lambda \approx 0.5\,\mathrm{GeV} \ll \Lambda_\chi \approx 1\,\mathrm{GeV}$.

It is pretty obvious that results for the T-matrix may depend sensitively on the regulator and its cutoff parameter. This is acceptable if one wishes to build models. For example, the meson models of the past [75, 79] always depended sensitively on the choices for the cutoff parameters which, in fact, were important for the fit of the NN data. However, the EFT approach wishes to be fundamental in nature and not just another model.

In field theories, divergent integrals are not uncommon and methods have been developed for how to deal with them. One regulates the integrals and then removes the dependence on the regularization parameters (scales, cutoffs) by renormalization. In the end, the theory and its predictions do not depend on cutoffs or renormalization scales.

So-called renormalizable quantum field theories, like QED, have essentially one set of prescriptions that takes care of renormalization through all orders. In contrast, EFTs are renormalized order by order.

As discussed, the renormalization of *perturbative* EFT calculations is not a problem. *The problem is nonperturbative renormalization.* This problem typically occurs in *nuclear* EFT because nuclear physics is characterized by bound states which are nonperturbative in nature. EFT power counting may be different for nonperturbative processes as compared to perturbative ones. Such difference may be caused by the infrared enhancement of the reducible diagrams generated in the LS equation.

Weinberg's implicit assumption [13, 16] was that the counterterms introduced to renormalize the perturbatively calculated potential, based upon naive dimensional analysis ("Weinberg counting"), are also sufficient to renormalize the nonperturbative resummation of the potential in the LS equation. In 1996, Kaplan, Savage, and Wise (KSW) [80] pointed out that there are problems with the Weinberg scheme if the LS equation is renormalized by minimally-subtracted dimensional regularization. This criticism resulted in a flurry of publications on the renormalization of the nonperturbative NN problem. See Ref. [81] for a thorough discussion, the bottom line of which is:

Crucial for an EFT are regulator independence (within the range of validity of the EFT) and a power counting scheme that allows for order-by-order improvement with decreasing truncation error. The purpose of renormalization is to achieve this regulator independence while maintaining a functional power counting scheme. After the comprehensive tries and errors of the past, it appears that there are two renormalization schemes which have the potential to achieve the above goals and, therefore, should be investigated systematically in the near future.

In *scheme one*, the LO calculation is conducted nonperturbatively (with $\Lambda \to \infty$ as in [82]) and subleading orders are added perturbatively in distorted wave Born approximation. Valderrama has started this in

S waves [83], but results in higher partial waves are needed to fully assess this approach. Even though at this early stage any judgement is speculative, we take the liberty to predict that this approach will be only of limited success and utility — for the following reasons. First, it will probably require about twice as many counterterms as Weinberg counting and, therefore, will have less predictive power. Second, this scheme may converge badly, because the largest portion of the nuclear force, namely, the intermediate-range attraction appears at NNLO. Third, as discussed in [84], this force may be problematic (and, therefore, impractical) in applications in nuclear few- and many-body systems, because of a pathologically strong tensor force that will cause bad convergence of energy and wave functions. Finally, in the work that has been conducted so far within this scheme by Valderrama, it is found that only rather soft cutoffs can be used.

The latter point (namely, soft cutoffs) suggests that one may then as well conduct the calculation nonperturbatively at all orders (up to N^3LO) using Weinberg counting, which is no problem with soft cutoffs. This is *scheme two* that we propose to investigate systematically. In the spirit of Lepage [85], the cutoff independence should be examined for cutoffs below the hard scale and not beyond. Ranges of cutoff independence within the theoretical error are to be identified using "Lepage plots". A very systematic investigation of this kind does not exist at this time and is therefore needed. However, there is comprehensive circumstantial evidence from the numerous chiral NN potentials constructed over the past decade [20, 31, 33–35, 86] indicating that this investigation will most likely be a success (cf. also Fig. 2, below). The potentials discussed in the following section are all based upon Weinberg counting.

4.3. *Constructing quantitative chiral NN potentials*

4.3.1. *What order?*

As discussed, the NN potential can be calculated up to various orders, cf. Eqs. (8)–(11), and the accuracy increases as the order increases. That triggers the obvious question: To what order do we have to go for an accuracy that we would perceive as necessary and sufficient for, e.g. reliable microscopic nuclear structure calculations?

To answer this question, we show in Table 1 the χ^2/datum for the fit of the world np data below 290 MeV for families of np potentials at NLO and NNLO constructed by the Juelich group [89]. The NLO potentials produce

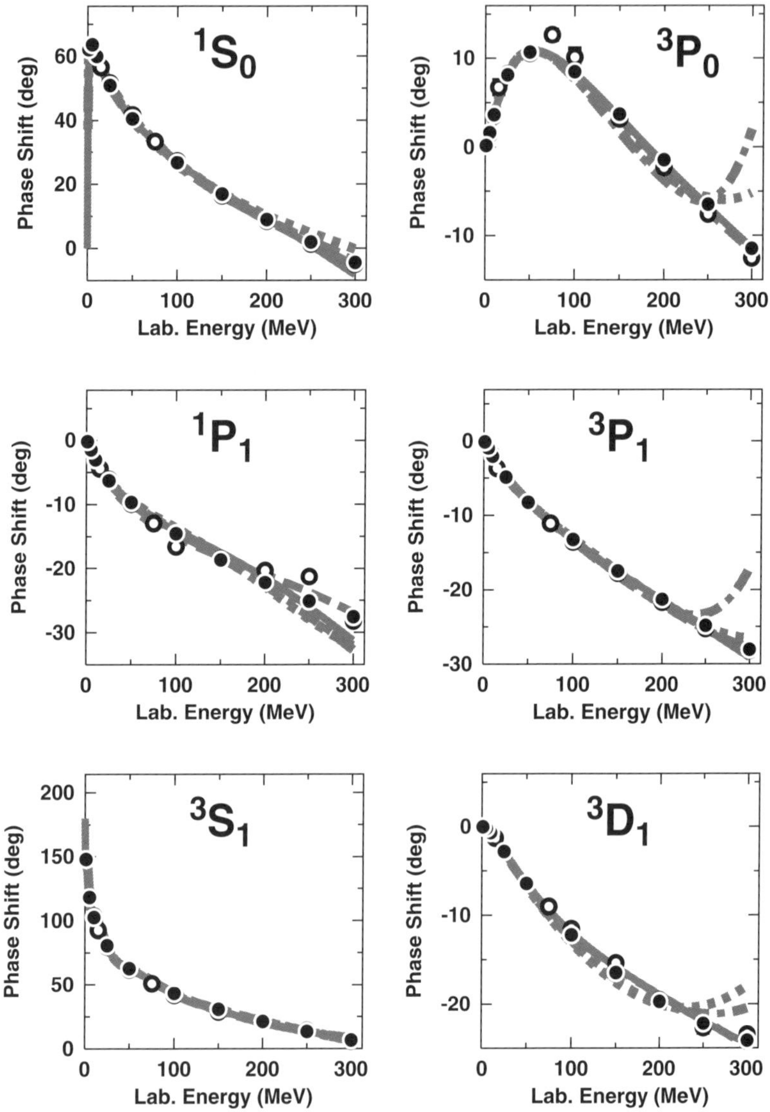

Fig. 2. Neutron-proton phase parameters as described by various chiral potentials at N^3LO. The solid and the dashed curves are calculated from Idaho N^3LO potentials [34] with $\Lambda = 500$ and 600 MeV, respectively; while the dash-dotted and the dotted curves are based upon Juelich N^3LO potentials [86] with cutoff combinations 600/700 and 450/500 MeV, respectively. Partial waves with total angular momentum $J \leq 2$ and laboratory energies up to 300 MeV are displayed. The solid dots and open circles are the results from the Nijmegen multi-energy *np* phase shift analysis [90] and the VPI/GWU single-energy *np* analysis SM99 [98], respectively.

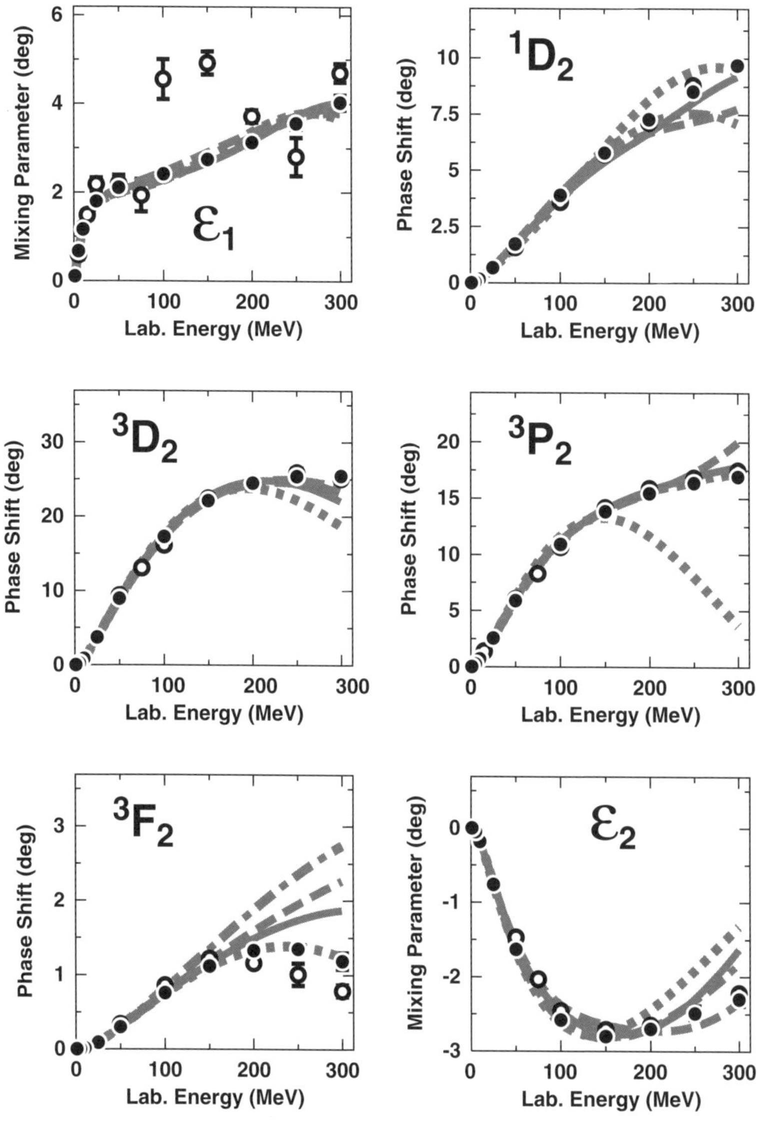

Fig. 2. (*Continued*)

the very large χ^2/datum between 67 and 105, and the NNLO are between 12 and 27. The rate of improvement from one order to the other is very encouraging, but the quality of the reproduction of the *np* data at NLO and NNLO is obviously insufficient for reliable predictions.

Table 1. Columns three and four show the χ^2/datum for the reproduction of the 1999 *np* database [87] (subdivided into energy intervals) by families of *np* potentials at NLO and NNLO constructed by the Juelich group [89]. The χ^2/datum is stated in terms of ranges which result from a variation of the cutoff parameters used in the regulator functions. The values of these cutoff parameters in units of MeV are given in parentheses. $T_{\rm lab}$ denotes the kinetic energy of the incident neutron in the laboratory system.

		— *Juelich np potentials* —	
$T_{\rm lab}$ bin (MeV)	# of *np* data	NLO (550/700–400/500)	NNLO (600/700–450/500)
0–100	1058	4–5	1.4–1.9
100–190	501	77–121	12–32
190–290	843	140–220	25–69
0–290	2402	67–105	12–27

Based upon these facts, it has been pointed out in 2002 by Entem and Machleidt [32,33] that one has to proceed to N^{3}LO. Consequently, the first N^{3}LO potential was published in 2003 [34].

At N^{3}LO, there are a total of 24 contact terms (24 parameters) which contribute to the partial waves with $L \leq 2$. These 24 LECs are essentially free constants which parametrize the short-ranged phenomenological part of the interaction. In Table 2, column "Q^4/N^{3}LO", we show how these terms are distributed over the partial waves. Most important for the improved reproduction of the NN phase shifts (and NN observables) at N^{3}LO is the fact that contacts appear for the first time in D-waves. D-waves are not truely peripheral and, therefore, 1PE plus 2PE alone do not describe them well. The D-wave contacts provide the necessary short-range corrections to get the D-phases right. Besides this, at N^{3}LO, another contact is added to each P-wave, which leads to substantial improvements, particularly, in $^3P_0$ and $^3P_1$ above 100 MeV.

In Table 2, we also show the number of parameters used in the Nijmegen partial wave analysis (PWA93) [90] and in the high-precision CD-Bonn potential [88]. The table reveals that, for S and P waves, the number of parameters used in high-precision phenomenology and in EFT at N^{3}LO are about the same. **Thus, the EFT approach provides retroactively a justification for the phenomenology used in the 1990's to obtain high-precision fits.**

Table 2. Number of parameters needed for fitting the *np* data in phase-shift analysis and by a high-precision NN potential *versus* the total number of NN contact terms of EFT based potentials to different orders.

	Nijmegen partial-wave analysis [90]	CD-Bonn high-precision potential [88]	— *Contact Potentials* —		
			Q^0 LO	Q^2 NLO/NNLO	Q^4 N^3LO
$^1S_0$	3	4	1	2	4
$^3S_1$	3	4	1	2	4
$^3S_1$-$^3D_1$	2	2	0	1	3
$^1P_1$	3	3	0	1	2
$^3P_0$	3	2	0	1	2
$^3P_1$	2	2	0	1	2
$^3P_2$	3	3	0	1	2
$^3P_2$-$^3F_2$	2	1	0	0	1
$^1D_2$	2	3	0	0	1
$^3D_1$	2	1	0	0	1
$^3D_2$	2	2	0	0	1
$^3D_3$	1	2	0	0	1
$^3D_3$-$^3G_3$	1	0	0	0	0
$^1F_3$	1	1	0	0	0
$^3F_2$	1	2	0	0	0
$^3F_3$	1	2	0	0	0
$^3F_4$	2	1	0	0	0
$^3F_4$-$^3H_4$	0	0	0	0	0
$^1G_4$	1	0	0	0	0
$^3G_3$	0	1	0	0	0
$^3G_4$	0	1	0	0	0
$^3G_5$	0	1	0	0	0
Total	35	38	2	9	24

At NLO and NNLO, the number of contact parameters is substantially smaller than for PWA93 and CD-Bonn, which explains why these orders are insufficient for a quantitative potential. The 24 parameters of N^3LO are close to the 30+ used in PWA93 and high precision potentials. Consequently (see following sections for details), at N^3LO, a fit of the NN data is possible that is of about the same quality as the one by the high-precision NN potentials [88, 91–93]. Thus, one may perceive N^3LO as the order of ChPT that is necessary and sufficient for a reliable NN potential.

4.3.2. *A quantitative NN potential at N^{3}LO*

We choose the Idaho N^{3}LO potential [34] as example. Note that for an accurate fit of the low-energy *pp* and *np* data, charge-dependence is important. We include charge-dependence up to next-to-leading order of the isospin-violation scheme (NLØ, in the notation of Ref. [94]). Thus, we include the pion mass difference in 1PE and the Coulomb potential in *pp* scattering, which takes care of the LØ contributions. At order NLØ we have pion mass difference in the NLO part of TPE, $\pi\gamma$ exchange [95], and two charge-dependent contact interactions of order Q^0 which make possible an accurate fit of the three different $^1S_0$ scattering lengths, a_{pp}, a_{nn}, and a_{np}.

For the cutoff parameter Λ of the regulator Eq. (14), we choose initially 500 MeV. Within a certain reasonable range, results should not depend sensitively on Λ (cf. discussion in Sec. 4.2). Therefore, we have also made a second fit for $\Lambda = 600$ MeV.

The fitting procedure starts with the peripheral partial waves because they involve fewer and more fundamental parameters. Partial waves with $L \geq 3$ are exclusively determined by 1PE and 2PE because the N^{3}LO contacts contribute to $L \leq 2$ only. 1PE and 2PE at N^{3}LO depend on the axial-vector coupling constant, g_A (we use $g_A = 1.29$), the pion decay constant, $f_\pi = 92.4$ MeV, and eight low-energy constants (LECs) that appear in the dimension-two and dimension-three πN Lagrangians. The LECs are listed in Table 3, where column "*NN* pot". shows the values

Table 3. Low-energy constants applied in the N^{3}LO *NN* potential (column "*NN* pot".). The c_i belong to the dimension-two πN Lagrangian and are in units of GeV^{-1}, while the $\bar{d}_i$ are associated with the dimension-three Lagrangian and are in units of GeV^{-2}. The column "πN analysis" shows values determined from πN data.

	NN pot.	πN analysis
c_1	-0.81	-0.81 ± 0.15[a]
c_2	2.80	3.28 ± 0.23[b]
c_3	-3.20	-4.69 ± 1.34[a]
c_4	5.40	3.40 ± 0.04[a]
$\bar{d}_1 + \bar{d}_2$	3.06	3.06 ± 0.21[b]
$\bar{d}_3$	-3.27	-3.27 ± 0.73[b]
$\bar{d}_5$	0.45	0.45 ± 0.42[b]
$\bar{d}_{14} - \bar{d}_{15}$	-5.65	-5.65 ± 0.41[b]

[a] Table 1, Fit 1 of Ref. [96].
[b] Table 2, Fit 1 of Ref. [97].

used for the present N^3LO potential. In the fitting process, we varied three of them, namely, c_2, c_3, and c_4. We found that the other LECs are not very effective in the NN system and, therefore, we left them at their central values as determined in πN analysis. The most influential constant is c_3, which — in terms of magnitude — has to be chosen on the low side (slightly more than one standard deviation below its πN determination), otherwise there is too much central attraction. Concerning c_4, our choice $c_4 = 5.4\,\text{GeV}^{-1}$ lowers the $^3F_2$ phase shift (and slightly the $^1F_3$) bringing it into closer agreement with the phase shift analysis — as compared to using the πN value $c_4 = 3.4\,\text{GeV}^{-1}$. The other F waves and the higher partial waves are essentially unaffected by this variation of c_4. Finally, the change of c_2 from its πN value of $3.28\,\text{GeV}^{-1}$ to $2.80\,\text{GeV}^{-1}$ (our choice) brings about some subtle improvements of the fit, but it is not essential. Overall, the fit of all $J \geq 3$ waves is very good.

We turn now to the lower partial waves. Here, the most important fit parameters are the ones associated with the 24 contact terms that contribute to the partial waves with $L \leq 2$. In addition, we have two charge-dependent contacts which are used to fit the three different $^1S_0$ scattering lengths, a_{pp}, a_{nn}, and a_{np}.

In the optimization procedure, we fit first phase shifts, and then we refine the fit by minimizing the χ^2 obtained from a direct comparison with the data. We start with *pp*, since the *pp* phase shifts and data are more accurate than the *np* ones. The *pp* fit fixes essentially the $I = 1$ potential. The $I = 1$ *np* potential is just the *pp* one modified by charge-dependence due to nucleon-mass difference, pion-mass splitting in 1PE, $\pi\gamma$ exchange, and omission of Coulomb. In addition to this, the non-derivative contact in the $^1S_0$ state is changed such as to reproduce the *np* scattering length. The *nn* potential is the *pp* one without Coulomb, using neutron masses, and fitting the *nn* scattering length in the $^1S_0$ state with the non-derivative contact.

The χ^2/datum for the fit of the *np* data below 290 MeV are shown in Table 4, and the corresponding ones for *pp* are given in Table 5. These tables reveal that at N^3LO a χ^2/datum comparable to the high-precision Argonne V_{18} [92] potential can, indeed, be achieved. The Idaho N^3LO potential [34] with $\Lambda = 500\,\text{MeV}$ produces a $\chi^2/\text{datum} = 1.1$ for the world *np* data below 290 MeV which compares well with the $\chi^2/\text{datum} = 1.04$ by the Argonne potential. In 2005, also the Juelich group produced several N^3LO NN potentials [86], the best of which fits the *np* data with a $\chi^2/\text{datum} = 1.7$ and the worse with 7.9 (Table 4).

Table 4. Columns three to five display the χ^2/datum for the reproduction of the 1999 ***np* database** [87] (subdivided into energy intervals) by various *np* potentials. For the chiral potentials, the χ^2/datum is stated in terms of ranges which result from a variation of the cutoff parameters used in the regulator functions. The values of these cutoff parameters in units of MeV are given in parentheses. T_{lab} denotes the kinetic energy of the incident nucleon in the laboratory system.

T_{lab} bin (MeV)	# of ***np*** data	*Idaho* N^{3}LO [34] (500–600)	*Juelich* N^{3}LO [86] (600/700–450/500)	Argonne V_{18} [92]
0–100	1058	1.0–1.1	1.0–1.1	0.95
100–190	501	1.1–1.2	1.3–1.8	1.10
190–290	843	1.2–1.4	2.8–20.0	1.11
0–290	2402	1.1–1.3	1.7–7.9	1.04

Table 5. Same as Table 4 but for ***pp***.

T_{lab} bin (MeV)	# of ***pp*** data	*Idaho* N^{3}LO [34] (500–600)	*Juelich* N^{3}LO [86] (600/700–450/500)	Argonne V_{18} [92]
0–100	795	1.0–1.7	1.0–3.8	1.0
100–190	411	1.5–1.9	3.5–11.6	1.3
190–290	851	1.9–2.7	4.3–44.4	1.8
0–290	2057	1.5–2.1	2.9–22.3	1.4

Turning to *pp*, the χ^2 for *pp* data are typically larger than for *np* because of the higher precision of *pp* data. Thus, the Argonne V_{18} produces a $\chi^2/\text{datum} = 1.4$ for the world *pp* data below 290 MeV and the best Idaho N^{3}LO *pp* potential obtains 1.5. The fit by the best Juelich N^{3}LO *pp* potential results in a $\chi^2/\text{datum} = 2.9$ and the worst produces 22.3.

Phase shifts of *np* scattering from two Idaho (solid and dashed lines) and two Juelich (dash-dotted and dotted lines) N^{3}LO *np* potentials are shown in Fig. 2. The phase shifts confirm what the corresponding χ^2 have already revealed. For the low-energy scattering and deuteron parameters, see Ref. [34].

5. Few-Nucleon Forces

The chiral 2NF discussed in the previous section has been applied in microscopic calculations of nuclear structure with, in general, a great deal of success [41–55]. However, from high-precision studies conducted in the 1990s, it is well-known that certain few-nucleon reactions and nuclear structure issues require 3NFs for their microscopic explanation. Outstanding examples are the A_y puzzle of N-d scattering [99,100] and the ground state of ^{10}B [101]. As noted before, an important advantage of the EFT approach to nuclear forces is that it creates two- and many-nucleon forces on an equal footing (cf. the overview given in Fig. 1).

For a 3NF, we have $A = 3$ and $C = 1$ and, thus, Eq. (5) implies

$$\nu = 2 + 2L + \sum_i \Delta_i. \tag{15}$$

This equation can be used to analyze 3NF contributions order by order. The lowest possible power is obviously $\nu = 2$ (NLO), which is obtained for no loops ($L = 0$) and only leading vertices ($\sum_i \Delta_i = 0$). This 3NF happens to vanish [18,77,78]. The first non-vanishing 3NF occurs at NNLO.

The power $\nu = 3$ (NNLO) is obtained when there are no loops ($L = 0$) and $\sum_i \Delta_i = 1$, i.e. $\Delta_i = 1$ for one vertex while $\Delta_i = 0$ for all other vertices. There are three topologies which fulfill this condition, known as the two-pion exchange (2PE), 1PE, and contact graphs [36,37] (Fig. 3).

The 2PE 3N-potential is given by

$$V^{\rm 3NF}_{\rm 2PE} = \left(\frac{g_A}{2f_\pi}\right)^2 \frac{1}{2} \sum_{i \neq j \neq k} \frac{(\vec{\sigma}_i \cdot \vec{q}_i)(\vec{\sigma}_j \cdot \vec{q}_j)}{(q_i^2 + m_\pi^2)(q_j^2 + m_\pi^2)} F^{ab}_{ijk} \, \tau^a_i \tau^b_j \tag{16}$$

with $\vec{q}_i \equiv \vec{p_i}' - \vec{p}_i$, where $\vec{p}_i$ and $\vec{p_i}'$ are the initial and final momenta of nucleon i, respectively, and

$$F^{ab}_{ijk} = \delta^{ab} \left[-\frac{4c_1 m_\pi^2}{f_\pi^2} + \frac{2c_3}{f_\pi^2} \vec{q}_i \cdot \vec{q}_j \right] + \frac{c_4}{f_\pi^2} \sum_c \epsilon^{abc} \, \tau^c_k \, \vec{\sigma}_k \cdot [\vec{q}_i \times \vec{q}_j]. \tag{17}$$

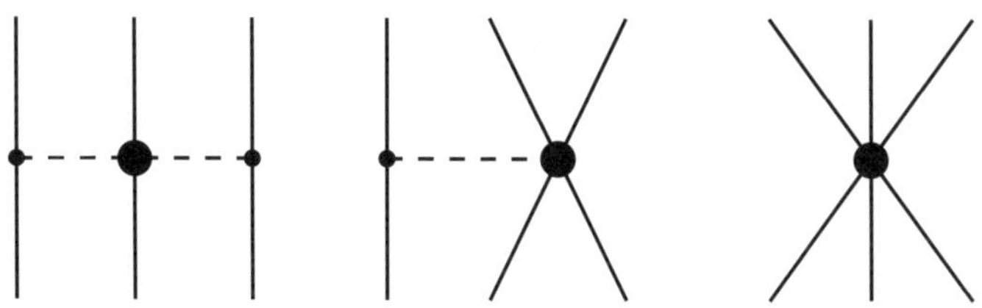

Fig. 3. The three-nucleon force at NNLO. From left to right: 2PE, 1PE, and contact diagrams. Notation as in Fig. 1.

There are great similarities between this force and earlier derivations of 2PE 3NFs, notably the 50-year old Fujita–Miyazawa [102], the Tucson-Melbourne (TM) [103], and the Brazil [104] forces. This is also the 3NF advocated by Gerry Brown in Refs. [2, 3].

The 1PE contribution is

$$V_{\rm 1PE}^{\rm 3NF} = -D \, \frac{g_A}{8f_\pi^2} \sum_{i \neq j \neq k} \frac{\vec{\sigma}_j \cdot \vec{q}_j}{q_j^2 + m_\pi^2} (\boldsymbol{\tau}_i \cdot \boldsymbol{\tau}_j)(\vec{\sigma}_i \cdot \vec{q}_j) \tag{18}$$

and the 3N contact potential reads

$$V_{\rm ct}^{\rm 3NF} = E \, \frac{1}{2} \sum_{j \neq k} \boldsymbol{\tau}_j \cdot \boldsymbol{\tau}_k . \tag{19}$$

The last two 3NF terms involve the two new parameters D and E, which do not appear in the 2N problem. There are many ways to pin these two parameters down. In Ref. [37], the triton binding energy and the nd doublet scattering length ${}^2a_{nd}$ were used. One may also choose the binding energies of ^{3}H and ^{4}He [65] or an optimal over-all fit of the properties of light nuclei [66]. Once D and E are fixed, the results for other 3N, 4N, etc. observables are predictions.

The 3NF at NNLO has been applied in calculations of few-nucleon reactions [37, 56–64], structure of light- and medium-mass nuclei [65–68], and nuclear and neutron matter [69, 70] with a good deal of success. Yet, the famous "A_y puzzle" of nucleon-deuteron scattering is not resolved [37, 61]. When only 2NFs are applied, the analyzing power in p-^{3}He scattering is even more underpredicted than in p-d [40, 105]. However, when the NNLO 3NF is added, the p-^{3}He A_y substantially improves (more than in p-d) [64] — but a discrepancy remains. Furthermore, the spectra of light nuclei leave room for improvement [66].

To summarize, the 3NF at NNLO is a remarkable contribution: It represents the leading many-body force within the scheme of ChPT; it includes terms that were advocated by Gerry Brown already some 40 years ago [2, 3]; and it produces noticeable improvements in few-nucleon reactions and the structure of light nuclei. But unresolved problems remain. Moreover, in the case of the 2NF, we have seen that one has to proceed to N^{3}LO to achieve sufficient accuracy. Therefore, the 3NF at N^{3}LO is needed for at least two reasons: for consistency with the 2NF and to hopefully resolve outstanding problems in microscopic structure and reactions. For further discussion of the 3NF (beyond NNLO), see Ref. [81].

6. Conclusions and Outlook

Forty years after Gerry Brown had stressed repeatedly the outstanding importance of chiral symmetry, we have finally arrived at a derivation of nuclear forces which takes this symmetry consistently into account.

The greatest progress occurred during the past 15 years. Key to this development was the realization that low-energy QCD is equivalent to a chiral effective field theory which allows for a perturbative expansion that has become known as chiral perturbation theory (ChPT). In this framework, two- and many-body forces emerge on an equal footing and the empirical fact that nuclear many-body forces are substantially weaker than the two-nucleon force is explained naturally.

In this chapter, we have shown how the two-nucleon force is derived from ChPT and demonstrated that, at N^3LO, the accuracy can be achieved that is necessary and sufficient for reliable microscopic nuclear structure predictions. First calculations applying the N^3LO NN potential [34] in the conventional shell model [41–43], the *ab initio* no-core shell model [44–46], the coupled cluster formalism [47–53], and the unitary-model-operator approach [54, 55] have produced promising results.

We also discussed nuclear many-body forces based upon chiral EFT. The 3NF at NNLO has been known for a while [36, 37] and applied in few-nucleon reactions [37, 56–64], structure of light- and medium-mass nuclei [65–68], and nuclear and neutron matter [69, 70] with good success; but some open issues remain, which are presently under investigation [81].

We are optimistic that the remaining outstanding problems will be resolved within the next few years, such that, after 80 years of desperate struggle, we may finally claim that the nuclear force problem is essentially under control — last not least, due to a thorough accounting of chiral symmetry.

Acknowledgments

It is our great pleasure to dedicate this chapter to Gerry Brown on the occasion of his 85th birthday. One of us (R.M.) has been a postdoc with Gerry during the years of 1976 and 1977 and stayed in contact with him ever since. R.M. has always been impressed by Gerry's liberating personality and inspiring physics intuition. This work was supported in part by the U.S. Department of Energy under Grant No. DE-FG02-03ER41270 (R.M.), and the Ministerio de Ciencia y Tecnología under Contract No. FPA2007-65748 and the Junta de Castilla y León under Contract Nos. SA-106A07 and GR12 (D.R.E.).

References

[1] G.E. Brown, *Comments Nucl. Part. Phys.* **4**, 140 (1970).
[2] G.E. Brown, in: Meson in Nuclei, Vol. I, edited by M. Rho and D.H. Wilkinson (North-Holland, Amsterdam, 1979) p. 330.
[3] G.E. Brown, A.M. Green and W.J. Gerace, *Nucl. Phys.* **A115**, 435 (1968).
[4] J. Schwinger, *Ann. Phys. (N.Y.)* **2**, 407 (1957).
[5] M. Gell-Mann and M. Levy, *Nuovo Cimento* **16**, 705 (1960).
[6] T. Ericson and W. Weise, *Pions and Nuclei* (Clarendon Press, Oxford, 1988).
[7] S. Weinberg, *Phys. Rev. Lett.* **18**, 188 (1967).
[8] S. Weinberg, *Phys. Rev. Lett.* **17**, 616 (1966).
[9] Y. Tomozawa, *Nuovo Cimento A* **46**, 707 (1966).
[10] S. Weinberg, *Phys. Rev.* **166**, 1568 (1968).
[11] S. Coleman, J. Wess and B. Zumino, *Phys. Rev.* **177**, 2239 (1969); C.G. Callan, S. Coleman, J. Wess and B. Zumino, *ibid.* **177**, 2247 (1969).
[12] S. Weinberg, *Physica* **96A**, 327 (1979).
[13] S. Weinberg, *Effective Field Theory, Past and Future*, arXiv:0908:1964 [hep-th].
[14] J. Gasser and H. Leutwyler, *Ann. Phys.* **158**, 142 (1984); *Nucl. Phys.* **B250**, 465 (1985).
[15] J. Gasser, M.E. Sainio and A. Švarc, *Nucl. Phys.* **B307**, 779 (1986).
[16] S. Weinberg, *Phys. Lett. B* **251**, 288 (1990).
[17] S. Weinberg, *Nucl. Phys.* **B363**, 3 (1991).
[18] S. Weinberg, *Phys. Lett. B* **295**, 114 (1992).
[19] C. Ordóñez, L. Ray and U. van Kolck, *Phys. Rev. Lett.* **72**, 1982 (1994).
[20] C. Ordóñez, L. Ray and U. van Kolck, *Phys. Rev. C* **53**, 2086 (1996).
[21] N. Kaiser, R. Brockmann and W. Weise, *Nucl. Phys.* **A625**, 758 (1997).
[22] N. Kaiser, S. Gerstendörfer and W. Weise, *Nucl. Phys.* **A637**, 395 (1998).
[23] N. Kaiser, *Phys. Rev. C* **61**, 014003 (2000).
[24] N. Kaiser, *Phys. Rev. C* **62**, 024001 (2000).
[25] N. Kaiser, *Phys. Rev. C* **63**, 044010 (2001).
[26] N. Kaiser, *Phys. Rev. C* **64**, 057001 (2001).
[27] N. Kaiser, *Phys. Rev. C* **65**, 017001 (2002).
[28] C.A. da Rocha and M.R. Robilotta, *Phys. Rev. C* **49**, 1818 (1994); **52**, 531 (1995); M.R. Robilotta, *Nucl. Phys.* **A595**, 171 (1995); M.R. Robilotta and C.A. da Rocha, *Nucl. Phys.* **A615**, 391 (1997); J.-L. Ballot, C.A. da Rocha and M.R. Robilotta, *Phys. Rev. C* **57**, 1574 (1998).
[29] R. Higa and M.R. Robilotta, *Phys. Rev. C* **68**, 024004 (2003).
[30] E. Epelbaum, W. Glöckle and U.-G. Meißner, *Nucl. Phys.* **A637**, 107 (1998).
[31] E. Epelbaum, W. Glöckle and U.-G. Meißner, *Nucl. Phys.* **A671**, 295 (2000).
[32] D.R. Entem and R. Machleidt, *Phys. Lett. B* **524**, 93 (2002).
[33] D.R. Entem and R. Machleidt, *Phys. Rev. C* **66**, 014002 (2002).
[34] D.R. Entem and R. Machleidt, *Phys. Rev. C* **68**, 041001 (2003).
[35] R. Machleidt and D.R. Entem, *J. Phys. G: Nucl. Phys.* **31**, S1235 (2005).
[36] U. van Kolck, *Phys. Rev. C* **49**, 2932 (1994).
[37] E. Epelbaum *et al.*, *Phys. Rev. C* **66**, 064001 (2002).

[38] S. Ishikawa and M.R. Robilotta, *Phys. Rev. C* **76**, 014006 (2007).
[39] V. Bernard, E. Epelbaum, H. Krebs and U.-G. Meißner, *Phys. Rev. C* **77**, 064004 (2008).
[40] A. Deltuva and A.C. Fonseca, *Phys. Rev. Lett.* **98**, 162502 (2007); *Phys. Rev. C* **76**, 021001 (2007); arXiv:1005.1308 [nucl-th].
[41] L. Coraggio *et al.*, *Phys. Rev. C* **66**, 021303 (2002).
[42] L. Coraggio *et al.*, *Phys. Rev. C* **71**, 014307 (2005).
[43] L. Coraggio *et al.*, arXiv:1005.2896 [nucl-th].
[44] P. Navrátil and E. Caurier, *Phys. Rev. C* **69**, 014311 (2004).
[45] C. Forssen *et al.*, *Phys. Rev. C* **71**, 044312 (2005).
[46] J.P. Vary *et al.*, *Eur. Phys. J. A* **25** s01, 475 (2005).
[47] K. Kowalski *et al.*, *Phys. Rev. Lett.* **92**, 132501 (2004).
[48] D.J. Dean and M. Hjorth-Jensen, *Phys. Rev. C* **69**, 054320 (2004).
[49] M. Wloch *et al.*, *J. Phys. G* **31**, S1291 (2005); *Phys. Rev. Lett.* **94**, 21250 (2005).
[50] D.J. Dean *et al.*, *Nucl. Phys.* **752**, 299 (2005).
[51] J.R. Gour *et al.*, *Phys. Rev. C* **74**, 024310 (2006).
[52] G. Hagen, T. Papenbrock, D.J. Dean and M. Hjorth-Jensen, *Phys. Rev. Lett.* **101**, 092502 (2008).
[53] G. Hagen, T. Papenbrock, D.J. Dean and M. Hjorth-Jensen, *Phys. Rev. C* **82**, 034330 (2010).
[54] S. Fujii, R. Okamoto and K. Suzuki, *Phys. Rev. C* **69**, 034328 (2004).
[55] S. Fujii, R. Okamoto and K. Suzuki, *Phys. Rev. Lett.* **103**, 182501 (2009).
[56] K. Ermisch *et al.*, *Phys. Rev. C* **71**, 064004 (2005).
[57] S. Kistryn *et al.*, *Phys. Rev. C* **72**, 044006 (2005).
[58] H. Witala, J. Golak, R. Skibinski, W. Glöckle, A. Nogga, E. Epelbaum, H. Kamada, A. Kievsky and M. Viviani, *Phys. Rev. C* **73**, 044004 (2006).
[59] J. Ley *et al.*, *Phys. Rev. C* **73**, 064001 (2006).
[60] E. Stephan *et al.*, *Phys. Rev. C* **76**, 057001 (2007).
[61] N. Kalantar-Nayestanaki and E. Epelbaum, *The Three-Nucleon System as a Laboratory for Nuclear Physics: The Need for 3N Forces*, *Nucl. Phys. News* **17**, 22 (2007), arXiv:nucl-th/0703089, and references therein.
[62] L.C. Marcucci, A. Kievsky, L. Girlanda, S. Rosati and M. Viviani, *Phys. Rev. C* **80**, 034003 (2009).
[63] A. Kievsky, M. Viviani, L. Girlanda and L.E. Marcucci, *Phys. Rev. C* **81**, 044003 (2010).
[64] M. Viviani, L. Giarlanda, A. Kievsky, L.E. Marcucci and S. Rosati, arXiv:1004.1306 [nucl-th].
[65] A. Nogga, P. Navratil, B.R. Barrett and J.P. Vary, *Phys. Rev. C* **73**, 064002 (2006).
[66] P. Navratil, V.G. Gueorguiev, J.P. Vary, W.E. Ormand and A. Nogga, *Phys. Rev. Lett.* **99**, 042501 (2007).
[67] G. Hagen *et al*, *Phys. Rev. C* **76**, 034302 (2007).
[68] T. Otsuka, T. Susuki, J.D. Holt, A. Schwenk and Y. Akaishi, *Phys. Rev. Lett.* **105**, 032501 (2010).

[69] S.K. Bogner *et al.*, *Nucl. Phys.* **A763**, 59 (2005); S.K. Bogner *et al.*, *Nuclear Matter from Chiral Low-Momentum Interactions*, arXiv:0903.3366 [nucl-th].
[70] K. Hebeler and A. Schwenk, *Phys. Rev. C* **82**, 014314 (2010).
[71] S.R. Beane, P.F. Bedaque, K. Orginos and M.J. Savage, *Phys. Rev. Lett.* **97**, 012001 (2006).
[72] N. Ishii, S. Aoki and T. Hatsuda, *Phys. Rev. Lett.* **99**, 022001 (2007).
[73] J. Goldstone, *Nuovo Cim.* **19**, 154 (1961).
[74] J. Goldstone, A. Salam and S. Weinberg, *Phys. Rev.* **127**, 965 (1962).
[75] R. Machleidt, K. Holinde and Ch. Elster, *Phys. Rep.* **149**, 1 (1987).
[76] M. Lacombe, B. Loiseau, J.M. Richard, R. Vinh Mau, J. Côté, P. Pires and R. de Tourreil, *Phys. Rev. C* **21**, 861 (1980).
[77] S.N. Yang and W. Glöckle, *Phys. Rev. C* **33**, 1774 (1986).
[78] S.A. Coon and J.L. Friar, *Phys. Rev. C* **34**, 1060 (1986).
[79] R. Machleidt, *Adv. Nucl. Phys.* **19**, 189 (1989).
[80] D.B. Kaplan, M.J. Savage and M.B. Wise, *Nucl. Phys.* **B478**, 629 (1996); *Phys. Lett. B* **424**, 390 (1998); *Nucl. Phys.* **B534**, 329 (1998).
[81] R. Machleidt and D.R. Entem, *J. Phys. G: Nucl. Part. Phys.* **37**, 064041 (2010).
[82] A. Nogga, R.G.E. Timmermans and U. van Kolck, *Phys. Rev. C* **72**, 054006 (2005).
[83] M.P. Valderrama, *Phys. Rev. C* **83**, 024003 (2011).
[84] R. Machleidt, P. Liu, D.R. Entem and E. Ruiz Arriola, *Phys. Rev. C* **81**, 024001 (2010).
[85] G.P. Lepage, *How to Renormalize the Schrödinger Equation*, nucl-th/9706029.
[86] E. Epelbaum, W. Glöckle and U.-G. Meißner, *Nucl. Phys.* **A747**, 362 (2005).
[87] The 1999 NN data base is defined in Ref. [88].
[88] R. Machleidt, *Phys. Rev. C* **63**, 024001 (2001).
[89] E. Epelbaum, W. Glöckle and U.-G. Meißner, *Eur. Phys. J.* **A19**, 401 (2004).
[90] V.G.J. Stoks, R.A.M. Klomp, M.C.M. Rentmeester and J.J. de Swart, *Phys. Rev. C* **48**, 792 (1993).
[91] V.G.J. Stoks, R.A.M. Klomp, C.P.F. Terheggen and J.J. de Swart, *Phys. Rev. C* **49**, 2950 (1994).
[92] R.B. Wiringa, V.G.J. Stoks and R. Schiavilla, *Phys. Rev. C* **51**, 38 (1995).
[93] R. Machleidt, F. Sammarruca and Y. Song, *Phys. Rev. C* **53**, 1483 (1996).
[94] M. Walzl *et al.*, *Nucl. Phys.* **A693**, 663 (2001).
[95] U. van Kolck *et al.*, *Phys. Rev. Lett.* **80**, 4386 (1998).
[96] P. Büttiker *et al.*, *Nucl. Phys.* **A668**, 97 (2000).
[97] N. Fettes *et al.*, *Nucl. Phys.* **A640**, 199 (1998).
[98] R.A. Arndt, I.I. Strakovsky and R.L. Workman, SAID, Scattering Analysis Interactive Dial-in computer facility, George Washington University (formerly Virginia Polytechnic Institute), solution SM99 (Summer 1999); for more information see, e.g. R.A. Arndt, I.I. Strakovsky and R.L. Workman, *Phys. Rev. C* **50**, 2731 (1994).
[99] W. Glöckle *et al.*, *Phys. Rep.* **274**, 107 (1996).

[100] D.R. Entem, R. Machleidt and H. Witala, *Phys. Rev. C* **65**, 064005 (2002), and references to earlier work therein.
[101] E. Caurier *et al.*, *Phys. Rev. C* **66**, 024314 (2002).
[102] J.-I. Fujita and H. Miyazawa, *Prog. Theor. Phys.* **17**, 360 (1957).
[103] S.A. Coon, M.D. Scadron, P.C. McNamee, B.R. Barrett, D.W.E. Blatt and B.H.J. McKeller, *Nucl. Phys.* **A317**, 242 (1979).
[104] H.T. Coelho, T.K. Das and M.R. Robilotta, *Phys. Rev. C* **28**, 1812 (1983); M.R. Robilotta and H.T. Coelho, *Nucl. Phys.* **A460**, 645 (1986).
[105] B.M. Fisher *et al.*, *Phys. Rev. C* **74**, 034001 (2006).

Chapter 15

Chiral Symmetry, Nuclear Forces and All That

Ulf-G. Meißner*

These are personal recollections of how Gerry Brown's work and thinking influenced the development of the chiral effective Lagrangian based theory of nuclear forces.

1. The Early Days at Stony Brook

I first met Gerry on the occasion of the Erice school in 1981, where he and Mannque Rho were giving quite interesting lectures on the chiral bag model and related issues. At that point, I was close to finishing my diploma thesis under the supervision of Manfred Gari at Bochum, working on two-meson exchange currents in the framework of unitary transformations. More specifically, the goal was to find out the relative contribution of the isoscalar $\rho\pi^2\gamma$ current compared to the well-established leading order isoscalar $\rho\pi\gamma$ current. This calculation turned out to be — to use Howard Georgi's words — an excellent exercise in self-torture, as one had to evaluate about 1000 time-ordered diagrams by hand. Also, the result that these corrections were already sizeable at momentum transfers below $q^2 \simeq 1\,\text{GeV}^2$ did not quite fit into the standard lore which was so eloquently summarized by Gerry and Mannque in their chiral filter hypothesis [1]. This hypothesis that grew out of explicit meson-exchange current calculations including heavier hadrons like the $\Delta(1232)$ or the vector mesons ρ, ω was clearly emphasizing the role of the pion. Whenever present, the one-pion exchange

*HISKP and Bethe Center for Theoretical Physics, Universität Bonn, D-53115 Bonn, Germany; Institut für Kernphysik, Institute for Advanced Simulation and JCHP, Forschungszentrum Jülich, D-52425 Jülich, Germany. E-mail: meissner@hiskp.uni-bonn.de

current was supposed to dominate the pertinent response of light nuclei to electromagnetic probes, even up to momentum transfers as large as $1\,\mathrm{GeV}^2$, way beyond the soft-pion limit. In fact, it was clear to Gerry and Mannque — and others — that the understanding of the nucleon structure from QCD and the nuclear interactions were two intimately connected issues, firmly rooted in the chiral symmetry of QCD. It is worth quoting from Ref. [1]: "... that there must exist an intimate connection between what makes up a hadron and what induces a rich variety of interactions between hadrons. ..., it will be rather unlikely one will gain a full understanding of one without the other." Only then too much emphasis was put onto the quark structure of the nucleon — as we understand now, low-energy QCD is well approximated by an hadronic effective Lagrangian. When doing my calculations at Bochum, I did not know that, but learned about it quickly when I arrived at Stony Brook in April of 1982. The Erice lectures on the little bag model had aroused my interest and, honestly, I was fed up calculating meson-exchange currents. Therefore, I contacted Gerry about becoming a graduate student at Stony Brook. It did not take him long to accept me — as it turned out, I spent a short but very productive time in Gerry's group until my graduation in December 1984. But let me briefly come back to the meson-exchange current (MEC) story. In fact, only after some strong support from Dan-Olof Riska, one of the MEC pioneers [2], my diploma work was finally published [3] but largely ignored by the community. The crux of the matter was that in those days one did not have a power counting that allowed one to systematically address the corrections to the leading order one-pion contribution either in the nuclear forces or in the corresponding exchange currents, though a tremendous amount of phenomenology had been built up over the years. In particular, the Stony Brook and Paris approach to the two-nucleon force that used dispersion relations to connect the πN scattering amplitudes to the 2π-exchange in the NN force supplemented by some heavy mass Yukawa-type functions with parameters determined from a fit to the data where already very close to what would become later the "modern theory of nuclear forces". In any case, I had to learn the basis of the fundamentals of the Stony Brook potential by working through the book of Gerry and Andy Jackson [4]. This turned out to be a very painful exercise as the notation would change from chapter to chapter, but made me familiar with dispersion theory and the many intricacies of nuclear interactions. In fact, one of the early projects Gerry assigned to me was the coupling of ρ-mesons to the little bag, which I worked out within a few weeks and which was supposed to be the starting point for a more microscopic investigation of vector meson exchanges in the nuclear force, thus further contributing to the sought-after unified picture of nucleons and the nuclear interactions. When I presented to Gerry my notes in the form of a handwritten draft — LaTeX was not yet available — he just told me that I was too fast for him and proposed to me that I work alone. So I teamed up with Andreas Wirzba — a fresh new graduate student from Münster — and Ismail Zahed, then a post-doc in Gerry's group to work on Casimir effects in chiral bag models and on many aspects of Skyrmions, which was a hot topic at that time. The Skyrme model promised to lead to the long wanted unification of nucleon and nuclear physics, and many impressive calculations revealed further deep connections, but again the lack of a power counting was the stumbling stone.

2. A First Step: Studying Pion-Nucleon Scattering in Chiral Perturbation Theory

During my stay as a Heisenberg fellow at Bern, I was lucky to learn chiral perturbation theory (CHPT) from Jürg Gasser and Heiri Leutwyler, who had transformed Weinberg's ideas into an effective machinery to analyze low-energy QCD. With Véronique Bernard and Norbert Kaiser we systematically analyzed electroweak reactions on the nucleon and in particular neutral pion photoproduction off the proton, $\gamma p \to \pi^0 p$, triggered by measurements at Saclay, Mainz and Saskatoon to test the then believed low-energy theorems based on soft-pion algebra (plus one extra assumption). The explanation of the data by a new one-loop effect helped to establish the method as a precision tool in the single nucleon sector. In the the late nineties and early years of the new millennium, with my students Sven Steininger and Nadia Fettes [5] we investigated pion-nucleon scattering in the framework of heavy baryon chiral perturbation theory — this later turned out to be an important ingredient in the construction of the few-nucleon forces based on effective field theory (EFT). The underlying chiral Lagrangian of pions and nucleons coupled to external currents can be written as

$$\mathcal{L}_{\rm eff} = \mathcal{L}^{(1)} + \mathcal{L}^{(2)} + \mathcal{L}^{(3)} + \mathcal{L}^{(4)} + \cdots \,, \tag{1}$$

where the superscript denotes the chiral dimension (the power in small external momenta and/or pion masses). Tree level computations are done with $\mathcal{L}^{(1)} + \mathcal{L}^{(2)}$, a complete one-loop calculation involves all loop graphs with insertions from $\mathcal{L}^{(1)}$ and at most one insertion from $\mathcal{L}^{(2)}$ and so on.

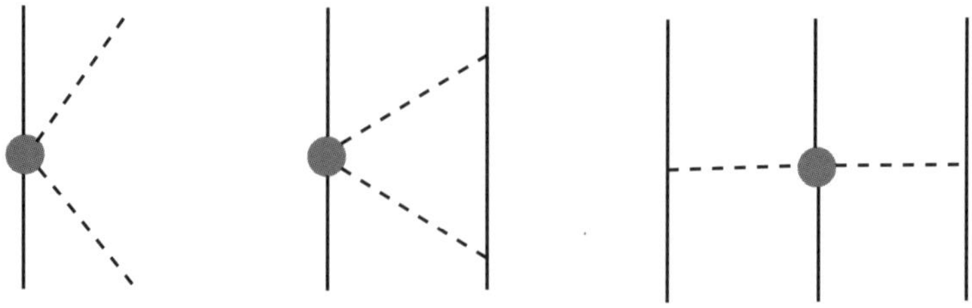

Fig. 1. The LECs c_i (circles) in pion-nucleon scattering (left), the two-nucleon (NN) interaction (center) and the three-nucleon (NNN) interaction (right).

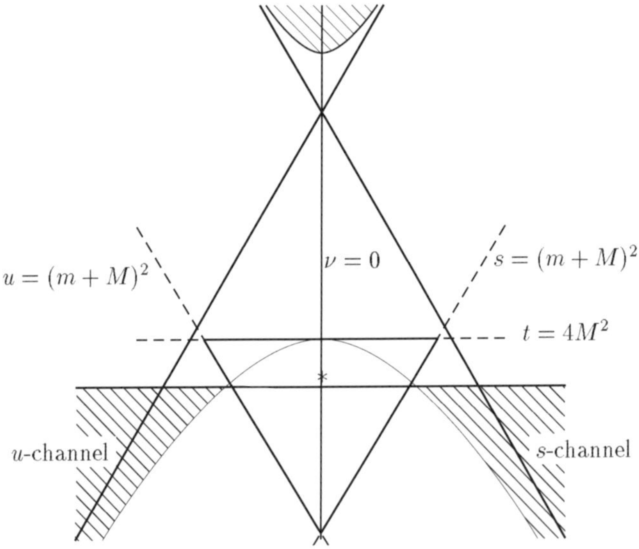

Fig. 2. Mandelstam plane. The Mandelstam triangle is the inside of the thick lines.

In its full glory, $\mathcal{L}_{\rm eff}$ can be found in Ref. [7]. Of particular importance in these studies are the finite dimension two low-energy constants (LECs) c_i, see Fig. 1, that also feature prominently in the description of the two- and three-nucleon forces — Gerry was one of the first who had understood this long before the advent of EFT. An even better determination of some of the c_i was possible by analytically continuing the amplitudes into the interior of the Mandelstam triangle, see Fig. 2 — the ability to use dispersion relations as I learned it in Stony Brook turned out to be very valuable. All this work can be summarized by the following values for the c_i in units of GeV^{-1}:

$$c_1 = -0.9^{+0.2}_{-0.5}, \quad c_2 = 3.3 \pm 0.2, \quad c_3 = -4.7^{+1.2}_{-1.0}, \quad c_4 = 3.5^{+0.5}_{-0.2}. \tag{2}$$

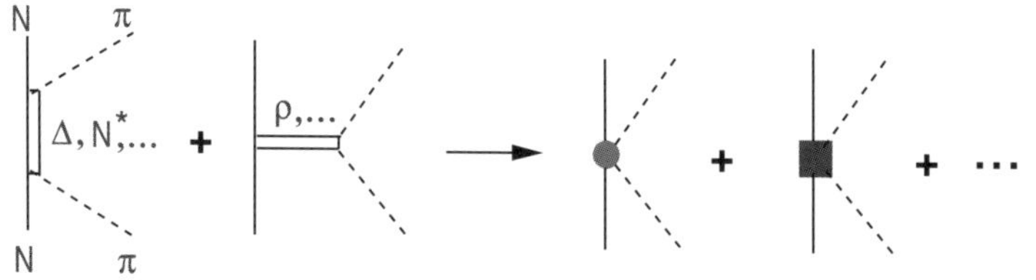

Fig. 3. Resonance saturation for the dimension two (circles) and three (squares) LECs.

The values for c_2, c_3 and c_4 are larger than the expected (natural) values: $c_i \sim g_A/\Lambda_\chi \simeq 1.1 \ldots 1.5$, with $g_A = 1.267$ the nucleon axial-vector coupling and $\Lambda_\chi \simeq 1\,\text{GeV}$ the scale of chiral symmetry breaking. This was understood early in terms of resonance saturation — the fact that the heavier states integrated out from the EFT leave their imprint in the values of the LECs. This method was pioneered in the meson sector by Gerhard Ecker and collaborators and also by John Donoghue and collaborators [8]. In 1997, with Véronique Bernard and Norbert Kaiser, we had already determined the c_i by a comparison to scattering lengths and subthreshold parameters and extended the resonance saturation scheme to the one-baryon sector, see Fig. 3. Not surprisingly, the LECs c_2 and c_3 are largely dominated by the close-by $\Delta(1232)$ resonance, whereas c_4 also receives an important contribution from the ρ-meson [9]. This leads to the following values of $c_{2,3,4}$ (in brackets the ranges obtained from resonance saturation) $c_2 = 3.9\,[2\ldots4], c_3 = -5.3\,[-4.5\ldots-5.3], c_4 = 3.7\,[3.1\ldots3.7]$, whereas c_1 is given by scalar meson exchange ($\pi\pi$ correlations).[a] These numbers are also consistent with the recent analysis of pion-nucleon scattering and subthreshold parameters, $c_1 = -1.2\ldots-0.9, c_2 = 2.6\ldots4.0, c_3 = -6.1\ldots-4.4$ [10]. Let me also stress that the uncertainties in the c_i are strongly correlated, as best seen from the combination $-2c_1 + c_2 + c_3 - g_A^2/(8m)$ that determines the small isoscalar pion-nucleon scattering length.

3. Flashback: The Texas Revolution

Already in the early nineties a decisive step towards a systematic theory of the nuclear forces had been done. This was initiated by a series of papers by Steven Weinberg, who had already pioneered the use of effective chiral Lagrangians decades earlier under the nowadays somewhat strange-reading

[a]Note that the central values from Ref. 8 are superseded by the values given in Eq. (2).

title "Phenomenological Lagrangians" [11]. In that marvelous paper the idea of the power counting was developed in detail and further a very illuminating application of the renormalization group to work out the so-called chiral logarithms was presented. Years later, while teaching the effective field theory of pions and nucleons, Weinberg suddenly realized that there are additional terms consistent with the power counting, terms quadrilinear in the nucleon fields.[b] He immediately realized that these must be related to the much discussed mechanism for the short-range repulsion of the nuclear forces. Here the fundamental principle of effective field theory works at it best — in the low-energy sector of a given theory one is not able to resolve the physics at large momentum scales but simply parameterizes it in terms of operators made of the fields one has at one's disposal — it just does not matter whether the nucleon-nucleon repulsion is due to vector-meson exchange, quark rearrangement energies or whatever model has been invented before. Weinberg then worked out the consequences of the spontaneous chiral symmetry breaking for the forces between two, three and four nucleons using the method of phenomenological Lagrangians [12], nowadays called chiral perturbation theory or chiral effective field theory. Remarkably, the first of these two papers contains only three references, one to the review Gerry had written in 1985 with Sven-Olaf Bäckman and Jouni Niskanen [13] and also, he explicitly thanks Gerry for "enlightening conversations on nuclear forces". The important step taken by Weinberg was to realize that in the nuclear force problem one can not apply the power counting directly to the S-matrix, but rather to the effective potential — these are all diagrams without N-nucleon intermediate states. Such diagrams lead to pinch singularities in the infinite nucleon mass limit (the so-called static limit), so that e.g. the nucleon box graph is enhanced as m/Q^2, with m the nucleon mass and $Q \ll \Lambda_\chi$ a small momentum. The beautiful power counting formula for the graphs contributing with the νth power of Q or a pion mass to the effective potential reads (considering only connected pieces):

$$\nu = 2 - N - 2L + \sum_i V_i \left[d_i + \frac{n_i}{2} - 2\right]. \qquad (3)$$

Here, N is the number of in-coming and out-going nucleons, L the number of pion loops, V_i counts the vertices of type i with d_i derivatives and/or pion mass insertions and n_i is the number of nucleons participating in this kind of vertex. Because of chiral symmetry, the term in the square brackets is larger

[b] I am grateful to Steve Weinberg for providing me with this recollection.

than or equal to zero and thus the leading terms contributing e.g. to the NN potential can easily be identified. These are the time-honored one-pion exchange and two four-nucleon contact interactions without derivatives. These contact interactions were indeed the missing part which were earlier modeled by heavy meson-exchanges or by some kind of fit function — but none of these approaches was controlled or systematic. Here again the awesome power of power counting becomes crystal-clear — it just took one person not biased by nuclear folklore to take this final step. The so-constructed effective potential is then iterated in the Schrödinger or Lippman-Schwinger equation, generating the shallow nuclear bound states as well as scattering states. The resulting contributions to the 2N, the 3N and the 4N forces are depicted in Fig. 4. I will come back to a detailed discussion of the various entries below, at this point it is important to observe that — consistent with phenomenological observations — three-nucleon forces appear only two orders after the dominant NN forces and four-nucleon forces are even further suppressed, appearing only at N^{3}LO. Let me stress that the beautiful analysis of few-nucleon forces in chiral EFT by Bira van Kolck [14] was an important ingredient to set up the power counting as

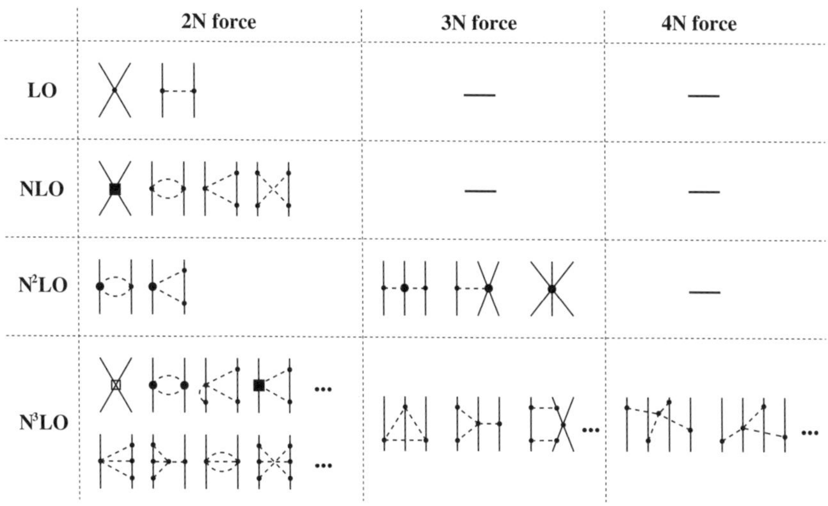

Fig. 4. Contributions to the effective potential of the 2N, 3N and 4N forces based on Weinberg's power counting. Here, LO denotes leading order, NLO next-to-leading order and so on. Dimension one, two and three pion-nucleon interactions are denoted by small circles, big circles and filled boxes, respectively. In the 4N contact terms, the filled and open box denote two- and four-derivative operators, respectively.

displayed in Fig. 4 and that this extended power counting has been the major development on the way to the chiral theory of nuclear forces.

The first large-scale numerical evaluation of this chiral potential was done by Bira van Kolck and collaborators a few years after Weinberg's seminal papers, see Refs. [15]. They worked up-to-and-including N^2LO in a theory of pions, nucleons and also the Δ isobar, fitting the LECs of the contact interactions to low-energy NN phases separately for isospin $I = 0$ and $I = 1$ and properties of the deuteron. This work was certainly groundbreaking but it was also not immediately accepted in the nuclear physics community — for the simple reason that the phenomenological approaches to the two-nucleon problem had achieved a much higher precision — at the expense of many more parameters and lack of systematicity. Still, the first important step was done. Before picking up this theme, let me stress that Weinberg extended his approach to pion reactions on light nuclei [16], using at that time deuteron wave functions from phenomenological potentials. This so-called "hybrid appraoch" laid the ground for systematic studies of pion scattering off light nuclei, pion photo- and electroproduction off light nuclei and of pion production in NN collisions.

4. The Chiral Theory of Nuclear Forces as a Precision Tool

4.1. *The nucleon-nucleon interaction*

When I was still at Bonn filling in for the vacant chair of Max Huber, then rector of Bonn University, I met a few times with Karl Holinde of the IKP Jülich. Karl had become famous for his work on the Bonn potential and was hired by Josef Speth at Jülich in 1984, which gave Karl a fantastic home base to continue his work. We talked about the chiral potential of the Texas group, and he was clearly interested but also aware that the precision was not yet competitive. So we decided to work together to try to improve the Texas potential. Unfortunately, before really getting started, this collaboration was terminated by Karl's untimely death in December of 1996. In fact, Gerry came to Jülich the next year and gave a wonderful talk about the roots, the physics and the achievements of the Bonn potential, focusing on Karl's contribution. Fortunately, a bit later Walter Glöckle from Bochum contacted me — he had acquired a brilliant young student named Evgeny Epelbaum, who had done some calculations of the chiral NN potential in his diploma thesis and was supposed to continue working on this topic for his doctoral thesis. Walter was (and still is) a world-leading expert in few-nucleon calculations but did not feel at ease with effective chiral Lagrangians. So we decided that we would team up to supervise Epelbaum, which turned out to be a true pleasure as we did not have to do much. Being aware of the short-comings of energy-dependent potentials in few-body calculations, we decided to recalculate the Texas potential using the method of unitary transformations — which I had not used for about 15 years. Based on Epelbaum's diploma thesis, we published a paper on the power counting adapted to this method in Ref. [17]. Before working out the phenomenological consequences of this, we decided to work out some simpler examples to get better acquainted with the many subtleties of the nuclear interaction and how they reflect on the corresponding effective theory. Interestingly, in two papers we developed a method to integrate out the high-momentum components from an effective theory based on a simple local potential of the Malfliet-Tjon type [18]. This preceeded the work of Achim Schwenk, Tom Kuo, Bengt Friman, Scott Bogner Gerry and others at Stony Brook on the so-called $V_{\rm low-k}$ potentials based on elegant renormalization group techniques, see e.g. the review [19] and references therein. Such parallel developments are a fine witness of Gerry's legacy — he has taught many generations of students a deep understanding of physics that will necessarily lead to progress in the field of nuclear physics. But back to the nuclear force problem. We published our N^{2}LO results in Ref. [20], which were markedly improved as compared to the Texas potential. The simple reason was that we did not make global fits but rather projected the chiral potential onto partial waves and determined the four-nucleons LECs by fitting to the S- and P-waves and the $^3S_1$-$^3D_1$ mixing parameter in the neutron-proton system. An important paper that had been published earlier was the work by the Munich group, where the role of the two-pion exchange to the peripheral partial waves was scrutinized [21]. They demonstrated that these peripheral waves are well described by one- and two-pion exchanges in the Born approximation, thus only the low partial waves were to be used in fitting the multi-nucleon LECs. However, the large values of the c_i, cf. Eq. (2), lead to a very strong isoscalar central potential at short distances that generates unphysical deep bound states for the larger values of the cut-off scale ($\Lambda \simeq 1\,$GeV) in the LS-equation. Again, earlier work done by Chemtob, Durso and Riska — that was taught at Stony Brook — came to our rescue, we simply adopted the so-called spectral function regularization method to deal with the unwanted short-distance behavior of the two-pion exchange [22]. So the ground was prepared for going to yet higher

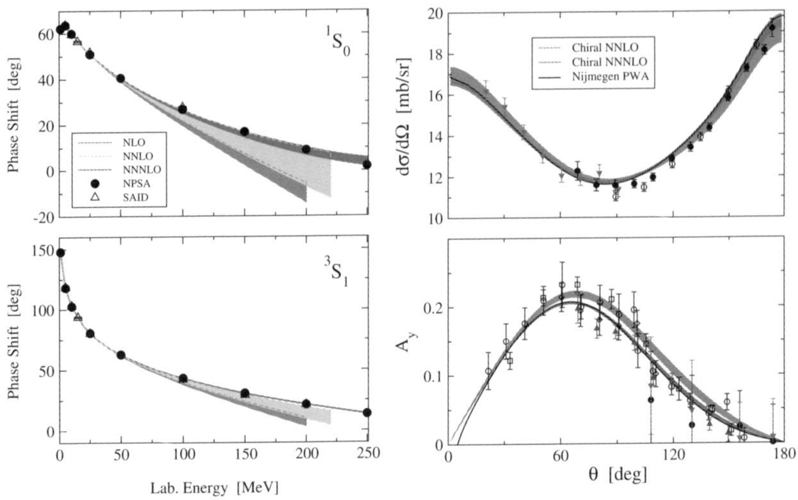

Fig. 5. Left: np S-waves at NLO (grid), N^{2}LO (light shaded) and N^{3}LO (dark shaded). Right: Differential cross section and vector analyzing power for np scattering at $E_{\rm lab} =$ 50 MeV compared to the N^{2}LO (light shaded) and N^{3}LO (dark shaded) predictions.

orders. While Evgeny had moved to Jefferson Lab as the first Nathan-Isgur-Distinguished-Fellow, we continued our collaboration and published the N^{3}LO results in [23]. Earlier on, Ruprecht Machleidt and David Entem had published N^{3}LO results using dimensional regularization in the two-pion exchange [24] based on a series of papers from Norbert Kaiser — this story is told in the contribution of Machleidt to this volume. Some characteristic results of our N^{3}LO calculation are displayed in Fig. 5 — these are not more accurate than the ones based on phenomenological potentials, however, they provide a measure of the theoretical uncertainty and, more importantly, they are based on the chiral symmetry of QCD. To close the circle, one can indeed show that the four-nucleon operators to a large extent are saturated by resonance excitations, as explicit calculations mapping boson-exchange models onto the chiral expansion of the effective potential showed, for details see Ref. 25. In a way, EFT provides a reason why the meson-exchange models of the nuclear force have been so successful.

4.2. *Three-nucleon interactions*

One of the biggest advantages of the chiral Lagrangian approach to the nuclear force problem — as stressed early in Weinberg's papers — is the

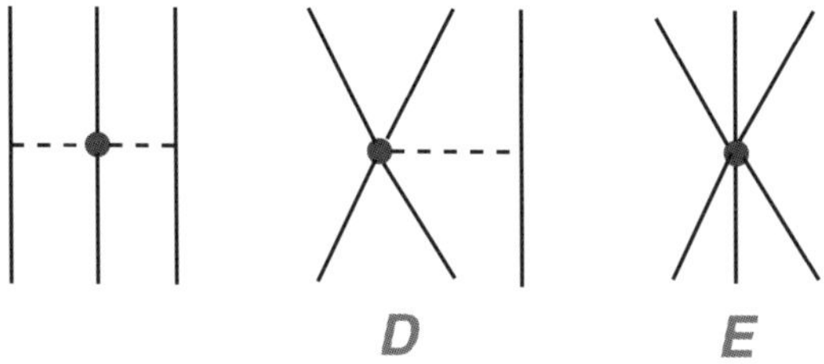

Fig. 6. Topologies of the leading contributions to the chiral 3NF. From left to right: Two-pion exchange, one-pion-exchange and 6N contact interaction.

consistent derivation of the 3- and 4-nucleon forces and also the meson-exchange currents (that is, the response to electroweak probes). It is worth mentioning that decades earlier Gerry was aware of the role of chiral symmetry in the description of three-body forces in nuclei — see his nice paper with Saul Barshay in 1972 [26]. But let me come back to the modern approach. As already noted, the three-nucleon force (3NF) only appears two orders after the leading NN interaction. At this order, there are only three topologies contributing, see Fig. 6. The two-pion exchange topology is given again in terms of the c_i, as discussed in detail in Ref. [27]. The so-called D-term, which is related to the one-pion exchange between a 4N contact term and a further nucleon, has gained some prominence over the last years, as many authors have tried to pin it down based on a cornucopia of reactions, such as $Nd \to Nd$, $NN \to NN\pi$, $NN \to d\ell\nu_\ell$ or $d\pi \to \gamma NN$. This demonstrates nicely the power of EFT - very different processes are related through the same LECs thus providing many different tests of chiral symmetry (as it is also the case with the LECs c_i, see Fig. 1). The LEC E related to the 6N contact interaction can only be fixed in systems with at least three nucleons, say from the triton binding energy. Remarkably, although there are many diagrams at N^{3}LO, there are no new LECs to be determined. The long-range terms of this force can be found in Ref. [28] and the shorter ranged ones will be published soon (see also Ref. [29]). Many applications of these forces and the testing of their structure can be found in the reviews [30, 31].

As a very nice example that these forces make their way into nuclear structure calculations, I show here the results from Petr Navrátil and collaborators [32], who performed large-basis no-core shell model (NCSM) calculations including the leading chiral 3N forces and demonstrated their necessity to describe the spectra of nuclei with $A = 10, \ldots 13$, see the entry NN+NNN in Fig. 7. Note, however, that in this calculation the 2N force

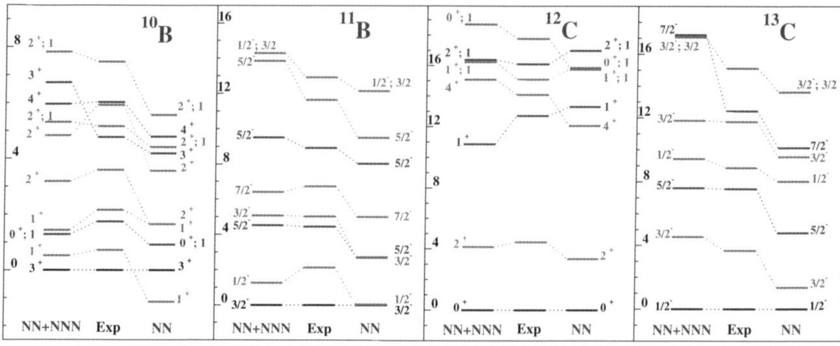

Fig. 7. Spectra of light nuclei using chiral NN and chiral NN+NNN forces in a NCSM calculation compared to the data. Figure adopted from Ref. [32] and courtesy of P. Navrátil.

was employed at N^{3}LO while the 3NF was only included at N^{2}LO. It will be interesting to see how the inclusion of the sub-leading 3NF terms will modify these results.

So far, I have only discussed the EFT with pions and nucleons. Clearly, the two-pion exchange diagram in Fig. 6 is a controlled approximation to the time-honored Fujita-Miyazawa force, whose 50th anniversary was celebrated in Tokyo in 2007. For a light reading of how the Fujita-Miyazawa force is related to the EFT description of the 3NF, I refer to Ref. [33] and references therein.

4.3. *The Nuclear Matter Problem*

Another important question in nuclear physics is to understand the saturation properties of nuclear matter — an idealized infinite system of nucleons in which all Coulomb effects are switched off. From the properties of heavy nuclei using some sophisticated mass formula, one can extrapolate to nuclear matter — and determine its saturation properties. The binding energy per nucleon E/A is approximately -16 MeV at a Fermi momentum of about $1.3\,\text{fm}^{-1}$. So what does chiral EFT have to say? A first important step was taken by Norbert Kaiser, Wolfram Weise and collaborators at Munich, who calculated the contribution of pion exchange(s) to the energy density of nuclear matter and showed that the energy density of isospin symmetric nuclear matter can be extremely well approximated by the simple form

$$E/A = \frac{3k_F^2}{10m} - \alpha\frac{k_F^3}{m^2} + \beta\frac{k_F^4}{m^3}. \tag{4}$$

They then calculated the coefficients α and β from chiral dynamics. With some fine-tuning of the regulator, one finds an astonishingly good description of the energy density of nuclear matter [34]. This was later improved by including e.g. higher orders (sensitive again to the LECs c_i) and isobar degrees of freedom [35]. Interestingly, all this was based on a loop expansion with no explicit power counting — a power counting that indeed explained the success of these calculations was only set up much later, see Ref. [36]. Space forbids a discussion of this power counting and the resulting physics in detail — I only would like to mention that it could be shown that for many reactions the contributions from the multi-nucleon interactions cancel to leading order which is at the heart of the success of the Munich group calculations.

Here, I briefly pursue a more "conventional" approach to the nuclear matter problem based on more standard nuclear many-body theory. For decades, the method of choice to analyze nuclear matter has been the G-matrix of Brueckner and others — leading to the problem that for phase-shift equivalent NN potentials the resulting saturation properties lie on the Coester line that does not pass the empirical values. Recently, Siegfried Krewald and collaborators have recalculated the properties of nuclear matter based on the chiral NN and 3N forces (see the v2 version of Ref. [38]) utilizing the R-matrix as proposed by Baker in 1971 [37]. As can be seen from the left panel of Fig. 8, the NN interaction alone

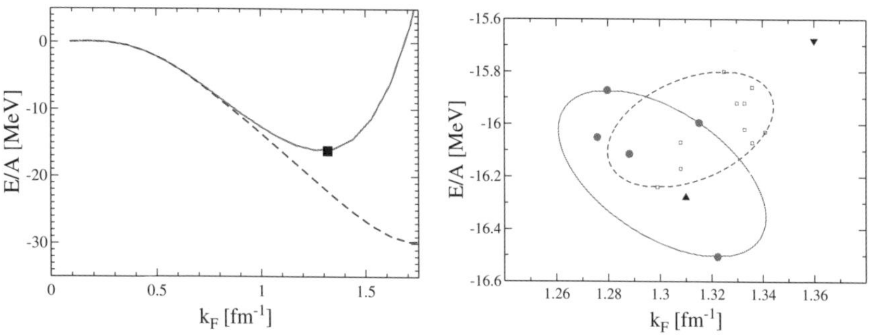

Fig. 8. Left panel: Binding energy per particle of nuclear matter as a function of the Fermi momentum k_F for the N^{2}LO potential using a Lippman-Schwinger cut off $\Lambda = 550\,\text{MeV}$ and a spectral function cut off $\tilde{\Lambda} = 600\,\text{MeV}$ (solid line). The binding energies obtained in the absence of three-nucleon interactions are shown by the dashed line. The black square gives the empirical nuclear matter properties. Right panel: Saturation points of nuclear matter. Downward triangle, upward triangle and rectangle: various phenomenological approaches. The dashed circle summarizes the phenomenological results. Solid circle: chiral EFT at N^{2}LO.

only binds for Fermi momenta larger than $1.7\,\mathrm{fm}^{-1}$. However, including the 3NF with natural values for D and E, one finds binding with the proper strength at the proper density. In the right panel, the sensitivity to the two cut-offs in the NN force as compared to phenomenological determinations is shown, it is of comparable size. The dependence on the 3-body LECs D and E is also weak, see Fig. 3 in Ref. [38]. It will be interesting to push these calculations to $\mathrm{N}^3\mathrm{LO}$ and to include the $\Delta(1232)$ in the EFT.

4.4. *Back to square one — meson-exchange currents*

As told in Sec. 1, I started my career calculating meson-exchange currents based on the method of unitary transformations. It is quite a nice turn of events that in chiral nuclear EFT one is now able to do this in a controlled and systematic manner, thanks to the power counting. In our approach, Stefan Kölling, Hermann Krebs and Evgeny Epelbaum have taken up the task — the two-pion exchange electromagnetic current based on this method is published in Ref. [39] and together with the Cracow group led by Henryk Witala and Jacek Golak a thorough investigation of electro-nuclear processes is underway, for a first application to deuteron photodisintegration see Ref. [40]. However, the group around Rocco Schiavilla, using old-fashioned time-ordered perturbation theory, has taken the lead and performed calculations of the MECs of one- and two-pion range and applied this to magnetic moments of light nuclei [41]. Most recently, they have calculated thermal neutron capture on d and ^{3}He [42]. The resulting radiative capture cross sections σ^{γ}_{nd} and $\sigma^{\gamma}_{n^3\mathrm{He}}$ and the photon circular polarization parameter R_c, resulting from the capture of polarized neutrons on the deuteron, are shown in Fig. 9 in comparison to two state-of-the-art conventional approaches. As the authors note, these processes are not the best or simplest to illustrate the convergence pattern of chiral nuclear EFT. So much more work is needed here and will be done.

5. Flash Forward: Combining Nuclear EFT with Lattice Simulations

Having now constructed nuclear forces between two, three and four nucleons, one has a new tool to address the nuclear many-body problem. One venue is to combine standard many-body techniques with these forces,

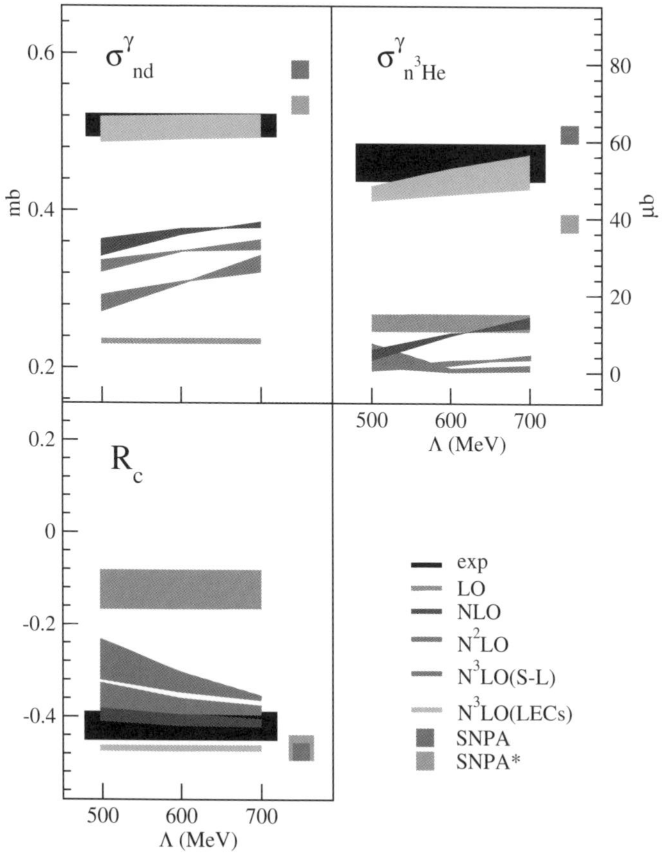

Fig. 9. Predictions for the *n*-*d* and *n*-^{3}He capture reactions. Chiral EFT: the curve labeled N3LO(S-L) includes pion loop corrections contact terms, while the curve labeled N3LO(LECs) includes in addition contributions from (higher order) pion exchanges and non-minimal contact currents. Pre-EFT approaches: SNPA and SNPA* denote up-to-date conventional calculations. Figure courtesy of Rocco Schiavilla.

as exemplified by the NCSM calculation shown in Sec. 4.2. or the discussion of nuclear matter in Sec. 4.3. (see also the contribution by Josef Speth, Siegfried Krewald and Frank Grümmer to this Festschrift). Here, I briefly want to outline a very different and novel approach that combines the chiral EFT for the forces with the method of Monte Carlo simulations, that are so successfully utilized in the lattice approach to QCD. The basic idea is to formulate the chiral EFT on a Euclidean space-time lattice as depicted in the left panel of Fig. 10 — here, the lattice spacing serves as an UV

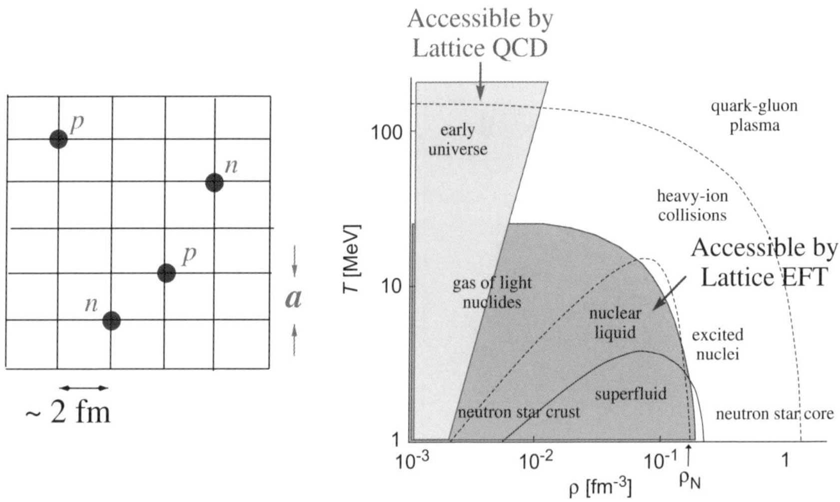

Fig. 10. Left panel: Euclidean space-time lattice with point-like nucleons on the lattice sites and a typical lattice spacing of $a = 2\,\mathrm{fm}$. Right panel: Nuclear phase diagram as accessible by lattice QCD and by nuclear lattice EFT. Figure courtesy of Dean Lee.

regulator and has to be chosen such that the nucleons — which are treated as non-relativistic point-like particles on the lattice sites — do not overlap. A lattice spacing of $a = 2\,\mathrm{fm}$ entails an UV cut-off $\Lambda = \pi/a \simeq 300\,\mathrm{MeV}$. The pion and nucleon propagators as well as the one- and two-pion exchanges and the multi-nucleon contact interactions are written in terms of lattice variables, making use of Hubbard-Stratonovich fields to bring the 4- and 6-nucleon interactions into a quadratic form (for more details, see Ref. [43] and references therein as well as the nice review by Dean Lee [44]). Given this Euclidean formulation of the action, the generating functional of the theory can then be evaluated with stochastic methods, such as the hybrid Monte Carlo approach. It is important to stress that the approximate Wigner SU(4) symmetry of the nuclear interactions strongly suppresses the sign oscillations and thus makes such nuclear lattice simulations easier to handle than lattice QCD calculations.

Within this scheme, a systematic study of nuclei up to $A \simeq 40$ will be possible using petascale computing as long as the total spin and isospin of the nucleus under consideration is zero. Also, the nuclear phase diagram can be studied for a wide range of temperatures and densities, as indicated in the right panel of Fig. 10. The state-of-the-art of these simulations is the calculation of the ground-state energy of ^{6}Li and ^{12}C [45]. In these works, for the

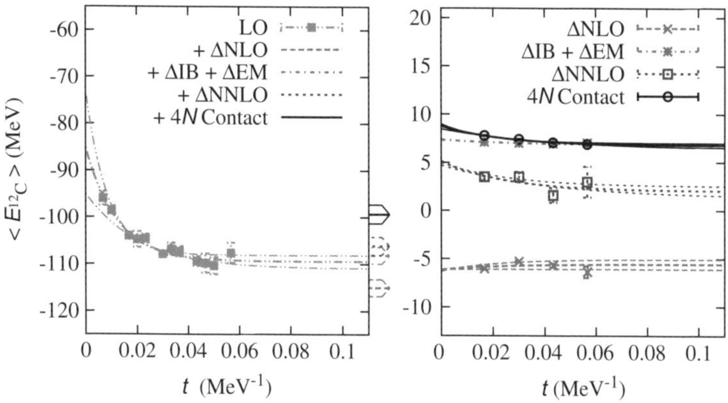

Fig. 11. Ground-state energy of ^{12}C as a function of Euclidean time. For details, see Ref. [45].

first time the contributions from the Coulomb interaction between protons and strong isospin-breaking effects were included. A parameter-free prediction for the energy difference between the triton and ^{3}He could be given,

$$E(^3\text{H}) - E(^3\text{He}) = 0.78(5)\,\text{MeV}, \tag{5}$$

in good agreement with the empirical value of 0.76 MeV. The ground-state energies of ^{6}Li and ^{12}C are calculated as $-32.9(9)$ MeV and $-99(2)$ MeV, respectively, not far from the empirical values of -32.0 and -92.2 MeV, cf. also Fig. 11. This accuracy is comparable to other so-called ab initio calculations (like the NCSM or Greens function Monte Carlo), that often are based on a less consistent formulation of the underlying nuclear forces. In fact, using this approach, it has recently been possible to perform the first ever *ab initio* calculation of the Hoyle state (the first excited 0^+-state) that plays such a central role in the generation of carbon in stars.[46] Taking ^{12}C as the benchmark, it is interesting to estimate the required computing time and storage needed for the calculation of larger nuclei.[c] With an improved algorithm, the required CPU time for the ^{12}C simulation on a BlueGene/P architecture and the necessary storage to make the configurations (about 3500) available is

$$X_{^{12}C}^{\text{CPU}} = 5 \times 10^{-5}\,\text{PFlop} - \text{yr}, \quad X_{^{12}C}^{\text{storage}} = 0.14\,\text{TB}. \tag{6}$$

[c]I am grateful to Dean Lee for providing me with these estimates.

For a nucleus with A nucleons and with total spin S and isospin I the required CPU time and data storage place can be estimated as

$$X^{\rm CPU} \approx X^{\rm CPU}_{^{12}C} \times \left(\frac{A}{12}\right)^{3.2} \exp[0.10(A-12) + 3(S \bmod 2) + 4I],$$

$$X^{\rm storage} \approx X^{\rm storage}_{^{12}C} \times \left(\frac{A}{12}\right)^{2} \exp[0.10(A-12) + 3(S \bmod 2) + 4I], \qquad (7)$$

so that e.g. the calculation of the ground-states energies of the magnesium isotopes ^{24,25,26}Mg would require $0.002, 0.064, 0.131$ PFlop-yr and one would need $1.9, 73.8, 145.5$ TB to store the configurations for the calculation of matrix elements for these nuclei. In many cases, neutron scattering off nuclei can also be calculated making use of Lüscher's formula that relates the continuum phase shift to the finite energy shift measured on the lattice. The CPU and storage requirements for neutron scattering of the isotopes ^{24,26}Mg would be $0.064, 5.43$ PFlop-yr and $73.8, 5473$ TB, respectively. All this work remains to be done but bears a lot of promise. The possibilities that would be offered in a future exascale era are simply breath-taking.

6. Some Final Words

The chiral effective Lagrangian of QCD offers a tool to systematically investigate the structure of the nucleon and the nuclear interactions with high accuracy — it appears that we are now on the right track to achieve what could only be dreamed about a few decades ago — see the discussion in Sec. 1. Nuclear physics will no longer be based on model-building and phenomenological approaches, although still quite a bit of work is ahead of us. But this work — even if hard and time-consuming — will be very rewarding at the end. Furthermore, strangeness nuclear physics can also be addressed within this framework [47], but here a sound data basis has yet to be established. Eventually, lattice QCD might also contribute significantly — but there are still quite a few obstacles to be overcome, see e.g. [48]. It is fair to say that nuclear physics as a field, and I personally, owe a lot to Gerry — thank you.

Acknowledgments

I am very much indebted to Gerry — as a physicist and a human. I am grateful to Evgeny Epelbaum, Dean Lee, Mannque Rho, Rocco Schiavilla, and Steven Weinberg for useful communications during the preparation of this contribution. I also thank Silas Beane and Véronique Bernard for a careful reading of the manuscript.

References

[1] M. Rho and G.E. Brown, *Comments Nucl. Part. Phys.* **10**, 201 (1981).
[2] D.O. Riska and G.E. Brown, *Phys. Lett.* **38B**, 193 (1972).
[3] U.-G. Meißner and M. Gari, *Phys. Lett. B* **125**, 364 (1983).
[4] G.E. Brown and A.D. Jackson, The nucleon-nucleon interaction, North-Holland Pub. Co., (Amsterdam and New York, 1976).
[5] N. Fettes, U.-G. Meißner and S. Steininger, *Nucl. Phys. A* **640**, 199 (1998) [arXiv:hep-ph/9803266]; N. Fettes and U.-G. Meißner, *Nucl. Phys. A* **676**, 311 (2000) [arXiv:hep-ph/0002162]; N. Fettes and U.-G. Meißner, *Nucl. Phys. A* **693**, 693 (2001) [arXiv:hep-ph/0101030].
[6] P. Buettiker and U.-G. Meißner, *Nucl. Phys. A* **668**, 97 (2000) [arXiv:hep-ph/9908247].
[7] N. Fettes, U.-G. Meißner, M. Mojzis and S. Steininger, *Annals Phys.* **283**, 273 (2000) [Erratum-ibid. **288**, 249 (2001)] [arXiv:hep-ph/0001308].
[8] G. Ecker, J. Gasser, A. Pich and E. de Rafael, *Nucl. Phys. B* **321**, 311 (1989); G. Ecker, J. Gasser, H. Leutwyler, A. Pich and E. de Rafael, *Phys. Lett. B* **223**, 425 (1989); J.F. Donoghue, C. Ramirez and G. Valencia, *Phys. Rev. D* **39**, 1947 (1989).
[9] V. Bernard, N. Kaiser and U.-G. Meißner, *Nucl. Phys. A* **615**, 483 (1997) [arXiv:hep-ph/9611253].
[10] J. Gasser, V.E. Lyubovitskij and A. Rusetsky, *Phys. Rept.* **456**, 167 (2008) [arXiv:0711.3522 [hep-ph]].
[11] S. Weinberg, *Physica A* **96**, 327 (1979).
[12] S. Weinberg, *Phys. Lett. B* **251**, 288 (1990); S. Weinberg, *Nucl. Phys. B* **363**, 3 (1991).
[13] S.O. Bäckman, G.E. Brown and J.A. Niskanen, *Phys. Rept.* **124**, 1 (1985).
[14] U. van Kolck, *Phys. Rev. C* **49**, 2932 (1994).
[15] C. Ordonez, L. Ray and U. van Kolck, *Phys. Rev. Lett.* **72**, 1982 (1994); C. Ordonez, L. Ray and U. van Kolck, *Phys. Rev. C* **53**, 2086 (1996) [arXiv:hep-ph/9511380].
[16] S. Weinberg, *Phys. Lett. B* **295**, 114 (1992) [arXiv:hep-ph/9209257].
[17] E. Epelbaum, W. Glöckle and U.-G. Meißner, *Nucl. Phys. A* **637**, 107 (1998) [arXiv:nucl-th/9801064].
[18] E. Epelbaum, W. Glöckle and U.-G. Meißner, *Phys. Lett. B* **439**, 1 (1998) [arXiv:nucl-th/9804005]. E. Epelbaum, W. Glöckle, A. Krüger and U.-G. Meißner, *Nucl. Phys. A* **645**, 413 (1999) [arXiv:nucl-th/9809084].
[19] S.K. Bogner, T.T.S. Kuo and A. Schwenk, *Phys. Rept.* **386**, 1 (2003) [arXiv:nucl-th/0305035].
[20] E. Epelbaum, W. Glöckle and U.-G. Meißner, *Nucl. Phys. A* **671**, 295 (2000) [arXiv:nucl-th/9910064].

[21] N. Kaiser, R. Brockmann and W. Weise, *Nucl. Phys. A* **625**, 758 (1997) [arXiv:nucl-th/9706045].
[22] E. Epelbaum, W. Glöckle and U.-G. Meißner, *Eur. Phys. J. A* **19**, 125 (2004) [arXiv:nucl-th/0304037].
[23] E. Epelbaum, W. Glöckle and U.-G. Meißner, *Nucl. Phys. A* **747**, 362 (2005) [arXiv:nucl-th/0405048].
[24] D.R. Entem and R. Machleidt, *Phys. Lett. B* **524**, 93 (2002) [arXiv:nucl-th/0108057]; D.R. Entem and R. Machleidt, *Phys. Rev. C* **68**, 041001 (2003) [arXiv:nucl-th/0304018].
[25] E. Epelbaum, U.-G. Meißner, W. Glöckle and C. Elster, *Phys. Rev. C* **65**, 044001 (2002) [arXiv:nucl-th/0106007].
[26] S. Barshay and G.E. Brown, *Phys. Rev. Lett.* **34**, 1106 (1975).
[27] J.L. Friar, D. Huber and U. van Kolck, *Phys. Rev. C* **59**, 53 (1999) [arXiv:nucl-th/9809065].
[28] V. Bernard, E. Epelbaum, H. Krebs and U.-G. Meißner, *Phys. Rev. C* **77**, 064004 (2008) [arXiv:0712.1967 [nucl-th]].
[29] S. Ishikawa and M.R. Robilotta, *Phys. Rev. C* **76**, 014006 (2007) [arXiv:0704.0711 [nucl-th]].
[30] E. Epelbaum, *Prog. Part. Nucl. Phys.* **57**, 654 (2006) [arXiv:nucl-th/0509032].
[31] E. Epelbaum, H.W. Hammer and U.-G. Meißner, *Rev. Mod. Phys.* **81** (2009) 1773 [arXiv:0811.1338 [nucl-th]].
[32] P. Navratil, V.G. Gueorguiev, J.P. Vary, W.E. Ormand and A. Nogga, *Phys. Rev. Lett.* **99**, 042501 (2007) [arXiv:nucl-th/0701038].
[33] U.-G. Meißner, *AIP Conf. Proc.* **1011**, (2008) 49.
[34] N. Kaiser, S. Fritsch and W. Weise, *Nucl. Phys. A* **697**, 255 (2002) [arXiv:nucl-th/0105057].
[35] S. Fritsch, N. Kaiser and W. Weise, *Nucl. Phys. A* **750**, 259 (2005) [arXiv:nucl-th/0406038].
[36] J.A. Oller, A. Lacour and U.-G. Meißner, *J. Phys. G* **37**, 015106 (2010) [arXiv:0902.1986 [nucl-th]]; A. Lacour, J.A. Oller and U.-G. Meißner, *Ann. Phys.* **326**, 241 (2011).
[37] G. A. Baker, *Rev. Mod. Phys.* **43**, 479 (1971).
[38] P. Saviankou, S. Krewald, E. Epelbaum and U.-G. Meißner, arXiv:0802.3782 [nucl-th].
[39] S. Kölling, E. Epelbaum, H. Krebs and U.-G. Meißner, *Phys. Rev. C* **80**, 045502 (2009) [arXiv:0907.3437 [nucl-th]].
[40] D. Rozpedzik, J. Golak, S. Kölling and E. Epelbaum, arXiv:1010.6269 [nucl-th].
[41] S. Pastore, R. Schiavilla and J.L. Goity, *Phys. Rev. C* **78**, 064002 (2008) [arXiv:0810.1941 [nucl-th]]; S. Pastore, L. Girlanda, R. Schiavilla, M. Viviani and R.B. Wiringa, *Phys. Rev. C* **80**, 034004 (2009) [arXiv:0906.1800 [nucl-th]].
[42] L. Girlanda, A. Kievsky, L.E. Marcucci, S. Pastore, R. Schiavilla and M. Viviani, arXiv:1008.0356 [nucl-th].

[43] B. Borasoy, E. Epelbaum, H. Krebs, D. Lee and U.-G. Meißner, *Eur. Phys. J. A* **31**, 105 (2007) [arXiv:nucl-th/0611087].
[44] D. Lee, *Prog. Part. Nucl. Phys.* **63**, 117 (2009) [arXiv:0804.3501 [nucl-th]].
[45] E. Epelbaum, H. Krebs, D. Lee and U.-G. Meißner, *Phys. Rev. Lett.* **104**, 142501 (2010) [arXiv:0912.4195 [nucl-th]]; E. Epelbaum, H. Krebs, D. Lee and U.-G. Meißner, *Eur. Phys. J. A* **45**, 335 (2010) [arXiv:1003.5697 [nucl-th]].
[46] E. Epelbaum, H. Krebs, D. Lee and U.-G. Meißner. arXiv:1101.2547 [nucl-th], *Phys. Rev. Lett.* (2011) in press.
[47] C.L. Korpa, A.E.L. Dieperink and R.G.E. Timmermans, *Phys. Rev. C* **65**, 015208 (2002) [arXiv:nucl-th/0109072]; H.W. Hammer, *Nucl. Phys. A* **705**, 173 (2002) [arXiv:nucl-th/0110031]; H. Polinder, J. Haidenbauer and U.-G. Meißner, *Nucl. Phys. A* **779**, 244 (2006) [arXiv:nucl-th/0605050]; H. Polinder, J. Haidenbauer and U.-G. Meißner, *Phys. Lett. B* **653**, 29 (2007) [arXiv:0705.3753 [nucl-th]]. J. Haidenbauer and U.-G. Meißner, *Phys. Lett. B* **684**, 275 (2010) [arXiv:0907.1395 [nucl-th]].
[48] S. Beane, *PoS C* **D09**, 076 (2009) [arXiv:0912.5404 [hep-lat]].

Chapter 16

Transport Properties of a Non-Relativistic Delta-Shell Gas with Long Scattering Lengths

Sergey Postnikov* and Madappa Prakash†

The coefficients of diffusion, thermal conductivity, and shear viscosity are calculated for a system of non-relativistic particles interacting via a delta-shell potential $V(r) = -v\,\delta(r-R)$ when the average distance between particles is smaller than R. The roles of resonances and long scattering lengths including the unitary limit are examined. Results for ratios of diffusion to viscosity and viscosity to entropy density are presented for varying scattering lengths.

1. Introduction

Recently, studies of transport properties, particularly viscosities of interacting particles in a many-body system, have attracted much attention as observed phenomena in diverse fields of physics (such as atomic physics and relativistic heavy ions physics) are greatly influenced by their effects [1]. In atomic physics, the relaxation of atomic clusters from an ordered state have been observed [2]. In relativistic heavy-ion physics, viscosities influence the observed elliptic flow versus transverse momentum of final state hadrons [3]. In both of these fields, the challenge of determining the transport properties at varying temperatures and densities has been vigorously pursued [1].

In this paper, we wish to study a simple system in which the high temperature properties of a dilute gas are significantly affected by the nature

*Instituto de Astronomía, Universidad Nacional Autónoma de México, Mexico D.F. 04510, Mexico; Department of Physics and Astronomy, Ohio University, Athens, Ohio 45701, USA. E-mail: sp315503@ohio.edu

†Department of Physics and Astronomy, Ohio University, Athens, Ohio 45701, USA. E-mail: prakash@harsha.phy.ohiou.edu

of two-body interactions. The two-body interaction chosen for this study — the delta shell interaction — allows us study the roles of long scattering lengths, finite-range corrections, and resonances. Our study is mainly pedagogical, but several universal characteristics are revealed. Our results stress the need to perform analyses beyond what we offer here, particularly with regard to contributions from higher than two-body interactions.

Gottfried's treatment [4, 5] of the quantum mechanics of two particles interacting through a delta-shell potential $V(r) = -v\,\delta(r-R)$, where v is the strength and R is the range, forms the basis of this work in which the transport properties of a dilute gas of non-relativistic particles are calculated for densities n for which the average inter-particle distance $d \sim n^{-1/3} \gg R$. The delta-shell potential is particularly intriguing as the scattering length a_{sl} and the effective range r_0 can be tuned at will, including the interesting case of $a_{sl} \to \infty$ that is accessible in current atomic physics experiments (see, for e.g. Kinast [2]). The delta-shell potential also allows us to understand the transport characteristics of nuclear systems in which the neutron-proton and neutron-neutron scattering lengths ($a_{sl} \simeq -23.8$ fm and -18.5 fm, respectively [6]) are much longer than the few fm ranges of strong interactions. In addition, the delta-shell potential admits resonances.

For small departures of the distribution function from its equilibrium, the net flows of mass, energy and momentum are characterized by the coefficients of diffusion, thermal conductivity and shear viscosity, respectively. Here we calculate these coefficients in first and second order of deviations (from the equilibrium distribution function) using the Chapman–Enskog method [7]. In particular, we examine these transport properties for varying physical situations including the case of an infinite scattering length, i.e. the unitary limit, and when resonances are present.

2. Two-body Interaction

We begin by collecting the two-particle quantum mechanical inputs [4, 5, 8] and cast them in forms suitable for the calculation of the transport coefficients. The Hamiltonian for two particles interacting via the delta-shell potential is

$$\hat{H} = -\hbar^2 \frac{\Delta}{2\mu} - v\,\delta(r-R), \tag{1}$$

where μ denotes the reduced mass, $\Delta = \frac{1}{r^2}\frac{\partial}{\partial r} r^2 \frac{\partial}{\partial r} - \frac{\hat{L}^2}{r^2}$ is the Laplacian in spherical coordinates ($\hat{L}$ is the orbital angular momentum operator),

r is the separation distance, v and R are the strength and range parameters of the potential, respectively. In terms of the dimensionless variable $\rho = kr$ (k is the wave number), $E = \frac{\hbar^2 k^2}{2\mu}$, and l the orbital quantum number, the general solution of Shrödinger equation is given by the spherical Bessel functions $j_l(\rho)$ and $n_l(\rho)$:

$$\psi(\rho) \equiv u(\rho)/\rho = A_l j_l(\rho) + B_l n_l(\rho). \tag{2}$$

The phase shifts δ_l are obtained through [4]

$$\tan(\delta_l) = \frac{g\, x\, j_l^2(x)}{1 + g\, x\, j_l(x)\, n_l(x)}, \tag{3}$$

where $x = kR$, and the single dimensionless parameter

$$g = 2\mu v\, R/\hbar^2 \tag{4}$$

controls *all* physical outcomes. At low energies ($l = 0$), the scattering length a_{sl}, the effective range r_0 and the shape parameter P, are defined through [9]:

$$k \cot(\delta_0) = -\frac{1}{a_{sl}} + r_0 \frac{k^2}{2} - P\, r_0^3 k^4 + O(k^6),$$

$$a_{sl} = \frac{Rg}{g-1}, \quad r_0 = \frac{2R}{3}\left(1 + \frac{1}{g}\right) \quad \text{and} \quad P = -\frac{3}{40}\frac{g^2(3+g)}{(1+g)^3}.$$

Note that as $g \to 1$, $a_{sl} \to \infty$, but $r_0/R \to 4/3$ and $P \to 3/80$. For $l > 0$, Newton's generalization of scattering lengths and effective range parameters reads as [9]

$$k^{2l+1} \cot(\delta_l) = -\frac{1}{a_{sl}^{(l)}} + r_0^{(l)} \frac{k^2}{2} + O(k^4). \tag{5}$$

For the delta-shell potential, the $l \geq 0$ results are

$$\frac{a_{sl}^{(l)}}{R^{2l+1}} = \frac{(2l+1)}{((2l+1)!!)^2}\left(\frac{g}{g-(2l+1)}\right), \tag{6}$$

$$\frac{r_0^{(l)}}{R^{1-2l}} = \frac{2((2l+1)!!)^2}{(2l+3)(2l-1)}\left(\frac{2l-1}{g} - 1\right). \tag{7}$$

The scattering lengths in Eq. (6) become very large for values of $g \to 2l+1$. In Fig. 1, we show a_{sl} and r_0 as functions of the strength parameter g for

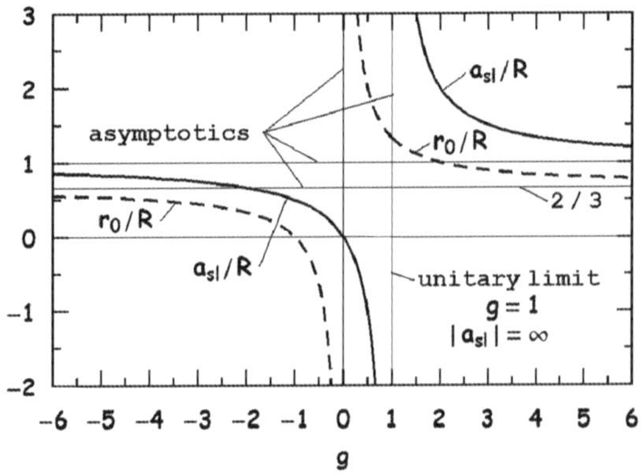

Fig. 1. Scattering lengths and effective ranges as functions of the strength parameter g in Eq. (4).

$l = 0$. The diverging tendencies of of $|a_{sl}|$ as $g \to 1$ and $|r_0|$ as $g \to 0$ are clearly evident. The case of $g \to -\infty$ corresponds to the hard-sphere case.

3. Transport Coefficients

We turn now to calculate the coefficients of diffusion, thermal conductivity and shear viscosity, based on the Chapman–Enskog approach [7, 10]. The central ingredient is the transport cross-section of order n:

$$\phi^{(n)} = 2\pi \int_{-1}^{+1} d\cos\theta (1 - \cos^n \theta) \left. \frac{d\sigma(k,\theta)}{d\Omega} \right|_{c.m.}, \tag{8}$$

where the scattering angle θ and the collisional differential cross-section $\left. \frac{d\sigma(k,\theta)}{d\Omega} \right|_{c.m.}$ are calculated in the center of mass reference frame of the two colliding particles. For indistinguishable particles, an expansion of the cross-section in partial waves $\sum_l' (2l+1)(e^{i2\delta_l} - 1)P_l(\cos\theta)$ and the orthogonality of the Legendre polynomials P_l renders the integral above to the infinite sums

$$q^{(1)} \equiv \frac{\phi^{(1)}}{4\pi R^2} = \frac{2}{x^2} \sum_l{}' (2l+1) \sin^2(\delta_l),$$

$$q^{(2)} \equiv \frac{\phi^{(2)}}{4\pi R^2} = \frac{2}{x^2} \sum_l{}' \frac{(l+1)(l+2)}{(2l+3)} \sin^2(\delta_{l+2} - \delta_l). \tag{9}$$

The prime on the summation sign indicates the use of even l for Bosons and odd l for Fermions. In Eq. (9), the low energy hard-sphere cross section

($g \to -\infty$) $4\pi R^2$ has been used to render the transport cross sections dimensionless. If the particles possess spin s, then the properly symmetrized forms are:

$$q^{(n)}_{(s)} = \frac{s+1}{2s+1}\, q^{(n)}_{\text{Bose}} + \frac{s}{2s+1}\, q^{(n)}_{\text{Fermi}}, \quad \text{for integer } s,$$

$$q^{(n)}_{(s)} = \frac{s+1}{2s+1}\, q^{(n)}_{\text{Fermi}} + \frac{s}{2s+1}\, q^{(n)}_{\text{Bose}}, \quad \text{for half-integer } s.$$

Here, we will present results for the case of spin-1/2 particles only. The transport coefficients are given in terms of the transport integrals

$$\omega^{(n,t)}_{\alpha}(T) \equiv \int_0^\infty d\gamma\, e^{-\alpha\gamma^2} \gamma^{2t+3} q^{(n)}(x), \tag{10}$$

where $\gamma = \frac{\hbar k}{\sqrt{2\mu k_B T}} = \frac{x}{\sqrt{2\pi}}\left(\frac{\lambda(T)}{R}\right)$ with the thermal de-Broglie wavelength $\lambda = (2\pi\hbar^2/m\, k_B T)^{1/2}$.

In what follows, the coefficients of self diffusion $\mathscr{D}$, shear viscosity η, and thermal conductivity κ are normalized to the corresponding hard-sphere-like values

$$\tilde{\mathscr{D}} = \frac{3\sqrt{2}}{32}\frac{\hbar}{mnR^3}, \quad \tilde{\eta} = \frac{5\sqrt{2}}{32}\frac{\hbar}{R^3}, \quad \text{and} \quad \tilde{\kappa} = \frac{75}{64\sqrt{2}}\frac{\hbar k_B}{mR^3}. \tag{11}$$

In the first order of deviations from the equilibrium distribution function, the transport coefficients are

$$\begin{aligned} \frac{[\mathscr{D}]_1}{\tilde{\mathscr{D}}} &= \left(\frac{R}{\lambda(T)}\right)\frac{1}{\omega_1^{(1,1)}(T)}, \\ \frac{[\eta]_1}{\tilde{\eta}} &= \frac{[\kappa]_1}{\tilde{\kappa}} = \left(\frac{R}{\lambda(T)}\right)\frac{1}{\omega_1^{(2,2)}(T)}, \end{aligned} \tag{12}$$

Equation (12) shows clearly that if $\omega_1^{(2,2)}$ is T-independent (as for hard-spheres with a constant cross section), the shear viscosity exhibits a $T^{1/2}$ dependence which arises solely from its inverse dependence with $\lambda(T)$. For energy-dependent cross sections, however, the temperature dependence of the viscosity is sensitive also to the temperature dependence of the omega-integral.

The second order results can be cast as

$$\frac{[\mathscr{C}]_2}{[\mathscr{C}]_1} = (1 + \delta_{\mathscr{C}}(T))(1 \pm n\lambda^3 \epsilon_{\mathscr{C}}(T)), \tag{13}$$

where $\mathscr{C}$ is $\mathscr{D}$ or η or κ, and the $\pm$ refers to Bose (+) and Fermi (−) statistics. For example,

$$\delta_\eta \equiv \frac{3\left(7\omega_1^{(2,2)} - 2\omega_1^{(2,3)}\right)^2}{2\left(\omega_1^{(2,2)}\left(77\omega_1^{(2,2)} + 6\omega_1^{(2,4)}\right) - 6\left(\omega_1^{(2,3)}\right)^2\right)},$$

$$\epsilon_\eta \equiv 2^{-7/2}\left[4 - \frac{128}{3^{3/2}}\frac{\omega_{4/3}^{(2,2)}}{\omega_1^{(2,2)}}\right]. \tag{14}$$

4. Asymptotic Behavior

It is useful to define a characteristic temperature

$$\tilde{T} \equiv \frac{2\pi\hbar^2}{k_B m R^2} \quad \text{or} \quad \frac{T}{\tilde{T}} = \left(\frac{R}{\lambda}\right)^2 \tag{15}$$

in terms of which limiting forms of the transport coefficients can be studied. For spin 1/2 statistics, Fig. 2 shows $[\mathscr{D}]_1$ from Eq. (12) normalized to $\tilde{\mathscr{D}}$ from Eq. (11). As expected, the diffusion coefficient grows steadily with temperature. For $g \neq 1, 3$, and for $T \ll \tilde{T}$, we find the asymptotic behavior

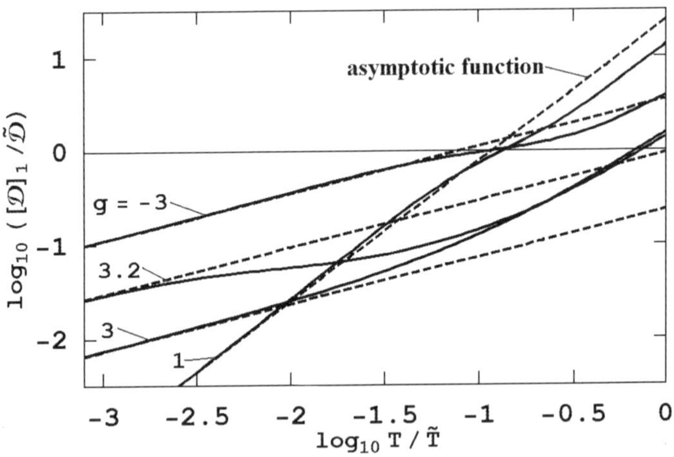

Fig. 2. Normalized (to $\tilde{\mathscr{D}}$ in Eq. (11)) diffusion coefficient as a function of normalized (to $\tilde{T}$ in Eq. (15)) temperature for various values of the strength parameter g in Eq. (4). Solid curves show the diffusion coefficient in the first approximation, $\mathcal{D}_1$ in Eq. (12), and the dashed lines are its asymptotic trends for $T \ll \tilde{T}$ in Eq. (15).

(the dashed lines in Fig. 2)

$$\frac{\mathscr{D}}{\tilde{\mathscr{D}}} \to 2\left(\frac{1-g}{g}\right)^2 \sqrt{\frac{T}{\tilde{T}}}. \tag{16}$$

The cases $g \to 1, 3$ require special consideration. In these cases, the asymptotic behavior for $T \ll \tilde{T}$ is obtained from a series expansion in $x = kR$ of the transport cross sections $q^{(n)}(x)$ in Eq. (9), and subsequent integrations of Eq. (10). We find the limiting forms:

$$\frac{\mathscr{D}}{\tilde{\mathscr{D}}} \to \begin{cases} 8\pi(T/\tilde{T})^{3/2} & \text{for } g = 1 \\ \dfrac{4}{17}(T/\tilde{T})^{1/2} & \text{for } g = 3. \end{cases} \tag{17}$$

The asymptotic behavior (for $T \ll \tilde{T}$) of the shear viscosity $[\eta]_1$ is similar to that of the diffusion coefficient:

$$\frac{\eta}{\tilde{\eta}} \to \begin{cases} \left(\dfrac{1-g}{g}\right)^2 (T/\tilde{T})^{1/2} & \text{for } g \neq 1, 3 \\ 6\pi(T/\tilde{T})^{3/2} & \text{for } g = 1 \\ \dfrac{1}{6}(T/\tilde{T})^{1/2} & \text{for } g = 3. \end{cases} \tag{18}$$

It is interesting that even at the two-body level, the coefficients of diffusion, thermal conductivity and viscosity acquire a significantly larger temperature dependence as the scattering length $a_{sl} \to \infty$ ($g \to 1$).

Figure 3 shows the normalized viscosity coefficient as a function of $\frac{R}{a_{sl}} = \frac{g-1}{g}$ for $T < \tilde{T}$. Enhanced cross sections at resonances produce significant drops in the viscosity as $g \to 2l + 1$. The widths of the dips in viscosity decrease with increasing values of l. For $T \leq \tilde{T}$, the dips become less prominent and disappear for $T \geq \tilde{T}$. For the densities and temperatures considered, the contribution from the second approximation is small as $|\delta_\eta| < 0.15$.

In Fig. 4, the ratio of the coefficients of diffusion (times mn) and viscosity are shown as functions of R/a_{sl} for $T \leq \tilde{T}$. As expected, the largest variations in this ratio occur as $g \to 2l + 1$, that is, as resonances are approached. For $T \ll \tilde{T}$, the ratio approaches the asymptotic value 6/5 away from resonances and 4/5 for the S-wave resonance (a point). With increasing T, resonances become progressively broader with diminishing strengths; for $T \approx \tilde{T}$ resonances disappear. For comparison, this figure also includes results for hard spheres.

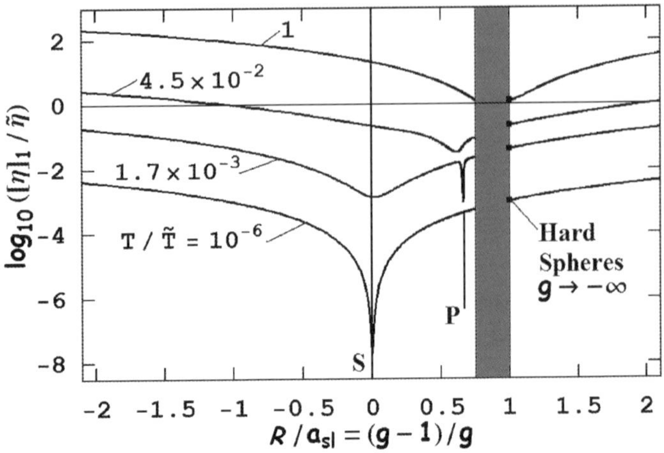

Fig. 3. The normalized viscosity coefficient in Eq. (12) as a function of the inverse scattering length. Effects due to resonances associated with the partial waves $l = 0$ and 1 are indicated by the letters S and P, respectively. In the vertical shaded region, a large number of partial waves are required to obtain convergent results.

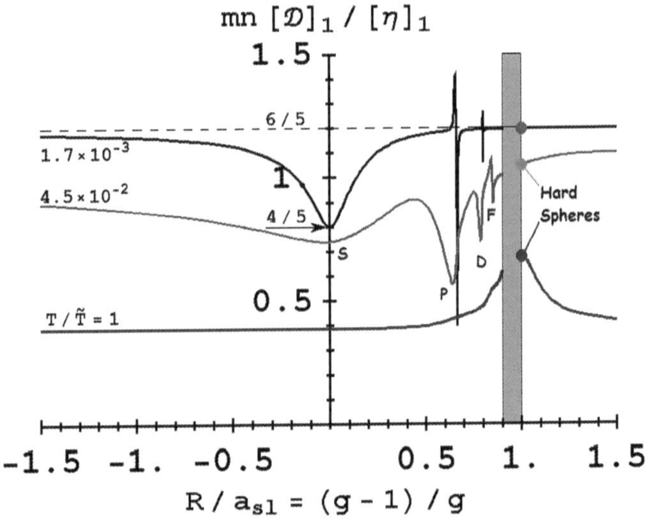

Fig. 4. Ratio of diffusion (times mn) to shear viscosity versus inverse scattering length. In the vertical shaded region, a large number of partial waves are required to obtain convergent results. The hard-sphere results ($g \to -\infty$) are shown by solid circles.

Table 1. First order coefficients of diffusion (times mn), shear viscosity, and their ratios for $T \ll \tilde{T}$ for select strength parameters g. The unitary limit ($g = 1$) result for η was obtained earlier in Massignan [11].

g	$mn\mathcal{D}$	η	$mn\mathcal{D}/\eta$
1	$\frac{3\sqrt{2}\pi}{4}\frac{\hbar}{\lambda^3}$	$\frac{15\sqrt{2}\pi}{16}\frac{\hbar}{\lambda^3}$	$\frac{4}{5} = 0.80$
3	$\frac{3}{68\sqrt{2}}\frac{\hbar}{\lambda R^2}$	$\frac{5}{96\sqrt{2}}\frac{\hbar}{\lambda R^2}$	$\frac{72}{85} = 0.85$
$\neq 1, 3$	$\frac{3\sqrt{2}}{16}\frac{\hbar}{\lambda a_{sl}^2}$	$\frac{5\sqrt{2}}{32}\frac{\hbar}{\lambda a_{sl}^2}$	$\frac{6}{5} = 1.20$

The coefficient of viscosity (as also mn times the coefficient of diffusion) has the dimension of action ($\hbar$) per unit volume. The manner in which the effective physical volume $\mathcal{V}$ changes as the strength parameter g is varied is illuminating as our results for $T \ll \tilde{T}$ in Table 1 shows. In the unitary limit ($g = 1$), the relevant volume is $\mathcal{V} \propto \lambda^3$. For $g = 3$, $\mathcal{V} \propto \lambda R^2$ (independent of a_{sl}), and for $g \neq 1, 3$, $\mathcal{V} \propto \lambda a_{sl}^2$ (independent of R).

5. Ratio of Shear Viscosity to Entropy Density

Recently, a lower limit to the ratio of shear viscosity to entropy density is being sought [13] with results even lower than $(4\pi)^{-1}(\hbar/k_B)$ first proposed by Kovtun [12]. We therefore examine $[\eta]_{1,2}/s$ using the entropy density

$$s = (5/2 - \ln(n\lambda^3) + \delta s(T)\, nR^3)nk_B,$$
$$\delta s(T) = \left(\frac{a_2(T)}{2} - T\frac{da_2(T)}{dT}\right)\left(\frac{\lambda}{R}\right)^3, \tag{19}$$

which includes the second virial correction [10] to the ideal gas entropy density. The second virial coefficient [10]

$$a_2(T) = \mp 2^{-5/2} - 2^{3/2}{\sum_l}'(2l+1)$$
$$\times\left(e^{-E_l/(k_BT)} + \frac{1}{\pi}\int_0^\infty dx\,\frac{\partial\delta_l}{\partial x}e^{-\xi(T)x^2}\right), \tag{20}$$

where the prime indicates summation over even l's for Bosons ($-$) and odd l's for Fermions ($+$), E_l is the energy of the bound state with angular

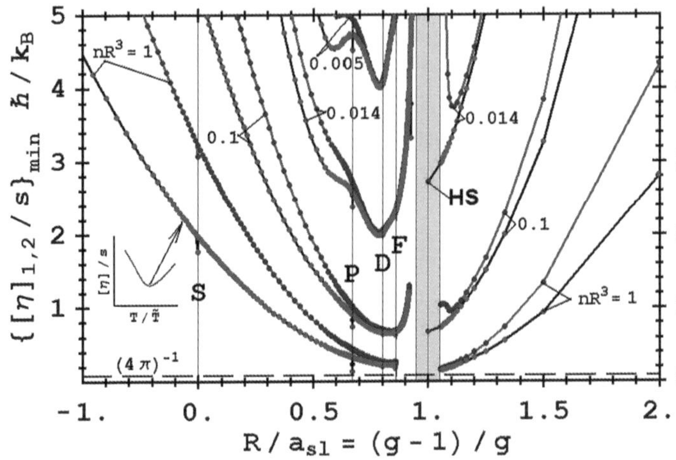

Fig. 5. The insert shows the occurrence of a minimum in the ratio viscosity to entropy density versus T. For each nR^3, this minimum value in the first (lower curve) and second (upper curve) order calculations of shear viscosity is plotted versus R/a_{sl} for the indicated nR^3. In the vertical shaded region, a large number partial waves are needed. Vertical lines with letters (S, P, D, and F, respectively) indicate resonances associated with the partial waves $l = 0, 1, 2$ and 3. The symbol HS denotes hard-spheres result for which $g \to -\infty$. The horizontal dashed line shows $1/(4\pi)$, the conjectured lower bound in Ref. [10].

momentum l and $\xi(T) = (\lambda/R)^2/(2\pi)$. The insert in Fig. 5 shows a characteristic minimum of η/s versus T for fixed dilution nR^3 and strength g. Values of $[\eta]_{1,2}/s$ at the minimum are also shown functions of R/a_{sl} for several nR^3. The lower (upper) curve for each nR^3 corresponds to the first (second) order calculations of η. The large role of improved estimates of η on the ratio η/s is noticeable. The case of $nR^3 = 1$ possibly requires an adequate treatment of many-body effects not considered here. We can, however, conclude that in the dilute gas limit η/s for the delta-shell gas remains above $(1/4\pi)\hbar/k_B$.

Our analysis here of the transport coefficients of particles subject to a delta-shell potential has been devoted to the dilute gas (non-degenerate) limit, in which two-particle interactions dominate, but with scattering lengths that can take various values including infinity. Even at the two-body level considered, a rich structure in the temperature dependence and the effective physical volume responsible for the overall behavior of the transport coefficients are evident. The role of resonances in reducing the transport coefficients are amply delineated. Matching our results to those of intermediate and extreme degeneracies [11, 14] which highlight the additional roles of superfluidty and superconductivity reveals the extent to which many-body effects play a crucial role.

Personal Note (Madappa Prakash)

This contribution to Gerry's Festschrift is in collaboration with my former graduate student Sergey Postnikov, who obtained his Ph.D. degree in 2009, and who would have impressed Gerry with his originality. For Gerry, it is important that researchers be passionate about physics. I know that Sergey is in the category of people of whom Gerry would have approved. Gerry is very fond of schematic models that captures the essential physics. Our study here is an offer to the spirit in which the famous and classic Brown and Bolsterli paper [15] on "Dipole State in Nuclei" was written. I always chide him with the remark "You should strive for that excellence". He always retorts "What do you mean? I always do".

Gerry cares a lot about both undergraduate and graduate students. During the years 1981–2005, I have known of many instances in which he helped students to achieve their career goals, helping them with guidance in science, as well as with financial support (from his own purse when funds could not be arranged officially) and with emotional backing when needed.

I first saw his uncanny abilities with students during a course he taught with Kevin Bedell in 1982–83 on the applications of Landau Fermi-liquid theory. Gerry gave out many ideas, as a result of which, even before the end of the course, there were three articles written for Physical Review Letters, which were quickly accepted for publication. I was impressed.

To this day, I wish I had Gerry's ways with students. Gerry and I taught many courses together at Stony Brook. When we got students' evaluations, I usually got a numerical score that was better than Gerry's. However, students invariably said "Gerry is a fun teacher", a compliment that I never got. When we talked about our teaching skills, he said "how come I don't get your scores?" My reply was "how come I don't get your compliments?"

As Gerry is recovering from his illness, I am now supervising graduate student Constantinos Constantinou, who began his research with Gerry. I hope I do as well for Constantinos as Gerry would have done.

Acknowledgments

This research was supported by the Department of Energy under the grant DE-FG02-93ER40756.

References

[1] T. Schafer and D. Teaney, *Rept. Prog. Phys.* **72**, 126001 (2009).
[2] J. Kinast, A. Turlapov and J.E. Thomas, *Phys. Rev. Lett.* **94**, 170404 (2005).
[3] U.W. Heinz, J.S. Moreland and H. Song, *Phys. Rev.* **C80**, 061901 (2009).
[4] K. Gottfried, *Quantum Mechanics, Vol. I: Fundamentals* (W.A. Benjamin, Inc., New York, 1966), Chapter 15.
[5] K. Gottfried and Tung-Mow Yau, *Quantum Mechanics, Vol. II* (Springer-Verlag, New York, 2004).
[6] W. von Witsch *et al.*, *Phys. Lett.* **B80**, 187 (1979); J. Deng, A. Siepe and W. von Witsch, *Phys. Rev.* **C66**, 047001 (2002).
[7] S. Chapman and T.G. Cowling, *The Mathematical Theory of Non-Uniform Gases* (Cambridge Univ. Press, Great Britain, 1970), Chapters 9, 10 and 17.
[8] J.-P. Antoine *et al.*, *J. Phys. A: Math. Gen.* **20**, 3687 (1987).
[9] R.G. Newton, *Scattering Theory of Waves and Particles* (McGRAW-HILL, New York, 1966), p. 311.
[10] J.O. Hirschfelder *et al.*, *Molecular Theory of Gases and Liquids* (John Wiley & Sons, Inc., New York, 1967), Chapters 6 and 10.
[11] P. Massignaan, G.M. Brun and H. Smith, *Phys. Rev.* **A71**, 033607 (2005); G.M. Brun and H. Smith, *Phys. Rev.* **A72**, 043605 (2005), *Phys. Rev.* **A75**, 043612 (2007).
[12] P. Kovtun, D.T. Son and A.O. Starinets, *Phys. Rev. Lett.* **94**, 111601 (2005).
[13] Y. Kats and P. Petrov, *JHEP* **0901**, 044 (2009); M. Brigante *et al.*, *Phys. Rev. Lett.* **100**, 191601 (2008); A. Buchel *et al.*, *JHEP* **03**, 084 (2009).
[14] G. Rupak and T. Schäfer, *Phys. Rev.* **A76**, 053607 (2007).
[15] G.E. Brown and M. Bolsterli, *Phys. Rev. Lett.* **3**, 472 (1959).

Chapter 17

Subtle is the Manifestation of Chiral Symmetry in Nuclei and Dense Nuclear Matter

Mannque Rho*

The history of how chiral symmetry has entered in nuclear physics, in which Gerry Brown and I have participated from 1970 up to today, is described from my personal viewpoint. The route of development we have traversed together goes from meson exchange currents, to "little chiral bag," to chiral effective field theory, to "Brown-Rho scaling" and then to dense matter and chiral phase transition. It has been a great fun and exciting, some right and some wrong in what we have done together, but none that can be dismissed as "not even wrong." We have found all along that whatever signal there may be for the manifestation of chiral symmetry in nuclear medium, be it at low density in meson exchange currents or at high density approaching the chiral phase transition, is rich in its intricacy and subtlety.

1. Introduction

Gerry Brown and I started to ask how chiral symmetry manifests itself in nuclear structure and dense nucleonic matter way back in late 1960's and have been chasing after it ever since. At that early time, quantum chromodynamics was not yet around and hence chiral symmetry was not considered in terms of nearly zero mass quarks. What was available then was a set of current-algebra relations associated with pions as Nambu-Goldstone bosons, and it was agreed that pions must play an important role in nuclear interactions. Gerry, at the time at Nordita in Copenhagen, was looking at the problem in connection with nuclear forces [1] and I, freshly tenured at Saclay after a year of post-doc there, was more interested in how chiral

*Institut de Physique Théorique, CEA Saclay, 91191 Gif-sur-Yvette Cédex, France & Department of Physics, Hanyang University, Seoul 133-791, Korea.

symmetry could be exploited in nuclear responses to external fields, such as electromagnetic and weak.

In this note I would like to describe the line of development I followed, some in collaboration with him and some with others but strongly influenced by Gerry's thinking.

2. How It All Came About

With Marc Chemtob, then a thesis student at Saclay working on exchange currents in nuclei, we were asking how the soft-pion theorems associated with the current algebras worked out by particle theorists could be used to constrain the pion-exchange currents which we thought would dominate the processes at not too large momentum transfers.

We were motivated to address this question from conversations with Benjamin Lee who was at that time on his sabbatical stay at Institut des Hautes Études Scientifiques at Bures-sur-Yvette, which is located very near Saclay, and was on his celebrated work on the renormalization of the sigma model — a prelude to the renomalization of Yang-Mills gauge theory. These stimulating and fruitful conversations with Ben Lee led us to write in 1969-70 a paper where exchange currents were derived using S-matrix formalism in which the soft-pion theorems were invoked in calculating the longest-range exchange currents [2].

It turned out that the result obtained in [2] was instrumental to Gerry and Dan Olof Riska for their pinning down, to a few% accuracy, the long-standing exchange-current contribution in the thermal $p + n \to d + \gamma$ process [3]. This calculation was later confirmed by chiral perturbation theory, an effective field theory of QCD, to which I will return below. This set in motion my long collaboration with Gerry on how chiral symmetry, later identified with an essential ingredient of QCD, figured in not just organizing corrections to nuclear matrix elements of the electroweak currents, but more importantly in playing an indispensable role in the basic structure of the nucleon itself and its vacuum changes in hot and dense background. Along the way, we have come up with the concept of "Little Chiral Bag," followed by Brown-Rho scaling and their place in nuclear chiral perturbation theory developed in 1990's. Many diverse phenomena observed in nuclei have been found naturally to fit in the scheme, influencing our later work on dense (and to some extent hot) matter.

When there is a novel concept involved, one would naturally like to check it by singling out experimentally a distinctive signal — a "smoking-gun" — for the effect. A naive thinking would be that if the proton mass of ~1 GeV were generated (mostly if not entirely) by spontaneous breaking of chiral symmetry as widely believed, with the quark mass being near zero, then one should be able to "unbreak" the symmetry in the laboratory by heating the system to high temperature or compressing it to high density at which the symmetry breakdown is supposed to take place. Such experiments have indeed been performed. What has come out of such searches turns out, however, to be unclear, both unsettling and intriguing. Indeed experiments done at the SPS/CERN and the RHIC/Brookhaven in heavy ion collisions have failed so far to isolate a "smoking-gun" signal for the effect. At first thought, this seems disappointing. At second thought, however, it is intriguing that chiral symmetry — which must be broken in nuclear systems as indicated, though indirectly, in a variety of nuclear processes — could hide from detection. In this note contributing to Gerry's 85th birthday celebration, I would like to describe how the chiral symmetry that shows up unambiguously in exchange currents — and also in various other ways — can hide from view in those experiments specifically looking for a direct evidence. It is significant that the same old pions that we started to take seriously in what we were doing a long time ago turn out to take up a surprising new role today on par with quarks and gluons in a modern effective theory of QCD [4]. This development seems to resuscitate the old idea of "BR scaling" in the original form Gerry and I proposed in 1991 although there are recent indications that this form should be modified at high density.

3. From Soft-Pion Theorems to Chiral Perturbation Theory

One of the pressing questions in nuclear physics in 1960's was how meson degrees of freedom figured in nuclear interactions. While it was generally accepted that the pion entered in giving the nuclear forces, it was not very well understood in what way. The question then was how one can "see" in a form of direct signal its role in nuclear dynamics. To put the question more specifically, consider first two-nucleon systems in strong interactions below the pion production threshold, the pion being the lowest-mass meson with mass $m_\pi \approx 140\,\text{MeV}$. Given a well-determined two-nucleon potential V_2, two-nucleon properties, such as scattering, binding energy etc. can be computed more or less accurately, with no explicit intervention of other degrees of freedom than the nucleons. However, the situation is changed when an external field is applied to the system, which is given by the matrix element

$$M = \langle f|\mathcal{M}|i\rangle \tag{1}$$

where i and f are, respectively, the initial and final wave functions of the states involved and $\mathcal{M}$ is the appropriate probe operator, weak or EM in our case. Expressed in terms of the nucleon variables, the operator $\mathcal{M}$ will consist of the sum of one body operators $\mathcal{M}_1+\mathcal{M}_2$ and a two-body operator $\mathcal{M}_{12}$. "Accurate" wave functions are given if the potential V_2 is accurately known. Assuming symmetries involved in the process, a fully "realistic" potential can be determined provided there are enough experimental data points available to fix the parameters figuring in the potential. Such a potential is also referred to as "sophisticated potential".[a]

Now suppose the matrix element of the operator $\mathcal{M}$ between single nucleons is known from on-shell experiments. This gives the single-particle operator. Then one may take as the first approximation for the two-nucleon system the sum $\mathcal{M}_1+\mathcal{M}_2$. In this case, M will be given entirely by on-shell nucleon quantities. This leads to what is called "impulse approximation." Given accurate wave functions, *any* deviation from the experimental value must then be attributed to a two-body contribution that comes from other degrees of freedom than the nucleons that figure in the interactions. In terms of the Yukawa meson-exchange picture for nuclear forces, this corresponds to the "meson-exchange-current" contribution to M.

The ~10% deviation of the impulse-approximation calculation from the experimental cross section in the radiative capture

$$n + p \to d + \gamma \tag{2}$$

clearly indicated the presence of meson degrees of freedom contributing to the current. How to calculate this correction had been a long-standing problem until early 1970's.

Given a reliable field theory technique to calculate the potential that fits the scattering data, the same theory should enable one to compute the exchange currents in consistency with the potential. But there was no such

[a]By "realistic" or "accurate" or "sophisticated" two-nucleon potential, one means a potential that is fit to a large number of scattering data up to lab energy $\lesssim 300$ MeV with χ^2/datum $\lesssim 1$. The present status of chiral perturbation calculation of the NN potential at the highest chiral order reached — which is next-to-next-to-next leading order or N^3LO — is that one has to go still higher in chiral order to reach the accuracy of the most "sophisticated" potential. It is unlikely that such a calculation is fully doable within the foreseeable future.

theory available then. The wave functions are what Nature is supposed to be giving to the extent that they are extracted from experiments but how to compute the exchange currents compatible with the wave functions was unknown. In [2], it was proposed that the currents associated with the operator $\mathcal{M}$ be calculated via an S-matrix method, involving necessarily on-shell quantities, constrained by the assumed symmetries of the current. For the EM current, there is the vector current conservation, and for the weak current, both the vector-current and axial-vector-current conservations. Such conservation laws applied to the total current matrix elements give certain relations between two-body currents and two-body potentials. A well-known example is the Siegert theorem for the vector current. One can write down an analogous relation for the axial current, but here the exchange current is not directly related to the NN potential.

The question addressed in [2] was how best to write down the total current compatible, if not fully consistent in the sense of the modern chiral counting as is done nowadays, with the potential used to get the wave functions. The key point of [2] was that the longest-range exchange current for low-momentum-transfer processes could be given reliably by a pion exchange and that the pion exchange be constrained by the low-energy theorems. Indeed the exchange currents so constructed — with an additional correction from the Δ resonance which turns out to be higher order in the chiral counting found later — were the source of the fit to the process (2) obtained by Riska and Brown [3]. What was surprising at the time was that these exchange currents derived in the soft-pion limit worked well also at momentum transfers as high as $q^2 \sim 15$ fm^{-2} in the electrodisintegration of the deuteron [5]

$$e + d \to e' + p + n. \tag{3}$$

It was not clear then why shorter-distance contributions, namely, the exchange of heavier mesons for which no systematic tools were available then, were not needed. The understanding of this result required an input from chiral perturbation theory. This story was described already in 1981 in a comment article by Gerry and me [6].

As far as I know, the first application of the so-called chiral perturbation theory (ChPT for short) to nuclear processes was made in a 1981 lecture in Erice [7].[b] What motivated me to resort to chiral perturbation theory was

[b]At this school, I suggested — and discussed how — to implement chiral counting in computing pion-exchange currents. It was, however, incomplete then but later completed

the article by Weinberg where his "folk theorem" on effective field theory was proposed [8]. The idea was that the soft-pion theorems were just the leading order terms in a field theory that allows one to calculate to all orders of low-energy processes involving pions in chiral perturbation theory. This meant that one should be able to compute systematically the full exchange currents going beyond the soft-pion theorems as Weinberg did for pion interactions. Nucleons were involved in nuclear processes in addition to the pions considered by Weinberg. However, if one were to confine to two-nucleon processes, then one simply could take over Weinberg's counting rule — with some tractable additional terms such as Δ — to the exchange current problem, as I did in the Erice lecture. The counting associated with the nucleons (internal and external) as well as others did not affect — apart from an overall factor — the power counting to the order discussed then. Later Weinberg wrote down the complete counting rule for nuclear processes that includes both nucleons and pions [9], which subsequently led to a more precise formulation of chiral perturbation approach to the exchange currents [10]. The power of chiral perturbation treatment was that it explained in a simple way how the soft-pion terms dominate the corrections in certain processes, while allowing to calculate higher-order terms.

The modern state-of-art calculations following the above strategy account for the process (2) within 1% [11] and (3) within a few % accuracy.

The same current algebra argument made an equally clear-cut prediction on axial charge transitions in nuclei [12]. The pertinent observation was that the nuclear beta-decay transition from nucleus A to A'

$$A(J^{\pm}) \to A'(J^{\mp}) + e^{+}(e^{-}) + \nu(\bar{\nu}), \quad \Delta T = 1 \tag{4}$$

where J and T are, respectively, the angular momentum and isospin and the superscript $\pm$ stands for the parity, is predominantly given by the time component of the axial current and that there the impulse approximation contribution is strongly suppressed [13]. Chiral perturbation theory straightforwardly explains why the two-body exchange axial-charge operator is also dominated by the soft-pion terms. This observation turns

as described below. Ulf Meissner was one of the graduate students in that Erice school. He probably does not recall that lecture; he has since then made significant contributions to the field of applying chiral perturbation theory to the physics of pion-nucleon systems and nuclear effective field theory.

out to have a nontrivial consequence on how chiral symmetry manifests itself in dense medium [14].

Looked from the present vantage point of effective field theory of nuclear physics, what I have described here is not strictly speaking consistent. The state-of-art wave functions are obtained with a sophisticated phenomenological potential (SPP) that we may consider to be of high order in chiral perturbation theory, say, ν^p, with $p \gg 1$. What this means is that the higher one goes in chiral counting in ChPT, the closer the potential comes closer to the SPP. However, the current is computed at some finite order ν^q with $q \ll p$. This necessarily entails a mismatch in chiral counting between the potential and the current. The mismatch is a practical one in that it can be avoided if one can go to high enough an order in chiral expansion. It is not the sort of mismatch that is present in the Weinberg scheme in which there is a separation between the "irreducible" graphs — calculated to some finite order — defining the potential and the "reducible" graphs that enter in solving the Schrödinger equation or Lippmann-Schwinger equation with the given potential entailing an infinite order [15]. Our philosophy here is that this strict order-by-order adherence to full consistency in chiral counting is not necessary for attaining the accuracy desired, given that the state-of-art wave functions are available. The power of this approach is that it can be applied not only to few-nucleon systems but also to heavy nuclei for which reliable wave functions are available such as in Pb nuclei [13]. This point was stressed in [16].

This strategy has proven to be successful in accurately describing responses to external fields in a variety of other processes. This matter will be described elsewhere in this volume [17], so let me just give a list of what has been accomplished. In addition to the radiative *np* capture and the deuteron electrodisintegration, calculations of unprecedented accuracy have been performed for, e.g., (a) the solar $p+p \to d+e^+ + \nu_e$ [18], (b) the solar *hep* (^{3}He $+p \to ^4$He $+ e^+ + \nu_e$) process [18, 19], (c) the *hen* (^{3}He $+n \to ^4$He $+ \gamma$) process [20, 21], and (d) the muon capture on deuteron and ^{3}He [22].

4. The Little (Chiral) Bag and the Cheshire Cat

In the development described so far in addressing nuclear interactions, there was no reference to the microscopic degrees of freedom of QCD, namely, quarks and gluons. The natural question raised is how the microscopic picture, QCD, maps to the macroscopic picture, nuclear EFT. This question was addressed in 1978.

In Fall 1978, I arrived at Stony Brook for an academic year. It was the second in my regular one-year visits to Stony Brook to work with Gerry — which lasted for many years afterwards. Returning from his visit to Caltech, Gerry suggested to look at how one could make the "bags" in which three quarks are confined smaller in nuclear environment. In Caltech, Gerry had conversations with Richard Feynman in which the size of the MIT bag was discussed. At the time, the MIT bag model was very popular. In this model, hadrons are described as three quarks for baryons and a quark and an anti-quark for mesons confined within a bag of size of typically ∼1 fm in radius. To our thinking, this size for the nucleon was too big. There were also indications from high-energy proton-proton scattering as well as experiments on high mass muon pairs for a nucleon with core size of ∼0.2 – 0.3 fm [23]. There are more recent indications of similar size in deep inelastic scattering off nucleon [24]. To us, a nucleon of the bag size ∼1 fm in a heavy nucleus, say, Pb nucleus, was like a "grapefruit" in a salad bowl. There was something basically bizarre about this. The puzzle that we were having was how can the picture of rigid balls of ∼1 fm radius substantially overlapping each other in the bag model in, say, Pb where the average inter-nucleon distance is ∼2.2 fm be reconciled with the highly successful shell model or more specifically for us, the effective field theory description given above where point-like nucleons interact via the exchange of "soft" mesons — which is essentially the old "standard nuclear physics approach" (SNPA for short) anchored on Yukawa theory.

We proposed in that year that the resolution of this puzzle was *in the way chiral symmetry is manifested* in the structure of the nucleon [25]. With only the MIT bags in many-nucleon systems, pions — accepted nowadays as the key ingredient of the spontaneously broken chiral symmetry of QCD — were missing. Pions had to be introduced in consistency with QCD. It was then recognized that pions could be brought in by imposing the conservation of the baryon axial current [26]. What we chose to do was that the pion field be coupled to the quarks at the bag boundary and do not penetrate into the bag. This procedure adopted by Callan, Dashen and Gross [27] was natural in that the interior of the bag was supposed to be in the Wigner-Weyl (Wigner for short) mode whereas the exterior was supposed to be in the Nambu-Goldstone (Goldstone for short) mode. This structure turns out to possess an intricate feature known as "the Cheshire Cat phenomenon." It may be taken as a particular *manifestation of chiral symmetry* in the guise of "quark-baryon continuity." I will have more to say on this below.

The main idea of [25] was as follows. Suppose that the bag were squeezed by an external pressure to a small size. Then the energy density of the confined quarks would increase making the quark condensate $\langle\bar{q}q\rangle$ — which is non-zero in the vacuum — go to zero, as expected in the Wigner phase. With the radius squeezed to a small size, $\sim$0.3 fm, the energy of the quarks would increase accordingly $\propto 1/R$, where R is the bag radius, due to the kinetic energy of the confined quarks. The assumption was that the self-energy effect due to the coupling to the pion outside would provide a large enough attraction to cancel the increase in mass.

Considerations of subtle quantum effects in the chiral bag structure indicated, however, that this picture was not correct in its original form. Apart from the fact that such a strong attractive self-energy effect was not obvious, there was a problem associated with an anomaly — a quantum effect — in the baryon number $U(1)$ current at the boundary which was not taken into account in the "little bag" scenario. While the color is confined within the bag, the baryon charge which is conserved classically is not, at the quantum level. The quantum anomaly generated by an induced axial gauge field arising from the continuity of the axial current at the boundary makes the baryon charge leak out into the outside of the bag. The pion field lodged outside has the mean field of the hedgehog form $\pi = \frac{1}{2}\tau^a\pi^a(\mathbf{r}) \sim \frac{f_\pi}{2}\vec{\tau}\cdot\hat{\mathbf{r}}\cos\theta$ where θ is known as the "chiral angle" and goes from 0 to π as a function of radius. This object has a topology and picks up the leaking baryon charge. It was first shown in [28] that for the chiral angle $\theta = \pi/2$ (dubbed as "magic angle"), the baryon charge is partitioned equally between the inside and outside of the bag, and then proven generally in [29] that the partition takes place for *any* angle. This led to the suggestion that physics should not depend on the bag radius, a phenomenon given the name "Cheshire Cat" [30]. This phenomenon can be rigorously established in (1+1) dimensions thanks to the exact bosonization technique. However, bosonization technique for (3+1) dimensions was — and still is — not available and hence the phenomenon could only be approximate. Highly approximative though it may be, the Cheshire Cat notion has met with success for a variety of nuclear phenomena studied in the model [31]. This picture has reappeared in a holographic dual QCD description of the baryon as described by Ismail Zahed in this volume [32].

The lesson from the Cheshire Cat phenomenon is that for long-wavelength probes, the bag could be shrunk to zero. When the bag is shrunk to zero, the resulting baryon is the skyrmion soliton [33], with the confinement size rendered unphysical.

What happens here illustrates even a subtler way chiral symmetry can be manifested in *nuclear* systems.

A similar analysis showed that the color could also leak out due to the well-known $U(1)_A$ anomaly [34]. In contrast to the baryon charge where the hedgehog pion absorbs the baryon charge, there is nothing outside of the bag that can absorb the leaking color charge. This required that the color-leakage be stopped by fiat, that is, by putting by hand a boundary condition in the Lagrangian. This implied that the color symmetry was broken at the classical level and restored at the quantum level. This procedure in some aspect resembles the Green-Schwarz anomaly cancelation mechanism in string theory. The resulting chiral bag Lagrangian was found to reproduce the Witten-Veneziano formula for the η' mass, i.e., the resolution of the $U(1)_A$ anomaly problem [35].

An interesting modern development is that the Little Bag structure seems to appear in a strong coupling lattice (SCL) QCD calculation of nuclear matter [36],[c] so it appears that somehow the small size core [23–25] refuses to disappear in the Cheshire Cat smile. I will come back to this matter when I describe the instanton structure of the nucleon in holographic dual QCD.

5. B(rown)R(ho) Scaling

In Fall 1990, I was spending a semester at Seoul National University giving a special topics course. Early in my stay at SNU, I received a fax from Gerry that contained a copy of the draft of Weinberg's paper on chiral Lagrangian approach to nuclear forces — published later in [9] — and a summary of his thoughts on Weinberg's approach concerning the role of the ρ meson in nuclear physics. Gerry mentioned in the fax that he was disagreeing with Weinberg on the necessity of having the ρ meson explicitly in the construction of the nuclear forces. In Weinberg's chiral perturbation approach, the ρ meson (and all other massive degrees of freedom than the nucleon) have been integrated out, with their effects, if any, appearing in higher chiral-order terms. Gerry was stressing the need of having the ρ *explicitly* because of what we found in our work on the tensor forces in nuclei where the cancelation between the pion tensor and the

[c]It is interesting — and perhaps significant — that a meson-exchange picture arises in this SCLQCD calculation in a manner resembling the old Heitler-London model of Cutkosky [37].

ρ tensor — which come with an opposite sign [38] — became important. The argument in [38] was that if one assumed that the "intrinsic" hadron masses dropped in increasing density as naive constituent quark model indicated, then the cancelation in the tensor forces would become greater as density increased and Gerry was arguing that such an effect — that depends directly on the ρ degree of freedom — would be hard to capture if the ρ meson were absent in the consideration.

This fax from Gerry led to two notable developments on my side, one with and the other without Gerry. The one without Gerry was that Weinberg's (draft) paper immediately — and almost trivially — indicated how to complete the (incomplete) formulation given in my 1981 Erice lecture of exchange currents in chiral perturbation theory [10]. The other with Gerry was the idea of in-medium scaling based on chiral Lagrangian, i.e. the BR scaling [39]. The former cleared my fuzzy understanding of the role of the soft-pion theorems in certain nuclear electroweak processes and paved the road to my part in the work described in the contribution of Kuniharu Kubodera and myself [17]. The latter put on a more formal ground the argument we developed in [38] for the in-medium weakening of the tensor forces. The special role of vector mesons in nuclear interactions will become more prominent in a modern holographic dual QCD model of the nucleon mentioned below.

The idea behind the BR scaling went as follows.

The Cheshire Cat principle suggested that baryons in nuclear matter could be approximated by skyrmions considered as chiral bags shrunk to a point. Since the skyrmion has well-defined charge (electric, baryonic etc.) distributions, it should be easy to treat its property in medium. Here the first question was: What is the appropriate Lagrangian for the skyrmion description? The second question was: Given such a Lagrangian, how does one treat nucleon properties in medium?

In addressing the first question, suggested by Weinberg's folk theorem, we needed to identify the degrees of freedom relevant to the low-energy scale involved. Then guided by the large N_c arguments [40], one would write an effective theory for QCD at low energy as a power series (in derivatives and the pion mass) in terms of the chosen degrees of freedom. For simplicity, we took the simplest Lagrangian with Goldstone pions only with the heavy degrees freedom such as vector mesons, scalar mesons, glueballs etc. assumed to be integrated out. Baryons should arise as topological solitons from such a Lagrangian, though strictly valid in the large N_c limit. At near zero energy, we knew what it should look like from the current algebras,

$$\mathcal{L}_{\mathbf{CA}} = \frac{f_\pi^2}{4}\mathrm{tr}(U^{-1}\partial_\mu U)^2. \tag{5}$$

By itself, this term does not support a soliton, so to get a stable soliton, one needs higher derivative terms. Skyrme in his path-breaking paper [33] put by hand a quartic term — called Skyrme term — of the form

$$\mathcal{L}_{\mathrm{Skyrme}} = \frac{1}{32e^2}\mathrm{tr}[U^{-1}\partial_\mu U, U^{-1}\partial_\nu U]^2. \tag{6}$$

This term does stabilize the soliton. The Lagrangian consisting of the two terms (5) and (6), called "Skyrme Lagrangian" in the literature, has been extensively studied for the structure of baryons [41] as well as finite nuclei and infinite nuclear matter [42]. It is, however, too simple to be a realistic model for the baryon, even more so for many-nucleon systems, although it embodies the basic concept of solitonic baryon. There were lots of objections raised against the Lagrangian, among which most serious was that there is no reason why the derivative expansion should stop at the quartic order. We ignored all these caveats at the time. Some of these objections were lifted later when vector mesons, in fact an infinite tower of them in holographic QCD,[d] were incorporated.

In [39], we decided to bypass dynamical contents but focus only on, or exploit, the chiral symmetry structure of the Skyrme Lagrangian for describing the properties of hadrons in dense medium. The immediate goal here was to justify the scaling behavior of the hadrons used in [38]. For this, we had to figure out how to do an effective field theory calculation with the Skyrme Lagrangian in a dense medium. We were principally motivated by the success achieved by Walecka mean field theory for nuclear matter [43]. This is anchored on the premise that such a mean-field approach corresponds to doing Fermi-liquid theory and that such a mean-field theory

[d]In the holographic dual QCD model to which I will return, the low-energy action in the large N_c limit is given by a 5-D $U(N_f)$ (where N_f is the number of flavors) Yang-Mills action $\propto (F_{AB})^2$ in curved space (ignoring higher order terms such as $(F_{AB})^3$). When all the vector fields that emerge in dimensional reduction to 4D à la Klein-Kaluza are set to equal to zero, one is left with precisely the Skyrme Lagrangian, with no other quartic terms. There is only one quartic term here, i.e., the Skyrme-term type. This is because the Skyrme term is the only quartic term with one time derivative coming from the reduction of the $\propto (F_{AB})^2$. It is worth noting that in the hQCD model, the Skyrme term has no role whatsoever in stabilizing the soliton (instanton). It is the Chern-Simons term involving the ω field that prevents the collapse of the soliton.

can be constructed in a chirally symmetric way as will be elaborated below. Therefore the strategy was to write a chiral Lagrangian in the form of the Skyrme Lagrangian and then do the mean field. For this, we felt that a scalar meson of the mass comparable to that used in Walecka model would be needed to address many-nucleon matter. The question was: What is this scalar meson? It was surely not the scalar meson σ in the linear σ model, because we were dealing with the nonlinear sigma model, for which the σ of the linear σ model, being heavy with mass $\gtrsim 1$ GeV, has been integrated out. The scalar needed in Walecka model had to be a chiral singlet with mass $\lesssim 600$ MeV. In [39] the scalar was implemented with a dilaton field χ that we assumed to interpolate only the soft — and not hard — component of the trace anomaly of QCD that is locked to the quark condensate, the order parameter of chiral symmetry.[e] At the time, quite a few particle theorists were interested in connecting conformal symmetry, more specifically, scale symmetry to chiral symmetry and we adopted their strategy. The idea is to make the action for the Goldstone bosons and the vector mesons — apart from the chiral symmetry-breaking mass term — scale-invariant and attribute the scale non-invariance, i.e., the trace anomaly, to a potential $V(\chi)$ [44,46]. This meant that one should multiply the Lagrangian $\mathcal{L}_{\mathbf{CA}}$ by $(\chi/f_\chi)^2$ where f_χ is the vev of χ, $f_\chi = \langle\chi\rangle_0$, and leave $\mathcal{L}_{\text{Skyrme}}$ (the action of which is scale-invariant) untouched. Similarly for the action with vector mesons. Such an action would lead to $\theta^\mu_\mu \sim \chi^4$ without involving other fields except for the pion mass term.

The next step is then to consider the soliton solution of the resulting Lagrangian as giving the baryon and construct an effective Lagrangian that has as the pertinent degrees of freedom the pion, the vector mesons, the scalar χ and the nucleons.[f] Taking the mean field à la Walecka, the Lagrangian will then give nuclear matter in which the in-medium vev $\langle\chi\rangle^*$ will figure as a parameter of the Lagrangian. This led to the scaling law

---

[e] The trace anomaly is given by $\theta^\mu_\mu = \partial_\mu D^\mu$ where D_μ is the dilatation current. In QCD, $\theta^\mu_\mu = \frac{\beta(g)}{2g}(G_{\mu\nu})^2$ in the chiral limit with vanishing quark mass. Defining the dilaton field χ as interpolating the gluon condensate, it is useful to decompose, following [44], $(G_{\mu\nu})^2 \sim (\chi^4)_{\text{soft}} + (\chi^4)_{\text{hard}}$ and associate the soft component tied to chiral symmetry breaking. This subtle point in doing the separation was clarified much later in a more satisfactory way in collaboration with Hyun Kyu Lee in the WCU program at Hanyang University [45]. The soft dilaton will be denoted simply as χ in what follows.

[f] The Skyrme Lagrangian does not have the vector mesons but we can recover their properties from the hidden local symmetry Lagrangian [47, 48] from which it can be considered to come from. This is simply because the two Lagrangians are "gauge-equivalent."

of [39]

$$\Phi(n) = m_N^*/m_N \approx m_V^*/m_V \approx m_\chi^*/m_\chi \approx f_\pi^*/f_\pi \tag{7}$$

where $f_\pi^* \approx f_\pi(\langle\chi\rangle^*/f_\chi)$ and the subscript V stands for the vector mesons ρ and ω. For this relation to make sense, it is crucial that the χ field be associated with the soft component of the gluon condensate, the hard component remaining unchanged as density increases. At the time this relation was proposed, there was a big misunderstanding because people took χ to have both components as it appears in the trace anomaly.

5.1. *"Deriving" BR scaling*

There are several different ways of interpreting, or arriving at, (7). Let me discuss a few here.

Chiral symmetry enters in (7) through the *locking* with the soft scalar condensate and hence manifests itself in the mean-field as formulated. Clearly the connection to QCD is not direct. Thus it is not clear how that relation is connected to QCD-motivated quantities such as in-medium QCD sum rules. It has been suggested that a Walecka-type mean field theory for nuclear matter is equivalent to Landau Fermi liquid fixed point theory [49]. Now assuming this equivalence, the relation (7) holding at nuclear matter density n_0 can be considered as a fixed-point relation valid at $n = n_0$. In fact such a BR-scaling Lagrangian at mean field is found to give a correct description of nuclear matter [50]. Now the density dependence of the parameters such as mass (7) and coupling constants simulate part of the higher-dimension operator terms added to the Walecka-type mean-field models in order to make good fits to finite nuclei as well as infinite matter [51]. This could be interpreted as (a part of) the rearrangement terms brought in through many-body (typically, three-body) forces. The scaling parameter $\Phi(n = n_0)$ can be related to the Landau parameter F_1 at the fixed point through the nuclear anomalous orbital gyromagnetic ratio [52] as we will see below. Gerry understood this connection before he started working on BR scaling in dilepton processes in heavy-ion collisions. This issue became later highly controversial with no paucity of misunderstandings.

The fact that the scaling factor (7) is tied to the Fermi liquid fixed-point parameter which encodes a part of quasiparticle interactions clearly warns of the subtlety in the way chiral symmetry can be manifested in many-nucleon systems. As it stands, it encompasses both the quasiparticle nature of mesons and baryons and their interactions. Obviously the absence of visible effects does not necessarily signify the absence of chiral symmetry effects. A good example of this is given by the *invisible* effect of (7) in the Dirac phenomenology that I will mention below. And a similar story can be told of the JLab result as I will explain.

If one is probing a system away from the fixed point as in the case of heavy-ion processes, the question arises in what way the relation (7) can figure in the process. In particular, one would like to know how the scaling works near the chiral phase transition.

The first possibility is to consider the pion decay constant f_π as an order parameter of chiral symmetry, presumably locked to the quark condensate $\langle\bar{q}q\rangle$. In this case, if one were to naively extrapolate (7) all the way to the chiral phase transition point at which $\langle\bar{q}q\rangle = 0$ (in the chiral limit), then all the masses would go to zero. However, this is probably too naive or even wrong. In hidden local symmetry theory where the nonlinear sigma model is elevated to a local gauge theory [47], it was shown by renormalization group at one-loop order [48] that when the quark condensate goes to zero, independently of how it is driven, the hidden gauge coupling g (which is related to the vector coupling $g_{\rho\pi\pi}$ or $g_{\rho NN}$) goes to zero, arriving at what is called "vector manifestation (VM) fixed point." There it is the hidden gauge coupling g that is the order parameter, not the pion decay constant. Now since the vector meson mass goes as $m_V^2 \sim g^2$, the vector meson mass must scale near the VM fixed point as[g]

$$m_\rho^*/m_\rho \simeq g^*/g \simeq \langle\bar{q}q\rangle^*/\langle\bar{q}q\rangle \to 0. \tag{8}$$

This is a clear prediction within the HLS theory framework. It is not known what happens if one includes more vector mesons and/or other degrees of freedom. There is an analysis with constituent quarks in addition to the pions and the vector mesons, which suggests that the constituent quark mass m_Q *also* scales very near the VM fixed point as $m_Q^*/m_Q \simeq g^*/g \simeq \langle\bar{q}q\rangle^*/\langle\bar{q}q\rangle \to 0$.

The situation is even less clear when there are an infinite tower of vector mesons as implied by holographic QCD models.

I should point out that (8) is the only rigorous demonstration known so far of what happens to the meson mass at the chiral transition, although it can be established strictly within the specified framework. I know of no

[g]Gerry had been arguing that the ρ mass was the order parameter long before the VM was discovered. In fact, the VM vindicated Gerry's point.

other statement on the property of the meson mass that can be taken as rigorous. Lattice QCD cannot offer anything yet on it either in temperature or in density. No other models that I know of do the RG analysis in a consistent way. Most of the models are hodge-podge mixtures and can hardly be trusted.

That this happens at high temperature was shown within the HLS theory without introducing fermions. In density, however, fermions are needed. One might try to generate baryons as skyrmions from the HLS mesonic Lagrangian, but one could work equally well with quasiquarks (or constituent quarks). There one can show that the vector meson mass does also go to zero at high density in a manner analogous to temperature. Considered in terms of the constituent quarks mentioned below, the nucleon, a bound state of three constituent quarks, is expected to become massless at the chiral transition point. So very near the transition point, the BR scaling (7) seems to be recovered with Φ scaling linearly in $\langle\bar{q}q\rangle$. Unfortunately, up to date HLS can say nothing definite away both from the vacuum and from the VM fixed point.

There is at present no unquenched lattice data which supports or refutes (7). However, there is an evidence for the relation (7) in strong coupling lattice QCD in the leading order in $1/g_c$ and $1/d$ where g_c is the color coupling and d is the space dimension [53]. It remains to be seen what happens when $1/g_c$ and $1/d$ corrections are included.

The simplest — and perhaps the most appealing — way to get at (7) is the constituent quark model. It leads to a similar relation as that from the skyrmion picture but in a much simpler way. A recent discussion on this approach is based on the Manohar-Georgi model [54] which was recently given a strong backing by Weinberg in the large N_c limit [4]. In this model, the relevant degrees of freedom are the constituent quarks, the pions and the gluons. They figure in the phase where chiral symmetry is broken, but quarks and gluons are deconfined. The essential idea in this work is that the constituent quark mass, which is of order of the renormalization scale Λ_{QCD}, is N_c-independent. If one supposes that the constituent quark mass comes from the spontaneous breaking of chiral symmetry — which is, however, not implied by the model itself, then it seems reasonable to think that the constituent quark mass will go to zero when the quark condensate goes to zero (in the chiral limit) in a manner analogous to the VM of hidden local symmetry. In the Manohar-Georgi model, the nucleon mass scales as $m_N \sim \langle\bar{q}q\rangle$. Now if the vector mesons and baryons are loosely bound states of constituent quarks and antiquarks, then as density increases, the universal scaling of the type (7) will follow with f_π^*/f_π replaced by $(\langle\bar{q}q\rangle^*/\langle\bar{q}q\rangle)^q$ where q is some positive power $\sim 1/2 - 1$.[h]

5.2. *"Seeing" BR scaling*

5.2.1. *Nuclei and nuclear matter*

Up to nuclear matter density n_0 to which ample experimental data are available, the relation (7) works fairly well. I know of no cases where the scaling (7) is at odds with nature.[i] Let me list just a few notable ones, in which Gerry participated directly and indirectly:

- *Nuclear tensor forces*
  The early suggestion [38] that the scaling (7), through increased cancelation between the pion and ρ tensor forces mentioned above, would modify the structure of the tensor forces was reconfirmed in a striking way in the C14 dating beta decay process [57]. This is an exceptional case in nuclear physics where dialing certain parameters amplifies enormously the effect of the "dropping" ρ-meson mass.
- *Proton-nucleus scattering*
  Among numerous applications of (7) to nuclear scattering processes, the one where Gerry actively participated was proton-nucleus scattering at

[h]Some models like NJL or QCD sum rules give $q = 1/2$, whereas HLS theory gives $q = 1$.

[i]Since this matter is outside of the context of this note, I put it in a footnote. I know of one case in which the effort to *directly see* the scaling (7) with the ρ meson in nuclear medium has "failed," i.e., the CLAS/JLab [55]. Given the background analysis and other details of experimental procedure, one should be cautious of the interpretation of the result. Let me assume for the sake of discussions that one is given an experimentally determined spectral function. Since one is dealing with a low density no greater than that of the nuclear matter, one can approach the spectral function in two ways. (1) Start with a theory with the parameters of the effective chiral Lagrangian defined at the matter-free vacuum and compute the spectral function in chiral perturbation theory to as high an order as feasible. Here the vacuum parameters, not the scaling parameters (7), should be used. (2) Start with a chiral Lagrangian with parameters defined at a "sliding vacuum" with the scaling (7) and then compute corrections as in (1). One possibility is to start from the Fermi-liquid fixed point and treat deviations from the fixed point as was done in [52]. These two ways should give the same result if done fully consistently. It is not unlikely that there are cancelations in the second approach which is not present in the first, as exemplified by the example of the Dirac phenomenology [58] mentioned below. Gerry has emphasized a sort of hidden scale invariance in nuclear processes that can mask the scaling behavior as it does in the Dirac phenomenology. Furthermore, the usual vector dominance (à la Sakurai) — adopted by most of the theorists — is violated in dense medium (more rapidly than in hot medium mentioned below), so the interpretation becomes highly tricky, and it would be difficult to sort out all (theoretical) backgrounds well enough to un-mistakenly "identify" the scaling (7) in this sort of experiments.

several hundreds of MeV [58]. Incorporating the scaling (7) removed certain persistent discrepancy in analyses with the nonrelativistic impulse approximation, a standard technique at the time, and improved agreement with experiment. There are other processes where similar improvements are obtained, that I will not go into.

- *Anomalous orbital gyromagnetic ratio in nuclei (δg_l)*
  Since BR-scaling chiral Lagrangian in the mean field can be identified with Landau-Migdal theory [52], one can derive the Migdal formula $\delta g_l = \frac{1}{6}\frac{F_1'-F_1}{1+F_1/3}\tau_3$ [59] in terms of Φ of Eq. (7). The scaling parameter Φ enters in place of the Landau-Migdal parameters F_1 and F_1', giving $\delta g_l = \frac{4}{9}[1/\Phi - 1 - \kappa_\pi]\tau_3$ where κ_π is given by the pion, so is accurately known. This formula gives in the Pb region $\delta g_l^p \approx 0.22$ for $\Phi(n_0) \approx 0.8$ given by QCD sum rules. The result from experiment on giant dipole resonances is 0.23 ± 0.03. This is a clear indication that the scaling, with an intricate, albeit indirect, link to chiral symmetry, is encapsulated also in the Landau parameters at the fixed point.
- *Warburton ratio in heavy nuclei*
  As mentioned above, the time component of the axial current in nuclei is kinematically suppressed, so the "impulse approximation" for the axial charge transition (4) does not dominate the transition matrix element. In fact, the soft-pion exchange current correction can be rather big [12]. At just about the same time as the scaling relation (7) was proposed, Warburton found [13] that his analysis of the ratio (that I refer to as "Warburton ratio") $\epsilon_{\rm MEC} = M_{\rm exp}/M_{\rm sp}$ where $M_{\rm exp}$ is the experimental matrix element of the axial charge operator and $M_{\rm sp}$ is the shell-model single-particle matrix element in the lead region, came out as $\epsilon_{\rm MEC} = 2.01 \pm 0.05$, considerably larger than the prediction ~ 1.4 based on the operator of [12]. This enhancement was quantitatively explained by the scaling (7) [14]. In fact the enhancement in $\epsilon_{\rm MEC}$ ranging from $A = 12$ to $A = 208$ observed in experiments is fairly well — if not entirely — explained in an extremely simple way by the scaling (7) [60].
- *Nuclear matter properties*
  One of the subtlest aspect of chiral symmetry manifestation is that it can be made a part of nuclear many-body interactions. When formulated in terms of the sliding vacuum as in the cases given above and further elaborated below in Sec. 6, the scaling (7) is embedded in the Landau parameters. Thus nuclear matter can be described in an economical way in a generalized mean-field formulation in terms

of a chiral Lagrangian with (7) [50], which corresponds to the "single-decimation" scheme mentioned below in Sec. 6. It is true that such a formulation is not yet made quantitatively accurate, but it seems highly feasible to make it so if there are enough data to fit the parameters of the effective chiral Lagrangian that one starts with. I should also mention what is called "double-decimation" scheme which belongs to the same class of approaches for nuclear matter. This will be discussed in Sec. 6.

The effects I have discussed thus far are all *indirect* in that the particles involved are off-shell. They are in fact closely connected with the meson-exchange currents that were clarified by effective chiral field theory as I described above. In this sense, one can only say that (7) is not inconsistent with what has been observed in Nature: They are clearly not a "snap-shot" of an on-shell particle propagating with the scaling mass, which is what one wants. I will turn to this issue next. But before doing so, one can, however, ask whether there are no evidences that (7) contradicts Nature, in other words, is falsified by experiments. If the degrees of freedom figuring in a particular process have masses different from the free-space values with which physics has been done up to now, are there no cases where the agreement obtained without scaling masses is upset by the scaling masses?

This question was addressed by Gerry. One of the most prominent successes of the original (linear) Walecka mean-field model was the Dirac phenomenology on the spin observables in proton-nucleus scattering. One would naively think that the scaling of the type (7) would upset the most spectacular fit of the data. It turns out, however, that this was not the case [58, 61]: There was no inconsistency over the whole range of kinematics measured. Gerry has argued that there is a hidden scale invariance in nuclear many-body process that preserves certain spin observables. This is yet another illustration that the absence of visible signals does not necessarily imply the absence of the scaling effects.

5.2.2. *Dileptons*

Suppose a ρ meson is produced inside a big chunk of dense matter and that one would like to "see" how it behaves inside the dense matter. Now suppose that the detector is put inside the matter. It will then detect the ρ meson propagating with its properties, such as its mass, couplings with other "particles" etc. determined by the "vacuum" affected by the density of the medium.

However, in all the experiments performed so far, the detector is outside of the medium. Given the situation, it was natural to think that the best means to take the "snapshot" of the properties of the ρ in medium would be to measure the spectral function of the dileptons produced via the process $\rho \to l^+l^-$ (where $l = e, \mu$) inside the matter. The question asked was: Can the scaling (7) be singled out in the dilepton production in nuclear collisions when measured outside of the medium?

In order for the relation (7) to be applicable, the large N_c approximation should be reliable. Recall that it is a mean-field relation and only in the large N_c limit can it represent a quasiparticle. Whether or not the vector meson can be treated reliably as a quasiparticle with a "narrow" width in density greater that of n_0 is of course totally unknown. If the residual (quasiparticle) interactions become too strong, then the quasiparticle picture will surely break down.

As emphasized, within the well-defined HLS scheme with the vector mesons ρ and ω and the pion as the only degrees of freedom (with nucleons arising as solitons), the RG arguments do predict that the masses will go to zero according to (8) with the widths also going to zero. So in that scheme, as $\langle\bar{q}q\rangle \to 0$, there will be a sharp peak at low invariant mass. Will this be detected in a measurement with the detector outside of the medium?

Gerry applied the scaling (7) to the dilepton measurements in heavy-ion collisions in mid 1990's assuming that a similar scaling holds in temperature,[j] and it seemed to explain the low-mass dileptons seen in the SPS/CERN experiments [62]. This scenario is called "dropping mass scenario (DropMS)." It turned out that these dilepton data could also be accounted for if one were to take into account such standard nuclear many-body effects as particle-hole excitations (or "sobars"), collisional broadening etc. Let me refer to this scenario where the strong mundane nuclear effects dominate as "melting scenario (MeltS)." These many-body effects are

[j]There is no strong theoretical argument to suggest that one can apply (7) to heavy-ion processes where high temperature is involved. Although, unlike in density, there is lattice information on temperature, there are no unquenched lattice data for light-quark hadrons near the critical temperature. The only theoretical statement one can make is that the vector manifestation is *shared* by both temperature and density in the framework of HLS. But at present, HLS can say little on what happens in between the vacuum and the transition point (both in T and n).

expected to be there anyway even if one has the quasiparticle picture as the first approximation. The question would be whether the quasiparticle picture makes sense in the heavy ion process in question. There have been long debates as to whether it is the DropMS *or* MeltS that is seen in the experiment. Gerry and I believe this is a wrong debate. The correct approach should be DropMS *and* MeltS together *treated consistently*, not DropMS *or* MeltS.

A succinct summary of both experimental and theoretical situations on dileptons is given in [63]. The upshot can be put in one line: *The dilepton data in heavy-ion collisions can be more or less entirely explained by mundane (standard) nuclear interactions.* Does this mean that Gerry was wrong in [62]? My answer is: Not entirely. Before I explain this, let me mention that what the present interpretation in terms of the MeltS mechanism is telling us is that what has been seen in the experiments is what I would call the "background" to what one is looking for as the *signal* for the working of chiral symmetry in dense/hot medium. That the background accounts for the observed result simply means that whatever is looked for has not been exposed. It does not necessarily mean that it is not there. It is currently said in some quarters that the "melting mechanism" is a signal for chiral symmetry restoration. This is as absurd as saying that the explanation of the anomalous gyromagnetic ratio or absence of the deviation in the Dirac phenomenology mentioned above *evidences* chiral symmetry at work. The background, the "gooey" stuff, could have been a consequence of just plain strong interactions regardless of whether there is chiral symmetry involved or not.

Now let me address the issue of whether or not one can rule out (7) or (8) from the dilepton results.[k] My point is that one cannot. The mechanism Gerry was advocating in [62] could very well be there but as we now know it should be accompanied by other intricate mechanisms contained in HLS formalism. For instance, his consideration did not take into account that there is "hadronic free regime" in which the photon is most likely decoupled from the dropping-mass vector mesons. The RG consideration in HLS suggests that because of the scaling in the gauge coupling g and the

[k]Both Gerry and I disown the mysterious green curve one sees in certain experimental papers and conference talks attributed to "BR scaling" and invoked to rule out (7). Such a curve has absolutely nothing to do with what was originally predicted in [39] or in later versions of the scaling.

parameter a,[l] in the temperature or density regime where (8) is applicable, the photon coupling to the BR-scaling vector mesons gets suppressed. There are several agents for this which individually may not be significant but become important together. One of them is that the naive Sakurai vector dominance picture breaks down in temperature and in density, so that the strength of the ρ coupling to the photon gets suppressed. This can be exacerbated by the infinite tower effect seen in holographic QCD where the tower of vector mesons can play an important role in photon coupling [64]. This means that the dileptons seen in the measurement carry no unambiguous imprint, free of the background, of the vector mesons whose intrinsic mass has undergone the shift (7) [65]. In other words, even if the dropping-mass vector meson were there, it would be like a needle in the haystack swamped by the background of strong nuclear interactions. What is observed in the experiments is just the lepton pair coming from hadrons in on-shell processes taking place at the "flash point"[m] and this process has little to do with the vacuum change that one is looking for. The same argument applies to the CLAS/JLab result on electro-produced ρ mesons in nuclei. Somewhat detailed discussion on this issue was given above in the footnote i.

In sum: What has been shown in this flurry of activities is simply that nucleons in nuclei do indeed interact strongly, which of course nobody would question. If this were all there is to it, then that would be the most boring situation. Maybe one is not doing the right thing in the search for the "smoking gun".[n] I should mention that the ambiguity and difficulty in extracting medium effects on the mass of the decaying particle as in dilepton

[l] a is a parameter in hidden local symmetry theory defined as the ratio f_s^2/f_π^2 where f_s is the decay constant of the scalar that is Higgsed to give the ρ the mass or the longitudinal component of the ρ. This parameter signals deviation from vector dominance in medium.

[m] The "flash point" in our language [56] is defined to be the density or temperature at which particles with decreased masses and weak coupling flowing from the hadronic free regime recover most (not necessarily all) of their free-space masses and strong interactions. It is difficult to precisely pin down the flash density or temperature. But to give a rough idea, let me give some rough guesses here. In the case of density, there is an indication from dense matter simulated on crystals that it could lie at $\sim(1.3-2)n_0$ and in temperature, not far from the critical temperature $\sim$170 MeV. Pinning down this point (either in density or in temperature) would be a challenge for the future.

[n] One cannot help but wonder, however, why in condensed matter physics, things could be equally complicated in a strongly-coupled system and yet there is beautiful order and simplicity as one gets to understand the system better, and phenomena with the same notions of quasiparticles and long-range order do not occur in the strong interactions.

processes, presumed to reflect the property of chiral symmetry in medium, was clearly pointed out by Yamazaki and Akaishi [66].

6. Double Decimation and V_{lowk}

In Fall 1999, I was asked by Gerry to come to Stony Brook and give a series of lectures on renormalization group approach to Landau-Migdal theory of nuclear matter. At the time I was working on something related to what I did with Bengt Friman in 1996 at GSI on connecting BR scaling to the Landau parameters [52]. I was attempting to use Shankar's renormalization group approach to Landau Fermi liquid theory [67], to identify the scaling at $n = n_0$ as the RG fixed point. The reasoning was that the chiral Lagrangian with BR scaling parameter, when treated in the mean field, corresponded to the Fermi liquid fixed point theory. My lecture was on this idea. What Gerry with his collaborators did was first to arrive at the V_{lowk} in the first decimation [68] and then go, via BR scaling, to the Fermi liquid fixed point using Babu-Brown induced interactions in the second decimation [69]. This was the idea of the double decimation [61]. What I did then was combining these two decimations into one. These two approaches more or less agree with each other, notably for δg_l which was difficult to get right in the double-decimation method *without* BR scaling.

The double-decimation approach that implements BR scaling finds an additional support in the calculation of nuclear-matter properties in a standard many-body approach [70]. Here the second decimation was made using a ring-diagram technique in place of field-theoretic Fermi-liquid theory of [69]. Again there is an equivalence between multi-body effects brought in by BR scaling and those found in empirical multi-body forces, e.g., of the Skyrme type.

7. Strangeness in Nuclear Matter

Kaon is massive because the strange (current) quark mass is about 20 times the mass of the light quark. But it is lighter than other hadrons, such as the ρ. If one puts the kaons on the same footing as the pions and consider them as Goldstone particles with, however, the quark masses taken into account as corrections, then as first shown by Kaplan and Nelson [71], negatively charged kaons (K^-'s) can condense in nuclear matter at high density. This happens because the density, through attractive kaon-nuclear interactions, eats into the kaon mass, bringing the kaon mass to zero at some density. At this point, the kaon field picks up the vev $\langle K \rangle$ and the kaon Bose-condenses. Such a high density could be reached in neutron-star matter, so it was thought to be relevant to neutron stars. It was also thought to be reachable in heavy-ion collisions in the laboratories.

Together with Kuniharu who was on sabbatical at Stony Brook and Vesteinn (Thorsson), Gerry's student, we proposed a mechanism that we found was more pertinent for neutron stars [72]. The scenario proposed by Kaplan and Nelson required that the pions condense first, which then trigger kaon condensation [73]. Our estimate showed that this scenario would take place at too high a density to be relevant to neutron stars. The reason was that pions were thought to be unlikely to condense at a density regime relevant to neutron stars due to highly nonlinear nuclear correlations at short distances. In the scenario that we came up with, the kaon mass need not fall all the way to zero before the kaon condenses. All that is required is that the kaon mass m_K drop at roughly the same rate as it does in the Kaplan-Nelson's at increasing density. The new ingredient in our scenario is to exploit that in neutron-star matter, there are electrons, and that the electron chemical potential μ_e increases as the matter density increases. If, at some density, the increasing μ_e crosses the dropping mass of the kaon, $m_K^* < m_K$, then the electrons will beta-decay into the kaons as

$$e^- \to K^- + \nu_e. \tag{9}$$

These kaons will Bose-condense. The density at which the condensation takes place turns out to be not so high, say, around 3–4 times the nuclear matter density. Gerry, together with Hans Bethe and Chang-Hwan Lee, made extensive studies of neutron star properties with this scenario, making some startling predictions on the maximum neutron star mass ($\sim 1.6 M_\odot$) and large number of small mass black holes in the Universe. This was summarized in a review article that he co-authored with Chang-Hwan Lee and me [74]. We also speculated what the consequence of the increased number of black holes would be on cosmology [75].

There are several ways to provide support to the early, somewhat crude, prediction of the kaon condensation threshold. One possibility is to invoke the scaling relation (7) in the framework of Walecka mean-field theory starting from nuclear matter density, i.e., the Fermi-liquid fixed point [76]. Another possibility is to exploit the sliding-vacuum method and start from the VM fixed point instead of the Fermi-liquid fixed point. This is in principle more astute since within the well-defined theoretical premise of hidden local symmetry, the VM fixed point is a robust starting point around which perturbation theory can be developed. Furthermore, the kaon condensation takes place nearer to the VM point than to the Fermi-liquid fixed point. This was done in [77] obtaining the same result as approaching from the Fermi-liquid fixed point. These two approaches illustrate the point I raised above, namely, that the scaling (7) makes sense only in a framework where the starting point is defined with a sliding vacuum.

8. The Multifaceted Skyrmion

The dual character between the "macroscopic" skyrmion picture and the "microscopic" QCD picture for the baryon clearly points to a fuzzy notion of topology that enters in the former. The chiral bag model with the Cheshire Cat phenomenon attests to this ambivalence. There is a trade-in between topology and quark/gluon degrees of freedom. When the Skyrme model was resurrected in the early 1980's, a large number of nuclear and particle theorists had jumped on the "bandwagon" and worked on it. However string theory re-emerged soon after with its "first revolution," which took away most of the particle theorists involved in the Skyrme model into the main stream of string theory. That left the problem mostly to nuclear theorists who then went on to further elaborations of the model and to applying it to many-baryon systems. The output unfortunately was limited and even poor. The reasons for the paucity of significant results were multi-fold. First of all, as stressed throughout this note, the Skyrme model as it stands — which was studied by most workers — is at best only a poor caricature of QCD. Other degrees of freedom, notably the vector mesons — possibly the infinite tower of them, the scalar mesons etc. need be included. There is a clear signal for this need in the recent development of the holographic dual model of the baryon. (This matter is briefly touched on in the next section.) Secondly, even at the level of the pure Skryme model, little is understood of what is contained in the soliton structure. Pure mathematicians and mathematical physicists, some as celebrated as Atiyah and Manton, are finding fascinating aspects of the model at the classical level, but nothing much is known at the quantum level. When the vector-meson degrees of freedom are included as clearly required by the Cheshire Cat principle of the chiral bag picture [32], even less is understood. It is undoubtedly a difficult problem to resolve. For these and other reasons, the model was "abandoned" by many workers in the field who then turned to more manageable and schematic models. The difficulty and the consequent abandonment were mistakenly identified by many of them as the defect in the soliton picture of the baryon itself, not as the failures or shortcomings of the workers themselves.

Gerry and I found this tendency of abandonment unjustified as well as disingenuous. We felt that what we have seen in the model for the baryon structure *per se* as well as for many-body systems, i.e., dense baryonic matter, was just a tip of a big iceberg and there were a wealth of hitherto unexplored fascinating phenomena to be uncovered. This led to our editing of the volume *The Multifaceted Skyrmion* [78] in which we gathered contributions ranging from nuclear physics to condensed matter physics to string theory, all dealing with the same notion of the skyrmion soliton ranging in dimensions from 3 to 5. The objective here was to illustrate by stunning developments in condensed matter physics and what is taking place in string theory how pervasive and rich the physics captured by the soliton structure can be. In fact, in strong interaction physics with the intricacy involved with the Cheshire Cat phenomenon that is not shared by condensed matter, there is a growing evidence for certain phenomena associated with topology that are not visible in other models, such as for instance a half-skyrmion phase that appears at high density [79] that resembles the merons or half-skyrmions that are found in condensed matter systems [78], e.g., bilayer quantum Hall phenomena, deconfined quantum critical point and thin chiral magnetic films.

9. Recent Developments

I would like to end this note with an account of recent developments in which Gerry did not participate but that are closely related to both what he has done and what we have done together.

There has been a long-standing puzzle why Sakurai's vector dominance works well for the pion (and mesons in general) but fails famously for the nucleon. In [80], together with Wolfram Weise, we proposed to resolve this problem by means of a little chiral bag defined at the "magic chiral angle" $\theta = \pi/2$ at which the baryon charge is shared equally inside and outside of the bag. In this case, the nucleon has a "core" of $\sim 0.2 - 0.3$ fm as seemed to be required by the experimental observations mentioned above [23, 24]. Recently this issue was revived by the discovery of an instanton picture for the baryon that arises from holographic dual QCD [81]. The baryon in this theory essentially corresponds to a skyrmion built of the pion and an infinite tower of vector mesons [82]. This picture turns out to give an extremely good description of the proton form factors in terms of a surprisingly simple formula derived from the infinite tower of the vector mesons [83]. The result, however, gives a structure of what corresponds to the "core" that is drastically different from what was found before in terms of the quark-bag description [80] and also from the holographic Cheshire Cat [32]. I have at present no clear idea what this means in terms of QCD variables[o] and I am wondering what Gerry will say about it.

Another new development is a new form of scaling which modifies (7) for $n \gtrsim n_0$ that comes from the structure of cold dense baryonic matter simulated on crystal lattice [84]. When skyrmions are put on an FCC lattice to simulate dense matter, one finds that there is a phase change from the skyrmion matter to a half-skyrmion matter in CC. At a certain density $n_{1/2}$ above that of the nuclear matter — the precise value of which cannot be pinned down at the moment, a skyrmion in a dense medium fractionizes into 2 half-skyrmions with the chiral symmetry putatively restored in the sense that $\mathrm{Tr}\, U \sim \langle \bar{q}q \rangle$ goes to zero at that density. When this happens, the nucleon mass stops dropping until it reaches the deconfined phase which takes place at a density $> n_{1/2}$ whereas the vector meson mass continues to drop with the dropping hidden gauge coupling g^*. Thus for $n \gtrsim n_{1/2}$ the scaling differs from (7)

$$m_N^*/m_N \sim \text{const}, \tag{10}$$

$$m_V^*/m_V \sim g^*/g. \tag{11}$$

This means that the tensor forces would look quite different below and above the density $n_{1/2}$, with a sudden change-over at that density. This would not affect the C14 dating process of [57] described above, since up to n_0, it remains the same, but it makes a dramatic effect on the tensor forces at $n \gtrsim n_0$. Specifically the ρ tensor is almost completely killed by $n \sim 3n_0$, leaving only the pion tensor effective. This is in strong contrast to what one expects with the scaling (7) which makes the whole tensor forces nearly vanish at that density. This would have a serious ramification on the symmetry energy which plays an essential role in the EOS of compact stars [84]. Again I wonder what Gerry will say about this.

[o]It may be that "integrating out" the high tower in the bulk (gravity) sector involving geometry gives a different physical picture from that of "integrating out" higher energy scale degrees of freedom in the boundary (QCD) sector. The meaning of the "core" in the QCD sector may not be directly mapped to what happens in the bulk sector.

Acknowledgments

What I described in this note may not be exactly what Gerry sees as what has happened. I only hope that it does not deviate too much from his version. Gerry and I have written a large number of papers, many of which with provocative ideas, working with Sunday transatlantic phone calls whenever I was not at Stony Brook. It was a wonderful journey together, full of fun and excitement — and never a dull moment — and I would like to thank him for that.

This work was partially supported by the WCU project of Korean Ministry of Education, Science and Technology (R33-2008-000-10087-0).

References

[1] Reviewed by G.E. Brown in *Mesons in Nuclei*, ed. by M. Rho and D.H. Wilkinson (North-Holland, Amsterdam, 1980).

[2] M. Chemtob and M. Rho, *Phys. Lett.* **B29**, 540 (1969); 'Meson exchange currents in nuclear weak and electromagnetic interactions', *Nucl. Phys.* **A163**, 1 (1971).

[3] D.O. Riska and G.E. Brown, 'Meson exchange effects in $n+p \to d+\gamma$', *Phys. Lett.* **B38**, 193 (1972).

[4] S. Weinberg, 'Pions in large-N quantum chromodynamics', *Phys. Rev. Lett.* **105**, 261601 (2010).

[5] J. Hockert, D.O. Riska, M. Gari and A. Huffman, 'Meson exchange currents in deuteron electrodisintegration', *Nucl. Phys.* **A217**, 14 (1973).

[6] G.E. Brown and M. Rho, 'The role of chiral invariance in nuclei', *Comments Nucl. Part. Phys.* **10**, 201 (1981).

[7] M. Rho, 1981 Erice Lecture, 'Pions and the chiral bag', *Prog. Part. Nucl. Phys.* **8**, 103 (1982).

[8] S. Weinberg, 'Phenomenological Lagrangians', *Physica* **A96**, 327 (1979).

[9] S. Weinberg, 'Nuclear forces from chiral Lagrangians', *Phys. Lett.* **B251**, 292 (1990); 'Effective chiral Lagrangians for nucleon — pion interactions and nuclear forces', *Nucl. Phys.* **B363**, 3 (1991).

[10] M. Rho, 'Exchange currents from chiral Lagrangians', *Phys. Rev. Lett.* **66**, 1275 (1991).

[11] T.S. Park, D.P. Min and M. Rho, 'Radiative neutron-proton capture in effective chiral Lagrangians', *Phys. Rev. Lett.* **74**, 4153 (1995); T.S. Park, K. Kubodera, D.P. Min and M. Rho, 'Effective field theory approach to $\vec{n}+\vec{p} \to d+\gamma$ at threshold', *Phys. Lett.* **B472**, 232 (2000).

[12] K. Kubodera, J. Delorme and M. Rho, 'Axial currents in nuclei', *Phys. Rev. Lett.* **40**, 755 (1978).

[13] E.K. Warburton, 'Mesonic enhancement of the weak axial vector current evaluated from beta decay in the lead region', *Phys. Rev. Lett.* **66**, 1823 (1991).

[14] K. Kubodera and M. Rho, 'Axial charge transitions in heavy nuclei and in-medium effective chiral Lagrangians', *Phys. Rev. Lett.* **67**, 3479 (1991).
[15] D.B. Kaplan, M.J. Savage and M.B. Wise, 'Two nucleon systems from effective field theory', *Nucl. Phys.* **B534**, 329 (1998).
[16] M. Rho, 'Predictiveness of effective field theory in nuclear physics', nucl-th/0610003.
[17] K. Kubodera and M. Rho, chapter 10 in this volume.
[18] T.S. Park *et al.*, 'Parameter free effective field theory calculation for the solar proton fusion and hep processes', *Phys. Rev.* **C67**, 055206 (2003).
[19] K. Kubodera and T.S. Park, 'The solar *Hep* process', *Ann. Rev. Nucl. Part. Sci.* **54**, 19 (2004).
[20] R. Lazauskas, Y.H. Song and T.S. Park, 'Heavy-baryon chiral perturbation theory approach to thermal neutron capture on ^{3}He', arXiv:0905.3119 [nucl-th].
[21] L. Girlanda, A. Kievsky, L.E. Marcucci, S. Pastore, R. Schiavilla and M. Viviani, 'Thermal neutron captures on d and ^{3}He', arXiv:1008.0356 [nucl-th].
[22] L.E. Marcucci, M. Piarulli, M. Viviani, L. Girlanda, A. Kievsky, S. Rosati and R. Schiavilla, 'Muon capture on deuteron and ^{3}He', arXiv:1008.1172 [nucl-th].
[23] J. Orear, 'Evidence for a proton core', *Phys. Rev.* **D18**, 2484 (1978).
[24] R. Petronzio, S. Vergata, S. Simula and G. Ricco, 'Possible evidence of extended objects inside the proton', *Phys. Rev.* **D67**, 094004 (2003); Erratum-ibid. **D68**, 099901 (2004). For a related observation, see R.B. Ferroli, S. Pacetti and A. Zallo, 'Pointlike baryons?', arXiv:0812.3283 [hep-ph].
[25] G.E. Brown and M. Rho, 'The little bag', *Phys. Lett.* **B82**, 177 (1979).
[26] A. Chodos and C. Thorn, 'Chiral hedgehogs in the bag theory', *Phys. Rev.* **D12**, 2733 (1975).
[27] C.G. Callan, R.F. Dashen and D.J. Gross, 'The structure of hadrons in QCD', *Phys. Lett. B* **78**, 307 (1978).
[28] V. Vento, M. Rho, E.M. Nyman and J.H. Jun, 'Chiral pion dynamics for spherical nucleon bags', *Nucl. Phys.* **A345**, 413 (1980); M. Rho, A.S. Goldhaber and G.E. Brown, 'Topological soliton bag model for baryons', *Phys. Rev. Lett.* **51**, 747 (1983).
[29] J. Goldstone and R.L. Jaffe, 'The baryon number in chiral bag models', *Phys. Rev. Lett.* **51**, 1518 (1983).
[30] S. Nadkarni, H.B. Nielsen and I. Zahed, 'Bosonization relations for bag boundary conditions', *Nucl. Phys.* **B253**, 308 (1964).
[31] M. Rho, 'The Cheshire Cat hadrons', *Phys. Rept.* **240**, 1 (1994); corrected and updated with errata 'The Cheshire Cat hadrons revisted', in hep-ph/0206003.
[32] I. Zahed, 'Holographic nucleons', contribution to this volume, arXiv: 1010.5980[hep-ph]; H.B. Nielsen and I. Zahed, 'The Cheshire Cat Principle from holography', arXiv:0904.0429 [hep-th].
[33] T.H.R. Skyrme, 'A unified field theory of mesons and baryons', *Nucl. Phys.* **31**, 556 (1962).
[34] H.B. Nielsen, M. Rho, A. Wirzba and I. Zahed, 'Color anomaly in a hybrid bag model', *Phys. Lett.* **B269**, 389 (1991).

[35] H.B. Nielsen, M. Rho, A. Wirzba and I. Zahed, 'The tale of the η' from the cheshire cat principle', *Phys. Lett.* **B281**, 345 (1992).
[36] P. de Forcrand, 'Nuclear matter from the strong coupling lattice QCD', Talk at ECT*, July 2010, http://www2.yukawa.kyoto-u.ac.jp/ fuku/cold_dense/ talks/Friday/Forcrand.pdf.
[37] R.E. Cutkosky, 'Meson exchange effects in two-nucleon states', *Phys. Rev.* **112**, 1027 (1958).
[38] G.E. Brown and M. Rho, 'In-medium stiffening of the nucleon nucleon spin-isospin interaction', *Phys. Lett.* **B237**, 3 (1990).
[39] G.E. Brown and M. Rho, 'Scaling effective Lagrangians in a dense medium', *Phys. Rev. Lett.* **66**, 2720 (1991).
[40] G. 't Hooft, 'A planar diagram theory for strong interactions', *Nucl. Phys. B* **72**, 461 (1974); E. Witten, 'Baryons in the 1/N expansion', *Nucl. Phys. B* **160**, 57 (1979).
[41] I. Zahed and G.E. Brown, 'The Skyrme model', *Phys. Rept.* **142**, 1 (1986).
[42] M. Rho, *Chiral Nuclear Dynamics II: From quarks to nuclei to compact stars* (World Scientific Co, Singapore, 2008)
[43] For recent review, B.D. Serot and J.D. Walecka, 'Recent progress in quantum hadrodynamics', *Int. J. Mod. Phys. E* **6**, 515 (1997).
[44] V.A. Miransky and V.P. Gusynin, 'Chiral symmetry breaking and nonperturbative scale anomaly in gauge field theories', *Prog. Theor. Phys.* **81**, 426 (1989).
[45] H.K. Lee and M. Rho, 'Dilatons in hidden local symmetry for hadrons in dense matter', *Nucl. Phys. A* **829**, 76 (2009).
[46] J. Ellis and J. Lanik, 'Is scalar gluonium observable?', *Phys. Lett.* **B150**, 289 (1985).
[47] M. Bando, T. Kugo and K. Yamawaki, 'Nonlinear realization and hidden local symmetries', *Phys. Rept.* **164**, 217 (1988).
[48] M. Harada and K. Yamawaki, 'Hidden local symmetry at loop: A new perspective of composite gauge boson and chiral phase transition', *Phys. Rept.* **381**, 1 (2003).
[49] T. Matsui, 'Fermi liquid properties of nuclear matter in a relativistic mean-field theory', *Nucl. Phys.* **A370**, 365 (1981).
[50] C. Song, G.E. Brown, D.P. Min and M. Rho, 'Fluctuations in 'Brown-Rho scaled' chiral Lagrangians', *Phys. Rev.* **C56**, 2244 (1997).
[51] See, e.g. G.A. Lalazissis, S. Karatzikos, R. Fossion, D. Pena Arteaga, A.V. Afanasjev and P. Ring, 'The effective force NL3 revisited', *Phys. Lett. B* **671**, 36 (2009).
[52] B. Friman and M. Rho, 'From chiral Lagrangians to Landau Fermi liquid theory of nuclear matter', *Nucl. Phys.* **A606**, 303 (1996).
[53] A. Ohnishi, N. Kawamoto and K. Miura, 'Brown-Rho scaling in the strong coupling lattice QCD', *Mod. Phys. Lett. A* **23**, 2459 (2008).
[54] A. Manohar and H. Georgi, 'Chiral quarks and the nonrelativistic quark model', *Nucl. Phys. B* **234**, 189 (1984).
[55] R. Nasseripour *et al.* [CLAS Collaboration], 'Search for medium modification of the ρ meson', *Phys. Rev. Lett.* **99**, 262302 (2007).

[56] G.E. Brown, J.W. Holt, C.H. Lee and M. Rho, 'Late hadronization and matter formed at RHIC: Vector manifestation, BR (Brown-Rho) scaling and hadronic freedom', *Phys. Rept.* **439**, 161 (2007).
[57] J.W. Holt, G.E. Brown, T.T.S. Kuo, J.D. Holt and R. Machleidt, 'Shell model description of the 14C dating beta decay with Brown-Rho-scaled NN interactions', *Phys. Rev. Lett.* **100**, 062501 (2008).
[58] G.E. Brown, A. Sethi and N.M. Hintz, 'Proton nucleus scattering and density dependent meson masses in the nucleus', *Phys. Rev. C* **44**, 2653 (1991).
[59] A.B. Migdal, *Theory of finite nuclear system and application to atomic nuclei* (Interscience Publishers/John Wiley, 1967).
[60] G.E. Brown and M. Rho, 'On the manifestation of chiral symmetry in nuclei and dense nuclear matter', *Phys. Rept.* **363**, 85 (2002).
[61] G.E. Brown and M. Rho, 'Double decimation and sliding vacua in the nuclear many-body system', *Phys. Rept.* **396**, 1 (2004).
[62] G.Q. Li, C.M. Ko and G.E. Brown, 'Enhancement of low mass dileptons in heavy ion collisions', *Phys. Rev. Lett.* **75**, 4007 (1995).
[63] R. Rapp, 'In-medium vector mesons, dileptons and chiral restoration', arXiv:1010.1719 [nucl-th].
[64] M. Harada, S. Matsuzaki and K. Yamawaki, 'Holographic QCD integrated back to hidden local symmetry', arXiv:1007.4715 [hep-ph].
[65] G.E. Brown, M. Harada, J.W. Holt, M. Rho and C. Sasaki, 'Hidden local field theory and dileptons in relativistic heavy ion collisions', *Prog. Theor. Phys.* **121**, 1209 (2009); M. Rho, 'Dileptons get nearly 'blind' to mass-scaling effects in hot and/or dense matter', arXiv:0912.3116 [nucl-th].
[66] T. Yamazaki and Y. Akaishi, 'Nuclear medium effects on invariant mass spectra of hadrons decaying in nuclei', *Phys. Lett. B* **453**, 1 (2000).
[67] R. Shankar, 'Renormalization group approach to interacting fermions', *Rev. Mod. Phys.* **66**, 129 (1994).
[68] S.K. Bogner, T.T.S. Kuo and A. Schwenk, 'Model-independent low momentum nucleon interaction from phase shift equivalence', *Phys. Rept.* **386**, 1 (2003).
[69] J.W. Holt, G.E. Brown, J.D. Holt and T.T.S. Kuo, 'Nuclear matter with Brown-Rho-scaled Fermi liquid interactions', *Nucl. Phys. A* **785**, 322 (2007).
[70] L.W. Siu, J.W. Holt, T.T.S. Kuo and G.E. Brown, 'Low-momentum NN interactions and all-order summation of ring diagrams of symmetric nuclear matter', *Phys. Rev. C* **79**, 054004 (2009).
[71] D.B. Kaplan and A.E. Nelson, 'Strange goings on in dense nucleonic matter', *Phys. Lett.* **B175**, 57 (1986).
[72] G.E. Brown, V. Thorsson, K. Kubodera and M. Rho, 'A novel mechanism for kaon condensation in neutron star matter', *Phys. Lett. B* **291**, 355 (1992).
[73] H.D. Politzer and M.B. Wise, 'Kaon condensation in nuclear matter', *Phys. Lett. B* **273**, 156 (1991).
[74] G.E. Brown, C.H. Lee and M. Rho, 'Recent developments on kaon condensation and its astrophysical implications', *Phys. Rept.* **462**, 1 (2008).
[75] G.E. Brown, C.H. Lee and M. Rho, 'Kaon condensation, black holes and cosmological natural selection', *Phys. Rev. Lett.* **101**, 091101 (2008).

[76] G.E. Brown and M. Rho, 'From chiral mean field to Walecka mean field and kaon condensation', *Nucl. Phys.* **A596**, 503 (1996).

[77] G.E. Brown, C.H. Lee, H.J. Park and M. Rho, 'Study of strangeness condensation by expanding about the fixed point of the Harada-Yamawaki vector manifestation', *Phys. Rev. Lett.* **96**, 062303 (2006).

[78] *The Multifaceted Skyrmion*, ed. G.E. Brown and M. Rho (World Scientific, Singapore, 2010).

[79] H.K. Lee and M. Rho, in *The Multifaceted Skyrmions*, ed. G.E. Brown and M. Rho (World Scientific, Singapore, 2010).

[80] G.E. Brown, M. Rho and W. Weise, 'Phenomenological delineation of the quark-gluon structure from nucleon electromagnetic form-factors', *Nucl. Phys.* **A454**, 669 (1986).

[81] T. Sakai and S. Sugimoto, 'Low energy hadron physics in holographic QCD', *Prog. Theor. Phys.* **113**, 843 (2005); 'More on a holographic dual of QCD', *Prog. Theor. Phys.* **114**, 1083 (2005).

[82] D.K. Hong, M. Rho, H.U. Yee and P. Yi, 'Dynamics of baryons from string theory and vector dominance', *JHEP* **0709**, 063 (2007); 'Chiral dynamics of baryons from string theory', *Phys. Rev. D* **76**, 061901 (2007); 'Nucleon form factors and hidden symmetry in holographic QCD', *Phys. Rev. D* **77**, 014030 (2008); K. Hashimoto, T. Sakai and S. Sugimoto, 'Holographic baryons: Static properties and form factors from gauge/string duality', *Prog. Theor. Phys.* **120**, 1093 (2008); K.Y. Kim and I. Zahed, 'Electromagnetic baryon form factors from holographic QCD', *JHEP* **0809**, 007 (2008) [arXiv:0807.0033 [hep-th]].

[83] M. Harada and M. Rho, 'Integrating holographic vector dominance to hidden local symmetry for the nucleon form factor', arXiv:1102.5489 [hep-ph].

[84] H.K. Lee, B.Y. Park and M. Rho, *Phys. Rev.* **C83**, 025206 (2011).

Chapter 18

Multi-W-Z-top Bags, and their Possible Role in Cosmological Baryogenesis

Edward Shuryak*

As an amusing incarnation of Nuclear Physics, the gauge bosons W, Z and the top quarks may form collective states bound by Higgs-based attraction. Since Higgs vacuum expectation value (VEV) is significantly modified inside them, they are called "bags". We argue that creation of such objects can explain certain numerical studies of cosmological baryogenesis. Using as an example a hybrid model, combining inflationary preheating with cold electroweak transition, we identify "spots of unbroken phase" found in numerical studies of this scenario with such $W-Z$ bags. We argue that the baryon number violation should happen predominantly inside these objects, and show that the rates calculated in numerical simulations can be analytically explained using finite-size pure gauge sphaleron solutions, developed previously in the QCD context by Carter, Ostrovsky and Shuryak (COS). Furthermore, we point out significant presence of the top quarks/antiquarks in these bags (which were not included in those numerical studies). We find that tops help to stabilize them for a longer time. Another enhancement of the transition rate comes from the "recycling" of the tops in the topological transition. Although we do not discuss it here, it is highly probable that this scenario would produce CP violation by the usual CKM matrix mechanism large enough to explain the observed baryon asymmetry.

1. Few Stories about Gerry

Having daily lunch discussions in our common room for about twenty years, dominated by Gerry Brown, I heard his favorite stories about people he met in his life more than once. As he repeatly told us (and also wrote in his memoirs), his life path has been deeply affected by three "eagles", with

*Department of Physics and Astronomy, State University of New York, Stony Brook, NY 11794, USA.

whom he tried to "fly": Gregory Breit, Rudolf Peierls and Hans Bethe. His attitude toward them goes well beyond the normal gratitude of a student towards his teachers, as Gerry very well realized that the main thing he learned from them were not some scientific ideas, or particular facts or tools, but the attitude. Indeed, if he had not met them, he would be quite a different person. Gerry never forgot the life story of his father, who had certain math talent, but due to lack of connections and some unfortunate turn of events was discouraged to pursue research and settled for teaching math in small South Dakota college.

From what I have learned, his relations with Breit were not warm or very personal, but it was very different with two others. Gerry lived in the Peierls house for a long time, run by the strong hand of Peierls' wife Eugenia (Genya). From his stories it looked like she extended her maternal instincts beyond her immediate family and took care of the affairs of the entire Peierls group of postdocs and students, from their basic needs (like a roof under their head) to even their personal lives.

I was not a witness of his time in UK, of course, but still I think retelling one story, from Gerry's own words, may be instructive. Gerry does not use or like any gadgets (other than his beloved 10-dollar worth calculator) or computers, as he is a master of how to organize the right calculation whenever needed. And yet, during his Peierls days, when he badly needed certain atomic theory integrals be numerically calculated, he managed to do it, in a curious way. He said he hitchhiked or took bus to Cambridge, where he knew he could use their first lamp computer on weekends. It was so, Gerry recalled, because that machine was very unreliable and full of bad electrical contacts, so any calculation could only proceed if little kicks were applied periodically, and sometimes he and his friends had to hold certain pieces by hand to keep it going.

His relation with Bethe was extremely intense, right till Hans' death in 2005: in fact Gerry spoke with him for the last time on the morning of that very day. Nearly daily we could see faxes from Bethe to Gerry on our fax machine, full of formulae with frightening 4 or 5 digit page numbers. I recall once taking one of them to Gerry: he pointed out some vertical line in the middle of the page which made no relation to the text: "see", he said, "Hans fall asleep here."

I witnessed it from a distance, of course, starting from 1993. In that year I was one of the organizers of Santa Barbara program, while Gerry went to his usual winter retreat with Bethe in Caltech: and I visited them there. Here is how it worked: Gerry was in charge of practical things like cooking (their wives later were also going with them, but not on earlier years of their collaboration when they wanted to get away from everything and focus on science). After breakfast he was picking up a problem and provided empirical information when needed; in the days before the internet made it easier Gerry's art of navigating in the library and getting the relevant facts was a great help. (As an aside, let me point out here that Gerry was one of the organizers of Physics Reports, today a journal with about the highest impact factor, as he always highly valued a subfield being cleverly summarized.) Then they thought about it separately till lunch, after which they discussed the results. Hans, the great problem solver as he was, often was able to give a complete solution.

One story had especially impressed Gerry so that he told it again and again. At one of their early meeting, in Copenhagen, Gerry proposed to discuss the famous supernova problem. The prevailing idea was that when matter implodes and density grows, nuclei are decomposed into nucleons. Bethe, being presented with it in the morning by Gerry, in the evening told Gerry, in his straightforward style that "... they are all wrong... because the entropy is too low." And sure enough, Hans was right. The point of the story is that in a few hours (or less) Hans had overthrown years of other people's work, and he did so by just one simple thought based on a general principle, the entropy conservation. This shows what, among other qualities Gerry valued in his "eagles", the main one still was the simple fact that all three were exceptionally clever.

On a lighter note, I cannot resist relating a funny episode of that 1993 meeting. The phone rang and Gerry answered it. The call was from the Russian embassy: Gerry passed phone to me, as the Russian speaking person in the room. The embassy however simply wanted to know if Bethe was there, since they had to forward a letter to him from Yeltsin (then Russian president). I confirmed that and they hung up. Gerry asked: "Hans, why does President Yeltsin write to you?" Bethe said: "Well, it must be a reply to my letter". "And what did you write to him about?" inquired Gerry. "One evening I had nothing else to think about and I thought about how one can solve the economic problems of Russia, and I wrote to him what my suggestions were." I said: "See Gerry, the problems you keep providing are apparently not hard enough". Everybody laughed. Yet Bethe did not tell us about his economic suggestions, steering the conversation back to the physics subjects we discussed. (Later, however, he told Gerry that his idea was to back currency by parcels of state land, an idea he perhaps recalled from the time of Weimar Republic, when he was a young boy.)

Gerry values highly phenomenologists who can look at some apparently unrelated empirical facts and then draw out important consequences from it, by pure deduction. In fact, he has some Sherlock-Holmes-type abilities himself. Let me give an example outside science: Once on a weekend during the conference at Tennessee we went hiking in the mountains for few hours. Going downhill in a big group was too slow for Ismail Zahed, who just ran down ahead of the others. When we found him waiting for us at the bottom, he said: "Look, I just met a deer." "Perhaps at the right turn near the small brook?" asked Gerry. "How do you know?" asked Ismail, with surprise. "Well, there were fresh deer droppings there, which looked larger than usual. Had you frightened the deer?" "Yes indeed", confirmed Ismail.

Needless to say, Gerry helped me a lot, especially during our settling in at Stony Brook. But soon I learned that Gerry also practiced random acts of kindness to people he just met. A typical story: spotting a student who seemed to be in distress in the corridor, he learned he flunked the comprehensive exam. (To make it a bit clearer: this student was not from our group at all, most likely Gerry had just met him for the first time, as he was not teaching large courses.) Many people would be satisfied by the explanation, but not Gerry: he wanted to know why he thought he flanked. It turned out student's wife (back in Europe) told him she decided to leave him a couple days before. Gerry did not say a word, he just went down to the Physics Office to find out how to organize an additional oral exam for him, which saved him from expulsion.

Let me end this memoir section with some comments on how Gerry works. At lunches he often presented certain facts, and then asked us to guess their explanations, usually providing rather provocative ones of his own to get the discussion going. The audience happily jumped on the occasion to disprove them, criticizing from every possible angle: it was precisely what Gerry wanted. "You take a position in the argument," he was instructing us, "and then let them shoot at you". This trick was very effective in getting other people become involved in problems he was interested in.

Gerry knows very well how important it is to attract the best students possible, and he is very good at that. In fact the number of Ph.Ds defended under him is unbelievably high, 72 according to his count, which may be some kind of a record. And, whatever they did later, many of them I met kept personal relationships with Gerry for decades.

2. Motivation and Previous Papers

For this collection I selected an issue which is well outside of the mainstream of Gerry's or my interests, and even outside from any well-trodden paths at all, being somewhat in between the nuclear and high energy physics. Most of it has been done during winter retreat to Auckland, New Zealand, in January 2010, with my collaborator and long-time friend Victor Flambaum and now is available as two papers [1,2]. (Yes, of course, we tried to emulate Brown-Bethe winter retreats...)

Let me start with a nuclear physics-based motivation, based on (much-simplified) Walecka model, in which the nuclear forces can be approximately described by the σ and ω meson exchanges. The σ is "the Higgs boson of the nuclear physics", obtaining VEV due to chiral symmetry breaking phenomenon. The σ and ω meson masses set a scale $1/m_{\sigma,\omega} \sim 0.3\,\mathrm{fm}$ for the range of nuclear forces. Note however that this scale is not at all the typical internuclear distances $n^{-1/3} \sim 1.5\,\mathrm{fm}$ in nuclei, which is about a factor 5 larger. Moreover, since the Yukawa forces fall off exponentially, the typical nuclear potential is much smaller than what one would naively get from these forces. This all happens because their masses ($m_\sigma \sim 600\,\mathrm{MeV}$, $m_\omega \sim 770\,\mathrm{MeV}$) as well as couplings are similar. As a result, the sigma-induced attraction and the omega-induced repulsion nearly completely cancel each other! As we tell at the beginning of nuclear physics courses, nuclear forces are so weak that they fail to bind two neutrons and barely bind a deuteron. And yet, they still lead to lots of bound/metastable nuclei with larger number of nucleons.

Standard model (SM) Higgs is quite similar to σ of the nuclear physics. Can Higgs exchange between top quarks or gauge bosons also create multi-quanta bound states? Unfortunately the strength of the Higgs exchange is about an order of magnitude weaker than that for nuclear forces: even for the heaviest of all, the top quark, the corresponding Yukawa coupling is not $O(1)$ but only

$$\frac{g_t^2}{4\pi} = 0.08 \tag{1}$$

This however may be helped by adding more particles, as Higgs field cannot be screened and is just additive.

What about the range of this interaction, in respect to masses of bound particles? We do not yet know the Higgs boson mass, and for now assume that it is about $M_H \approx 120\,\mathrm{GeV}$, the first non-excluded island. If so, it has roughly the same relation to the top quark mass ($M_t = 172\,\mathrm{GeV}$) as the sigma mass has to the nucleon mass (600 and 939 MeV, respectively). And, last but not least, we do *not* know or expect existence of any vectors (analogs of the ω) inducing comparable repulsion. (I am not an expert but I asked around and people told me that the existing models for e.g. a "techni-rho" place their mass rather high, above the 1 TeV.)

So, can many tops be bound together by a common Higgs potential, providing "a second incarnation" of the nuclear physics? Before we try to answer this question, let me point out two more useful analogies. Because the Higgs exchange is an *universal attraction* between all kinds of particles with the strength proportional to their masses, it is similar to gravity. Gravity, feeble as it is, is able to hold together planets, stars and even create closed systems (black holes), provided its rather weak coupling is compensated by a huge number N of participating particles acting together. True, unlike a graviton the Higgs boson is not massless: but its mass is strongly reduced near the electroweak phase transition: can that help to bind?

The last question leads to another analogy (for me, the most important one). Second order phase transitions are associated with infinite correlation lengths (or massless modes) which in practice is cut off by the IR cutoff available such as the system size. As suggested by Stephanov, Rajagopal and myself [3] some time ago, such example *may* be found in heavy ion collision, known as the QCD (tri)critical point. A long-range attraction induced by the *massless* mode (a mixture between the sigma and omega, in this case) may lead to interesting effects discussed in [4].

3. The Bags

3.1. *The twelve tops*

The interest in the issue originated from the question whether a single sufficiently heavy SM-type fermion should actually be accompanied by its own "bag" [5], depleting the Higgs VEV around. Although classically this seemed to be possible, it was shown (by Farhi *et al.* [6] and others) that quantum (one loop) effects destabilize such bags. Unfortunately, at strong enough Yukawa coupling the SM itself gets apparently sick, with an instability of its ground state.

The issue rest dormant for some time till Nielsen and Froggatt [7] suggested to look at many tops. The first magic number is 12 tops+antitops, corresponding to the lowest $l = 0$, $j = 1/2$ orbital, with 3 colors and 2 from $t + \bar{t}$. Using simple formulae from atomic physics these authors suggested that such system already forms a bound state. In fact by some coincidence of numbers they even get extremely bound state, with a near-zero total mass. We have checked [8] this interesting claim and found that, unfortunately, this is *not* the case. While for a massless Higgs there are indeed bound states of 12 tops, for a realistic Higss mass, $m_H > 100\,\text{GeV}$ they disappear. Further variational improvement of the binding conditions for the 12-quark system were discussed by Richard [9] who confirmed that 12 tops are unbound for Higgs mass $m_H > M_c(12) \sim 50\,\text{GeV}$.

The role of quantum corrections may be crucial to the existence of *multi*-quanta bags. This led us [10] to study of the magnitude of one-loop quantum corrections to such bags in a scalar approximation (which still used some simplifications, we ignored the spin of the top quark and the Pauli principle, using a scalar field instead) as a function of N and g. Specifically, by writing the quantum-corrected energy of such states as a loop expansion

$$E(N) = E_{\text{cl.}}\left[1 + \sum_k C_k(N)\left(\frac{g^2}{4\pi}\right)^k\right], \tag{2}$$

we calculated the one-loop coefficient C_1 in [10], following both the specific model and the practical methods developed in [6]. The central issue we investigated was the N-dependence of quantum corrections for different couplings: calculation of $C_1(N)$ for $g = 1.8$ showed that the for $g \sim 1$, the correction for large bags is dominated by the scattering phases of vacuum fields on the bag, and that $C_1 \sim O(1)$ for large N, without a strong N-dependence. For top quarks, the Yukawa coupling is $g^2/4\pi \approx .08$ and one-loop quantum corrections are about $\Delta E = 0.06\,E_{\text{cl.}}$. We thus concluded that although for hypothetical fermions with a coupling a few times larger that the top mass (i.e. mass$\,>1\,$TeV) quantum corrections are indeed important. And yet for top-bags they are still under good theoretical control.

3.2. *Two types of bags*

According to the SM, the dynamics of the Higgs field Φ is described by the Lagrangian density

$$\mathcal{L}_{\text{Higgs}} = \frac{1}{2}|D_\mu\Phi|^2 - \frac{\lambda}{4}(\Phi\Phi^\dagger - v^2)^2, \tag{3}$$

where D_μ is the corresponding (gauge) covariant derivative. The usual vacuum state therefore corresponds to a homogeneous Higgs field $\langle\Phi\rangle = v$ permeating all of space. In the conventional unitary gauge, the Higgs field

Φ is represented by the real, dimensionless, field ϕ

$$\Phi = v \begin{pmatrix} 0 \\ \phi \end{pmatrix}, \tag{4}$$

where v is the vacuum expectation value (VEV) which the Higgs achieves when $\phi = 1$. Assuming spherical symmetry, the energy reads

$$E_{\text{Higgs}} = 2\pi v^2 \int_0^\infty dr\, r^2 \left[\phi'^2 + \frac{1}{4} m_H^2 (\phi^2 - 1)^2\right] \tag{5}$$

where $m_H^2 \equiv 2\lambda v^2$ is the Higgs physical mass. We will take $v = 246\,\text{GeV}$ and $m_H = 100\,\text{GeV}$ and use units of $100\,\text{GeV}$ throughout the paper.

Consider now the addition of fields with a global internal symmetry, leading to a conserved particle number N associated to each field. If these N particles (fermions or bosons), couple strongly to the Higgs field, or $N \gg 1$, the Higgs could be strongly distorted in the vicinity of these particles. To describe this situation, we adopt a mean-field approximation, in which all the particles are described by the *same* wavefunctions in the background of a non-trivial Higgs field. Corrections to the mean-field description, such as, many-body, recoil and retardation of the Higgs field are expected to be suppressed by factors v/m, m_H/m and $1/N$. In the semiclassical approximation, the total energy of the system will therefore be

$$E_{\text{cl.}} = E_{\text{Higgs}} + \sum_a n_a \varepsilon_a, \tag{6}$$

where $\{\varepsilon_a\}$ is the spectrum of the corresponding field in the Higgs background, n_a is the occupation number.

We can crudely estimate the order of magnitude of N for which such bags could exist. If we were to deplete a certain large volume of the Higgs VEV (surface terms neglected for now), it would require an energy

$$Vol \cdot \frac{m_H^2}{8} v^2.$$

For a bag of radius, say, $R \cdot 100\,\text{GeV} = 4$, this energy is about $20\,\text{TeV}$. Thus, if the lowest W-boson energy level has a binding energy of the order of $20\,\text{GeV}$ per W or Z, an order of $O(1000)$ of them would be needed to compensate for the bag energy and to make binding possible. Although top quarks are heavier and one may hope to get binding of the order of $100\,\text{GeV}$, the situation with fermions is much more delicate: like in atomic and nuclear physics, the states should be populated shell by shell.

If the Higgs VEV is distorted nearly to zero in some volume of space, particles become nearly massless and we shall refer to such bags as "no-Higgs" bags. These are analogous in spirit to the various bags used in the 1970's to model QCD confinement (e.g. the famous MIT bag), although there is no need in our case to take the limit of infinitely massive quanta outside the bag.

There is another, maybe less obvious, configuration which we shall refer to as an "inverted bag". In this configuration, we let the Higgs reverse its sign inside the bag, crossing zero at some finite radius r_0, and asymptotically taking the usual Higgs VEV v. If the Higgs field gets close to $-v$ inside the sphere, such a bag is simply filled with another Higgs vacuum, so there is no volume energy but only a surface contribution. As we shall see below, fermions are localized around the node at $r \approx r_0$, within a shell of width $O(1/m_H)$. Therefore, for large N we expect the system to be described by a 2-dimensional gas of volume $\propto r_0^2$.

3.3. *Pure W-Z bags*

The gauge bosons, as massive spin-1 particles, have 3 polarizations, and thus 3 different types of modes in the bag: the electric (e), longitudinal (l) and magnetic (m) ones. Assuming stationary fields and spherical symmetry, one can define

$$\boldsymbol{W}^{(e,l,m)} = \boldsymbol{Y}_{jm}^{(e,l,m)} f_{(e,l,m)}(r)/r, \tag{7}$$

where $\boldsymbol{Y}_{jm}$ are spherical harmonic vectors and $f_e(r), f_l(r)$ and $f_m(r)$ are the radial wavefunctions for each mode. In a static, spherically symmetric, background the magnetic mode satisfies simple Klein-Gordon equation for the radial wavefunction

$$\left(\frac{d^2}{dr^2} + \omega^2 - M_W^2\phi^2 - \frac{j(j+1)}{r^2}\right) f_m(r) = 0 \tag{8}$$

with $j \geq 1$. However, the Laplacian mixes the electro-longitudinal modes, leading to the set of coupled equations

$$\left(\frac{d^2}{dr^2} + \omega^2 - M_W^2\phi^2 - \frac{j(j+1)}{r^2}\right) f_e(r) - \frac{2\sqrt{j(j+1)}(1 - r\frac{d\ln\phi}{dr}) f_l(r)}{r^2} = 0. \tag{9}$$

$$\left(\frac{d^2}{dr^2} + \omega^2 - M_W^2\phi^2 - \frac{j(j+1)}{r^2}\right) f_l(r)$$

$$+ 2r\frac{d}{dr}\left(\frac{f_L(r)}{r}\frac{d\ln\phi}{dr}\right) - \frac{2\sqrt{j(j+1)}f_e(r)}{r^2} = 0. \qquad (10)$$

The last term (not present for scalars or spinors) is singular where the Higgs ϕ crosses zero, effectively repelling the electro-longitudinal modes. However, it vanishes identically for the magnetic modes causing, these energy levels to be below the electro-longitudinal ones, and so we concentrate on magnetic levels.

The total energy of the system with N W-bosons in the lowest energy level ε_0 simply reads

$$E_{\text{cl.}} = E_{\text{Higgs}} + N\varepsilon_0. \qquad (11)$$

To prove the existence of local minima with finite N, we took a variational approach based on the Higgs ansatz

$$\phi(r) = 1 - \alpha \exp\left(-r^2/w^2\right), \qquad (12)$$

where the two parameters, α and w describe the depth and the width of the bag potential. We can parametrize the Higgs energy (5) in terms of α and w, which results in

$$E_{\text{Higgs}} = \frac{(2\pi\alpha v)^2}{16\sqrt{2\pi}}\left[3w + w^3 M_H^2\left(1 - \frac{4}{3\sqrt{6}}\alpha + \frac{1}{8\sqrt{2}}\alpha^2\right)\right]. \qquad (13)$$

We then solve numerically for the levels (see examples in Fig. 1) and then searched for local minima of the total energy both in the α and w direction. As expected, for the Higgs mass of about 100 GeV, it only happens when N is of the order of few thousends.

3.4. *Top quarks in a Higgs bag*

Consider a system of N heavy fermions interacting with a background Higgs field Φ, described by the SM Lagrangian density

$$\mathcal{L} = \mathcal{L}_{\text{Higgs}} + \bar{\psi}(i\gamma^\mu\partial_\mu - g\Phi)\psi \qquad (14)$$

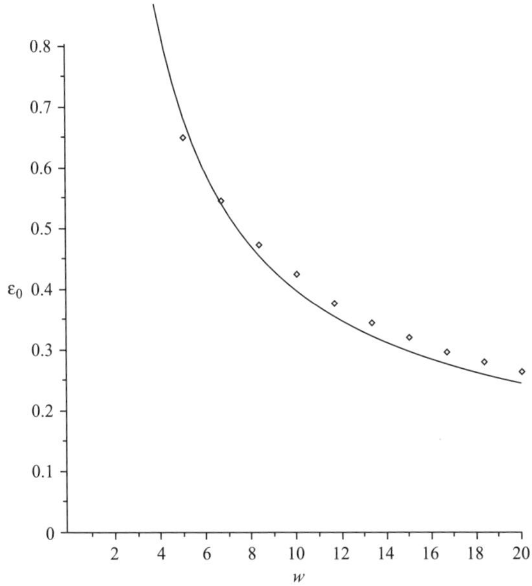

Fig. 1. Lowest magnetic mode ($j = 1$) levels for a Gaussian bag with $\alpha = 1.3$. The numerical results (points) are in good accordance with the analytical result we do not discuss here.

where Φ is complex scalar in the fundamental representation of $SU(2)_L$ and ψ is a Dirac spinor in 3+1 dimensions and

$$N \equiv \int d^3x \psi^\dagger i\partial_0 \psi \tag{15}$$

is the conserved fermion number. For completeness, let us remind the reader the standard notations for Dirac spinors in spherical coordinates

$$\psi = \frac{1}{r}\begin{pmatrix} F(r)\Omega_{jlm} \\ (-1)^{1/2(1+l-l')}G(r)\Omega_{jl'm} \end{pmatrix}, \tag{16}$$

where Ω_{jlm} are spherical 2-component spinors and we take normalization $\int dr\,(F^2 + G^2) = 1$. The so-called Dirac parameter κ is defined as

$$\kappa = \begin{cases} -(l+1) & \text{for } j = l + 1/2 \\ l & \text{for } j = l - 1/2 \end{cases} \tag{17}$$

and runs over all nonzero integers, being positive for anti-parallel spin and negative for parallel spin. Dirac's equation reads

$$(\varepsilon - m\phi)F = -G' + (\kappa/r)G \qquad (18)$$
$$(\varepsilon + m\phi)G = F' + (\kappa/r)\ F$$

In order to analytically justify the physical picture of inverted bags advocated for earlier, we calculate the energy of surface-fermions for an inverted bag of large radius. We assume that $\phi(r)$ has a node at large r_0, at which the derivative is *positive*, $\phi'(r_0) > 0$, we search for solutions of the Dirac equation in the near-Gaussian form

$$F(r) = A(r)\, e^{-S(r)}, \qquad (19)$$
$$G(r) = B(r)\, e^{-S(r)},$$
$$S(r) = m \int_{r_0}^{r} \phi(r')dr' \simeq \frac{1}{2} m\phi'(r_0)(r - r_0)^2.$$

The large mass m here favors the localization of $F(r)$ and $G(r)$ in the vicinity of r_0, which justifies the approximation. Omitting further details, we found analytic answer for the energy

$$\varepsilon_{\pm}^2 = m\phi'(r_0) + \kappa^2/r_0^2 \pm \sqrt{m^2\phi'^2(r_0) + \frac{1}{r_0^4}}. \qquad (20)$$

We see from this expression that the levels ε_+ are finite for $r \to \infty$, while ε_- goes to zero. The latter correspond to the would-be zero modes in the 1 dimensional case. These will be lower than the former for large enough bags and in what follows we consider these levels only. Note that to order $O(\frac{1}{mr_0})$,

$$\varepsilon_- \simeq \frac{|\kappa|}{r_0}, \qquad (21)$$

hence in this approximation the spectrum is independent of the mass of the fermion and the value for the first level ($\kappa = -1$) is less than 2.04, which appears for the volume bag. Thus, not only the Higgs energy, but the fermionic energy benefits as well.

Solving Dirac's equation numerically is rather straightforward and examples of the levels for a particular bag are shown in Fig. 2.

Let me spare the details of how it works, and go to conclusions. We showed in [1] that pure top bound systems cannot exist for any N if the Higgs is as heavy as experimental bound indicate, $m_H > 100\,\text{GeV}$: they

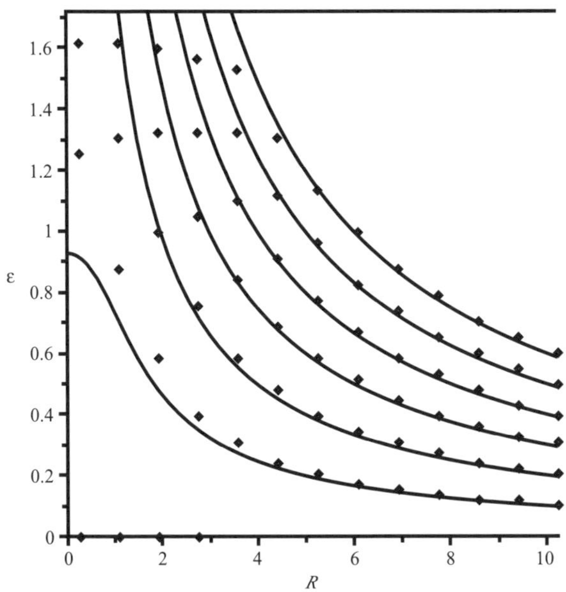

Fig. 2. Spectrum of the Dirac operator in the background of an inverted bag ($\eta = 1$) as a function of the bags's size R. These levels correspond to increasing l (from the bottom up) for $\kappa < 0$ only. The analytical ε_- results for large R (solid lines) are in good accordance with the numerical results (points) for even rather small R.

would be there in large numbers if say the Higgs mass be about 50 GeV. But tops can still be well bound in the W-Z -dominated bags: and in the next section we will argue that this may happen in cosmology, with important consequences for baryosinthesis.

4. Baryogenesis Beyond the Standard Cosmology

4.1. *The hybrid scenario*

The question how the observed baryonic asymmetry of the Universe was produced is among the most difficult open questions of physics and cosmology. Sakharov [11] had formulated three famous necessary conditions: the baryon number and the CP violations, with obligatory deviations from thermal equilibrium. Although all of them are present in the Standard Model (SM) and standard Big Bang cosmology, the known mechanisms creating the baryon asymmetry can only produced effects many orders smaller than the observed amount, usually expressed as the ratio of the baryon density to that of the photons $n_B/n_\gamma \sim 10^{-10}$. To find a scenario in which this puzzle can at least in principle be solved is one of the motivation of this paper. It follows from general anomaly relation and, in particularly, from known electroweak instanton and sphaleron solutions, that baryon number can be violated by certain processes providing a change in topology of the gauge field. In the broken phase of the SM we live in today these processes are basically frozen, by a very high barrier associated with Higgs VEV. On the other hand, in the symmetric phase — the electroweak plasma — the sphaleron rates are suppressed only by the powers of the coupling: those rates are high enough to wipe out any primordial asymmetry, if it has been generated cosmologically prior to the electroweak transition era.

The bubbles associated with the expected 1-st order electroweak phase transition had attracted significant attention in 1980's–1990's. Moving walls of such bubbles may supply the necessary local deviations from thermal equilibrium. The rate of thermal sphaleron transitions, both in equilibrium and near the bubble walls, has been studied extensively, see [15–18] and many more. However, as the experimental limits on the Higgs mass have over time evolved upward, to $M_H > 116\,\mathrm{GeV}$ today, it became clear that the first order transition is actually impossible in the SM. (Transition is a crossover for $M_H > 80\,\mathrm{GeV}$, see again [12] for details and the original references). After this has fact has been acknowledged, people either looked at various phenomena *beyond* the SM such as its supersymmetric extensions, which may still allow some window for the 1st order transitions.

An alternative is the modification of the standard cosmology. Instead of the usual Big Bang scenario, with its adiabatically slow crossing of the phase transition and tiny deviations from equilibrium, the so called "hybrid" scenario has been proposed [22, 23] It combines the end of the inflation era with the establishment of the electroweak broken phase. In it the nonperturbative production of long-wavelength gauge and Higgs fields happens far from equilibrium, returning to equilibrium later, at a temperature well below the critical temperatures T_c of the electroweak transition; therefore it is sometimes called the "cold" electroweak scenario. While based on some fine tuning of the unknown physics of the inflation, it avoids many pitfalls of the standard cosmology, such as "erasure" of asymmetries generated before the electroweak scale.

(i) One important finding of the simulations is that the initial coherent oscillations of scalars soon give way to the usual broken phase. The most important feature is persistence of "no-Higgs spots" in which Higgs VEV is very far from the equilibrium value v and is instead close to zero. The gauge fields in them have however rather high magnitude. Figure 3 (from [24]) show an example of a snapshot of the Higgs field modulus. Typically the volume fraction occupied by such "no-Higgs spots" is of the order of several percents and is decreasing with time.

(ii) The second important findings is that of topologically nontrivial fluctuations of the gauge fields. As shown in the lower Fig. 3, those are well localized. According to the anomaly relation, changes in topology leads to violation of $B + L$ (baryon and lepton) numbers, via emission/absorption of fermions, 12 in the SM.

It was observed that topological fluctuations happen only *inside* the "no-Higgs spots" mentioned above. Indeed, this becomes apparent from the distribution of the topological charge shown in the lower part of Fig. 3, for the same time configuration of the Higgs as shown in the upper part of the Fig. 3. Of course, it is only one snapshot, but the authors found from the simulations that it is true for the whole sample.

The fraction of no-Higgs spots which induced topological transitions is also in the range of a percent. More precise measure is the so called sphaleron rate is defined by the mean square deviation from zero of the Chern-Simons number

$$\Gamma(t) = \frac{1}{m^4 V}\frac{d\Delta N_{CS}^2}{dt} \tag{22}$$

(here and below all quantities are defined via one characteristic mass parameter m: for the simulations its value is about 264 GeV, close to v. Since the process only exists for finite period of time, Ref. [24] reports its time integrated value defined by some time integral

$$I(mt) = \int_{t_i}^{t} d(mt)\Gamma(t) \tag{23}$$

during which it basically happens. (Too early there are no gauge fields and too late there are no no-Higgs spots.) By the end of the process ($mt < 45$) it is in the range

$$I \sim 10^{-4} \tag{24}$$

for more details about different parameters sets see Table II of [24].

Although we will not go into details of our paper [2], one can understand these numerical results using two ideas. One is that the "no-Higgs spots" are nothing else but mechanically stabilized W-Z bags. The other is that the sphalerons appearing inside those spots (and thus having a particular

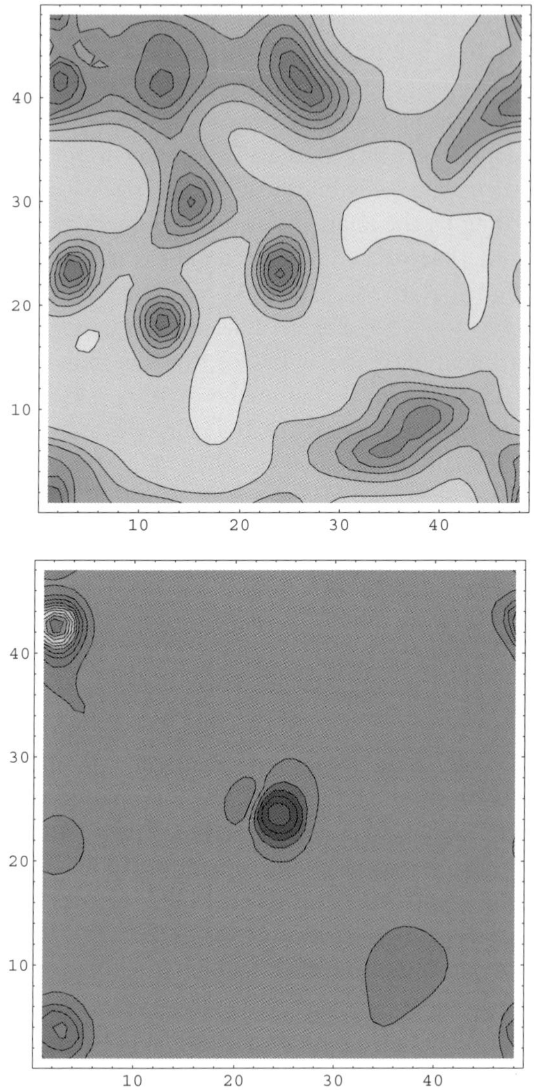

Fig. 3. The contour plots from [24] of the modulus of the Higgs field $|\phi(x)|^2$ (upper plot) and the topological charge density $Q(\vec{x}, t)$ at time $mt = 19$, for the model A1, $N_s = 48$. Dark areas in the upper plot correspond to small VEV, while light bulk corresponds to the broken phase. On the lower plot lumps of the topological charge density appear as dark regions. While many of the no-Higgs spots do not have the topological transitions, all transitions seem to be inside the spots.

size ρ) can be described by the so called COS (Carter-Ostrovsky-Shuryak) sphaleron [21], which has fixed size and is a solution to pure Yang-Mills equations, without Higgs. Using both one is able to explain the sphaleron rate found in the numerical simulations under consideration.

4.2. *A cosmological role of the top quarks*

Since the original numerical simulations have included the gauge fields but ignored fermions, we have to discuss first, at quite qualitative level, what their effects can be, relative to that of the gauge fields. Those tops would be added to the metastable bubbles of the symmetric phase, the no-Higgs spots, like the W discussed above. However, there are at least three important differences:

(i) their fugacity is expected to be larger, due to more rapid production rate of t relative to W
(ii) the degeneracy factor $g_{\bar{t},t} = 12$ is twice larger than for gauge bosons (6).
(iii) their binding to Higgs bags is several times larger than for W, which makes it easier to satisfy the mechanical stabilization of the bag. (For details on that see [1].)

The first question is at what time and how the top quarks would be produced. Top quark has the largest coupling to Higgs: so it is produced most copiously, via the $HH \to \bar{t}t$ process. Let us crudely compare its rate relative to that of $HH \to WW$

$$\frac{\Gamma(HH \to \bar{t}t)}{\Gamma(HH \to WW)} \sim \left(\frac{m_t}{m_W}\right)^4 \sim 20 \tag{25}$$

where in the first estimate follows from the fact that the (vacuum) masses are proportional to the corresponding coupling constants, which appears in the power 4 in the rate. This simple estimate suggests that top production may be by about an order of magnitude more rapid than that of Ws, and thus they would in fact dominate in the formation of "spots".

The next step is concentration of top quarks into the bags, happening because they are massive in the broken phase and massless in the symmetric one. As discussed in detail in [1], their binding can be as large as 100 GeV or so. Less trivial reason is that a "1d kink" of the Higgs field (a 2-d surface where Higgs VEV crosses zero) possesses fermion zero modes. In practice it means that top quarks can glue themselves to the $\phi = 0$ surfaces and be transported along them, eventually into the remaining island of the symmetric phase. The binding energy in such case is nearly all the top mass, 172 GeV, several time the bulk temperature.

The main role of the top quarks is their "recycling" by sphalerons. The well known Adler-Bell-Jackiw anomaly require that a change in gauge field topology by $\Delta Q \pm 1$ must be accompanied by a corresponding change in baryon and lepton numbers, B and L. More specifically, such topologically nontrivial fluctuation can thus be viewed as a "t'Hooft operator" with 12 fermionic legs. Particular fermions depend on orientation of the gauge fields in the electroweak SU(2): since we are interested in utilization of top quarks, we will assume it to be "up". In such case the produced set contains $t_r t_b t_g c_r c_b c_g u_r u_b u_g \tau \mu, e$, where r, b, g are three quark colors. We will refer below to this process as the $0 \to 12$ reaction. Of course, in matter with a nonzero fermion density many more reactions of the type $n \to (12 - n)$ are allowed, with n (anti)fermions captured from the initial state.

To our knowledge, the only (analytic) solution to Dirac eqn of the "expansion stage" was obtained in [19], it describes motion from the sphaleron zero mode at $t = 0$ and all the way to large time $t \to +\infty$. As we know the outgoing physical state of fermions, one can calculate e.g. their mean energy or momentum distribution. A new element is that its time-reflection can also describe the compression stage, from free fermions captured by a convergent spherical wave of gauge field at $t \to -\infty$ and ending at the sphaleron zero mode at $t = 0$.

How the corresponding fermionic levels move is schematically shown in Fig. 4. In the Dirac sea notations, the level moves from some negative initial energy at early time to positive one at late time. For COS sphalerons the mean energy is $E_q = -3/\rho-$ to $E_q = +3/\rho$. By symmetry, the level is crossing zero at zero time $t = 0$: the corresponding fermionic zero energy

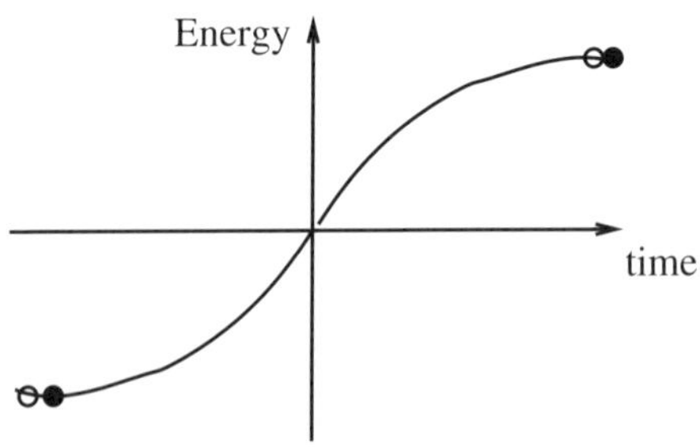

Fig. 4. Schematic picture of fermionic level motion as a function of time. Open and closed circles indicate unoccupied and occupied states.

state is at this moment nothing but the well known sphaleron zero mode. If the level was occupied (as most of levels below the surface of the Dirac sea are) it remains occupied and one gets new fermions produced. If there was antiquark (an unoccupied state or a hole, indicated by the open circle) it will remain unoccupied, thus having no meaning after the level moves to positive energy, where most levels are unoccupied. So, antiquarks — holes at negative energy sea — are actually *decelerated* by the radial electric field to zero energy modes, their energy being transmitted to the gauge field.

This conclusion is very important: it implies that energy of the initial fermions can be incorporated and used in the sphaleron transition. One finds that the optimum way to generate sphaleron transition is to use 3 initial top quarks, considering the $3 \to 9$ fermion process instead of the original $0 \to 12$ one. This process "economize" an amount of energy compable with the total hight of the COS sphaleron barrier (which, we remind, is about 2000 GeV) making the wholse scenario woring.

5. Summary and Discussion

In summary, we have first discussed the very existence of the $W-Z$-top Higgs-based bags, and found that with (admittedly very large) number of quanta those can be mechanically stable or even bound. We are all waiting for LHC to discover the Higgs boson: its mass is the key parameter of the problem needed to make it quantitative. It may also happen that LHC would reveal new heavier gauge bosons or fermions: and, if so, their binding would even happen at smaller number of them. The opposite thing which may happen would be discovery of vectors (like techni-rho) which may partly cancel Higgs-based attraction, as we know happens in nuclear physics with σ, ω exchanges.

We have also shown that top quarks may play an important role in baryogenesis. We have used, as a specific example of the so called hybrid scenario, for which numerical simulations are available. We first reproduced the main results of those simulation using the $W-Z$ bags and COS sphaleron, including qualitatively the sphaleron size and rate found numerically. We then discussed the role of the top quarks and found them to be quite important. They are produced via $HH \to \bar{t}t$ more effectively than W. They are collected into the bags together with the gauge quanta, presumably producing bags of larger size, with improved lifetime. In addition, we pointed out an interesting effect of "top quark recycling". This effect helps the transition, providing significant ($\sim$1/3) fraction of the energy needed to reach the sphaleron mass. And, last but not least, we argue — contrary to well known textbooks statements — that known CKM-base mechanism alone can provide the CP violation of the magnitude needed to explain the observed baryon asymmetry of the Universe. The only nontrivial element of the whole scenario is the "hybrid" cosmology, in which the end of inflation/creation of entropy of the Universe happens *below* the electroweak phase transition.

Right now there is a work in progress, by Yannis Burnier and myself, devoted to a calculation of the CP violation near the W bags and also sphalerons. Although the calculation is not yet finished, CP asymmetry of the order of 10^{-7} (needed to solve the baryon asymmetry puzzle in this scenario) seems to be reachable.

Let us finish this paper with a historical analogy. We do see stars on the sky which shine: yet basic nuclear physics was not enough to explain why they can do so, falling short by many orders of magnitude. It took decades of patient work by very clever people — Hans Bethe especially — to uncover highly nontrivial ways in which the star cycles work. It turned out that those crucially depend on some *catalysts*, whose presence is very small but crucial. Perhaps the $W-Z$-top bags are also a kind of a catalyst, or rather a "doorway state", helping to increase the probability of the baryon number violation and of the CP violation by a large factor.

Acknowledgments

Obviously the main gratitude goes to Gerry Brown, who has taught us, by example, not only how to do physics, but how effectively organize the process and to keep both the lines of personal interest and the group going. I am also indebted to Holger Nielson whose talk triggered this work, and obviously to my collaborators Victor Flambaum, Michail Kuchiev and Marcos Crichigno, with whom those results were obtained.

References

[1] M.P. Crichigno, V. Flambaum, M. Kuchiev and E. Shuryak, $W-Z$-Top Bags, arxiv:1006.0645, accepted to *Phys. Rev. D*.
[2] V.V. Flambaum and E. Shuryak, "Possible Role of the WZ-Top-Quark Bags in Baryogenesis," arXiv:1006.0249 [hep-ph]. accepted to *Phys. Rev. D*.
[3] M.A. Stephanov, K. Rajagopal and E.V. Shuryak, *Phys. Rev. Lett.* **81**, 4816 (1998) [arXiv:hep-ph/9806219].

[4] E. Shuryak, "Interactions between hadrons are strongly modified near the QCD (tri)critical point," arXiv:hep-ph/0504048.
[5] R. Johnson and J. Schechter, *Phys. Rev. D* **36**, 1484 (1987).
[6] E. Farhi, N. Graham, R.L. Jaffe, V. Khemani and H. Weigel, *Nucl. Phys. B* **665**, 623 (2003), [arXiv:hep-th/0303159].
[7] C.D. Froggatt, L.V. Laperashvili, R.B. Nevzorov, H.B. Nielsen and C.R. Das, (2008), 0804.4506; C.D. Froggatt and H.B. Nielsen, *Surveys High Energ. Phys.* **18**, 55 (2003), hep-ph/0308144; C.D. Froggatt, H.B. Nielsen and L.V. Laperashvili, *Int. J. Mod. Phys.* **A20**, 1268 (2005), hep-ph/0406110.
[8] M.Y. Kuchiev, V.V. Flambaum and E. Shuryak, *Phys. Rev. D* **78**, 077502 (2008), [arXiv:0808.3632 [hep-ph]].
[9] J.M. Richard, *Few Body Syst.* **45**, 65 (2009), [arXiv:0811.2711 [hep-ph]].
[10] M.P. Crichigno and E. Shuryak, arXiv:0909.5629 [hep-ph].
[11] A.D. Sakharov, *JETP Lett.* **6**, 24 (1967).
[12] V.A. Rubakov and M.E. Shaposhnikov, *Usp. Fiz. Nauk* **166**, 493 (1996) [*Phys. Usp.* **39**, 461 (1996)], [arXiv:hep-ph/9603208].
[13] M. Trodden, *Rev. Mod. Phys.* **71**, 1463 (1999), [arXiv:hep-ph/9803479]. [arXiv:hep-ph/0411301].
[14] M. Dine and A. Kusenko, *Rev. Mod. Phys.* **76**, 1 (2004), [arXiv:hep-ph/0303065].
[15] F.R. Klinkhamer and N.S. Manton, *Phys. Rev. D* **30**, 2212 (1984).
[16] P. Arnold and L.D. McLerran, *Phys. Rev. D* **36**, 581 (1987).
[17] L. Carson, X. Li, L.D. McLerran and R.T. Wang, *Phys. Rev. D* **42**, 2127 (1990).
[18] G.D. Moore, *Phys. Rev. D* **53**, 5906 (1996), [arXiv:hep-ph/9508405].
[19] E. Shuryak and I. Zahed, *Phys. Rev. D* **67**, 014006 (2003), [arXiv:hep-ph/0206022].
[20] Y. Burnier, M. Laine and M. Shaposhnikov, *JCAP* **0602**, 007 (2006), [arXiv:hep-ph/0511246].
[21] D.M. Ostrovsky, G.W. Carter and E.V. Shuryak, *Phys. Rev. D* **66**, 036004 (2002), [arXiv:hep-ph/0204224].
[22] J. García-Bellido, D. Grigoriev, A. Kusenko and M. Shaposhnikov, *Phys. Rev. D* **60**, 123504 (1999).
[23] L.M. Krauss and M. Trodden, *Phys. Rev. Lett.* **83**, 1502 (1999).
[24] J. Garcia-Bellido, M. Garcia-Perez and A. Gonzalez-Arroyo, *Phys. Rev. D* **69**, 023504 (2004), [arXiv:hep-ph/0304285].
[25] A. Tranberg and J. Smit, JHEP **0311**, 016 (2003), [arXiv:hep-ph/0310342]. Jon-Ivar Skullerud, Jan Smit, Anders Tranberg, *JHEP* **08**, 045 (2003).

Chapter 19

Theory of Finite Fermi Systems — The Stony Brook Jülich Interaction

J. Speth*, S. Krewald* and F. Grümmer*

The Landau-Migdal theory describes the linear response of strongly interacting Fermi systems by weakly interacting quasi particles. Gerry Brown initiated the first successful derivation of the quasi particle interaction from the ground state energy of nuclear matter. He realized the importance of the pion, which determines the long range part of the quasi-particle interaction and suggested to treat the pion exchange as an explicit part of the quasi particle interaction, which has to be complemented by Landau's contact interaction. The resulting quasi particle interaction has a strong momentum dependence which proved to be essential for a quantitative description of nuclear excitation modes with unnatural parity which are discussed in this review.

1. Personal Recollections by Josef Speth

I met Gerry for the first time 1972 in Osaka at a conference on magnetic moments. I gave my first talk at an international meeting and Gerry made some rather nasty comments after my talk which were even included in the proceedings. Our conversations during the meeting were much more pleasant and he invited me in 1973 to Copenhagen and in the following year to Stony Brook. I spent my first sabbatical in 1975 in his institute and from that time on we met several time a year in Stony Brook, in Jülich and at various conferences. He also spent the time of his *Humboldt award* in our institute and became an expert for walking trails on the *Sophien Höhe*, the artificial mountain near Jülich. Last but not least I had the pleasure to be his tennis partner and instructor. The photograph shows him in 1979 in his

*Institut für Kernphysik, Forschungszentrum Jülich, 52425 Jülich, Germany.

Gerry in Jülich in front of the *Kaiserhof* (1979).

best suit in front of the *Kaiserhof* in Jülich. He participated at the *first communion* of our youngest son.

2. Introduction

The Landau-Migdal Theory of *Finite Fermi Systems* (TFFS) [1] is based on Landau's *Fermi Liquid Theory* [2]. It is one of the most successful semi-empirical approaches to the structure of nuclei. The derivation of the basic formulas in the framework of many-body Green functions (GF) is fully microscopic. All quantities which appear in the final formulas, like the effective ph-interaction and the effective operators, are well defined and can in principle be calculated within the approach. It is important to mention that the derivation of the basic formulas is very general and the corresponding equations are not restricted to the TFFS. We will see in the next section, for the numerical application one needs single particle energies, single particle wave functions, an effective (ph)-interaction and for magnetic properties also effective operators. If pairing is included one also needs an effective particle-particle (pp) interaction. In the TFFS, and this is the phenomenological aspect, the single particle quantities are deduced from phenomenological single-particle models and the residual interactions as well as the effective operators are parametrized. In self consistent approaches the single particle data are obtained from a mean field calculated with an effective Hamiltonian, Lagrangian or energy functional and the residual ph-interaction is given in such an approach as second derivative of the energy. As we will see, Migdal's effective ph-interaction is in leading order a density-dependent zero-range interaction. The next to leading order, a momentum dependent zero-range force has never been used in numerical applications. In our contribution, we review a generalization of Landau's zero-range ansatz, which was initiated by Gerry. It started with the observation that in the spin-isospin channel, the long range one pion exchange will give an important contribution, which is not well described by a zero-range ansatz. For that reason, he suggested that the one-pion exchange contribution should be considered explicitly. This gives rise to a strong momentum dependence in the spin-isospin ph-channel. Actually, Migdal [3] was the first one, who considered the contribution of the pion in connection with pion condensation in nuclei. However, as shown by Gerry [4], Migdal neglected *his own* strongly repulsive g_0' parameter in the spin-isospin channel. The complete interaction does not allow pion condensation in nuclei. In the next section the basic formulas are given together with Migdal's zero range ansatz and the Stony-Brook Jülich generalization. In the fourth section we report on numerical applications that demonstrate the importance of the momentum dependent ansatz. In connection with the investigation of the Gamow-Teller resonances in (p,n)-reactions one observed a strong energy-dependence of the strength of the central spin-independent component of the (p,n) coupling potential, whereas the spin-isospin interaction remains constant. The explanation of this phenomenon is given in the fourth section. In the fifth section we report on microscopic calculations of the force parameters that also show a strong momentum dependence in the spin-isospin channel.

3. Theory of Finite Fermi Systems

3.1. *Basic formulas*

In the following we present a few steps of the derivation of the RPA equation within the Green function formalism, which is the basis not only for the

TFFS but for many other microscopic approaches. More details may be found in [5].

The one-particle and two-particle Green functions are defined as

$$g_{\nu_1\nu_2}(t_1;t_2) = (-i)\langle A0|T\{a_{\nu_1}(t_1)a^+_{\nu_2}(t_2)\}|A0\rangle \tag{1}$$

$$g_{\nu_1\nu_3\nu_2\nu_4}(t_1t_3;t_2t_4) = (-i)^2\langle A0|\, T\{a_{\nu_1}(t_1)a_{\nu_3}(t_3)a^+_{\nu_4}(t_4)a^+_{\nu_2}(t_2)\}\,|A0\rangle \tag{2}$$

with the time-dependent creation and annihilation operators of nucleons in the state φ_ν

$$a_\nu(t) = e^{iHt}a_\nu\, e^{-iHt} \quad a^+_\nu(t) = e^{iHt}a^+_\nu e^{-iHt}. \tag{3}$$

The symbol T denotes the time-ordering operator of Wick, which means that the operators should be taken in time-ordered form with the latest time to the left and the earliest to the right. In addition, the product has to be multiplied by a factor ± 1 depending on whether an even or odd number of permutations is needed to come from the given to the time-ordered form. The central object of the approach is the response function L which is defined as:

$$L(13,24) = g(13,24) - g(1,2)g(3,4). \tag{4}$$

The response function obeys an integral equation

$$L(13,24) = -g(1,4)g(3,2) - i\int d5d6d7d8g(1,5)K(57,68)L(83,74)g(6,2) \tag{5}$$

where K is an effective two-body interaction. A graphical representation of Eq. (5) is given in Fig. 1. One notices that the kernel K of the integral equation is irreducible with respect to the particle-hole propagator.

In Eq. (5) the kernel as well as the one-particle Green functions are unknown. Following the quasi-particle concept of Landau one separates the

Fig. 1. Graphical representation of the integral equation for the response function in the ph channel.

one-particle Green function into a singular part and a remainder:

$$g_{\nu_1\nu_2}(\omega) = \frac{\delta_{\nu_1\nu_2} z_{\nu_1}}{\omega - \epsilon_{\nu_1} + i\eta sign(\epsilon_{\nu_1} - \mu)} + g^{(r)}_{\nu_1\nu_2}(\omega). \tag{6}$$

In this equation ϵ_{ν_1} are the single particle energies measured from the Fermi surface and z_{ν_1} are the single particle strengths. The main goal is to obtain an equation for the response function that can be solved in practice. With the ansatz in Eq. (6) one writes the product of two GF as a singular part S and a rest B:

$$g(\epsilon, \Omega)g(\epsilon) = S(\epsilon, \Omega) + B(\epsilon, \Omega). \tag{7}$$

The singular part has the form:

$$A_{\nu_1\nu_2,\nu_3\nu_4}(\epsilon, \Omega) = 2\pi i z_{\nu_1} z_{\nu_2} \delta_{\nu_1\nu_3} \delta_{\nu_2\nu_4} \frac{n_{\nu_1} - n_{\nu_2}}{\epsilon_{\nu_1} - \epsilon_{\nu_2} - \Omega} \delta\left(\Omega - \frac{\epsilon_{\nu_1} + \epsilon_{\nu_2}}{2}\right) \tag{8}$$

here the n_ν are the occupation numbers for quasi particles: 1 and 0 for particles and holes, respectively. With this ansatz one can rewrite Eq. (5) and obtains after some analytical transformations (see [5]) the renormalized RPA equation for nuclei in the single-particle configuration space. The ph-interaction F^{ph} is energy dependent (which has to be neglected) and a function of the configuration space.

$$(\epsilon_{\nu_1} - \epsilon_{\nu_2} - \Omega_m)\chi^m_{\nu_1\nu_2} = (n_{\nu_1} - n_{\nu_2}) \sum_{\nu_3\nu_4} \mathrm{F^{ph}}_{\nu_1\nu_4\nu_2\nu_3} \chi^{\mathrm{m}}_{\nu_3\nu_4}. \tag{9}$$

Here Ω_m are the excitation energies and χ^m the corresponding quasi-particle quasi-hole transition matrix elements. The quantities n_ν are the occupation numbers 1 and 0 for particles and holes, respectively. Equation (9) is a short-hand notation of the well known RPA equation, e.g. Ref. [6]:

$$\begin{pmatrix} A & B \\ B^* & A^* \end{pmatrix} \begin{pmatrix} X^m \\ Y^m \end{pmatrix} = \Omega_m \begin{pmatrix} 1 & 0 \\ 0 & -1 \end{pmatrix} \begin{pmatrix} X^m \\ Y^m \end{pmatrix} \tag{10}$$

with the usual definitions of the matrix elements A and B see also Ref. [6]. The transition amplitudes between the excited state $|Am\rangle$ and the ground state $|A0\rangle$ of a nucleus with A nucleons for a Hermitian one-body operator

Q are given as:

$$\langle Am|Q|A0\rangle = \sum_{\nu_1\nu_2} Q^{\text{eff}}_{\nu_1\nu_2}\, \chi^m_{\nu_1\nu_2}. \tag{11}$$

Note: Due to the renormalization procedure [1,5] the one-particle operators are also renormalized. The renormalized one-particle operators Q^{eff} include the z_ν factors and the *regular* part B. In the case of electric multipole operators Q^{eff} corresponds to the bare operators, due to charge conservation, whereas the magnetic operators are parametrized, with universal parameters [1, 5]. The response function includes also an equation for moments and transitions in the neighboring odd-mass nuclei [1, 5]. These quantities are connected with the *vertex operator* τ

$$\langle A \pm 1, \alpha|Q|A \pm 1, \beta\rangle - \delta_{\alpha,\beta}\langle A0|Q|A0\rangle = \tau_{\alpha\beta}(\epsilon_{\alpha\beta}, Q). \tag{12}$$

The corresponding equation for the vertex operator $\tau_{\alpha\beta}(\epsilon_{\alpha\beta}, Q)$ has the form:

$$\tau_{\nu_1\nu_2}(\epsilon_{\alpha\beta}, Q) = Q^{eff}_{\alpha\beta}\delta_{\nu_1\alpha}\delta_{\nu_2\beta} + \sum_{\nu_3\nu_4} F^{ph}_{\nu_1\nu_4\nu_2\nu_3}\frac{n_{\nu_3} - n_{\nu_4}}{\epsilon_{\nu_3} - \epsilon_{\nu_4} - \epsilon_{\alpha\beta}}\tau_{\nu_3\nu_4}(\epsilon_{\alpha\beta}, Q). \tag{13}$$

Here $\epsilon_{\alpha\beta}$ is the energy difference between the two states α and β. From Eq. (12) we see that in the case of moments one can only calculate the difference between the even and odd mass nuclei. This, however, allows a very precise determination of the differences of corresponding charge distributions (isotope shifts).

3.2. *Make the equations solvable*

In order to solve the basic equations one needs as input single particle-wave functions, single-particle energies, the ph-and pp-interaction and for magnetic properties also an effective magnetic operator. Migdal has designed his theory in close connection to Landau's Fermi liquid theory. Therefore one takes as far as possible the input data from experiment or from models which reproduce the needed experimental data as good as possible. The single-particle wave functions are taken from a single-particle model and the single-particle energies are taken as far as possible from experiment. If the RPA equation is derived within the many-body Green function formalism [5] one obtains an explicit form for the ph-interaction F^{ph} that depends on the single-particle model and the size of the configuration space. In addition it is nonlocal and energy-dependent. In practice the ph-interaction is not calculated from that formula but it is parametrized. Following Landau's procedure F^{ph} is transformed into momentum space and one considers the interaction on the Fermi surface of nuclear matter. Here one can replace the energies by the Fermi energy and the magnitude of the momenta by the Fermi momentum. In this approximation F^{ph} depends only on the angle between the ph-momenta $\mathbf{P}$ and $\mathbf{P}'$ before and after the collision; it reads (we suppress the spin and isospin dependence)

$$F^{\rm ph}\left(\frac{\mathbf{P}\cdot\mathbf{P}'}{P_F^2}\right)=\sum_{l=0}^{\infty}F_l P_l\left(\frac{\mathbf{P}\cdot\mathbf{P}'}{P_F^2}\right). \tag{14}$$

Here $P_l(x)$ is the Legendre polynomial of order l and the constants F_l are the famous Landau-Migdal parameters. One introduces dimensionless parameters by defining:

$$F_l = C_0 f_l. \tag{15}$$

where $C_0 = 300\,[\mathrm{MeV\,fm}^3]$ is proportional to the inverse density of states at the Fermi surface. The Fourier transforms of the terms with $l = 0$ and $l = 1$ yield δ-functions and derivatives of δ-functions. The finite size of the nuclei is taken into account by density dependent parameters $F_l(\rho)$. The form of the interaction in leading order can be written as:

$$\begin{aligned}F^{\rm ph}(1,2) = C_0\delta(\mathbf{r}_1-\mathbf{r}_2)\cdot\big[&f_0(\rho)+f_0'(\rho)\tau_1\cdot\tau_2+g_0(\rho)\sigma_1\cdot\sigma_2\\&+g_0'(\rho)\sigma_1\cdot\sigma_2\tau_1\cdot\tau_2\big].\end{aligned} \tag{16}$$

By definition the parameters include the direct and exchange contribution of the ph-interaction, as the direct and exchange matrix elements of a zero-range interactions are identical. For that reason only the spin dependent components of the interaction contribute to magnetic moments and transitions.

3.3. *The one-pion-exchange in the ph-interaction*

For magnetic properties it has been shown [7] that one has also to include the effects of the one-pion-exchange. The direct part of the one-pion-exchange is long-ranged and can not be approximated by a zero-range force. For that reason on has to replace in the spin-isospin channel the conventional Landau-Migdal parametrization by the *Stony Brook Jülich ansatz*, which has the following form in momentum space:

$$G'(q) = \int \frac{d^3k}{(2\pi)^3}[V_\pi(\mathbf{k}) + V_\rho(\mathbf{k})]\Omega_c(\mathbf{q}-\mathbf{k}) + \delta G'_0 \sigma \cdot \sigma' \tau \cdot \tau' \tag{17}$$

with

$$\Omega_c(\mathbf{q}) = (2\pi)^3\delta(\mathbf{q}) - \frac{2\pi^2}{q^2}\delta(|\mathbf{q}| - q_c). \tag{18}$$

Here V_π and V_ρ denote the one-pion and one-rho exchange, $\Omega_c(\mathbf{q})$ a parametrization of the nuclear correlation function [8] with $q_c = 3.93$, which is the inverse of the Compton wavelength of the ω meson, and $\delta G'_0$ a small (universal) correction parameter that has to be adjusted to the data. The ρ exchange has been added because it introduces a natural cut-off for the tensor part of the one-pion exchange. A detailed discussion can be found in [7].

In the upper part of Fig. 2 the q-dependence of G' is plotted. One realizes that at small momentum transfers the central part of the

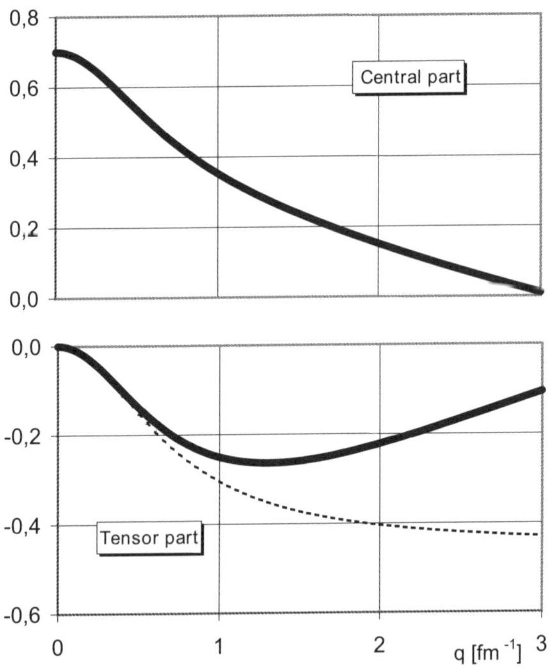

Fig. 2. Momentum dependence of the spin-isospin parameter $G'(q)$ in units of [300 MeV fm^3]. The full line is the completee model of Eq. (17), the dashed line in the lower part is the correlated one-pion exchange only.

spin-isospin interaction is strongly repulsive and for larger momentum transfers it is weak. This is the reason why, for example, the Gamow-Teller resonances, where small momenta are involved, are strongly shifted to higher energies whereas high-spin states are not [7,9]. For the central part π-meson and ρ-meson contributions have the same sign and for the tensor part the opposite sign. The ρ-meson exchange therefore acts as a natural cut-off for the strong tensor component of the one-pion exchange. All details of the calculations are given in the original publications.

4. Results

4.1. *Unnatural states in light nuclei*

Two questions which had to be answered were: (I) What is the magnitude of g_0' and (II) how can one test the q-dependence of Eq. (17). Here one was especially interested in the behavior of $g_0'(q)$ between $q \geq 1$ and $q \leq 3[\mathrm{fm}^{-1}]$ because this regime is important in connection with pion condensation. As we can see from Fig. 3, the unnatural parity states in ^{16}O are especially appropriate to investigate this momentum region. In this Figure the form factors $(\sigma \cdot \nabla)$ of various magnetic states in ^{16}O are shown that are peaked in the above mentioned momentum region. The effect of the momentum dependence of the force on a given state can be discussed qualitatively by considering these form factors, as the largest (diagonal) contributions to the ph-force are simply a double convolution of the transition density with the interaction. As one example, the effect of the π- and ρ- exchange on the 2^- states in ^{16}O is demonstrated in Fig. 4. The major part of the 2^- strength is concentrated around 24 MeV (δ) if one uses a zero-range force as F^{ph}-interaction, and it is shifted to 20 MeV with the Stony Brook Jülich ansatz, indicated by $(\delta + \pi + \rho)$ in good agreement with the data (Exp). In the second row the uncorrelated spectrum is shown ($F^{\mathrm{ph}} = 0$). If one compares the uncorrelated spectrum with the data one realizes that a repulsive interactions is needed in the RPA calculation. The pure δ-force would be too strong, whereas the inclusion of the π and ρ contributions reduces the strong repulsion and gives a theoretical result in agreement with the data. The important message, of that investigation, however, was that in the relevant momentum region where pion condensation should appear, the interaction is still repulsive and not attractive.

4.2. *Unnatural states in heavy mass nuclei*

In heavy mass nuclei the form factors of corresponding multipole states are shifted to smaller momenta, as the nuclei a larger. In Fig. 5 the form

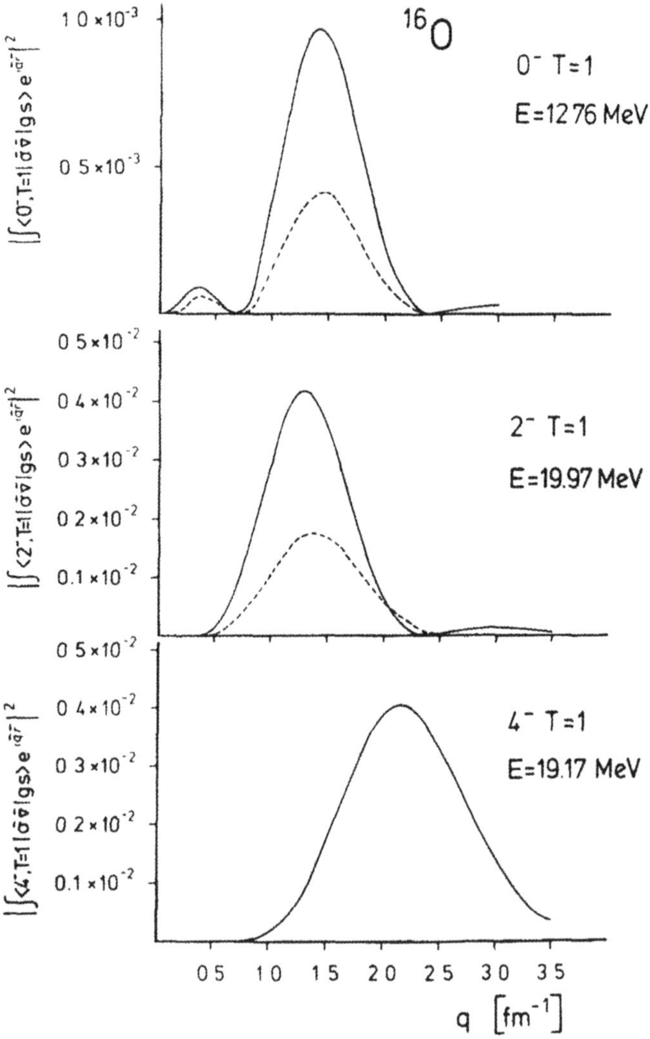

Fig. 3. Fourier transform of the $(\sigma \cdot \nabla)$ transition density of various unnatural parity states of ^{16}O. The dashed lines are the proton contributions of the strongest ph configuration.

factors of the transition density of the Gamow-Teller (GT)-resonance, the 0^-, 1^- and 2^- spin-isospin modes in ^{208}Pb are shown. From here one expects that the 0^-, 1^- and the GT-resonances are sensitive to the low momentum behavior of the interaction, i.e. the force is strongly repulsive. In those cases one obtains in a RPA calculation one collective state which

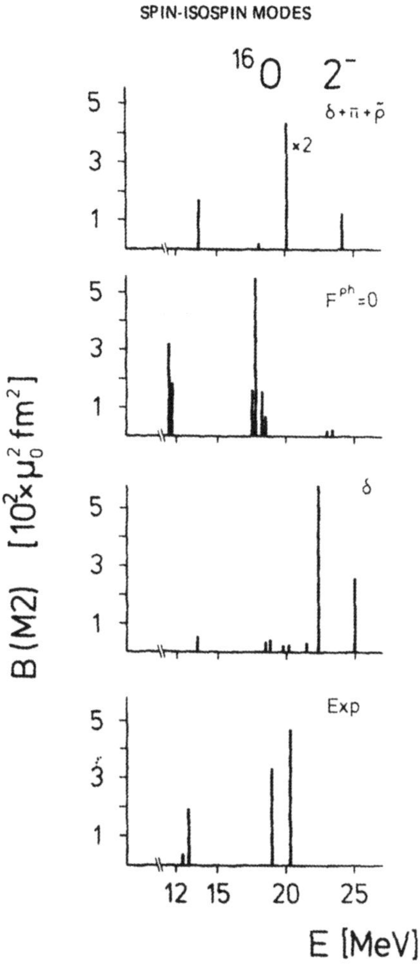

Fig. 4. Strength distribution of the 2^- states in ^{16}O are calculated for various choices of the ph interaction. In the lowest part the experimental distribution is shown.

is strongly shifted to higher energies. This is demonstrated in Table 1, where the averaged ph-energies ϵ_{ph} are compared with the RPA results E_{RPA}. In all three cases, the shift ΔE is of the order of 5 MeV. The 2^- result is qualitatively different which one already can expect from the form factor in Fig. 5. Here the ph-force is weak, so there exists no longer a collective state, but one obtains four states which are only little shifted from the uncorrelated ph-energies. The comparison with the data shows a fair agreement as far as the mean energies are concerned.

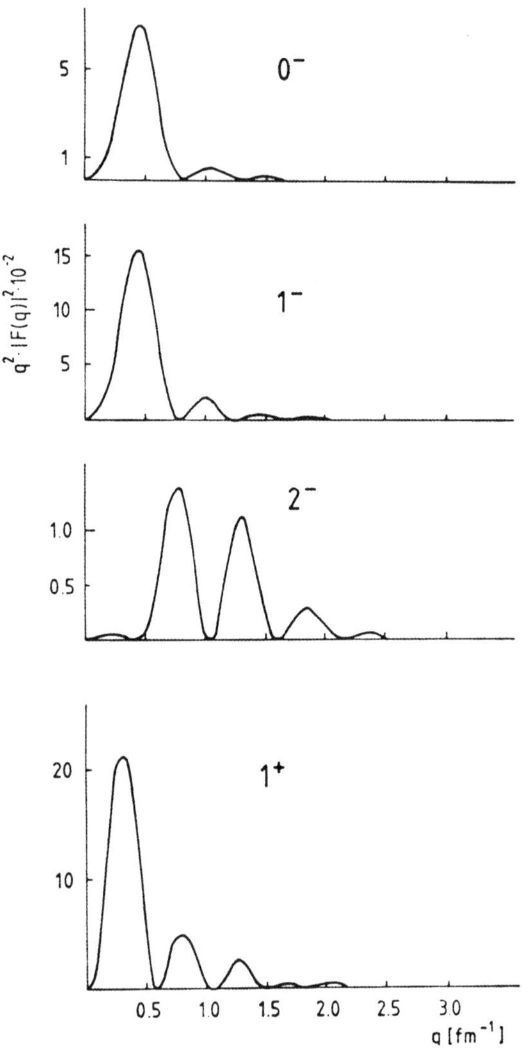

Fig. 5. Fourier transform of the transition density of the GT-resonance and the 0^-, 1^- and 2^- modes in ^{208}Pb.

The charge exchange resonances have relatively large widths which can not be obtained within a 1p1h approach but one has to include higher configurations. In Fig. 6 an example for such more involved calculations is shown. The authors [10] extended the conventional RPA approach and include 1p1h as well as 2p2h-configurations in a consistent way. This

Table 1. Charge-exchange resonances in ^{208}Pb.

J^π	ϵ_{ph} [MeV]	E_{RPA} [MeV]	ΔE [MeV]	E_{exp} [MeV]
0^-	21.8	26.6	4.8	25.1 ± 1.0
1^+	13.1	18.9	5.8	19.2 ± 0.2
1^-	21.5	26.3	4.8	25.1 ± 1.0
2^-	20.7	22.6	1.9	
	20.1	21.2	1.1	25.1 ± 1.0
	23.3	24.5	1.2	
	28.0	28.8	0.8	

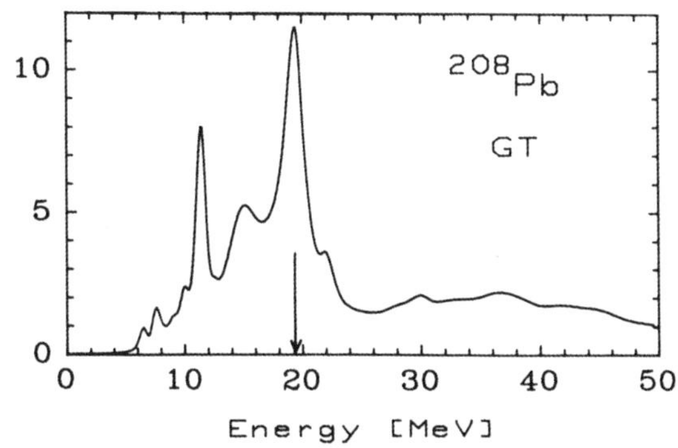

Fig. 6. Gamow-Teller resonance in ^{208}Pb calculated in an extended RPA model [10]. The arrow indicated the experimental mean energy.

calculation not only reproduced the known experimental energy and width but it also predicted a long tail up the 50 MeV, where nearly half of the strength is hidden. Several years later the theoretical prediction has been proven correct by experiment and explained the *missing* GT strength in nuclei.

4.3. *Magnetic high spin states in ^{208}Pb*

The most striking example for the Stony Brook Jülich interaction within an extended RPA model are the unnatural 12^- and 14^- high spin states [12] in ^{208}Pb the cross sections of which are peaked around $q \approx 2$ [fm^{-1}]. The energies of the two 12^- and the one 14^- states are close to the experimental ph-energies but the cross sections are only half of the shell model prediction. The explanation for this surprising result was given in Ref. [9]. As the spin-isospin interaction in the relevant momentum range is essentially zero, the

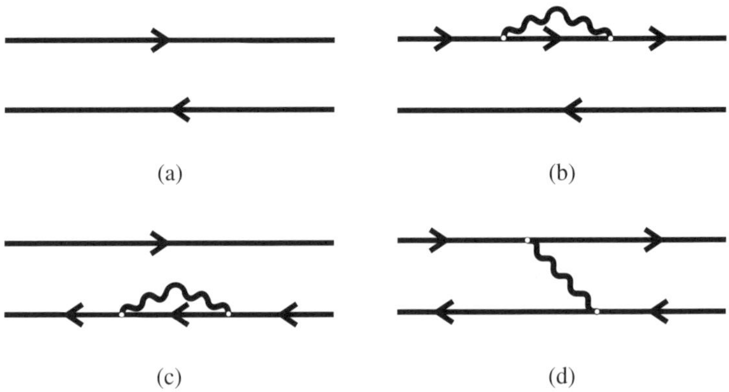

Fig. 7. Gamow-Teller resonance in ^{208}Pb calculated in an extended RPA model [10]. The arrow indicated the experimental mean energy.

RPA solutions are close to the unperturbed ph-energies. In their extended model the authors included also the effects of the low-lying phonons. The various contributions of the so called *core coupling* RPA [11] are given in Figs. 7(b)–7(d). In the conventional RPA only the shell model ph propagator (a) is considered. In the extended model the contributions (b) and (c) give rise to the fragmentation of the single particle strength and (d) modifies the ph-interaction. As the single particle states which couple to these high spin states are *intruder states* from the next higher shell these states are especially strongly effected by the low-lying 3^- and 5^- phonons which reduces the single particle strength by about 40%. In Fig. 8 the theoretical results are compared with the experimental data.

4.4. *Energy dependence of the coupling potentials in (p,n) reaction*

The above mentioned Gamow Teller resonances were predicted long time ago as very collective giant resonances in heavy mass nuclei. Experimentally it seemed difficult to observe these *spin-isospin* resonances with the existing accelerators with proton energies below 50 Mev. Here the *electric* charge-exchange resonances like the well known isobaric-analog resonances, which are connected with the isospin operator τ only, dominated the experimental cross sections.

The situation is reversed if the energy of the protons is increased. C. Goodman *et al.* [13] used 160 MeV protons and at this energy the spin-isospin resonance is the dominant structure in the differential cross sections

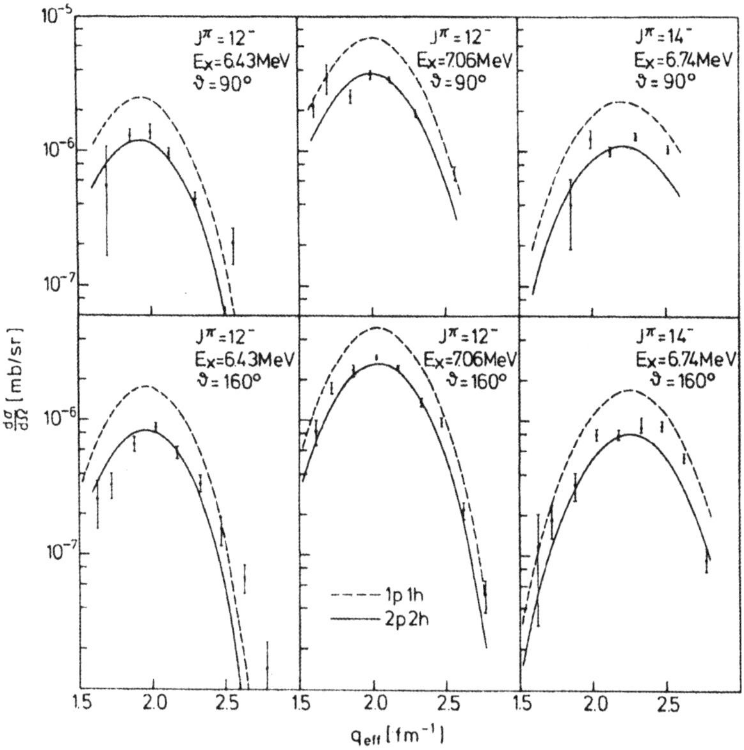

Fig. 8. Inelastic electron scattering cross sections at $\Theta = 90°$ and $\theta = 160°$ of the magnetic high spin states in ^{208}Pb. Experiments are compared with the RPA results (dashed lines) and the extended model. Calculations have been done in DWBA.

at zero degree in (p,n) reactions. From the analysis of the cross sections one deduces that the ratio of the corresponding coupling potential $[V_{\sigma\tau}/V_\tau]^2$ increases by about a factor 10 if one changes the proton energy from 40 to 200 MeV. Petrovich and Love obtained this unexpected energy dependence of the coupling potentials from their G-matrix interaction [14] in good qualitative agreement with the experimental finding. The physical origin of this surprising effect follows immediately from the Stony Brook Jülich interaction if one includes the next to leading order [15] as shown in Fig. 9.

The main results of the calculation can be summarized as: The $V_{\sigma\tau}$ coupling potential is only weakly dependent on the energy because is dominated by the direct part of the one-pion and one-rho exchange which is as a first order process energy independent (Fig. 9(a)). First order contributions to the V_τ coupling potential are small. The pure isovector

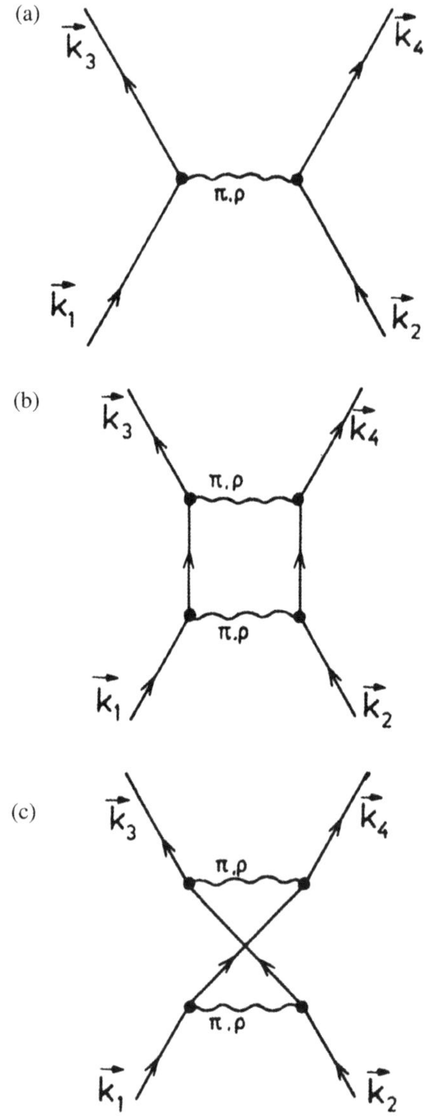

Fig. 9. Leading and next to leading contributions of the Stony Brook Jülich interaction.

potential is dominated by second order processes which are strongly energy dependent (Figs. 9(b) and 9(c)). A detailed discussion is given in Ref. [15]. The result of the calculation is shown in Fig. 10 where the model calculation is compared with the G-matrix result of Ref. [14].

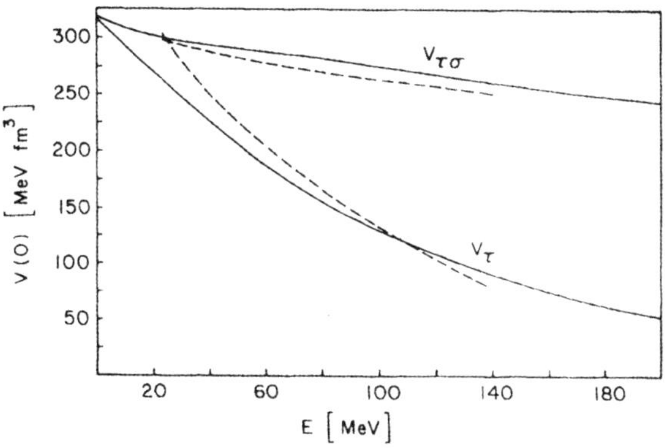

Fig. 10. Dependence of the isovector coupling potentials $V_\tau(q = 0)$ and $V_{\sigma\tau}(q = 0)$ on the energy of the incoming proton. The dashed lines give the Love-Petrovic [14] interaction.

From the various applications of the Stony Brook Jülich ansatz one may conclude that for magnetic properties, which are dominated by the spin isospin interaction, it is crucial to consider long range one-pion exchange. Pure zero range effective interactions like the Skyrme ansatz are inappropriate.

5. Derivation of the Landau Parameters

Gerry Brown and S. O. Bäckmann had shown how to derive the Landau parameters by functional differentiation of the ground state energy of nuclear matter with respect to the quasi particle occupation numbers, using the Brueckner-Bethe theory of nuclear matter, see Refs. [16, 17]. A major discrepancy between the calculated Landau parameter f_0 and the empirical value was found. Gerry realized the important role played by the collective modes in modifying the quasi particle interaction. In collaboration with S. Babu, he developed a theory of the phonon-induced interaction which provided a dramatic renormalization of the spin- and isospin independent part of Landau's quasi particle interaction [18]. The medium-dependent modification of the Landau parameters can be summarized by the Babu-Brown equations:

$$f_0 = f_0^G - \frac{1}{4}(f_0^I + 3f_0'^I + 3g_0^I + 9g_0'^I)$$

$$f_0' = f_0'^G - \frac{1}{4}(f_0^I - 3f_0'^I + 3g_0^I + 3g_0'^I)$$

$$g_0 = g_0^G - \frac{1}{4}(f_0^I + 3f_0'^I - g_0^I - 3g_0'^I)$$

$$g_0' = g_0'^G - \frac{1}{4}(f_0^I - f_0'^I - g_0^I + g_0'^I), \tag{19}$$

where the quantities f^I are generated by the phonon-induced interaction as follows:

$$f_0^I = \frac{f_0^2 \bar{U}}{1 - f_0 \bar{U}}. \tag{20}$$

Here, the angle averaged Lindhard function is abbreviated by $\bar{U}$.

The resulting Landau parameter obtained from the standard Brueckner theory based on the bare two-nucleon interaction HEA [19] (first row in Table 2) violates the stability criterion $f_0 > -1$, which has to be expected because standard Brueckner theory is known to lead to saturating Fermi momenta larger than $k_F = 1.5\,\text{fm}^{-1}$ for meson-exchange interactions. Despite this shortcoming, the Brueckner result for the parameter g_0' is close to the empirical value. This is due to the long range of the pion exchange which makes the spin-isospin dependent part of the quasi particle interaction insensitive to short-range medium dependent modifications. In the late seventies, Walecka's ideas of relativistic effects in nuclear structure had led to the observation that a medium dependent modification of the Dirac spinors used in deriving the bare meson-exchange potentials would lead to saturation of nuclear matter close to the empirical saturation point [20]. Indeed, the corresponding Landau parameter f_0 shown in the second row

Table 2. Landau parameters of nuclear matter for a Fermi momentum of $k_F = 1.36\,\text{fm}^{-1}$, obtained by using the one-boson exchange potential HEA as the two-nucleon interaction and different versions of Brueckner theory to derive the binding energy of nuclear matter. First row: standard Brueckner Hartee-Fock. Second row: The Dirac spinors of the bare two-nucleon interaction are modified by the nuclear medium. Third row: The phonon-induced interaction is considered in addition to the relativistic effects shown in the second row. Third row: the empirical parameters determined from linear response of finite nuclei. All parameters are given in units of $302\,\text{MeV}\,\text{fm}^3$.

method	f_0	f_0'	g_0	g_0'
BHF, non-relativistic	-1.17	0.32	0.2	0.63
BHF, relativistic	-0.72	0.42	0.12	0.63
Babu-Brown	-0.32	0.35	0.13	0.67
empirical	$+0 \pm 0.2$	0.8	0.2	0.9

in Table 2 satisfies Landau's stability condition [21]. The effect of the phonon-induced interaction, summarized by Eqs. (19), is to reduce the values shown in the second row by more than a factor 2. Again, one may note that the spin-isospin dependent part of the quasi particle interaction remains essentially unaffected by many-body effects. Therefore, the simple prescription to construct a quasi-particle interaction by starting from the longest range part of the two-nucleon interaction, i.e. the one-pion exchange, and to combine all complicated medium effects by a simple set of contact interactions summarizes the essential physics relevant for evaluating the linear response of nuclei.

The quasi particle interaction suggested by Gerry proved to be very useful for many applications, but yet, research was not finished. The remaining discrepancy between the theoretically derived Landau parameters and the empirical ones was too large to be ignored. Landau had related the parameter f_0 to the compression modulus of nuclear matter,

$$K = 3\frac{\hbar^2 k_F^2}{m}(1 + F_0).$$

Inserting the empirical value $K = 210 \pm 30\,\mathrm{MeV}$, one finds that F_0 cannot be negative. Moreover, the isomer shifts in heavy nuclei clearly rule out negative values of F_0 [22]. A lot of research was necessary to make further progress. In the 1980's the then new field of hadron physics emerged and improved our understanding of hadronic interactions. Effective Field Theories for pion-pion scattering, pion-nucleon interactions, and eventually nucleon-nucleon scattering were developed and provided a systematic expansion scheme which restricts the vast number of three-nucleon interactions allowed in older theories. A recent summary is given in Ref. [23]. In next-to-next-to-leading order, there are exactly three types of three-nucleon interactions. First theoretical investigations of nuclear matter based on those interactions have become available and are expected to make a major impact on the theoretical command of the Landau parameter [24, 25]. A successful description of neutron matter based on chiral three-nucleon interactions has been published and promises a solid basis for the theory of neutron-rich nuclei [26].

Acknowledgment

It is a pleasure to thank S. Drożdż, V. Klemt, J. Meyer-ter-Vehn, K. Nakayama, P. Ring, G. Tertychny+ and J. Wambach for the collaboration in developing and extending the Landau-Migdal theory in the past,

S. Avdeenkov, S. Kamerdziev, N. Lyutorovich and V. Tselyaev for their collaboration up to present. We thank the Deutsche Forschungsgemeinschaft for the grants DFG: 436 RUS 113/994/0-1 and DFG: 436 RUS 113/806/0-1. JS thanks the Foundation for Polish Science for financial support through the Alexander von Humboldt Honorary Research Fellowship.

References

[1] A.B. Migdal, *Theory of Finite Fermi Systems and Application to Atomic Nuclei*, Wiley, New York, 1967.
[2] L.D. Landau, *JETP* **3**, 920 (1957); **5**, 101 (1957); **8**, 70 (1959).
[3] A.B. Migdal, *Phys. Lett.* **45B**, 448 (1973).
[4] S. Barshay and G.E. Brown, *Phys. Lett.* **47B**, 107 (1973).
[5] J. Speth, E. Werner and W. Wild, *Physics Reports* **33**, 127 (1977).
[6] P. Ring and P. Schuck, *The Nuclear Many-Body Problem*, New York: Springer-Verlag, 1980.
[7] J. Speth, W. Klemt, J. Wambach and G.E. Brown, *Nucl. Phys.* **A343**, 382 (1980).
[8] M.G. Huber, S.T. Tuan and L.E. Wright, *Phys. Rev. Lett.* **23**, 174 (1969).
[9] S. Krewald and J. Speth, *Phys. Rev. Lett.* **45**, 458 (1980).
[10] S. Drożdż, V. Klemt, J. Speth and J. Wambach, *Phys. Lett.* **166B**, 18 (1986).
[11] J.S. Dehesa, S. Krewald, J. Speth and A. Fässler, *Phys. Rev.* **C15**, 1858 (1977).
[12] J. Lichtenstadt, J. Heisenberg, C.N. Papanicolas, C.P. Sarget, A.N. Courtemanche and J.S. McCarty, *Phys. Rev. Lett.* **40**, 1127 (1978).
[13] C.D. Goodman *et al. Phys. Rev. Lett.* **44**, 1755 (1980).
[14] W.G. Love, in *The (p,n)Reaction and the Nucleon-Nucleon Force*, edited by C.D. Goodman *et al.* Plenum, New York, 1980, p. 30; F. Petrovich, *ibid.*, p. 135.
[15] G.E. Brown, J. Speth and J. Wambach, *Phys. Rev. Lett.* **46**, 1057 (1981).
[16] G.E. Brown, *Rev. Mod. Phys.* **43**, 1 (1971).
[17] S.O. Bäckmann, *Nucl. Phys.* **A120**, 593 (1968).
[18] S. Babu and G.E. Brown, *Ann. Phys.* **78**, 1 (1973).
[19] K. Holinde, K. Erkelenz and R. Alzetta, *Nucl. Phys. A* **198**, 598 (1972).
[20] B.D. Serot and J.D. Walecka, *Adv. Nucl. Phys.* **16**, 1 (1986).
[21] S. Krewald, K. Nakayama and J. Speth, *Phys. Rep.* **161**, 103 (1988).
[22] V. Klemt, J. Speth and K. Goeke, *Phys. Lett.* **33B**, 331 (1970).
[23] E. Epelbaum, H.-W. Hammer and Ulf-G. Meißner, *Rev. Mod. Phys.* **81**, 1773 (2009).
[24] S.K. Bogner, R.J. Furnstahl and A. Schwenk, *Prog. Part. Nucl. Phys.* **65**, 94 (2010).
[25] J.W. Holt, G.E. Brown, J.D. Holt and T.T.S. Kuo, *Nucl. Phys.* **A785**, 322 (2007).
[26] K. Hebeler and A. Schwenk, *Phys. Rev.* **C82**, 014314 (2010).

Chapter 20

Separation Energy, Rearrangement Energy and Single Nucleon Wave Functions in Nuclei

Igal Talmi*

In a common method of constructing a single nucleon wave function in a given potential well, its energy is taken to be equal to its separation energy. This prescription is examined in view of Hartree-Fock theory where the potential well results from mutual interactions of the nucleons. It is shown that a more appropriate single nucleon energy of a valence nucleon is obtained by adding to the separation energy the rearrangement energy. This leads to a better prescription for constructing single nucleon wave functions in a given potential well.

1. Introduction

Some of the following considerations were developed in the fall of 1978 while I was staying in Stony Brook as a guest of Gerry Brown. They arose from fruitful discussions with Gerry, for which I am grateful. As far as I know, they were never published. This is an opportunity to present them here as a token of friendship and appreciation for Gerry.

Nuclei are systems of many interacting protons and neutrons (nucleons). The internal structure of nucleons in terms of quarks and gluons makes the situation even more complicated. Even in the usual approximation of dealing with nucleons as building blocks of nuclei, the situation is complex. The strong short range interaction between nucleons makes nuclei a very complex system where no simple features are expected to arise. Still, in spite of their complexity, nuclei exhibit some simple regular features which found a clear description in the shell model of the nucleus.

*Weizmann Institute of Science, Rehovot, Israel.

The most important simple feature of nuclei is associated with motion of nucleons within the nucleus as if they were independent particles moving in a common spherical and static potential. In the shell model, nucleons occupy the lowest orbits in the potential well. When proton and neutron orbits within a major shell are completely filled, the nucleus is particularly stable. The numbers of protons and the number of neutrons in such magic nuclei are called magic numbers.

The current version of the shell model was introduced in 1949 and was accepted by physicists only due to the overwhelming experimental evidence for magic numbers and orbits which give rise to them. It was not clear then, as it is also today, how can the motion of individual nucleons be reconciled with the observed strong and short range interaction between free nucleons. The latter was not very well determined in the early 1950s. Soon enough it was discovered that in addition to the short range attraction, the interaction between free nucleons has a strong repulsive core at shorter distances. It became very clear that such interactions cannot be simply reconciled with independent motion of nucleons.

An approach which attempts to reconcile between short range strong interactions and the shell model was introduced by Brueckner. In it, the effect of the short range correlations due to the interaction is transformed from the shell model wave functions onto the Hamiltonian. Shell model wave functions may be used for calculating energies provided a renormalized or effective interaction is used. The important step in applying these ideas to actual nuclei was done by Tom Kuo and Gerry Brown [1]. Since then there was a big effort to derive effective interactions from the bare interaction between free nucleons. Prescriptions for the latter were developed over the years but still, none of them follows from the more basic interaction between quarks and gluons. Only in the last few years, results of such efforts yield matrix elements of the effective interaction which may be used, without modifications, in shell model calculations of energies of nuclear states [2].

In the last few years, more ambitious calculations have been attempted. They do not start with given shell model states nor do they try to derive the effective interaction. They do not start from a central potential well but try to obtain it *ab initio*. This means using a certain version of the bare interaction to calculate nuclear energies. A typical attempt is the No Core Shell Model (NCSM). In it, harmonic oscillator wave functions are used only as a basis and the calculated wave functions are linear combinations of them. This expansion is limited by computational complexity but the aim is to include in it as many oscillator states as possible. This implies including states in higher and higher harmonic oscillator shells.

The success of such attempts critically depends on starting from a good bare interaction, making good approximations and using as many basis functions as necessary. If such an approach will yield accurate results, it will be very useful. It could be used for calculating energies which are not accessible to experiment. The resulting wave functions may be used for calculating other observables. These include electromagnetic moments and transitions as well as rates of beta decays and in particular, exact predictions of double beta decays. Still, the question may be asked whether the shell model will emerge from such calculations as a good approximation. Will we understand what kind of bare interactions lead to it, where it is valid and where it becomes useless. The shell model has a very successful record over the last 60 years and is still used as the main model for calculating and describing nuclear structure physics.

The success of shell model calculations using empirical effective two body interactions gave much credence to the shell model [3, 4]. Still, the successful use of interactions which are strongly renormalized, indicates strong admixtures to simple shell model wave functions. This could mean that the real wave functions of the nucleus may be very different from shell model wave functions. For the success of energy calculations, there is no need to assume the reality of single nucleon wave functions. Still, it could be asked why such calculations succeed, why magic numbers are evident if configurations are as mixed as they seem to be in some calculations of nuclear many body theory. There are, however, many successful applications of the shell model in which explicit use is made of single nucleon wave functions. From the many successful examples of such applications, only a few will be presented in the following.

2. On the Reality of Shell Model Wave Functions

In the theory of direct reactions, when one nucleon is removed from or added to a nucleus, the quantum numbers of this nucleon play an important role. The magnitude of the radial function at the nuclear boundary is important for such reactions. Since these are rather complicated processes, these calculations do not form a very solid evidence for the nature of single nucleon wave functions.

Elastic scattering of high energy electrons yielded detailed information on charge distributions in nuclei. By using ingenuous and accurate measurements of the charge distributions of ^{206}Pb and ^{205}Tl, the charge distribution of the least bound proton in ^{206}Pb was extracted [5]. It agrees remarkably well with the probability distribution due to the wave function of a proton in the $3s$ orbit in a potential well. It seems that the "wounds" inflicted on the single nucleon wave function by the short range correlations, amount only to a small fraction of the volume occupied by it. Such direct determinations, however, have not been carried out in other cases.

Coulomb energy differences have been calculated by using the wave function of a single valence proton. The difference in binding energies of two mirror nuclei, is due to the electrostatic interaction of the valence proton with core protons. The latter are taken to occupy the lowest orbits in a potential well. The dimensions of the well are determined by the measured charge radius of the core. The determination of its depth will be discussed in the following. Given the approximations involved, the results agree fairly well with experimental data. No detailed calculations will be presented here. In the example which will be discussed, only the difference between various orbits will be considered.

In the ^{17}O ground state, the valence neutron is in the $1d_{5/2}$ orbit. In an excited state at .875 MeV, the valence neutron is in the $2s_{1/2}$ orbit. In the mirror nucleus ^{17}F, the same states are the lowest but the $2s_{1/2}$ orbit is only .495 MeV above the ground state. This implies that the repulsive Coulomb interaction in the excited state is *lower* than in the ground state. The difference can be traced to the difference in wave functions of the valence nucleons. The $2s$ wave function is spread out more than the $1d$ one [6]. This effect is enhanced by the small separation energy of the $2s$ proton. Hence, its single nucleon energy is less affected by the electrostatic interaction. More examples may be found in Refs. [7,8]. A clear indication of the accurate description of the $1/2^+$ state of ^{17}F in terms of a single proton in the $2s$ orbit is offered by its spatial extension. The measured value of its root mean square radius is 5.333 fm as compared to 3.693 fm of the $1d_{5/2}$ orbit [9].

The situation presented above, is a special case of a general behavior of single nucleon wave functions. As recently clearly stated by Hamamoto [10], if the depth of the potential well is reduced, the higher the orbital angular momentum of a single nucleon orbit, the higher its energy is pushed up. This is due to the strong repulsion of the $l(l+1)/r^2$ term in the radial potential. Raising the bottom of the potential well, reduces the attractive part of it to a larger extent for higher l values.

There are also other ways to reduce the depth of the nuclear potential well. As realized many years ago [11], the attractive part of that potential well is due to the proton-neutron interaction. For neutrons it is due to the protons and *vice versa*. In ^{13}C, the $1p_{1/2}$ orbit is well below the $2s_{1/2}$ orbit. Removing from this nucleus two protons, the position of the $2s_{1/2}$ orbit in ^{11}Be was predicted to be somewhat lower than the $1p_{1/2}$ one [12]. This was based on the fact that the attractive interaction of a $1p_{3/2}$ proton with a $1p_{1/2}$ neutron, both with $l = 1$, is stronger than its interaction with a $2s_{1/2}$ neutron. The actual values of matrix elements of these interactions were taken from the observed levels of ^{12}B.

Another such case was presented together with the ^{11}Be shell model calculation [12]. As mentioned above, in ^{17}O the $2s$ orbit lies well above the $1d_{5/2}$ one. Removing two $1p_{1/2}$ protons, the ^{15}C nucleus is obtained, where the order of these orbits is reversed. This agrees with a shell model calculation based on matrix elements taken from levels of nearby nuclei. The $1p_{1/2}$ attraction to $1d_{5/2}$ neutrons turned out to be stronger than that with $2s_{1/2}$ neutrons. In this case, as well as in the ^{11}Be one, good agreement with the shell model is obtained by using two-body interactions whose matrix elements are taken from experiment. Still, a common feature of these cases is the fact that removing protons reduces the depth of the potential well of the neutrons. Hence, s orbits are less affected than p or d orbits. Such an argument, although highly qualitative, still gives a simple explanation based on properties of single nucleon wave functions in a potential well.

There is clear evidence for the reality of the s orbit of the valence neutron in the ground state of ^{11}Be. Due to its small separation energy and absence of a "centrifugal barrier" its wave function is considerably spread out. This provides a simple explanation of the large matter radius observed in this "halo nucleus".

Another important application of single nucleon wave functions is in the calculation of electromagnetic moments and transitions.

3. How Single Nucleon Wave Functions are Constructed

As explained above, much use is made of single nucleon wave functions which are often determined in the following way. The nucleus with A nucleons is replaced by a central potential, like the Wood-Saxon potential well. Its dimensions are usually determined by measurement of the extension of the charge distribution of that nucleus. Its depth is adjusted so that the single nucleon energy in the state considered, is equal to the energy needed to remove the valence nucleon from the nucleus with $A+1$ nucleons. This *separation energy* is equal to the difference between the binding energies of the nucleus with $A+1$ nucleons and the one with A nucleons. Accordingly, the single nucleon wave function will have the correct asymptotic behavior. Its exponential tail is determined by the measured separation energy of the valence nucleon considered. This procedure is based on the following considerations. When the valence nucleon is bound, the energy of the system is the binding energy of the nucleus with $A+1$ nucleons, E_{A+1} (binding energies in this paper are *negative*). When the valence nucleon is *adiabatically* removed, for very large distances, the *core* and the removed nucleon are independent of each other. The core, the nucleus with A nucleons, is in its ground state with energy E_A. The valence nucleon has in this situation, the energy $E = E_{A+1} - E_A$. Its wave function is

$$[R(r)/r]Y_{lm}(\vartheta, \varphi) \tag{1}$$

and, for very large r, $R(r)$ is the solution of the radial part of the Schröedinger equation in which the interaction, as well as the centrifugal potential, no longer appear

$$-(h^2/2m)R'' = ER = (E_{A+1} - E_A)R \tag{2}$$

From this equation follows that the asymptotic form of R is given by

$$R(r) = Cr^k \exp(-[(2m/h^2)|E_{A+1} - E_A|]^{1/2}r) \quad \text{for } r \to \infty \tag{3}$$

The values of C and of k are determined by the solution of the equation for small values of r.

This procedure is usually applied to a potential well which is fixed and independent of the position of the valence nucleon. The energy of the latter, E, in the potential well is, accordingly, equal to the separation energy $E_{A+1} - E_A$. The following discussion is concerned with this prescription and the conditions for its validity will be investigated. The question is whether the model in which the core is replaced by such a fixed potential well is accurate enough for the calculation of single nucleon wave functions. In the following, no distinction will be made between the state of the valence nucleon in the ground state and in an excited state of the nucleus with $A+1$ nucleons. In the latter case, the separation energy will be still denoted by $E_{A+1} - E_A$. In this case, E_{A+1} denotes the binding energy of the $A+1$ nucleus to which the energy of the excited state is added. In this case, E_{A+1} has a lower absolute value.

The discussion below is based on the assumption that the valence nucleon occupies an orbit with definite values of orbital and total angular momenta, l and j. Its wave function may have "wounds" inflicted on it by short range two-body correlations. Those are due to the strong interaction between nucleons. We assume that these wounds form a rather small part of the wave function. We also assume that the effect of these wounds may be replaced by using an effecive two-body interaction instead of the one between free nucleons. These conditions find a natural description in Hartree-Fock (HF) theory to which the following considerations apply. The results, however, will be derived without use of its formalism and will be shown to have wider applications. In the following, only spherical nuclei will be considered. To avoid complications due to several nucleons in the same orbit, a nucleus with a single nucleon outside closed shells (one *valence* nucleon) will be considered. This will be sufficient to make the point of this note.

The nucleus with closed shells (the *core*) is spherical also in Hartree-Fock theory. The wave function of the A nucleons will be denoted by Φ_A. It may be expressed as the product of two determinants, one of the protons and the other of the neutrons. Element ik of each determinant of order $A/2$, is defined by $\varphi_i(m_k)$, and hence, the wave function is fully antisymmetric. The indices i and k stand for l_i, l_k and j_i and j_k. The normalization factor of each determinant is $(A/2)^{-1/2}$. The A nucleons occupy the lowest orbits in a spherical potential well. In it, there are usually also higher orbits which are not occupied. They arise naturally in HF where they are determined by the interaction of the core nucleons with a valence nucleon. In the following, the valence nucleon will be taken to be a neutron. Due to the spherical symmetry of the core, the unoccupied orbits, like the occupied ones, have definite values of the angular momentum j.

The ground state energy of the nucleus with closed shells (core) is given by the lowest value of the variational integral

$$E_A = \int \Phi_A^*(\Sigma_i \mathbf{t}_i + \Sigma_{ij} V(i,j))\Phi_A d(1)d(2)\cdots d(A)$$

$$= \int \Phi_A^* H_C \Phi_A d(1)d(2)\cdots d(A) \tag{4}$$

In (4), $\mathbf{t}_i$ is the kinetic energy of the i-th nucleon and $V(i,j)$ is the effective interaction between nucleons i and j. The energy of the lowest *unoccupied* orbit in the (spherical) field of the core is obtained by calculating the lowest value of the integral

$$\begin{aligned} E^0_{A+1} &= \int (\Phi^*_A u^*_0)_a(\Sigma_i \mathbf{t}_i + \Sigma_{ij} V(i,j))(\Phi_A u_0)_a d(1)d(2)\cdots d(A)d(A+1) \\ &= E_A + \int (\Phi^*_A u^*_0)_a(\mathbf{t}_{A+1} + \Sigma_i V(i, A+1)) \\ &\quad \times (\Phi_A u_0)_a d(1)d(2)\cdots d(A)d(A+1) \\ &= E_A + \varepsilon_0. \end{aligned} \tag{5}$$

In (5), u_0 is the wave function of the valence nucleon and $(\Phi_A u_0)_a$ is a product of two normalized determinants. One of order $A/2$ of protons and the other, of order $A/2+1$, of the neutrons, including the valence neutron. The single nucleon wave function u_0 is the solution of the Schröedinger equation where the potential energy is equal to the expectation value

$$\int \Phi^*_A(\Sigma_i V(i, A+1))\Phi_A d(1)d(2)\cdots d(A) \tag{6}$$

and the energy eigenvalue is ε_0.

In (5) Φ_A is determined by (4) and is not varied. Hence, due to the variational principle, the integral (5) is higher than the energy of the nucleus with $A+1$ nucleons, $E^0_{A+1} \geq E_{A+1}$. From (5) then follows the inequality

$$\varepsilon_0 \geq E_{A+1} - E_A \tag{7}$$

The r.h.s. of (7) is the *calculated* separation energy of the valence neutron.

It is important to realize that this inequality does not follow from the fact that the calculated core energy in (4) may not yield the experimental binding energy. Even if in a better calculation yielding a spherical nucleus, the calculated core energy in (4) will agree with experiment, the inequality (7) will still hold. It is due to the fact that the *polarization* of the core by the valence nucleon is not taken into account. In the variational calculation in (5), the core wave function is not varied. Hence, in the inequality (7), *the separation energy may be taken from experiment.*

To obtain the correct value of the separation energy, the core wave function should also be varied. The interaction of the valence nucleon with the core (the nucleus with A nucleons) affects the orbits of the nucleons in it. Similarly, the wave function of the valence nucleon u_0 will also be changed. There are situations in which the valence nucleon still occupies a well defined orbit, with a wave function u, which differs from u_0 only in its radial part. The effect of the interaction between the valence and the core nucleons may lead just to modifications of the radial parts of the single nucleons wave functions. This is usually the case in constrained (spherical) HF theory and then the wave function of the core is still a determinant Φ'_A which may be different from Φ_A. The wave function of the nucleus with $A+1$ nucleons is then $(\Phi'_A u)_a$, defined as $(\Phi_A u_0)_a$ above. The interaction considered here, however, may lead to excited states of the core, all coupled to single nucleon states. The following formulae express the simpler situation, the discussion about its validity will be presented in the following.

Thus, we write

$$\begin{aligned} E_{A+1} &= \int (\Phi'^*_A u^*)_a (\Sigma_i \mathbf{t}_i + \Sigma_{ij} V(i,j))(\Phi'_A u)_a d(1)d(2)\cdots d(A)d(A+1) \\ &= \int \Phi'^*_A (\Sigma_i \mathbf{t}_i + \Sigma_{ij} V(i,j))\Phi'_A d(1)d(2)\cdots d(A) + \int (\Phi'^*_A u^*)_a \\ &\quad \times (\mathbf{t}_{A+1} + \Sigma_{iA+1} V(i, A+1))(\Phi'_A u)_a d(1)d(2)\cdots d(A)d(A+1) \\ &= E'_A + \varepsilon \end{aligned} \tag{8}$$

The single nucleon energy ε is *defined* by (8). Due to the variational principle, $E'_A \geq E_A$. The difference $E_A - E'_A$ is the energy of the core nucleons gained by adiabatic removal of the valence nucleon. The amount $E'_A - E_A$ is called the *rearrangement energy*. It will be denoted RE and is always non negative, $RE \geq 0$. Subtraction of E_A leads to

$$\varepsilon = E_{A+1} - E_A + (E_A - E'_A) \tag{9}$$

and hence, to the inequality

$$\varepsilon \leq E_{A+1} - E_A \tag{10}$$

The inequalities (7) and (10) may be combined into

$$\varepsilon \leq E_{A+1} - E_A \leq \varepsilon_0 \tag{11}$$

The relation (11) may become an equality if the rearrangement energy $E_A - E'_A$ vanishes. In that case $\Phi'_A = \Phi_A$ and $E^0_{A+1} = E_{A+1}$. This is the condition for the validity of the Koopmans Theorem [13] which deals with Hartree-Fock theory. It states that the single nucleon energy of the highest orbit is equal to the separation energy of that nucleon. If the state of A nucleons is not changed by the removal of nucleon $A+1$, then, indeed (11) is reduced to

$$\varepsilon = E_{A+1} - E_A = \varepsilon_0 \tag{12}$$

In his paper [13], Koopmans clearly stated the condition for his result to be valid. Unfortunately, this is not the case for some authors of textbooks, published decades after his paper.

The main concern of the present paper is the result (9) which may be written as

$$\varepsilon = E_{A+1} - E_A - RE \tag{13}$$

The following discussion is about its validity and its consequences.

The relations (7) and (10) are exactly obeyed in HF theory. It was argued above, that (7) holds also for the exact (experimental) separation energy. The valence nucleon may be, to a good approximation, in a well defined orbit with energy ε. Then the calculated binding energy in (8) with a more complicated E'_A, may yield the experimental E_{A+1} to a good approximation. The relations (9) and (10) will then hold also for the *exact* separation energy. These inequalities are due to the polarization of the core by the valence nucleon. Then, due to the variational principle, its binding energy, the expectation value of H_C in the state Φ'_A, is higher than its ground state binding energy E_A. This effect leads to the inequality (10). This effect may be demonstrated if perturbation theory may be used in the calculations above, as will be shown below.

As stated above, the wave function of the $A+1$ nucleus may well include excitations of the core. The wave function Φ_{A+1} may be expanded in a set of orthogonal states. In addition to Φ_A, It may include several orthogonal excited states Φ^i_A, coupled with u_0 or with other single nucleon states u_i, to produce states with the same values of total J and isospin T. The ground state of the $A+1$ nucleus is a linear combination of these states. The unnormalized wave function may be written, without specifying angular momentum couplings, as

$$\Phi_{A+1} = (\Phi_A u_0)_a + \Sigma_i \alpha_i (\Phi^i_A u_i)_a \tag{14}$$

where the states on the r.h.s. are antisymmetrized. The notion of a single nucleon wave function u, is strictly valid only for terms in which u_i has the same values of l and j as u_0 and Φ^i_A has zero angular momentum. If the coefficients α_i in the expansion of Φ_{A+1} are small, $\alpha_i^2 \ll 1$, then the description of the nucleus as a valence nucleon bound to a core, by $(\Phi'_A u)_a$, may be a good approximation. The derivations above, with some change in the notation, will be shown to be still valid.

The separation energy $E_{A+1} - E_A$ may be obtained by removing adiabatically the valence nucleon. In the asymptotic region, r, the distance of the valence nucleon from the core, becomes larger and larger. The interaction of this nucleon with core nucleons becomes weaker and weaker. In the *exact* wave function of the ground state, coefficients of Φ^i_A of excited core states become smaller and smaller. Hence, the core wave function Φ'_A approaches Φ_A, the energy E'_A approaches E_A and the value of ε, according to (5), approaches the separation energy $E_{A+1} - E_A$. This was shown above as the reason why the asymptotic tail of the exact wave function is determined by the separation energy. This does not happen in the case of HF wave functions. There, the core wave function Φ'_A is determined mainly by the interaction between nucleons in the region where it is strong. The core wave function Φ'_A is independent of the position of the valence nucleon. The energy of the valence nucleon is ε which may well be lower than the separation energy. Such a wave function may not have the correct asymptotic behavior but it may be a good approximation to the main bulk of the exact wave function. It is well known that HF wave functions do not have the correct asymptotic tail. This point will be further considered in the following.

If, however, the valence nucleon is removed instantanously, the situation is quite different. The core is left in either the ground state Φ_A or in any of the states Φ^i_A which were coupled with u_i in the expansion (14) of the ground state of the $A+1$ nucleus. The probability that the core is in any of these states, is proportional to the square of its amplitude in the ground state of the $A+1$ nucleus. This holds if the removal efficiency of the valence neutron is the same for all u_i orbits. The excitation energy of that state is $E_i - E_A$. The *average excitation energy* is equal to

$$RE = \Sigma_i \alpha_i^2 (E_i - E_A) \tag{15}$$

The value of the rearrangement energy (15) may be obtained by many runs of an experiment in which very high energy projectiles hit the nucleus. The excitation energy is measured in every successful hit and the average is computed. A detailed discussion of rearrangement energies in nuclei is presented by Hodgson [14].

4. Results in Perturbation Theory

If the coefficients α_i in (14) are small, $\alpha_i^2 \ll 1$, perturbation theory may be used. There is a problem with (14) since many of the states in the expansion are in the continuum. The discussion here will avoid these complications and excited states will be considered as if they are all bound states. The unperturbed state $(\Phi_A u_0)_a$ is an eigenstate of the unperturbed Hamiltonian. This is the core Hamiltonian, H_C, and a certain part due to the valence nucleon. The latter is the sum of the kinetic energy of the valence nucleon and its potential energy. The latter consists of an external potential, if there is any, plus the interaction energy between core nucleons in their ground state Φ_A and the valence nucleon. Since the core has spherical symmetry, this interaction is due to the monopole part of the direct term plus the corresponding exchange terms. Hence, the ground state eigenvalue of the unperturbed Hamiltonian thus defined, contains the following contribution to the energy, according to (6), namely

$$\langle 0 \,|\, V \,|\, 0 \rangle = \int (\Phi_A^* u_0^*)_a (\Sigma_i V(i, A+1))(\Phi_A u_0)_a d(1)d(2)\cdots d(A)d(A+1) \tag{16}$$

If the interaction between the valence nucleon and core nucleons could be considered as a perturbation, the expectation value (16) would be the contribution to the energy to first order in perturbation theory. Thus, (16) contains the contribution of the monopole part and is equal to both zero order and first order in perturbation theory. In the derivations above, the term (16) was included without any assumptions on its magnitude. In HF theory this is the only term which contributes to the single valence nucleon energy ε_0. It is included here in the unperturbed Hamiltonian and hence, may be rather large.

The monopole part of the interaction has also matrix elements leading from the unperturbed state to excited ones. Such matrix elements do not contribute to first order perturbation of the energy. These matrix elements, as well as higher multipoles of the interaction between core nucleons and the valence nucleon, are included in the perturbation. They do not contribute in the first order but only contribute to second order of perturbation theory. In first order of perturbation theory, only the contribution due to the *unperturbed* wave function is included. First order corrections to the wave function, due to polarization of the core by the valence nucleon contribute only in second order perturbation theory to which we now turn.

The second order contribution to the energy arising from H_C is obtained from the expectation value of H_C in the wave function (14). It is given by

$$\begin{aligned}(E_A/[1+\Sigma_i\alpha_i^2] + \Sigma_i\alpha_i^2 E_i) - E_A &= (E_A[1-\Sigma_i\alpha_i^2] + \Sigma_i\alpha_i^2 E_i) - E_A \\ &= \Sigma_i\alpha_i^2(E_i - E_A)\end{aligned} \tag{17}$$

where the normalization was included up to second order in the perturbation. Thus, the core energy in the nucleus with $A+1$ nucleons is *higher* than its ground state energy by the rearrangement energy (15).

In second order perturbation theory, the values of the expansion coefficients are given by the well known formula. Each is equal to the non-diagonal matrix element between the unperturbed lowest state, the state under consideration, and the i-th state, divided by their energy difference. The energy of the lowest state, $|0\rangle$, is $E_A + \varepsilon_0$. The energy of the i-th state is similarly given by $E_i + \varepsilon_i$ and thus, their difference is $E_A - E_i + \varepsilon_0 - \varepsilon_i$. Hence,

$$\alpha_i = \langle 0 \,|\, V \,|\, i \rangle / (E_A - E_i + \varepsilon_0 - \varepsilon_i) \tag{18}$$

Substituting this expression into (15), the average rearrangement energy becomes equal to

$$RE = \Sigma_i \,|\, \langle 0 \,|\, V \,|\, i \rangle \,|^2 (E_i - E_A)/(E_A - E_i + \varepsilon_0 - \varepsilon_i)^2 \tag{19}$$

To obtain the contribution in second order due to the complete *unperturbed* Hamiltonian, the contribution due to the single nucleon part of it should be added to (19). Like (17), it is obtained in the form

$$\begin{aligned}(\varepsilon_0/[1+\Sigma_i \alpha_i^2] + \Sigma_i \alpha_i^2 \varepsilon_i) - \varepsilon_0 &= (\varepsilon_0[1-\Sigma_i \alpha_i^2] + \Sigma_i \alpha_i^2 \varepsilon_i) - \varepsilon_0 \\ &= \Sigma_i \alpha_i^2 (\varepsilon_i - \varepsilon_0)\end{aligned} \tag{20}$$

Substituting the values (20) into the r.h.s. of (17) and adding the result to (19), the following expression is obtained

$$\Sigma_i \,|\, \langle 0 \,|\, V \,|\, i \rangle \,|^2 / (E_i - E_A + \varepsilon_i - \varepsilon_0) \tag{21}$$

This expression is equal to the full second order contribution to the energy multiplied by -1. The value of (21) is repulsive (positive) whereas the full second order contribution is always attractive (negative). To obtain the complete second order contribution, the amount due to the perturbing part of the Hamiltonian should be added. It is always equal to (21) multiplied by -2.

Thus, to second order perturbation theory, we obtain the result

$$\begin{aligned}E_{A+1} &= E_A + \varepsilon_0 + \Sigma_i \,|\, \langle 0 \,|\, V \,|\, i \rangle \,|^2 / (E_A - E_i + \varepsilon_0 - \varepsilon_i) \\ &= E_A + \Sigma_i \,|\, \langle 0 \,|\, V \,|\, i \rangle \,|^2 (E_i - E_A)/(E_A - E_i + \varepsilon_0 - \varepsilon_i)^2\end{aligned}$$

$$
\begin{aligned}
&+\varepsilon_0 + \Sigma_i \,|\, \langle 0 \,|\, V \,|\, i \rangle \,|^2 (\varepsilon_i - \varepsilon_0)/(E_A - E_i + \varepsilon_0 - \varepsilon_i)^2 \\
&+ 2\Sigma_i \,|\, \langle 0 \,|\, V \,|\, i \rangle \,|^2/(E_A - E_i + \varepsilon_0 - \varepsilon_i) \\
&= E_A + RE + \varepsilon
\end{aligned} \tag{22}
$$

If the state $(\Phi'_A u)_a$ is a good approximation of the nucleus with $A+1$ nucleons, then the single nucleon energy should be taken to be ε. From (22) follows a relation in which the single nucleon energy is expressed in terms of the separation energy and rearrangement energy. It is identical with (9) obtained above, namely,

$$\varepsilon = E_{A+1} - E_A - RE \tag{23}$$

which leads to the inequality (10). It is however, more precise. From (22) follows also an expression for ε_0 in second order perturbation theory. This relation leads to the inequality (7) obtained above. The derivations in this section are somewhat more general than those obtained from HF theory. Still, they were obtained in second order perturbation theory.

5. Summary and Conclusions

In many applications of the simple shell model, the actual single nucleon wave functions are used. For these applications, the main result of the discussion above, is the relation (9) or as expressed by (23), as

$$\varepsilon = E_{A+1} - E_A - RE \tag{24}$$

This relation was derived above by using several approximations. Still, it was argued that it is valid if the simple shell model is a good approximation for calculating nuclear energies. Thus, for construction of single nucleon wave functions, the binding energies appearing in (24) may be taken from experiment. The value of the rearrangement energy RE, however, cannot be determined with reasonable accuracy. Still, *it is always present if the core is polarized by the valence nucleon.*

There is ample evidence for polarization of the core by valence nucleons. This is evident also in spherical nuclei where such polarizations are considered as small perturbations. Such effects can be clearly seen in electromagnetic phenomena associated with valence neutrons. Charge radii of nuclei change when neutrons are added to a core nucleus. Electromagnetic transitions between levels of pure shell model *neutron* configurations are abundant. They are usually calculated by assigning to the neutron an "effective charge" which is of the same order of magnitude as the proton effective charge. The latter is usually larger than the charge of a single proton. Due to the variational principle, polarization of the core implies an increase of its energy. In the absence of an exact theory for such polarizations, it is difficult to calculate their magnitude. Nor is there a reliable method to determine the rearrangement energy by experiment.

Single nucleon wave functions are often determined by the method described above. A potential well is chosen whose range is determined by the size of the measured radius of the nucleus (core) considered. Its depth is adjusted so that the single nucleon energy is equal to the separation energy. This is due to the fact that the assumed potential well is fixed and independent of the coordinates of the valence nucleon. Hence, it follows that the single nucleon energy is equal to the separation energy. Looking at (24) it is evident that in this case, the rearrangement energy is ignored. This is so if the potential well is fixed but it should hardly be the case for a core of A nucleons. For actual nuclei, the single nucleon energy, given by (24), is equal to the *separation energy minus the rearrangement energy.*

In the *exact* solution, as r is increased, the reduction in the core-nucleon interaction leads to a reduced core polarization. The core energy is lowered to E_A and the rearrangement energy is transferred to the valence nucleon. Its energies is then raised to the separation energy $E_{A+1} - E_A$.

In the situation considered here, the potential well is not a function of the coordinates of the single nucleon. This is the case for a fixed potential well, as well as in Hartree-Fock theory. In the latter, the core energy is raised, making the valence nucleon more tightly bound. In the case of a fixed potential, the change in its depth or its radius or both, must be introduced "by hand". In both cases, the rearrangement energy is more than compensated by the lowering of the single nucleon energy according to (24). Thus, the wave function is more concentrated in the potential well and hence, its integral outside the well is reduced. The asymptotic behavior is given by $\varepsilon = E_{A+1} - E_A - RE$, which makes the exponential "tail", $\exp(-[(2m/h^2)|E_{A+1} - E_A - RE|]^{1/2}r)$, considerably smaller than the *exact* result given by (3) as $\exp(-[(2m/h^2)|E_{A+1} - E_A|]^{1/2}r)$. An extreme example of this effect is presented in the Appendix, taken from molecular physics.

Ignoring the rearrangement energy may lead to a wave function whose exponential tail is correct but it may be a poor approximation of the main body of the exact wave function. In many applications, it is the latter part of the wave functions which is important. Jokingly, it is the dog who should wag its tail rather than the tail wagging the whole body of the dog.

Appendix: The Hydrogen Molecular Ion

The point discussed above is that a wave function which is a very good approximation of the exact one, need not have the exact asymptotic tail. A rather extreme example of this situation is offered by the hydrogen molecular ion. The standard treatment of this system is based on the observation that the motion of the two protons is much slower that the motion of the electron. Their kinetic energy is not included in the approximate Hamiltonian. The protons are considered to be at a distance R apart and the Schröedinger equation is solved for the wave function of the electron. The energy eigenvalues thus obtained depend on R. To these energies $E(R)$, the Coulomb potential energy of the protons, e^2/R is added. Of these sums for various values of R, the lowest is adopted as the ground state energy. The motions of the two protons at R_0 for which $E(R)+e^2/R$, has its minimum value, turn out to have a negligible effect on the energy of the system.

The Schröedinger equation for this system is discussed in detail by Kotani *et al.* [15]. They show that it is separable for a certain set of variables, and hence, the wave functions are products of three functions of these variables. Of the latter, $\xi = (r_a + r_b)/R_0$ is the only variable which becomes large when the electron moves away from the protons. The r_a and r_b are the distances of the electron from the two protons. In the asymptotic region, where $\xi \to \infty$, it can be replaced by r which is the distance between the electron and the middle point between the two protons. The equation for $X(\xi) \sim u(r)$ is then expressed by

$$d^2u/dr^2 + (2/r)du/dr = (2m/h^2)(e^2/R_0 - E)u(r) \qquad \text{(A.1)}$$

where E is the binding energy of the ion. It is equal to the separation energy of the electron. The radial function $u(r)$ which satisfies this equation is of the form

$$r^k \exp(-[(2m/h^2)|E - e^2/R_0|]^{1/2} r) \qquad \text{(A.2)}$$

where r^k is the highest power of r in the exact wave function.

This asymptotic behavior was obtained for the wave function which gives a very good description of the H_2^+ molecule. This wave function is a very good approximation to the main bulk of the exact wave function of the electron. Still, this approximate wave function does not give the correct exponential tail of the exact solution of the three body problem.

The exponential decay of the exact wave function is given by

$$r^k \exp(-[(2m/h^2)|E|]^{1/2} r) \tag{A.3}$$

The difference between these two expressions has a clear origin. When r becomes very large, the force which holds the protons at their positions, becomes very weak. As a result, if $r \to \infty$ also $R \to \infty$ and $e^2/R \to 0$. The energy which determines the asymptotic dependence of the approximate (but very good) solution is the *separation energy* E from which the *rearrangement energy* e^2/R_0 is subtracted.

The value of R_0 in Ref. [15] turns out to be very close to 2 "atomic units" which makes e^2/R_0 equal to 13.6 eV. The separation energy of the electron from H_2^+ is equal to the total binding energy of H_2^+ since the ground state energy of two protons is zero. This energy is equal to −16.4 eV. Using these values, the exponential decay of the approximate (but very good) wave function is given by $r^k \exp(-1.48\, r/a_0)$ compared with the decay of the exact solution which is given by $r^k \exp(-1.10\, r/a_0)$, where a_0 is the Bohr radius (0.05292 nm). There is also some experimental measurement of the rearrangement energy of H_2^+ [16]. Accelerated ions hitting a thin foil, loose electrons in a rather fast process. The protons are left at a measured distance $\langle R^2 \rangle^{1/2}$ apart, which was found to be fairly close to the calculated value of R_0. It is clear that this case is rather extreme. Removal of the electron affects the state of the protons in a major way. They simply fly apart from their distance R_0.

References

[1] T.T.S. Kuo and G.E. Brown, *Nucl. Phys.* **85**, 40 (1966).

[2] L. Coraggio, A. Covello, A. Gargano. N. Itaco and T.T.S. Kuo, *Prog. Part. Nucl. Phys.* **62**, 135 (2009).

[3] I. Talmi, *Simple Models of Complex Nuclei*, Harwood 1993.

[4] I. Talmi, In *Advances of Nuclear Physics*, J.W. Negele and E. Vogt eds., Vol. **27**, 1 (2003).

[5] B. Frois, In *Proc. Intl. Conf. Nucl. Phys.* Florence 1983, P. Blasi and R.A. Ricci eds., Tipografia Compositori, Bologna 1983, 221.

[6] J.A. Nolen and J.P. Schiffer, *Ann. Rev. Nucl. Sci.* **19**, 471 (1969).

[7] I. Talmi and I. Unna, *Ann. Rev. Nucl. Sci.* **10**, 353 (1960).

[8] I. Talmi, In *Proc. 9th Intl. Spring Seminar Nucl. Phys.* 2007, A. Covello ed., World Scientific 2008, 231.

[9] R. Morlock *et al.*, *Phys. Rev. Lett.* **79**, 3837 (1997).

[10] I. Hamamoto, In *Proc. 10th Intl. Spring Seminar* 2010, A. Covello ed. (to be published).

[11] I. Talmi, *Rev. Mod. Phys.* **34**, 704 (1962).
[12] I. Talmi and I. Unna, *Phys. Rev. Lett.* **4**, 469 (1960).
[13] T. Koopmans, *Physica* **1**, 104 (1933).
[14] P.E.H. Hodgson, *Rep. Nucl. Phys.* **28**, 847 (1975).
[15] M. Kotani *et al.*, In *Encyclopedia of Physics* Vol. **XXXVII/2**, p. 54, Springer 1961.
[16] E.P. Kanter *et al.*, *Nucl. Instr. and Meth.* **170**, 87 (1980).

Chapter 21

Holographic Nucleons

Ismail Zahed*

Recent developments in holography have provided a new vista to the nucleon composition. A strongly coupled core nucleon tied with vector mesons emerge in line with the Cheshire cat principle. The cat is found to hide in the holographic direction. We discuss the one, two and many baryon problem in this context and point at the striking similarities between the holographic results and recent lattice simulations at strong coupling.

1. Dedication

This paper is dedicated to Gerry Brown who has been my mentor and colleague for the past three decades. I have met Gerry while in graduate school way back at MIT, in Feshbach's office on a sunny fall morning. After the meeting, Gerry asked me when I would graduate as he was prepared to hire me for the next three years. I did the following spring and have been with Gerry since then. Gerry is an outstanding scientist and humanist that has contributed immensely to our field. I thank him for his guidance and friendship, and wish him well for the years to come.

2. Introduction

Back in the eighties, quark bag models were proposed as models for hadrons that capture the essentials of asymptotic freedom through weakly interacting quarks and gluons within a bag, and the tenets of nuclear physics through strongly interacting mesons at the boundary. The delineation or bag radius was considered as a fundamental and physically measurable scale that separates ultraviolet from infrared QCD. Two competing pictures

*Department of Physics and Astronomy, SUNY Stony-Brook, NY 11794, USA.

emerged: The original MIT bag model with a large radius surrounded by a bare vacuum and the Stony-Brook bag model with a small radius surrounded by pions [1]. In fact, at low energy this delineation is unphysical as stated in the Cheshire cat principle [2]. Quantum effects and anomalies cause most of the charges (fermionic, axial, etc.) to leak making the bag boundary immaterial [3], much like the smile of the Cheshire cat in "Alice in wonderland" [4]. The Skyrme model typifies the extreme realization of the Cheshire cat principle whereby the immaterial bag radius is reduced to zero size [5].

The Skyrme model realizes QCD baryons as chiral solitons in the limit of large number of colors N_c. Recently, the same model was found to emerge from holographic QCD in the dual limit of large N_c and strong coupling t'Hooft coupling $\lambda = g^2 N_c$ [6, 7]. In the holographic construction, the Skyrmion is the holonomy of a flavor instanton tied to Witten's vertex in bulk [8]. Baryon number at the boundary is dual to instanton number in bulk. Although the exact holographic dualities are only known for a restricted set of string theories in bulk with mostly conformal field theories at the boundary [9], we will assume that such a correspondence holds for holographic QCD which is not conformal.

The present paper review some aspects of the holographic baryons following recent work in [10]. It is dedicated to Gerry Brown. In Sec. 3, we review the holographic baryon construction from a bulk instanton, and emphasize the emergence of the Cheshire cat principle. In Sec. 4 a top-down holographic model is briefly summarized and the bayonic current derived. In Sec. 5, the 2-nucleon problem is discussed using the ADHM 2-instanton configuration. In Sec. 6, cold and dense holographic matter is argued to be a crystal of instantons at low densities, and a crystal of dyons at higher densities. In Sec. 7, we estimate the melting temperature of these crystals into liquids. Our conclusions are summarized in Sec. 8.

3. The Skyrmion from the Instanton

In holographic QCD, a baryon is initially described as a flavor instanton in an $R^{1,3} \times R_Z$ dimensional space where the holographic direction R_Z is warped by gravity. At large t'Hooft coupling $\lambda = g^2 N_c$, the instanton is forced by gravity to drop to the bottom of R_Z. Topological (Coulomb) repulsion causes the instanton to lump at a size $Z \approx 1/\sqrt{\lambda} \ll 1$. In this limit, the effects of the gravitational warping on the instanton can be neglected. Thus the instanton SU(2) flavor configuration $\mathbb{A}_M$ and its supporting U(1)

Coulomb potential $\widehat{\mathbb{A}}_M$ read

$$\widehat{\mathbb{A}}_0 = -\frac{1}{8\pi^2 a\lambda}\frac{2\rho^2+\xi^2}{(\rho^2+\xi^2)^2}, \quad \mathbb{A}_M = \eta_{iMN}\frac{\sigma_i}{2}\frac{2x_N}{\xi^2+\rho^2}, \tag{1}$$

with all other gauge components zero. The size is $\rho \sim 1/\sqrt{\lambda}$. We refer to [6] (last reference) for more details on the relevance of this configuration for baryons. The ADHM configuration has maximal spherical symmetry and satisfies

$$(\mathbb{R}\mathbb{A})_Z = \mathbb{A}_Z(\mathbb{R}\vec{x}), \quad (\mathbb{R}^{ab}\mathbb{A}^b)_i = \mathbb{R}^T_{ij}\mathbb{A}^a_j(\mathbb{R}\vec{x}), \tag{2}$$

with $\mathbb{R}^{ab}\tau^b = \Lambda^+\tau^a\Lambda$ a rigid SO(3) rotation, and Λ is SU(2) analogue.

The holographic baryon is just the holonomy of (1) along the gravity bearing and conformal direction Z,

$$U^{\mathbb{R}}(x) = \Lambda\mathbf{P}\mathrm{exp}\left(-i\int_{-\infty}^{+\infty} dZ\,\mathbb{A}_Z\right)\Lambda^+ . \tag{3}$$

The corresponding Skyrmion in large N_c and leading order in the strong coupling λ is $U(\vec{x}) = e^{i\vec{\tau}\cdot\vec{x}\,\mathbf{F}(\vec{x})}$ with the profile

$$\mathbf{F}(\vec{x}) = \frac{\pi|\vec{x}|}{\sqrt{\vec{x}^2+\rho^2}} . \tag{4}$$

The holonomy (33) is a heavy flavor but colorless fermion "propagating" along the holographic direction with the instanton as a background field. The result is a small size Skyrmion map $U(x)$ at the boundary.

The baryon emerges from a semiclassical organization of the quantum fluctuations around the point-like source (33). To achieve this, we define

$$\begin{aligned} A_M(t, rx, Z) = \mathbb{R}(t)(\mathbb{A}_M(x - X_0(t), Z - Z_0(t)) \\ + C_M(t, x - X_0(t), Z - Z_0(t))). \end{aligned} \tag{5}$$

The collective coordinates $\mathbb{R}, X_0, Z_0, \rho$ and the fluctuations C in (5) form a redundant set. The redundancy is lifted by constraining the fluctuations to be orthogonal to the zero modes. This can be achieved either rigidly [11] or non-rigidly [12]. We choose the latter as it is causality friendly. For the collective iso-rotations the non-rigid constraint reads

$$\int_{x=Z=0} d\hat{\xi} C\, G^B \mathbb{A}_M , \tag{6}$$

with $(G^B)^{ab} = \epsilon^{aBb}$ the real generators of $\mathbb{R}$.

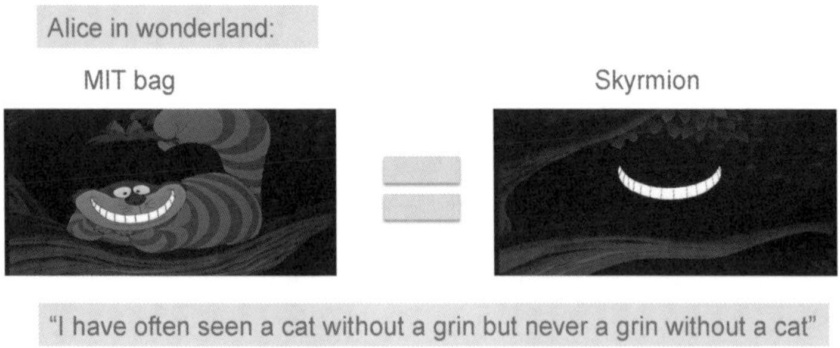

Fig. 1. Alice's Cheshire cat.

For Z and ρ the non-rigid constraints are more natural to implement since these modes are only soft near the origin at large λ. The vector fluctuations at the origin linearize through the modes

$$d^2\psi_n/dZ^2 = -\lambda_n\psi_n, \tag{7}$$

with $\psi_n(Z) \sim e^{-i\sqrt{\lambda_n}Z}$. In the spin-isospin 1 channel they are easily confused with $\partial_Z\mathbb{A}_i$ near the origin as we show in Fig. 3. Using the non-rigid constraint, the double counting is removed by removing the origin from the vector mode functions

$$\psi_n'(Z) = \theta(|Z| - Z_C)\psi_n(Z), \tag{8}$$

with $Z_C \sim \rho \sim 1/\sqrt{\lambda}$ which becomes the origin for large λ. In the non-rigid semiclassical framework, the baryon at small $\xi < |Z_C|$ is described by a flat or uncurved instanton located at the origin of R^4 and rattling in the vicinity of Z_C. At large $\xi > |Z_C|$, the rattling instanton sources the vector meson fields described by a semi-classical expansion with non-rigid Dirac constraints. Changes in Z_C (the core boundary) are reabsorbed by a residual gauge transformation on the core instanton. This is a holographic realization of the Cheshire cat principle [2] where Z_C plays the role of the Cheshire cat smile. In a way, Alice's Cheshire cat of Fig. 1 has gone out of sight in the holographic or 5th direction.

4. The Baryonic Current

To illustrate the Cheshire cat principle depicted in Fig. 2 more quantitatively, we now summarize the holographic Yang-Mills-Chern-Simons action in 5D curved background. This is the leading term in a $1/\lambda$ expansion of the D-brane Born-Infeld (DBI) action on D8 [6],

$$S = S_{YM} + S_{CS}, \tag{9}$$

$$S_{YM} = -\kappa \int d^4x dZ \,\mathrm{tr}\left[\frac{1}{2}K^{-1/3}\mathcal{F}_{\mu\nu}^2 + M_{\mathrm{KK}}^2 K\mathcal{F}_{\mu Z}^2\right], \tag{10}$$

$$S_{CS} = \frac{N_c}{24\pi^2}\int_{M^4\times R} \omega_5^{U(N_f)}(\mathcal{A}), \tag{11}$$

where $\mu,\nu = 0,1,2,3$ are 4D indices and the fifth(internal) coordinate Z is dimensionless. There are three things which are inherited by the holographic dual gravity theory: M_{KK}, κ, and K. M_{KK} is the Kaluza-Klein scale and we will set $M_{\mathrm{KK}} = 1$ as our unit. κ and K are defined as

$$\kappa = \lambda N_c \frac{1}{216\pi^3} \equiv \lambda N_c a, \quad K = 1 + Z^2. \tag{12}$$

$\mathcal{A}$ is the 5D $U(N_f)$ 1-form gauge field and $\mathcal{F}_{\mu\nu}$ and $\mathcal{F}_{\mu Z}$ are the components of the 2-form field strength $\mathcal{F} = \mathrm{d}\mathcal{A} - i\mathcal{A}\wedge\mathcal{A}$. $\omega_5^{U(N_f)}(\mathcal{A})$ is the Chern-Simons 5-form for the $U(N_f)$ gauge field

$$\omega_5^{U(N_f)}(\mathcal{A}) = \mathrm{tr}\left(\mathcal{A}\mathcal{F}^2 + \frac{i}{2}\mathcal{A}^3\mathcal{F} - \frac{1}{10}\mathcal{A}^5\right), \tag{13}$$

We note that S_{YM} is of order λ, while S_{CS} is of order λ^0. These terms are sufficient to carry a semiclassical expansion around the holonomy (33) with $\hbar = 1/\kappa$ as we now illustrate it for the baryon current.

To extract the baryon current, we source the reduced action with $\hat{\mathcal{V}}_\mu(x)$ a $U(1)_V$ flavor field on the boundary in the presence of the vector fluctuations ($C = \hat{v}$). The tree level baryonic current reads

$$J_B^\mu(x) = -\kappa K \widehat{\mathbb{F}}^{Z\mu}(x,Z)\,(1 - \sum_{n=1}^{\infty} a_{v^n}\psi_{2n-1})\Big|_{Z=B}$$
$$- \sum_{n,m} m_{v^n}^2 a_{v^n}\psi_{2m-1}\int d^4y\,\kappa K\widehat{\mathbb{F}}_{Z\nu}(y,Z)\Delta_{mn}^{\nu\mu}(y-x)\Bigg|_{Z=B}. \tag{14}$$

Fig. 2. The Cheshire cat principle.

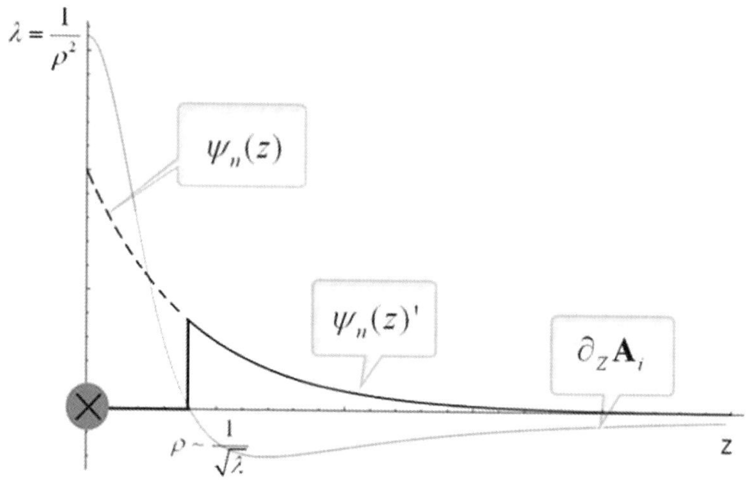

Fig. 3. The Z-mode in the non-rigid gauge vs $\partial_Z \mathbb{A}_i$.

The massive vector meson propagator in Lorentz gauge is

$$\Delta^{mn}_{\mu\nu}(x) = \int \frac{d^4p}{(2\pi)^4} e^{-ipx} \left[\frac{-g_{\mu\nu} - p_\mu p_\nu / m^2_{v^n}}{p^2 + m^2_{v^n}} \delta^{mn} \right], \tag{15}$$

The first contribution in (14) is the direct coupling between the core instanton and the $U(1)_V$ source as displayed in Fig. 4 (left). The second contribution sums up the omega, omega', ... contributions as displayed in Fig. 4 (right). We note that the direct or core coupling drops by the exact sum rule

$$\sum_{n=1}^{\infty} a_{v^n} \psi_{2n-1} = 1, \tag{16}$$

Fig. 4. Gerry Brown 50/50 scenario: Direct (left) + VMD (right). See text.

following from closure in curved space

$$\delta(Z - Z') = \sum_{n=1}^{\infty} \kappa\psi_{2n-1}(Z)\psi_{2n-1}(Z')K^{-1/3}(Z') \,. \tag{17}$$

after integrating over the tower of omega meson trajectory. Vector Meson Dominance (VMD) is exact in holography. A similar argument holds for the pion electromagnetic form factor in [6]. The results presented in this section were derived in [10] using the Cheshire cat descriptive. They were independently arrived at in [13] using the strong coupling source quantization. They also support, the bottom up effective approach described in [14] using the heavy nucleon expansion.

For many years Gerry Brown has been advocating the 50/50 scenario for the baryon form factor using both phenomenology and his democratic principle. In many ways, the present unwinding of the baryon current in holography supports that. Indeed, if we were to truncate the resonance contributions to the lowests, say the omega, then the core contribution is non-zero. In this case, the deleniation of the Cheshire cat smile is no longer arbitrary. A specific position of the smile, will garentee optimal rearrangement between the truncated cloud and the core contributions, thereby vindicating Gerry's 50/50 scenario.

5. 2-Skyrmions from 2-Instantons

The procedure from the 1-nucleon sector to the 2-nucleon sector in holography relies on substituting the 1-instanton by the 2-instanton configuration. The latter is encoded in the ADHM data Δ which is a $(1 + k) \times k$ matrix [15, 16]

$$\Delta = \begin{pmatrix} \lambda_1 & \lambda_2 \\ D - x & u \\ u & -D - x \end{pmatrix}, \quad \Delta^\dagger \equiv \begin{pmatrix} \lambda_1^\dagger & (D - x)^\dagger & u^\dagger \\ \lambda_2^\dagger & u^\dagger & (-D - x)^\dagger \end{pmatrix}, \tag{18}$$

where the coordinates x_M are defined as $x = x_M\sigma^M$, and the moduli parameters are encoded in the free parameters λ_1, λ_2, D: $|\lambda_i| \equiv \rho_i$ are the size parameters, $\lambda_1^\dagger\lambda_2/(\rho_1\rho_2) \in SU(2)$ is the relative gauge orientation, and

$\pm D$ is the location of the constituents. u is a parameter fixed by the ADHM constraint $\Delta^\dagger\Delta = f^{-1}\otimes 1$,

$$f^{-1} = \begin{pmatrix} \rho_1^2 + (x_M - D_M)^2 + \dfrac{\rho_1^2\rho_2^2 - (\lambda_1\cdot\lambda_2)^2}{4D_M^2} & \lambda_1\cdot\lambda_2 + 2x\cdot u \\ \lambda_1\cdot\lambda_2 + 2x\cdot u & \rho_2^2 + (x_M + D_M)^2 + \dfrac{\rho_1^2\rho_2^2 - (\lambda_1\cdot\lambda_2)^2}{4D_M^2} \end{pmatrix}, \tag{19}$$

where the notation $q\cdot p$ for two quaternions q and p is used,

$$q\cdot p \equiv \sum_M q_M p_M\,. \tag{20}$$

$\rho_i = \sqrt{\lambda_i\cdot\lambda_i}$ are the size parameters, $\pm D_M$ the relative positions of the instantons, and

$$\begin{aligned} 2x\cdot u = {} & \frac{1}{D\cdot D}[(\lambda_2\cdot D)(\lambda_1\cdot x) - (\lambda_1\cdot D)(\lambda_2\cdot x) \\ & - \epsilon^{MNPQ}(\lambda_2)_M(\lambda_1)_N D_P x_Q]. \end{aligned} \tag{21}$$

Without loss of generality, we may choose the moduli parameters to be

$$\lambda_1 = \rho_1(0,0,0,1), \quad \lambda_2 = \rho_2(\widehat{\theta}_a \sin|\theta|, \cos|\theta|), \quad D = \left(\frac{d}{2}, 0, 0, 0\right), \tag{22}$$

with $a = 1,2,3$, $|\theta| \equiv \sqrt{(\theta_1)^2 + (\theta_2)^2 + (\theta_3)^2}$ and $\widehat{\theta}_a \equiv \frac{\theta_a}{|\theta|}$. The spatial x^1 axis is chosen as the separation axis of two instantons at large distance d. The flavor orientation angles (θ_a) are relative to the λ_1 orientation. We assign an $SU(2)$ matrix U to the relative angle orientations in flavor space

$$U \equiv \frac{\lambda_1^\dagger\lambda_2}{\rho_1\rho_2} = e^{i\theta_a\tau^a} \in SU(2). \tag{23}$$

which is associated with the orthogonal $SO(3)$ rotation matrix R as

$$\begin{aligned} R_{ab} &= \frac{1}{2}\mathrm{tr}\,\left(\tau_a U \tau_b U^\dagger\right) \\ &= \delta_{ab}\cos 2|\theta| + 2\widehat{\theta}_a\widehat{\theta}_b \sin^2|\theta| + \epsilon_{abc}\widehat{\theta}_c \sin 2|\theta|\,. \end{aligned} \tag{24}$$

For instance R_{ab} reads

$$\begin{pmatrix} \cos 2\theta_3 & \sin 2\theta_3 & 0 \\ -\sin 2\theta_3 & \cos 2\theta_3 & 0 \\ 0 & 0 & 1 \end{pmatrix}, \quad \begin{pmatrix} 1 & 0 & 0 \\ 0 & \cos 2\theta_1 & \sin 2\theta_1 \\ 0 & -\sin 2\theta_1 & \cos 2\theta_1 \end{pmatrix}, \tag{25}$$

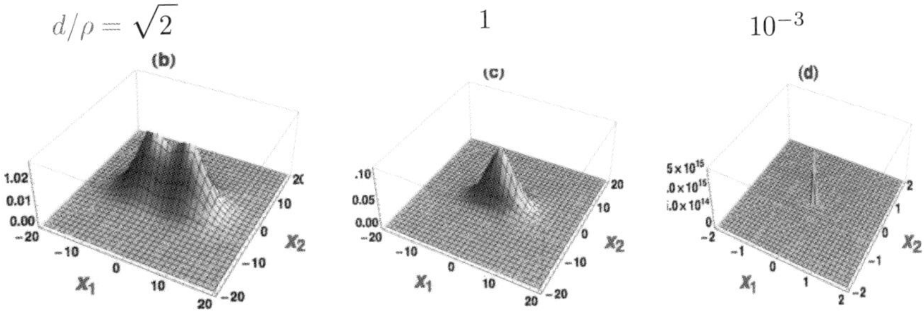

Fig. 5. 2-Skyrmions from 2-Instantons: Defensive.

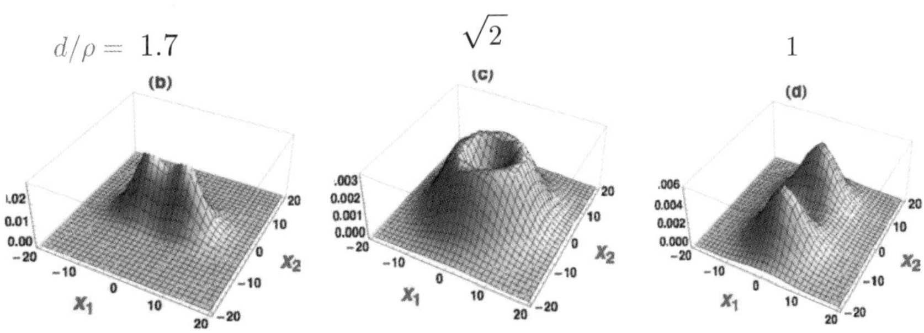

Fig. 6. 2-Skyrmions from 2-Instantons: Combed.

for $\theta_1 = \theta_2 = 0$ and $\theta_2 = \theta_3 = 0$ respectively. Note the double covering in going from $SU(2)$ to $SO(3)$.

In Fig. 5 we show the behavior of $\mathrm{Tr}(F_{\mu\nu}^2)$ for two parallell or defensive Skyrmions with $|\theta| = 0$ and $z = x_3 = 0$ and equal size cores $\rho = 9.64$ for several separations d/ρ. We recall that the physical core size in units of the KK-scale M_K is $\rho M_K \equiv \rho M_K/\sqrt{\lambda}$. In Fig. 6 we show the behavior for two antiparallel or combed Skyrmions with $\theta_1 = \theta_2 = 0$ and $\theta_3 = \frac{\pi}{2}$ or $|\,\theta| = \pi/2$. This is a π rotation along x_3 in the $SO(3)$ notation (24). For large separation two lumps form along the x^1 axis. For smaller separation the two lumps are seen to form in the orthogonal or x_2 direction. In between a hollow baryon 2 configuration is seen which is the precursor of the donut seen in the baryon number 2 sector of the Skyrme model [17]. The concept of $\widetilde{d}$ as a separation at small separations is no longer physical given the separation taking place in the transverse direction. What is physical is the dual distance u in the transverse plane.

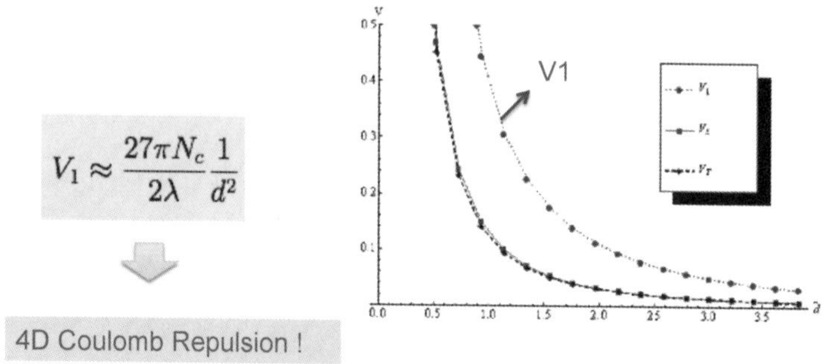

Fig. 7. The central potential: Core contribution.

At large separation, the nucleon-nucleon core interaction can be readily extracted from the Skyrmion-Skyrmion core interaction as it is linear in the $SO(3)$ rotation R. Indeed, the NN-potential can be decomposed as [18].

$$V_{NN} = V_1^+ + \vec{\tau}_1 \cdot \vec{\tau}_2 \, V_1^- + \vec{\sigma_1} \cdot \vec{\sigma_2} \left(V_S^+ + \vec{\tau}_1 \cdot \vec{\tau}_2 \, V_S^-\right) \tag{26}$$

$$+ (3(\vec{\sigma}_1 \cdot \widehat{d})(\vec{\sigma}_2 \cdot \widehat{d}) - \vec{\sigma}_1 \cdot \vec{\sigma}_2)(V_T^+ + \vec{\tau}_1 \cdot \vec{\tau}_2 \, V_T^-) \tag{27}$$

with the core contribution

$$V_{1,\mathrm{core}}^+ \approx \frac{27\pi N_c}{2\lambda}\frac{1}{d^2}, \tag{28}$$

as shown in Fig. 7. This repulsion is Coulomb-like in 5-dimensions, a hallmark of holography. The cloud contributions are meson-mediated. To lowest order they read [18]

$$V_{1,\widehat{V}}^+ \approx \sum_n G_{1\widehat{V},2n-1}^2 \frac{e^{-m_{2n-1}d}}{4\pi\, d}, \qquad G_{1\widehat{V},2n-1} \equiv \frac{N_c}{2}\psi_{2n-1} \sim \sqrt{\frac{N_c}{\lambda}}, \tag{29}$$

$$V_{S,A}^- \approx \sum_n G_{SA,2n}^2 \frac{e^{-m_{2n}d}}{4\pi d}, \qquad G_{SA,2n} \equiv -\frac{g_A \psi_{2n}}{\sqrt{6}\psi_0} \sim \sqrt{\frac{N_c}{\lambda}}, \tag{30}$$

$$V_{T,A}^- \approx \sum_n G_{TA,2n}^2 \frac{e^{-m_{2n}d}}{4\pi d}, \qquad G_{TA,2n} \equiv \frac{g_A \psi_{2n}}{\sqrt{12}\psi_0} \sim \sqrt{\frac{N_c}{\lambda}}, \tag{31}$$

$$V_{T,\Pi}^- \approx \frac{1}{16\pi}\left(\frac{g_A}{f_\pi}\right)^2 \frac{1}{d^3} \sim \frac{N_c}{\lambda}. \tag{32}$$

To order N_c/λ we note $V_1^- = V_S^+ = V_T^+ = 0$.

The present description of the 1- and 2-baryon configurations resemble in many ways the 1- and 2-nucleon structure emerging from strong coupling lattice QCD at finite density [19].

6. Dyonic Salt

At finite density, baryons crystalize at large N_c irrespective of the 't Hooft coupling λ since the Coulomb-like ratio is $\Gamma \approx N_c^2 \gg 1$. The crystal translates to a crystal of instantons in $T^3 \times R_Z$ with R_Z the holographic direction. Periodic directions are accompanied by twists or holonomies. Twisted instantons or instantons with holonomies are known to topologically split, if the twist or the holonomy are strong enough. An example is the KvLL instanton in $T^1 \times R^3$ [16,20,21] which is found to split into dyons.

Could such a splitting take place for an instanton arrangement in $T^3 \times R$? In [22] we suggested that it does, provided that the flavor gauge-symmetry is Higgsed, say the longitudinal "rho"

$$\langle \mathbf{A}_3^3 \rangle = \frac{2\pi}{2L}\, v\, T^3 \tag{33}$$

develops a non-vanishing expectation value. If that is the case, and proceeding by analogy with the KvLL instanton at finite temperature [16,20,21,23], the periodic instanton array in space splits into a dyonic array. In Fig. 8 we show how an initial crystal arrangement of fcc dyons split into a crystal arrangement of bcc dyons under the action of the flavor holonomy. The dyons are oppositely charged $e = g = \pm 1$ in units of T^3 along x^3. The dyon masses are $M_+ = MB_+ = Mv$ and $M_- = MB_- = M(1-v)$ — where v is the Higgs vev — with $B_\pm$ their topological charges respectively [23]. We

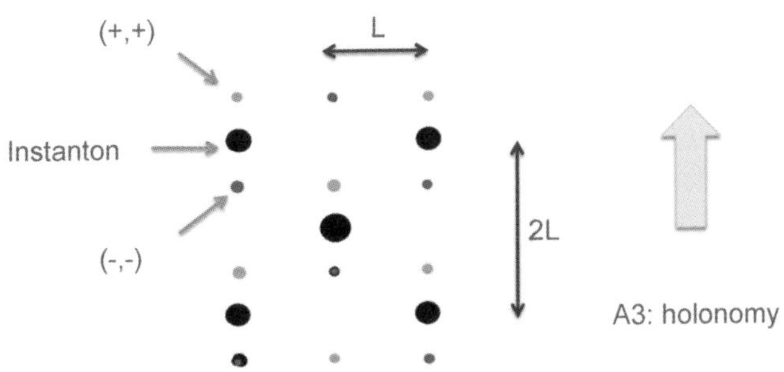

Fig. 8. Instantons (black) Splitting into oppositely charged Dyons $(e, g) = (\pm, \pm)$.

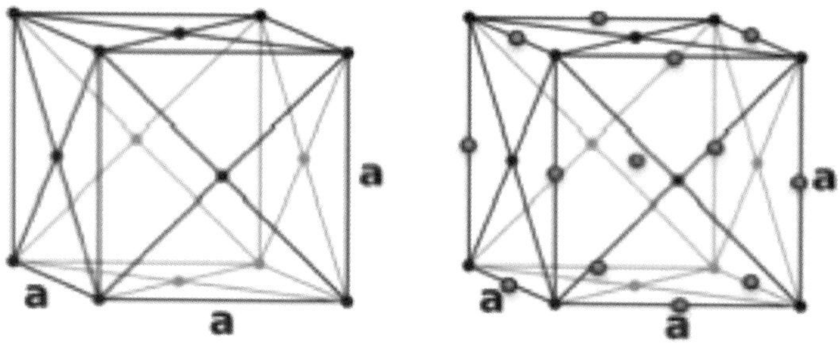

Fig. 9. (a) Instantons in fcc; (b) Dyons in bcc.

recall that $B_+ + B_- = v + (1 - v) = 1$ is the instanton number. Here $2L$ is the cell size of the initial fcc instanton arrangement.

To order $N_c\lambda \approx \kappa$ the instantons and dyons are BPS with an arbitrary value of the vev $0 \leq v < 1$. The dyonic crystal is salt-like with intertwined lattices of topological charges v and $(1 - v)$ at the vertices. In Fig. 9 we display the fcc instanton crystal (left) as it splits to a bcc crystal of dyons under the action of the spatial holonomy along x^3.

The instantons cease to be BPS at next to next to leading order (NNLO). Indeed, at NNLO the core instantons repel at short distances as in (28). In general, the exact many-dyon interaction for all ranges is involved [20]. Fortunately, for our dyonic crystal the details of the dyonic interactions are not important in the non-BPS regime. Indeed, once the instantons split into $e = g = \pm 1$ dyons as in Fig. 8, the Coulomb nature of the underlying charges will cause them to arrange in a salt-like configuration to maximally screen the + and − charges, and therefore balance the Coulomb forces. The Coulomb balance leaves the vev v arbitrary. However, the topological repulsion (28) wil maintain the Coulomb balance if $v = (1 - v)$. Thus $v = 1/2$ resulting into dyons of equal masses $M/2$ and equal charges $e = g = \pm 1$. This assumption is supported by dynamical calculations using colored instantons in SYM [24] and colored and thermal instantons in QCD in [25, 26].

The split dyon crystal arrangement is again salt-like with a unit cell $2L$. This is a bcc crystal of half-instantons or dyons per cubic cell L. The instanton or baryon density is

$$n_B = \frac{1/2}{L^3} = \frac{4}{(2L)^3} \tag{34}$$

which is commensurate with the initial density of fcc instantons, namely a cell unit of $(2L)$ with 4 instantons. Hence our initial choice of the fcc configuration for the instantons at low density. (34) reflects on the half-instanton symmetry of the bcc dyonic salt, which is dual to the half-skyrmion symmetry on the boundary. The density at which the splitting takes place can be estimated using the KvLL instanton. Indeed, in the latter dyon separation $R_{+,-}$ is [16]

$$R_{+-} = 2\pi \frac{\rho^2}{2L} \tag{35}$$

with ρ the KvLL instanton size with zero holonomy. Using (35) for our fcc crystal of cell size $(2L)^3$ and setting $R_{+-} = L$ at the transition to the bcc, yields $L = \sqrt{\pi}\rho$ or a critical density $n = 1/2/(\sqrt{\pi}\rho)^3$. In hQCD, $\rho \approx \sqrt{2/5}$ fm and $L \approx 1$ fm [22]. So the fcc to bcc transition takes place at $n_B \approx 1/2\,\mathrm{fm}^{-3}$ or 3 times nuclear matter density.

The energy density for which the fcc to bcc transition occurs can be estimated using only the Coulomb crystal (ignoring the core repulsion and the meson exchange attraction). The energy per baryon $E/N = M - \Delta$ is then [22]

$$\Delta = (e^2 + g^2)\,(T_3)^2\, \frac{\pi}{L}\, M_D \tag{36}$$

in the limit $\rho/L \ll 1$. Δ is the energy to bring a dyon in a bcc configuration and $M_D \approx 1.75$ is Madelung constant for salt. For $L \approx 1$fm, $\Delta \approx 350\,\mathrm{MeV}$. This estimate is on the high side since the Madelung constant for our 4-dimensional salt is smaller than that for the 3-dimensional salt, and $\rho/L \approx 1/\sqrt{\pi}$.

Does the dyonic salt configuration in bulk correspond to a chirally restored phase at high density? In [22] we have suggested that the Witten vertex maybe re-organized from the fcc to bcc configuration as shown in Fig. 10. In bulk, the probe flavor branes $D8 + \overline{D8}$ (left) split into separate $D8$ and $\overline{D8}$, each of which supporting its own crystal of 1/2-instantons or dyons $(L, R$ crystals). As a result, the right and left Wilson lines decouple

$$U^R_{1/2}(x) = P\exp\left(i\int_0^{+\infty} \mathbf{A}_Z(x,Z)\,dZ\right),$$
$$U^L_{1/2}(x) = P\exp\left(i\int_{-\infty}^{0} \mathbf{A}_Z(x,Z)\,dZ\right) \tag{37}$$

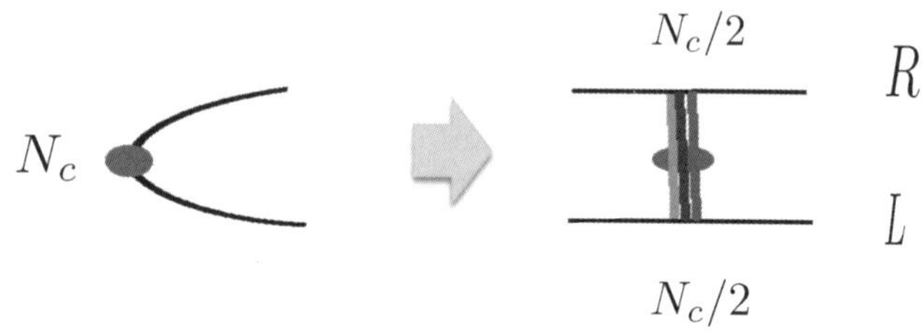

Fig. 10. Geometrical reorganization of Witten's vertex from fcc (left) to (bcc) right.

The L, R crystalline structures are commensurate with the $e = g = \pm 1$ dyonic structures as both are interchangeable by parity.

7. Dyonic Liquid

The cold dyonic crystal discussed in the preceding section can be dissociated by quanum fluctuations and/or thermal effects, both of which are subleading in the holographic counting. This notwithstanding, we may qualitatively ask how much temperature will be necessary to melt the instanton or dyon crystals. For that, Lindemann criterion tells us that when the vibration amplitude of the lattice ion reaches 10% of the nearest neighbor distance a_{NN}, the classical lattice melts [28]

$$\sqrt{\langle x^2 \rangle} \approx 10\%\, a_{NN} \tag{38}$$

The mean-square vibration can be estimated from the Einstein formulae

$$M\omega_E^2 \langle x^2 \rangle = k_B T \tag{39}$$

with $\omega_E \approx c_S k_{\max} = c_S \pi / a_{NN}$ the Einstein frequency. The speed of sound c_S in the crystal is related to its bulk compressibility $\mathbf{K}$,

$$\frac{c_S^2}{c^2} = \frac{\mathbf{K}}{n\, Mc^2} \approx (0.2 - 0.3) \frac{n_{NM}}{n} \tag{40}$$

with n_{NM} again the nuclear matter density and n the crystal baryon density. The last estimate borrows the typical nuclear matter compressibility estimate $\mathbf{K}/n_{NM} \approx 200\text{–}300\,\text{MeV}$ [29]. Inserting (39) and (40) in the Lindemann criterion (38) yields the estimated melting temperature

$$\frac{k_B T}{Mc^2} \approx \pi^2\, (10\%)^2\, (0.2 - 0.3)\, \frac{n_{NM}}{n} \tag{41}$$

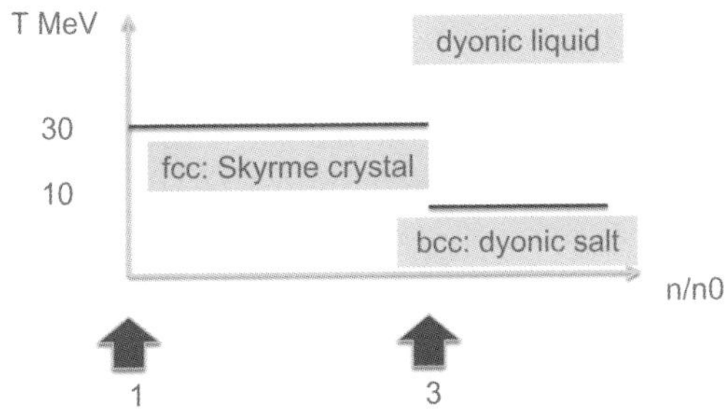

Fig. 11. Phase diagram at large N_c and low temperature.

The higher the crystal density, the easier to melt by this estimate. We show in Fig. 11 a typical structure of the phase diagram at low temperature and large N_c. The fcc crystals melts to a Skyrmion liquid at about 30 MeV, while the dyonic salt melts into a dyonic liquid at about 10 MeV. The low nature of these melting temperatures is an early indication that these crystals are easily dissociated by quantum vibrations due to kinetic energy for instance.

8. Conclusions

In hQCD baryons emerge as holograms of instantons in a curved five-dimensional space. The baryons embody the essentials of the Cheshire cat principle, with the Cheshire cat found "hiding" in the holographic direction. A baryon with a small core and a rich cloud of vector mesons is unravelled. Exact vector dominance emerges when the entire Regge trajectory of the vectors is kept in the cloud.

The two-baryon configuration is naturally obtained from the two-instanton configuration in bulk using the ADHM construction. Many features of the Skyrmion-Skyrmion interactions are this way recovered. The adventage of hQCD over the Skyrme model is the fact that the boundary action for the baryons is entirely fixed by the gauge-gravity in bulk using the large N_c and large λ bookeeping.

Cold dense nucleonic systems crystallize at large N_c. The crystals are found to be an arrangement of instantons in the fcc configuration at low densities, and a bcc arrangement of dyons at higher densities. The transition occurs at about 3 times nuclear matter densities in hQCD. Quantum kinetic effects or thermal effects are likely to turn each of the crystal into a strongly interacting liquid of instantons for the former and dyons for the latter.

Acknowledgments

I thank K.Y. Kim, M. Rho and S.J. Sin for the many discussions during our long term collaboration on the issues discussed here. This work was supported in part by US-DOE grants DE-FG02-88ER40388 and DE-FG03-97ER4014.

References

[1] A. Chodos, R. Jaffe, K. Johnson and C. Thorn, 'Baryon structure in the bag theory', *Phys. Rev. D* **10**, 2599 (1974); G.E. Brown and M. Rho, 'The little bag', *Phys. Lett. B* **82**, 177 (1979).
[2] S. Nadkarni, H.B. Nielsen and I. Zahed, 'Bosonization relations as bag boundary conditions', *Nucl. Phys. B* **253**, 308 (1985).
[3] H.B. Nielsen, M. Rho, A. Wirzba and I. Zahed, 'Color anomaly in hybrid bag model', *Phys. Lett. B* **269**, 389 (1991); M. Rho, 'The Cheshire cat hadrons revisited', *Phys. Rep.* **240**, 1 (1994).
[4] L. Caroll, 'Alice in Wonderland' Ed.
[5] I. Zahed and G.E. Brown, 'The Skyrme model', *Phys. Rep.* **142**, 1 (1986).
[6] T. Sakai and S. Sugimoto, 'Low energy hadron physics in holographic QCD', *Prog. Theor. Phys.* **113**, 843 (2005); T. Sakai and S. Sugimoto, 'More on a holographic dual of QCD', *Prog. Theor. Phys.* **114**, 1083 (2006); H. Hata, T. Sakai, S. Sugimoto and S. Yamato, 'Baryons from instantons in holographic QCD', arXiv:hep-th/0701280.
[7] D. Hong, M. Rho, H. Yee and P. Yi, 'Chiral dynamics of baryons from string theory' *Phys. Rev.* **D76**, 061901 (2007).
[8] E. Witten, 'Baryons and branes in anti de Sitter space', JHEP07, 006 (1998).
[9] J. Maldacena, 'The large N limit of superconformal field theories and supergravity', *Adv. Theor. Math. Phys.* **2**, 231 (1998).
[10] K.Y. Kim and I. Zahed, 'Electromagnetic baryon form factors from holographic QCD', *JHEP* **0809**, 007 (1998).
[11] C. Adami and I. Zahed, 'Soliton quantization in chiral models with vector mesons', *Phys. Lett. B* **215**, 387 (1988).
[12] H. Verschelde and H. Verbeke, 'Nonrigid quantization of the skyrmion', *Nucl. Phys. A* **495**, 523 (1989).
[13] K. Hashimoto, T. Sakai and S. Sugimoto, 'Holographic baryons : static properties and form factors from gauge/string duality', arXiv:0806.3122 [hep-th].
[14] D.K. Hong, M. Rho, H.U. Yee and P. Yi, 'Nucleon form factors and hidden symmetry in holographic QCD', arXiv:0710.4615 [hep-ph]; M. Rho, 'Baryons and vector dominance in holographic dual QCD', arXiv:0805.3342 [hep-ph].
[15] M.F. Atiyah, N.J. Hitchin, V.G. Drinfeld and Yu.I. Manin, 'Construction of instantons', *Phys. Lett. A* **65**, 185 (1978).
[16] T.C. Kraan and P. van Baal, 'Periodic instantons with non-trivial holonomy', *Nucl. Phys. B* **533**, 627 (1998) [arXiv:hep-th/9805168].

[17] E. Braaten and L. Carson, 'The deuteron as a toroidal Skyrmion', *Phys. Rev.* **D38**, 3525 (1988).
[18] K.Y. Kim and I. Zahed, 'Nucleon-nucleon potential from holography', *JHEP* **0903**, 131 (2009) [arXiv:0901.0012 [hep-th]].
[19] P. de Forcrand and M. Fromm, in talk 'Nuclear Matter from strong coupling lattice QCD', Trento, July 2010; [arXiv:0907.1915], [arXiv:0912.2524].
[20] K.M. Lee and P. Yi, 'Monopoles and instantons on partially compactified D-branes', *Phys. Rev. D* **56**, 3711 (1997) [arXiv:hep-th/9702107].
[21] K.M. Lee and C.H. Lu, 'SU(2) calorons and magnetic monopoles', *Phys. Rev. D* **58**, 025011 (1998) [arXiv:hep-th/9802108].
[22] M. Rho, S.J. Sin and I. Zahed, 'Dense QCD: A holographic dyonic salt', *Phys. Let.* **B689**, 23 (2010).
[23] D. Diakonov, 'Topology and confinement', arXiv:0906.2456 [hep-ph].
[24] N.M. Davies, T.J. Hollowood, V.V. Khoze and M.P. Mattis, 'Gluino condensate and magnetic monopoles in supersymmetric gluodynamics', *Nucl. Phys. B* **559**, 123 (1999) [arXiv:hep-th/9905015].
[25] A.R. Zhitnitsky, 'Confinement-deconfinement phase transition in hot and dense QCD at large N_c', *Nucl. Phys. A* **813**, 279 (2008) [arXiv:0808.1447 [hep-ph]]; A. Parnachev and A.R. Zhitnitsky, 'Phase Transitions, theta Behavior and Instantons in QCD and its Holographic Model', *Phys. Rev. D* **78**, 125002 (2008) [arXiv:0806.1736 [hep-ph]]; A.S. Gorsky, V.I. Zakharov and A.R. Zhitnitsky, 'On Classification of QCD defects via holography', *Phys. Rev. D* **79**, 106003 (2009) [arXiv:0902.1842 [hep-ph]].
[26] M.N. Chernodub, A. D'Alessandro, M. D'Elia and V.I. Zakharov, 'Thermal monopoles and selfdual dyons in the quark-gluon plasma' , arXiv:0909.5441 [hep-ph].
[27] C. Kittel, *Introduction to solid state physics (8th ed.)* (John Wiley & Sons, 2005).
[28] P. Hofmann, *Solid state physics: an introduction.* (Wiley-VCH, 2008).
[29] J.P. Blaizot, *Phys. Rep.* **64**, 171 (1980).